Advances in Digital Signal Processing

Advances in Digital Signal Processing

Editor: Emily McLaughlin

MURPHY & MOORE
www.murphy-moorepublishing.com

Published by Murphy & Moore Publishing
1 Rockefeller Plaza,
New York City, NY 10020, USA
www.murphy-moorepublishing.com

Advances in Digital Signal Processing
Edited by Emily McLaughlin

International Standard Book Number: 978-1-63987-017-2 (Hardback)

Cataloging-in-Publication Data

Advances in digital signal processing / edited by Emily McLaughlin.
p. cm.
Includes bibliographical references and index.
ISBN 978-1-63987-017-2
1. Signal processing--Digital techniques. 2. Signal processing. 3. Digital electronics. I. McLaughlin, Emily.
TK5102.9 .A38 2022
621.382 2--dc23

Contents

Preface

This book has been an outcome of determined endeavour from a group of educationists in the field. The primary objective was to involve a broad spectrum of professionals from diverse cultural background involved in the field for developing new researches. The book not only targets students but also scholars pursuing higher research for further enhancement of the theoretical and practical applications of the subject.

Digital signal processing (DSP) is a sub-field of signal processing. It refers to the processing of discrete-time sampled signals using digital circuits or general purpose computers. Sonar, radar, speech recognition, digital image processing, biomedicine, audio signal processing, video compression, financial signal processing, etc. are some of the various areas that require digital signal processing. Digital signal processing is superior to analog processing as it allows data compression, and error detection and correction in transmission. Digital signal processing may be done in the time domain, frequency domain, spatial domain, or wavelet domain. This choice is made depending on the domain that best represents the essential characteristics of the signal and the processing required. The ever growing need of advanced technology is the reason that has fueled the research in the field of digital signal processing in recent times. The topics included in this book on this discipline are of utmost significance and bound to provide incredible insights to readers. It is appropriate for students seeking detailed information in this area as well as for experts.

It was an honour to edit such a profound book and also a challenging task to compile and examine all the relevant data for accuracy and originality. I wish to acknowledge the efforts of the contributors for submitting such brilliant and diverse chapters in the field and for endlessly working for the completion of the book. Last, but not the least; I thank my family for being a constant source of support in all my research endeavours.

Editor

1

Compressive Background Modeling for Foreground Extraction

Yong Wang,[1] Qian Lu,[2] Dianhong Wang,[1,2] and Wei Liu[1]

[1]*Faculty of Mechanical and Electronic Information, China University of Geosciences, Wuhan 430074, China*
[2]*Department of Geological Science and Engineering, Wuhan University of Engineering Sciences, Wuhan 430200, China*

Correspondence should be addressed to Yong Wang; wycug2011@gmail.com

Academic Editor: Jiri Jan

Robust and efficient foreground extraction is a crucial topic in many computer vision applications. In this paper, we propose an accurate and computationally efficient background subtraction method. The key idea is to reduce the data dimensionality of image frame based on compressive sensing and in the meanwhile apply sparse representation to build the current background by a set of preceding background images. According to greedy iterative optimization, the background image and background subtracted image can be recovered by using a few compressive measurements. The proposed method is validated through multiple challenging video sequences. Experimental results demonstrate the fact that the performance of our approach is comparable to those of existing classical background subtraction techniques.

1. Introduction

Foreground extraction is often the first step of many visual surveillance applications such as object tracking, recognition, and anomaly detection. Background subtraction [1] is the most frequently used method to detect and extract objects automatically in video sequences. The basic principle can be formulated as a technique that builds the background model and compares this model with the current image frame in order to distinguish foreground, that is, moving objects from static or slow moving background. Many pixel-based methods have been investigated in the past decades. Among these, Gaussian mixture model (GMM) [2] is a representative method for robustly modeling complicated backgrounds with slow illumination changes and small repetitive movements. This method models the distribution of the values observed over time at each pixel by a weighted mixture of Gaussians. Several modified methods concerning the number of Gaussian components, learning rate, and parameters update [3–5] have been proposed. However, the main drawback of these approaches is being computationally intensive. In [6], a nonparametric kernel density estimation (KDE) has been proposed to model the background, but this method consumes too much memory. Besides, in [7], each pixel is represented by a codebook which is able to capture background motion over a long period of time with a limited amount of memory. The codebooks can evolve with illumination variations and moving backgrounds. Nevertheless, the codebook update will not allow the creation of new codewords once codebooks have been learned from a typically training sequence. Recently, a universal background subtraction method called ViBe has been presented in [8], where each pixel is modeled with a set of real observed pixel values. It is reported that this method outperforms current mainstream methods in terms of both computation speed and detection accuracy.

Most of pixel-level background subtraction techniques have considerable computation costs; thus some other methods are proposed to improve the computation efficiency. Our earlier work [9] proposed a block-based image reconstruction to reduce the video frame size and then applied GMM to model the background for the constructed frame. In recent years, there has been a growing interest in compressive sensing (CS) and the idea of CS has also been exploited for background subtraction. In [10], an image is divided into small blocks and random projections based on CS are then computed for each block to reduce the data dimensionality. After this, each projection value is modeled by GMM to determine whether the block belongs to foreground or not. The work in [11] describes a CS reconstruction method to directly recover background subtracted images from compressive measurements. The main idea is that the background subtracted images can be represented sparsely in the spatial image domain. Based on this, an improved method called adaptive rate compressive sensing (ARCS) is presented in [12]. Furthermore, a dynamic group sparsity (DGS) recovery algorithm [13] is proposed based on the extended CS theory to perform background subtraction. According to both sparsity and dynamic group clustering priors, DGS can recover stably background subtracted image using fewer measurements and computations.

Recently, the approach proposed in [14] exploits sparse representation (SR) to perform classification, where a test image can be sparsely represented by a subspace spanned from training images. CS has shown the ability to efficiently compress signals using SR. Motivated by this idea, we present a new background subtraction method combining CS and SR techniques in this paper. In our solution, the background modeling can be cast as a greedy sparse recovery problem. This paper is organized as follows. In Section 2, we briefly review the CS theory. This review presents the framework developed for compressive background modeling. Section 3 extensively describes our solution and the implementation process: sparse representation, background initialization, background modeling, and update mechanism. Section 4 compares our method with several state-of-the-art background subtraction techniques and discusses the performances in terms of accuracy and computation efficiency. Section 5 concludes this paper.

2. Compressive Sensing

The CS theory [15] states that a signal can be reconstructed from a small number of measurements with high probability, provided that the signal is sparse in the spatial domain or some transform domains, for example, wavelets. Assume that a signal x can be represented as $x = \Psi\theta$, where Ψ denotes a basis and θ is the coefficients corresponding to the basis. The signal x is said to be k-sparse if all other elements in θ vanish except for k nonzero coefficients. According to CS, for a sparse signal $x \in R^n$, compressive measurements can be collected by the following random projections:

$$y = \Phi x + \varepsilon, \quad (1)$$

where $\Phi \in R^{m\times n}$, $m \ll n$, is the measurement matrix, $y \in R^m$ contains m measurements, and ε is the measurement noise. Specifically, a high dimensional vector x is converted into a much lower dimensional measurement vector y. Moreover, the compressive measurements in y contain almost all the information of the sparse vector x. This means that we can work with data of significantly lower dimension so as to achieve computation efficiency as well as accuracy.

Since $m \ll n$, the recovery of the sparse signal x from y is underdetermined. However, the following two additional assumptions make the recovery possible [16, 17]. First, Φ is required to satisfy the restricted isometry property (RIP). Some randomly generated matrices, for example, Gaussian, Bernoulli distribution of ±1, obey the RIP and can be used for random projections. Second, m is greater than $\mathbf{O}(k\log(n/k))$; that is, the number of measurements must be large enough with respect to the number of nonzero coefficients in θ. Under these circumstances, the sparse recovery can be formulated by solving the following ℓ_0 minimization problem:

$$\hat{x} = \operatorname{argmin} \|x\|_0 \quad \text{s.t.} \ \|y - \Phi x\|^2 < \varepsilon. \quad (2)$$

It is known that (2) is an NP hard problem. In general, the sparse solution can be obtained by ℓ_1 convex optimization or greedy iterative algorithms.

3. Compressive Background Modeling

3.1. Sparse Representation of Background. To improve the computation efficiency of sparse recovery, the image frame is first divided into M blocks of pixels. We describe the sparse representation of background on a block because the operations to be carried out on the vector of each block are identical. For the ith block, $i = 1, \ldots, M$, it is vectorized into a column vector x_i of size $n \times 1$, where n denotes the product of the height and width of the block. The vectorized image block x_i is assumed to consist of both background b_i and background subtracted image f_i; that is,

$$x_i = b_i + f_i. \quad (3)$$

According to SR, the current background can be represented as a linear combination of all preceding background images. Due to the memory requirement blast for long video sequences, we use l preceding backgrounds to sparsely represent the new background, where l should be large enough to comprise a sufficient number of background samples. Thus the background b_i can be measured as follows:

$$b_i = A_i\alpha_i = [b_{i,1}, b_{i,2}, \ldots, b_{i,l}]\,\alpha_i, \quad (4)$$

where $A_i \in R^{n\times l}$ is a subspace spanned from preceding backgrounds $b_{i,j}$, $j = 1, \ldots, l$. $\alpha_i \in R^l$ is a sparse coefficient vector whose most elements are either zeros or very close to zeros. Substituting (4) into (3), we have

$$x_i = A_i\alpha_i + f_i = [A_i, I]\begin{bmatrix}\alpha_i \\ f_i\end{bmatrix} = \Psi_i s_i, \quad (5)$$

where $I \in R^{n\times n}$ is an identity matrix, $\Psi_i \in R^{n\times(l+n)}$ is called the dictionary, and $s_i \in R^{l+n}$ is the corresponding coefficient vector. In fact, the sizes of the foreground objects are relatively smaller than the size of the background image. Only a small part of pixels in the background subtracted image has nonzero values; that is, f_i exhibits sparsity in the spatial domain. The fact that both α_i and f_i are sparse implies that s_i is sparse. In this case, x_i can be well approximated using nonzero coefficients of s_i under Ψ_i, which is critical to the sparse recovery base on CS.

3.2. Background Initialization. As described before, the background subspace is characterized by A_i containing l preceding backgrounds. Thus the first problem we face is the initialization of A_i. Normally, most mainstream approaches need tens or hundreds of image frames to initialize background by estimating probability density functions or statistical parameters of background pixels. For instance, GMM initializes every background pixel by a mixture of K Gaussian distributions, where the weight, mean, and variance of each distribution are learned and updated from a sequence of frames. Nevertheless, the model initialization takes up large amounts of computation resources. Besides, the shape of a probability density function is sensitive to outliers and the evaluation of the statistical parameters is largely dependent upon the number of samples to be considered. As a matter of fact, it is not imperative to estimate the temporal distribution of background pixels using a large amount of pixel samples. Each background pixel can be initialized with a set of real observed pixel values instead of with an explicit pixel model.

Like the authors of [8], we initialize A_i from the first frame of the video sequence based on the assumption that neighboring pixels share a similar temporal distribution. Given a pixel located at u in ith image block, its value and the spatial neighborhood are denoted by $v(u)$ and $N_G(u)$, respectively. Therefore, the background sample value of the pixel u, $b_{i,j}(u)$, is equal to the value of a pixel v randomly selected in $N_G(u)$ in the first frame:

$$b_{i,j}(u) = v(v \mid v \in N_G(u)) \quad j = 1, \ldots, l. \tag{6}$$

Based on this, the background model of pixel u is filled with a collection of l sample values. Thus A_i can be initialized by background models of all pixels in the ith block:

$$A_i = [\{b_{i,1}(u)\}, \{b_{i,2}(u)\}, \ldots, \{b_{i,l}(u)\}], \quad u \in \text{block } i. \tag{7}$$

It is noteworthy that this strategy can perform foreground extraction from the second frame, which is beneficial for a short video sequence or memory-constrained embedded devices. Furthermore, the spatial neighborhood of moderate size is preferred. A large size leads to the degradation in the statistical correlation between pixels at different locations. Conversely, the diversity of background samples cannot be guaranteed under a small size. In our solution, samples are selected randomly in the 8-connected neighborhood of each pixel.

3.3. Compressive Background Modeling. Likewise, we take the ith image block for example. As mentioned earlier, due to the sparsity of s_i under the dictionary Ψ_i, x_i can be projected into compressive measurements without losing much information. Combining (5) with (1), random projections are executed for each block:

$$y_i = \Phi x_i + \varepsilon = \Phi [A_i, I] \begin{bmatrix} \alpha_i \\ f_i \end{bmatrix} + \varepsilon = \Phi \Psi_i s_i + \varepsilon. \tag{8}$$

Let Φ be an $m \times n$ random matrix whose entries $\varphi_{p,q}$ are independent realizations of ±1 Bernoulli random variables:

$$\varphi_{p,q} = \begin{cases} +\dfrac{1}{\sqrt{m}}, & \text{with probability } \dfrac{1}{2} \\ -\dfrac{1}{\sqrt{m}}, & \text{with probability } \dfrac{1}{2}. \end{cases} \tag{9}$$

Note that once Φ is generated in the beginning, the same matrix is used for each block throughout the video sequence.

Assume that $\Psi_{t,i}$ is the dictionary of the ith block at time t. For the image at time $t+1$, let $x_{t+1,i}$ be the vector of the ith block. Given Φ described in (9), the compressive measurements $y_{t+1,i}$ can be obtained by the matrix multiplication of Φ and $x_{t+1,i}$. According to (8), the background modeling at time $t+1$ is thus formulatedas the following sparse recovery problem:

$$(\alpha_{t+1,i}, f_{t+1,i}) = \operatorname{argmin} \|s_{t,i}\|_0, \quad \text{s.t. } \|y_{t+1,i} - \Phi \Psi_{t,i} s_{t,i}\|^2 < \varepsilon. \tag{10}$$

In our method, the sparse recovery in (10) is solved by a greedy algorithm called DGS [13]. The reason for this is that DGS can recover stably sparse data with less measurement requirement and lower computation complexity, because it considers not only sparsity but group clustering priors of sparse data as well. Compared to existing greedy recovery algorithms, the difference is that the estimation pruning uses DGS approximation rather than k-sparse approximation. To reduce the computational cost, background modeling can be performed at specified intervals instead of every frame.

3.4. Background Update. To adapt to gradual changes that occur in the background such as environment disturbances and slow lighting changes, the background modelneeds to be updated over time. In this paper, we tackle this problem by dynamically updating background model once every T frames, and the frame interval is set at 3 to 7. The reason for this is threefold. Firstly, it is acceptable that background model remains the same for a certain period of time, which does not materially deteriorate the results of background subtraction. Secondly, the continuous sparse recovery results in a relatively heavy computation overhead. Furthermore, too often update introduces small errors into background model at each time, and the accumulative errors have negative effects on the accuracy of foreground extraction.

At initialization, A_i is built using (7) from the first video frame. Subsequently, A_i is updated according to the sparse recovery result. Assume that the background subspace at time t is denoted as $A_{t,i}$. The sparse coefficient vectorat time $t+1$, $\alpha_{t+1,i}$, can be obtained via (10); thus the background at time $t+1$ is calculated by $b_{t+1,i} = A_{t,i}\alpha_{t+1,i}$. We can see from (4) that the background update is equivalent to replacing a certain sample $b_{i,j}$, $j = 1, \ldots, l$, in $A_{t,i}$ using $b_{t+1,i}$. The traditional approach to background update is to substitute the new values for old ones in turn. However, there is no reason to naively remove a valid sample if it corresponds to a background. Unlike this idea, we adopt a probability strategy to update $A_{t,i}$ where a sample in $A_{t,i}$ is chosen at random according to a uniform law and replaced by the new background $b_{t+1,i}$. Therefore the background subspace at time $t+1$ is denoted by

$$A_{t+1,i} = A_{t,i} \left\{ \forall b_{i,j} \mid b_{i,j} = b_{t+1,i}, \text{with probability } \frac{1}{l} \right\}. \tag{11}$$

Furthermore, the background update should also take sudden illumination changes into account. As stated earlier,

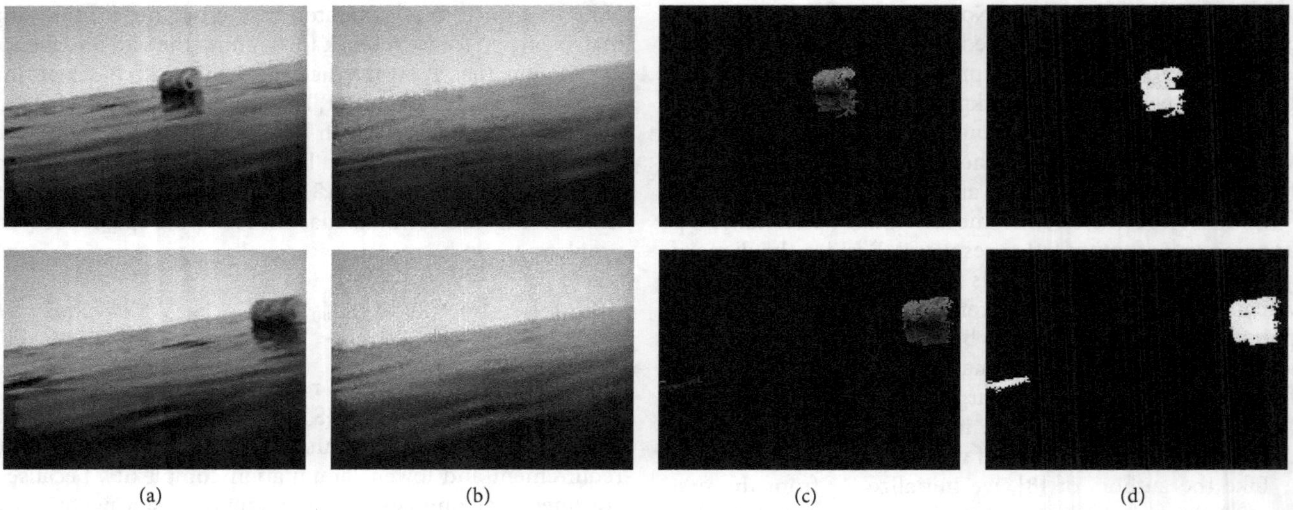

FIGURE 1: Results on a video sequence of 150 frames with background disturbances. (a) Original frames, (b) background, (c) background subtracted images, and (d) foreground masks.

A_i is initialized from a single frame. This technique provides a rapid response to sudden illumination changes; that is, the existing background model is discarded directly and a new model is initialized right now. Obviously, a sudden illumination change leads to the great change of pixel value. We count the pixels with great value change; if the ratio of these pixels to the total exceeds a threshold Th, the background is reinitialized according to the current frame in order to accomplish foreground extraction under the new illumination condition.

4. Experimental Results

4.1. Effectiveness Analysis. This experiment is designed to demonstrate the effectiveness of compressive background modeling. We implement the proposed method in MATLAB and test it on two complicated scenarios involving background disturbances and illumination changes. The same parameters are used for these sequences in the experiment. The values of parameters are as follows: $M = 4$, $l = 15$, $T = 5$, $\mathrm{Th} = 0.6$, and the measurement rate $m/n = 1/16$.

Figure 1 shows examples of background modeling and foreground extraction for two typical frames of the first sequence of 176×144 pixels, which concerns a floating tin can under condition of waves. Figure 1(a) is the original frames, Figures 1(b) and 1(c) show the recovered background images and background subtracted images, respectively. To facilitate the observations, Figure 1(d) gives the binary foreground masks that are directly obtained via background subtracted images. All results are the solutions of the optimization problem in (10) with a simple postprocessing based on morphological operations. As can be seen, the proposed method can almost accurately recover the background and extract foreground objects, except in the case of severe disturbances; for example, the background subtracted image shown in the second row contains a small piece of background corresponding to a heavy wave. The results show that the background modeling based on CS and SR techniques can handle dynamic scenes well. Moreover, the spatial consistency of neighboring pixels is absorbed into the initialization of background model, which can mitigate the impact of waves to some extent.

Figure 2 shows the background subtraction results on the other video sequence of 160×120 pixels, which concerns three different illumination conditions, that is, normal; the light is switched off and switched on again. As described in Section 3.4, when the light is switched off/on, a new background model is initialized instantaneously to keep track of the sudden change in illumination. After this, the recovery of the background and background subtracted images can proceed as before and continue to improve. From Figure 2, it is clear that the resulting images are clean even undergoing repetitive sudden illumination changes. This means that the proposed method is able to handle random illumination conditions. In addition, our method overcomes effectively the impact of the flickering monitors.

4.2. Qualitative Comparison. To evaluate the performances of the proposed method, we compare the results of our method with those of five representative background subtraction algorithms: (1) ARCS [12]; (2) CS-MoG [10]; (3) ViBe [8]; (4) KDE [6]; and (5) GMM [2]. We implement the five algorithms by ourselves, and all the parameters in these algorithms use the proposed default values according to the original papers. These algorithms are tested on seven different video sequences, and all the experiments give similar conclusions as described below. Due to space constraints, only five out of seven sequences are illustrated in Figure 3. The first three sequences involving pedestrians and vehicles detection under sunny and overcast days contain 500 frames of 320×240 pixels. The last two sequences concern the foreground extraction in the presence of background disturbances such as swaying tree and waves on the water, where the former is a standard sequence consisting of 286

(a) (b) (c) (d)

FIGURE 2: Results on a video sequence of 750 frames with sudden illumination changes. (a) Original frames, (b) background, (c) background subtracted images, and (d) foreground masks.

frames of 160 × 120 pixels and the latter comprises 150 frames of 176 × 144 pixels. According to the image size, the number of image blocks is set to $M = 16$ for the first three sequences and $M = 4$ for the last two sequences. Figure 3(a) shows one typical frame of each sequence, and Figures 3(b)–3(g) correspond to the foreground extraction results of the aforementioned six methods. Note that all the foreground masks are immediate results without any morphological postprocessing operations. Thus the comparisons based on these results are objective and reasonable.

There are several points we can learn from the experimental results. First, the results of our method and ARCS are similar due to the fact that background subtraction is cast as a sparse recovery problem. Both of them have relatively stable performance under various scenarios, as shown in Figures 3(b) and 3(c). Second, CS-MoG is a block-based background subtraction algorithm. The image is segmented into blocks of 8 × 8 pixels, and projections based on CS are computed for each block. Each projection value is modeled as a GMM to determine if the block is classified as foreground. This makes CS-MoG able to effectively eliminate background disturbances and correctly identify foreground regions. Meanwhile, it also leads to some problems. For example, the foreground masks are not accurate enough because of many holes and fragments. Besides, we can see from the third foreground mask in Figure 3(d) that vehicles with small size are omitted because of the block-based processing. Third, for the first two sequences, the results of ViBe look better. However, the foreground extraction of ViBe evidently deteriorates when there exist severe background disturbances. Fourth, KED and GMM are more sensitive to illumination changes and cannot deal with nonstatic scenes. From Figures 3(f) and 3(g), there are a large number of background pixels incorrectly classified as foreground under the condition of swaying tree and waves. Comparatively, our method is stable and effective under different illumination conditions and cluttered backgrounds.

Two metrics, the percentage of correct classification (PCC) [8] and the average processing time (APT), are utilized to quantitatively evaluate the six background subtraction algorithms. Naturally, the higher the PCC, the more accurate the foreground extraction; in the meanwhile, the less the APT, the higher the computation efficiency. The comparison results are computed as the mean of 50 consecutive frames. Figures 4 and 5 illustrate the PCCs and APTs of the six methods for five sequences, respectively. We can see from Figure 4 that the PCCs confirm the results illustrated in Figure 3. The PCC of ARCS is very close to that of our method under different scenarios. In most cases, our method has the best performance and its PCC is higher than those of the others. For one thing the PCC of ViBe is slightly greater than that of our method for the first two sequences and for another the PCC of ViBe reduces by an average of 7.48% in comparison with our method for the rest sequences. Moreover, the PCC of our method is 4.51 to 10.04% greater than that of CS-MoG.

Though our method and ARCS perform foreground extraction by solving sparse recovery that is relatively time-consuming, our method is faster than ARCS, as shown in Figure 5. The reason for this is twofold. First, in our solution, the image frame is processed in the form of subblocks of pixels instead of a whole image so that the computation efficiency of sparse recovery is improved. Second, we solve the sparse recovery using DGS with lower computation complexity and fewer measurements. As can be seen from Figure 5, CS-MoG which uses random projections to reduce

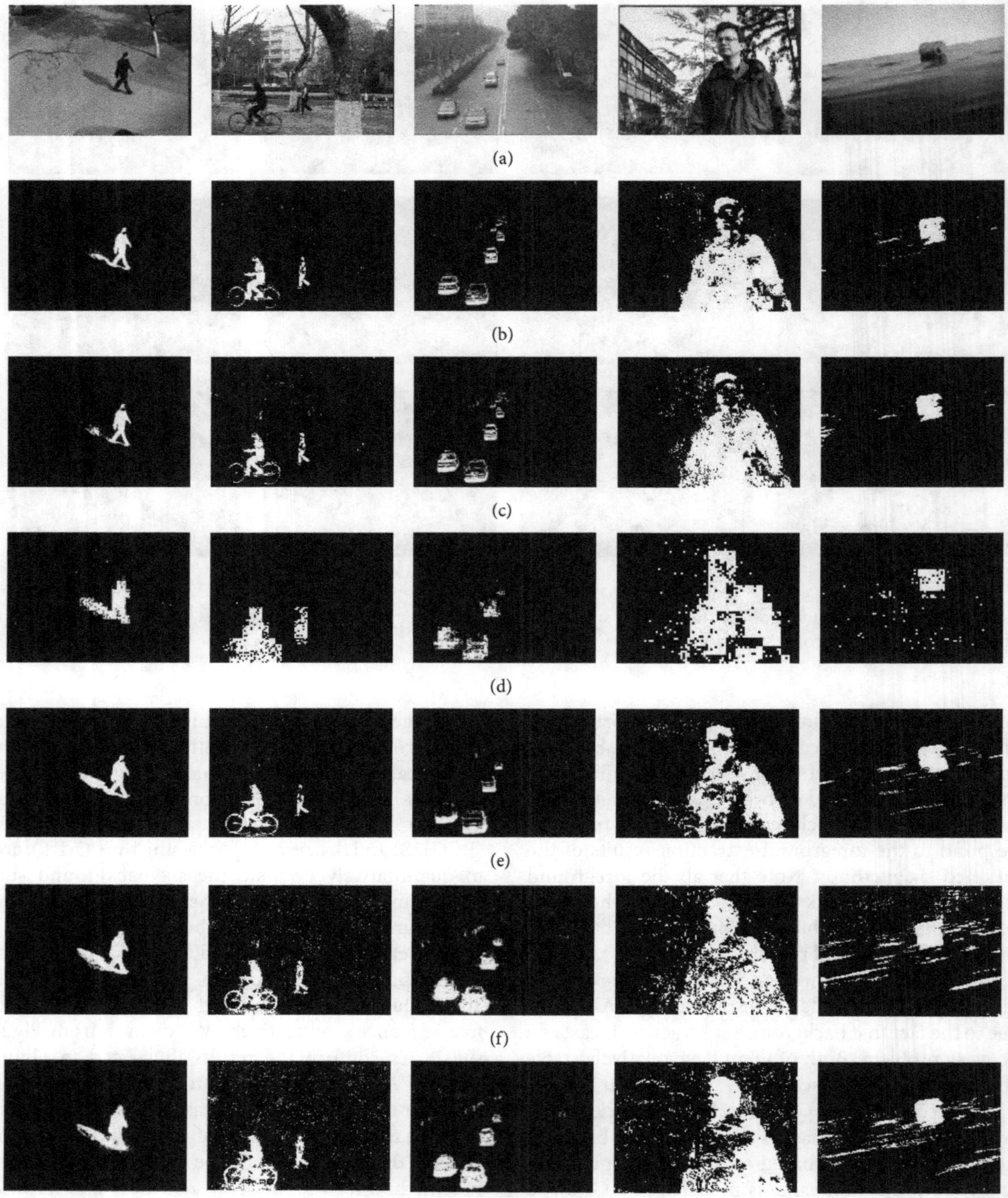

FIGURE 3: Comparative foreground masks of six methods. (a) Original frames, (b) our method, (c) ARCS, (d) CS-MoG, (e) ViBe, (f) KDE, and (g) GMM.

data dimensionality is significantly faster than other pixel-level techniques, that is, ViBe, KDE, and GMM. For KDE and GMM, background models maintain complete information on each pixel, which results in a considerable computation cost. For instance, for the first sequence, the APTs of our method, ARCS, CS-MoG, ViBe, KDE, and GMM, are 90.91, 129.87, 52.37, 79.07, 154.46, and 122.78 ms, respectively. Specifically, our method is approximately 42.7%, 69.7%, and 34.9% faster than ARCS, KDE, and GMM, respectively. On the other side, our method obtains foreground by solving optimization problem but CS-MoG and ViBe do not; therefore our method is 73.4% and 14.9% slower than CS-MoG and ViBe, respectively. Furthermore, we further observe the time consumption of our method. The test shows that the sparse

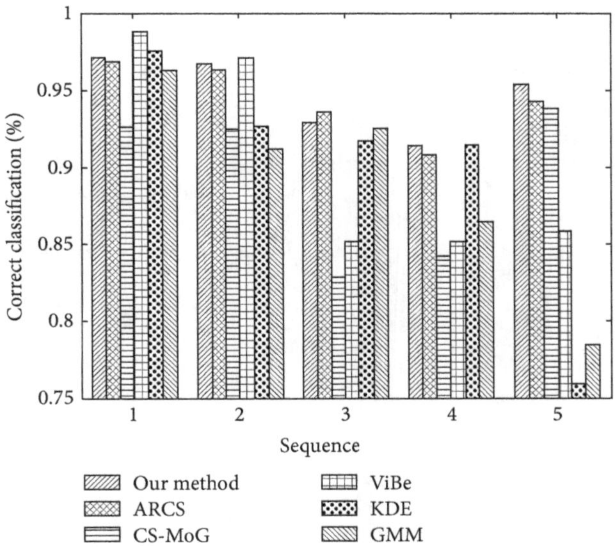

FIGURE 4: PCCs of six methods.

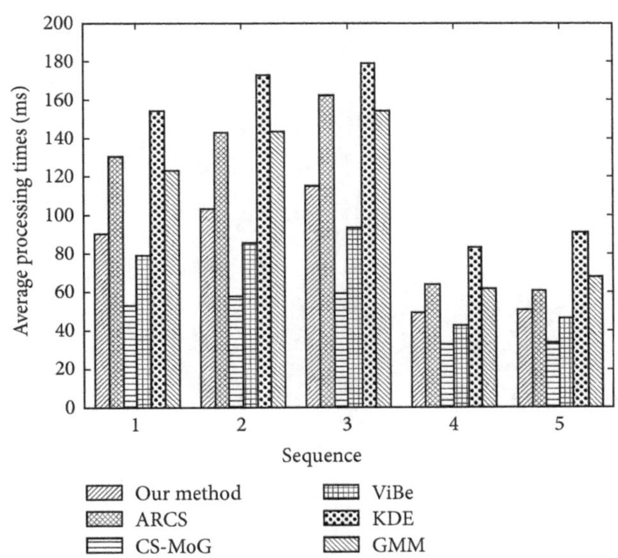

FIGURE 5: APTs of six methods.

recovery dominates the computation time and consumes as much as approximately 85.2% of total computation time. In comparison, the overhead introduced by background update is 10.6% of total computation overhead.

From the foregoing, our method is slower than CS-MoG and ViBe, while the foreground extraction accuracy of our method is superior to those of two others. It can be concluded that the proposed method achieves the best tradeoff between the accuracy and computation efficiency.

5. Conclusions

In this paper, we propose a background subtraction method based on CS and SR. Background is modeled as a sparse representation of preceding backgrounds. Combining the sparse representation of background with the sparsity of foreground, we use a few compressive measurements to recover background and the background subtracted image within the CS framework. Moreover, we provide the background initialization and update scheme which improve the robustness against the changes in the scenarios. According to comparisons and analyses involving several challenging sequences and five other state-of-the-art background subtraction algorithms, the experimental results clearly demonstrate the effectiveness of the proposed method.

Conflict of Interests

The authors declare that there is no conflict of interests regarding the publication of this paper.

Acknowledgments

This work was supported in part by Technology Research Program of Hubei Province, China (Grants 2012FFA108 and 2013BHE009), Wuhan Youth Chenguang Program of Science and Technology (Grant 2014070404010209), and Fundamental Research Funds for the Central Universities, China University of Geosciences (Wuhan).

References

[1] Y. Benezeth, P. M. Jodoin, B. Emile, H. Laurent, and C. Rosenberger, "Review and evaluation of commonly-implemented background subtraction algorithms," in *Proceedings of the 19th International Conference on Pattern Recognition (ICPR '08)*, pp. 1–4, Tampa, Fla, USA, December 2008.

[2] C. Stauffer and W. E. L. Grimson, "Adaptive background mixture models for real-time tracking," in *Proceedings of the IEEE Computer Society Conference on Computer Vision and Pattern Recognition (CVPR '99)*, pp. 246–252, Fort Collins, Colo, USA, June 1999.

[3] Y. Wang, J. W. Tian, and Y. H. Tan, "Effective Gaussian mixture learning and shadow suppression for video foreground segmentation," in *MIPPR 2007: Automatic Target Recognition and Image Analysis; and Multispectral Image Acquisition, 67861D*, vol. 6786 of *Proceedings of SPIE*, Wuhan, China, November 2007.

[4] Z. Zivkovic, "Improved adaptive Gaussian mixture model for background subtraction," in *Proceedings of the 17th International Conference on Pattern Recognition (ICPR '04)*, vol. 2, pp. 28–31, Washington, DC, USA, 2004.

[5] J. Cheng, J. Yang, Y. Zhou, and Y. Cui, "Flexible background mixture models for foreground segmentation," *Image and Vision Computing*, vol. 24, no. 5, pp. 473–482, 2006.

[6] A. Elgammal, D. Harwood, and L. Davis, "Non-parametric model for background subtraction," in *Computer Vision—ECCV 2000: 6th European Conference on Computer Vision Dublin, Ireland, June 26–July 1, 2000 Proceedings, Part II*, vol. 1843 of *Lecture Notes in Computer Science*, pp. 751–767, Springer, Berlin, Germany, 2000.

[7] K. Kim, T. H. Chalidabhongse, D. Harwood, and L. Davis, "Real-time foreground-background segmentation using codebook model," *Real-Time Imaging*, vol. 11, no. 3, pp. 172–185, 2005.

[8] O. Barnich and M. Van Droogenbroeck, "ViBe: a universal background subtraction algorithm for video sequences," *IEEE*

Transactions on Image Processing, vol. 20, no. 6, pp. 1709–1724, 2011.

[9] Y. Wang, D. H. Wang, and W. Fang, "Automatic node selection and target tracking in wireless camera sensor networks," *Computers & Electrical Engineering*, vol. 40, no. 2, pp. 484–493, 2014.

[10] Y. Shen, W. Hu, J. Liu, M. Yang, B. Wei, and C. T. Chou, "Efficient background subtraction for real-time tracking in embedded camera networks," in *Proceedings of the 10th ACM Conference on Embedded Networked Sensor Systems (SenSys '12)*, pp. 295–308, Toronto, Canada, November 2012.

[11] V. Cevher, A. Sankaranarayanan, M. Duarte, D. Reddy, R. Baraniuk, and R. Chellappa, "Compressive sensing for background subtraction," in *Proceedings of the European Conference on Computer Vision (ECCV '08)*, pp. 155–168, Marseille, France, October 2008.

[12] G. Warnell, D. Reddy, and R. Chellappa, "Adaptive rate compressive sensing for background subtraction," in *Proceedings of the IEEE International Conference on Acoustics, Speech, and Signal Processing (ICASSP '12)*, pp. 1477–1480, Kyoto, Japan, March 2012.

[13] J. Huang, X. Huang, and D. Metaxas, "Learning with dynamic group sparsity," in *Proceedings of the 12th International Conference on Computer Vision (ICCV '09)*, pp. 64–71, jpn, October 2009.

[14] J. Wright, A. Y. Yang, A. Ganesh, S. S. Sastry, and Y. Ma, "Robust face recognition via sparse representation," *IEEE Transactions on Pattern Analysis and Machine Intelligence*, vol. 31, no. 2, pp. 210–227, 2009.

[15] D. L. Donoho, "Compressed sensing," *IEEE Transactions on Information Theory*, vol. 52, no. 4, pp. 1289–1306, 2006.

[16] E. J. Candès, J. K. Romberg, and T. Tao, "Stable signal recovery from incomplete and inaccurate measurements," *Communications on Pure and Applied Mathematics*, vol. 59, no. 8, pp. 1207–1223, 2006.

[17] E. J. Candès, J. Romberg, and T. Tao, "Robust uncertainty principles: exact signal reconstruction from highly incomplete frequency information," *IEEE Transactions on Information Theory*, vol. 52, no. 2, pp. 489–509, 2006.

2

Intermediate Frequency Digital Receiver based on Multi-FPGA System

Chengchang Zhang[1,2] and Lihong Zhang[2]

[1]*Department of Electrical Engineering, College of Electronic Engineering, Chongqing University of Posts and Telecommunications, Chongqing, China*

[2]*Electrical and Computer Engineering, Faculty of Engineering and Applied Science, Memorial University, St. John's, NL, Canada*

Correspondence should be addressed to Chengchang Zhang; zhangcc@cqupt.edu.cn

Academic Editor: Bin-Da Liu

Aiming at high-cost, large-size, and inflexibility problems of traditional analog intermediate frequency receiver in the aerospace telemetry, tracking, and command (TTC) system, we have proposed a new intermediate frequency (IF) digital receiver based on Multi-FPGA system in this paper. Digital beam forming (DBF) is realized by coordinated rotation digital computer (CORDIC) algorithm. An experimental prototype has been developed on a compact Multi-FPGA system with three FPGAs to receive 16 channels of IF digital signals. Our experimental results show that our proposed scheme is able to provide a great convenience for the design of IF digital receiver, which offers a valuable reference for real-time, low power, high density, and small size receiver design.

1. Introduction

In the field of aerospace TTC system, a ground station is mostly used to capture and track aerial vehicles with the aid of a large high gain narrow beam. Such a searching and tracking process works when the angle error detector of an antenna servo system can detect the angle error (including azimuth error and pitch error) between laser gyroscope inertial axis and antenna axis. After amplification and certain operations, this angle error signal can drive a servo motor to make the electric antenna axis aim at the laser gyroscope inertial axis so as to realize automatic tracking to the aircraft [1, 2]. This traditional method is mainly based on analog devices and technologies, which lead to some major disadvantages, for example, expensive laser gyroscope and large-size turntable antenna.

With the advancement of digital devices and digital signal processing technology, a new concept of digital array radar (DAR), which can perfectly combine digital technology and antenna technology, emerges to adopt DBF to replace the traditional analog beam forming in both transmitting and receiving mode [3–5]. The key techniques of DAR mainly include digital transmitter and receiver (T/R) module, multichannel digital receiving, high speed and large capacity data transmission, wide band DBF, and high performance software signal processing. Multichannel digital receiver is the core of DAR because of its high hardware/software complexity, high integration, and high performance index.

Taking into account the strict requirements for DAR regarding correct identification of amplitude and phase between multiple receivers, DBF receiver adopts the thought of software defined radio (SDR) [6–8]. The central idea of SDR is to construct an open, standardized, and modular platform, which will use software to accomplish modulation and demodulation functions. And the bandwidth of the high speed broadband analog to digital converter (A/D) is highly close to the receiver or even to the antenna. The basic structures of SDR can be roughly divided into three types: radio frequency (RF) low pass sampling, RF band pass sampling, and IF band pass sampling.

In order to improve the selectivity, suppress out-band interferences, and reduce the processing rate of the subsequent signal processor, radars usually adopt the IF band pass sampling, which is an IF digital receiver usually adopting super heterodyne system [9]. Since the sampling takes place directly in intermediate frequency, the subsequent processing can all be handled in the digital form. This will overcome the shortcomings of the traditional receivers that are based on analog devices and analog signal processing technique, such as gain variation, DC temperature drift, and nonorthogonality between I and Q. Therefore, the stability and consistency can be significantly improved.

With the development and application of wideband high resolution radar, the bandwidth of digital receivers gets increasingly wider, and the sampling rate is increasingly higher. The design of wideband IF digital receivers is more complicated than that of the narrow band IF digital receivers [10–13], which are mainly reflected in the following aspects:

(1) With the increase of sampling data rate, the follow-up signal processing speed cannot be kept up with. In addition, the high data rate also leads to the problems of high speed data transmission and data synchronization.

(2) Combinational frequency interference within a receiver is greatly increased. Therefore, an optimized design considering a variety of factors (e.g., the working frequency, low phase noise, low spurious frequency source, and electromagnetic compatibility) is highly essential.

(3) RF front-end analog circuits of a receiver inevitably have amplitude and phase discordance, especially for the wideband receivers. In addition, there is inconsistency of the channel frequency response, which would affect the side lobe level, the output signal to noise ratio (SNR), or zero depth of adaptive beam forming. Therefore, the channel equalizer [14, 15] must be adopted for correction.

It can be seen that the multiple channel digital receivers based on DBF are much more complex than the traditional ones. As a matter of fact, a DBF receiver may include many channels, even up to thousands of channels. However, there exist some stringent requirements for the receiver size and weight of airborne radar, space borne radar, or other radars.

To resolve the abovementioned problems, in this paper we propose a DAR receiver based on Multi-FPGA system [16], which can provide a large number of digital I/O to meet the need of digital receivers for digital interface. The high speed parallel processing capability based on Multi-FPGA hardware, mature IP cores, and digital processing technology can greatly facilitate the design of our digital receiver, making it feasible for real-time application with the features of low power consumption, high density, and physical size miniaturization.

Our contributions of this work lie in the following aspects:

(1) We propose that the multichannel ADC output digital signal is directly connected to FPGA, and the multichannel digital down conversion (DDC) is also realized in FPGA. This method is simpler, cheaper, and more flexible compared to using the programmable digital down converter ASIC chip, such as GC4016 or AD6654. This improvement is especially significant for digital array receivers with hundreds of channels.

(2) Based on the principle and analysis of digital beam forming, we propose and implement CORDIC algorithm scheme for digital beam forming in FPGA.

(3) We realize our digital receiver scheme of circular array with 16 channels in a Multi-FPGA system. This Multi-FPGA system comprises 3 pieces of FPGA: two FPGAs are directly connected with the digital signal output from ADC, while the third FPGA is to implement DBF and terminal display driver.

2. Analysis for Array Antenna

Array antenna is a synthetic antenna composed of a plurality of radiating elements, which generates a strong direction radiation. Antenna array has two forms, one is a line array [17] where all the elements of the array are arranged in a straight line; the other is a planar array [18] where all the array elements are arranged on a plane. An array element might be a simple weak directional antenna, such as dipole, half wave dipole, and waveguide slot, or might be a more complex antenna, such as parabolic antenna and Yaga antenna. The array antenna forms narrow directivity beam, whose performance is determined by position, amplitude, and phase of the radiating elements. Figure 1 shows the geometric relationship of array antenna elements.

Let us assume an element is located at $P_1(x_1, y_1, z_1)$ and the phase reference point is $O(0,0,0)$. The electric field strength measured at the far field point P is

$$E(\theta,\phi) = I_0 \frac{e^{-jkR_1}}{R_1} f(\theta,\phi), \tag{1}$$

where I_0 is the complex amplitude, k is the wave number that equals $2\pi/\lambda$, and $f(\theta,\phi)$ is the radiation pattern.

Considering that the radiation source is composed of a number of array elements, the coordinate of any element P_i is (x_i, y_i, z_i) and the vector from point O to P_i is

$$r_i = \vec{a}_x x_i + \vec{a}_y y_i + \vec{a}_z z_i, \tag{2}$$

where $\vec{a}_x$, $\vec{a}_y$, and $\vec{a}_z$ are unit vectors of axes x, y, and z, respectively.

The far field component of the total electric field is

$$E_i(\theta,\phi) = I_i \frac{e^{-jkR_i}}{R_i} f(\theta_i,\phi_i), \tag{3}$$

where I_i is the complex amplitude and $f(\theta_i,\phi_i)$ is the radiation pattern. R_i can be represented as follows:

$$\begin{aligned} R_i &= \sqrt{(x-x_i)^2 + (y-y_i)^2 + (z-z_i)^2} \\ &= r\sqrt{1 + \frac{\left(x_i^2 + y_i^2 + z_i^2\right)}{r^2} - \frac{2\left(xx_i + yy_i + zz_i\right)}{r^2}}, \end{aligned} \tag{4}$$

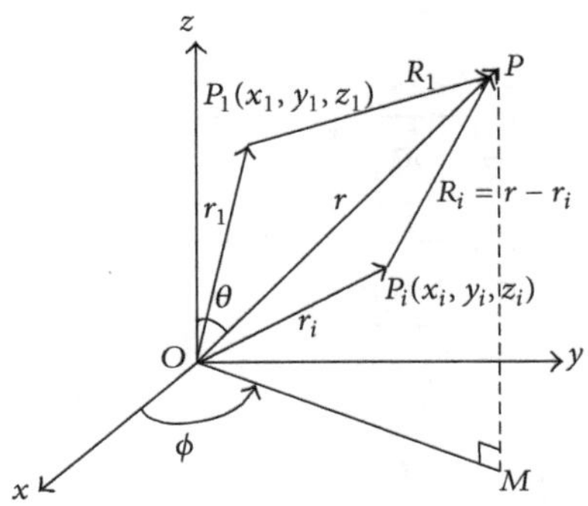

FIGURE 1: Geometry of array antenna.

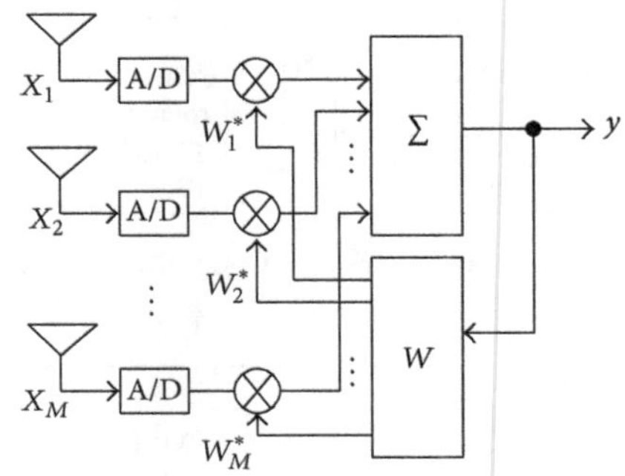

FIGURE 2: Structure of DBF.

where $x = r\sin\theta\cos\phi$, $y = r\sin\theta\sin\phi$, $z = r\cos\theta$, and

$$\frac{x_i^2 + y_i^2 + z_i^2}{r^2} = \frac{|r_i|^2}{r^2} \ll 1. \quad (5)$$

Based on binomial expansion, a better estimate for (4) is obtained using

$$R_i = r - r\left(x_i \sin\theta\cos\phi + y_i \sin\theta\sin\phi + z_i\cos\theta\right). \quad (6)$$

Relative to the reference point, the phase contribution of P_i in the far field is

$$e^{-jkR_i} = e^{-jkr} e^{jk(x_i\sin\theta\cos\phi + y_i\sin\theta\sin\phi + z_i\cos\theta)}. \quad (7)$$

The unit vector along the direction r is r_0 as shown in the following equation:

$$r_0 = \vec{a}_x \sin\theta\cos\phi + \vec{a}_y \sin\theta\sin\phi + \vec{a}_z\cos\theta. \quad (8)$$

Thus, (7) can be rewritten as

$$e^{-jkR_i} = e^{-jkr} e^{jk(r_i \cdot r_0)} = e^{-jkr} e^{j\psi_i(\theta,\phi)}. \quad (9)$$

For an equal amplitude excitation array, the radiation pattern of each array element is isotropic. Without loss of generality, we can assume $f(\theta,\phi) = 1$. Thus, the total electric field is

$$E(\theta,\phi) = \sum_{i=1}^{N} I_i e^{j\psi_i(\theta,\phi)}. \quad (10)$$

This is called the matrix factor expression. If $f(\theta,\phi)$ is equal to another value, only the coefficient of the expression changes, and the amplitude of the composed beam is changed to another value. However, this will not affect the shape and direction of the beam.

3. CORDIC-Based DBF

3.1. Model of DBF. Although each element pattern of antenna array is omnidirectional, the array's direction gain can be gathered in the direction of the desired signal by weighting the sampled data of each channel, which is equivalent to forming a high gain beam in the desired direction. The principle of selecting the weighting coefficients is to make the output signal maximal.

As shown in Figure 2, DBF for an array of M elements is achieved by adjusting the weighting coefficient of each element, the output of the array is the weighted sum of each component from the received signal vector $x(n)$. Let the weighting vector be $W = [W_1, \ldots, W_M]^T$; the output can be written as

$$y(n) = W^H x(n) = \sum_{m=1}^{M} W_M^* X_M(n), \quad (11)$$

where superscript $*$ represents complex conjugate.

Under narrowband model, the signal amplitude of each element at any instant is the same. Equation (11) can be realized only by adjusting weighting coefficient of phase shifter, which is only for signal phase, without a need for changing signal amplitude. If there is only one signal from direction θ_k and guidance vector is $a(\theta_k)$, then $y(n) = a^H(\theta_k)a(\theta_k) = M$ when weighting vector w is equal to $a(\theta_k)$. This is the largest, which is used to achieve the role of oriented positioning. At this time, the weights of the various signals are for the same phase superposition, which is also for the spatial matching filter.

3.2. DBF Module Based on CORDIC Algorithm [19]. Let the signal received by the ith array element be

$$X_i = x_{I_i} + jx_{Q_i}. \quad (12)$$

The complex weighting coefficient is

$$w_i = w_{I_i} + jw_{Q_i} = \exp\left(j\alpha_i\right). \quad (13)$$

The output after complex weighted is

$$X_i' = x_{I_i}' + jx_{Q_i}' = \left(x_{I_i} + jx_{Q_i}\right) * \exp\left(-j\alpha_i\right), \quad (14)$$

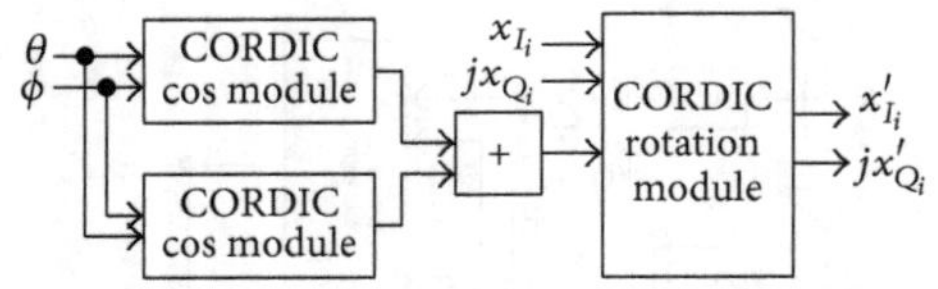

FIGURE 3: Single complex weighting module.

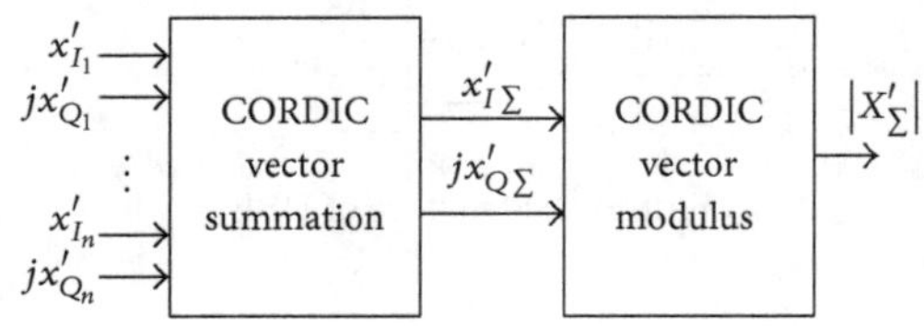

FIGURE 4: DBF module.

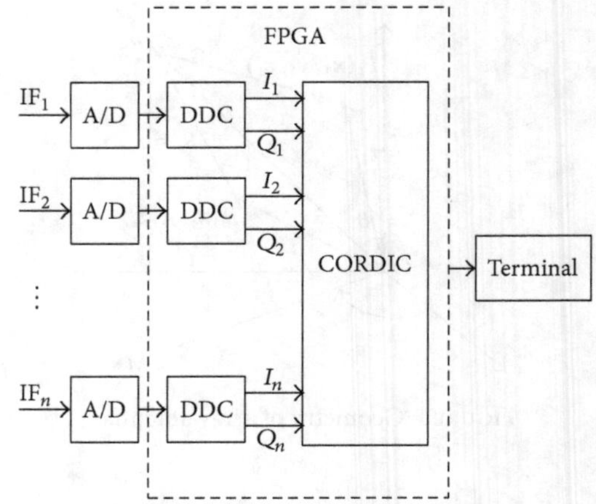

FIGURE 5: Block diagram of IF digital receiver.

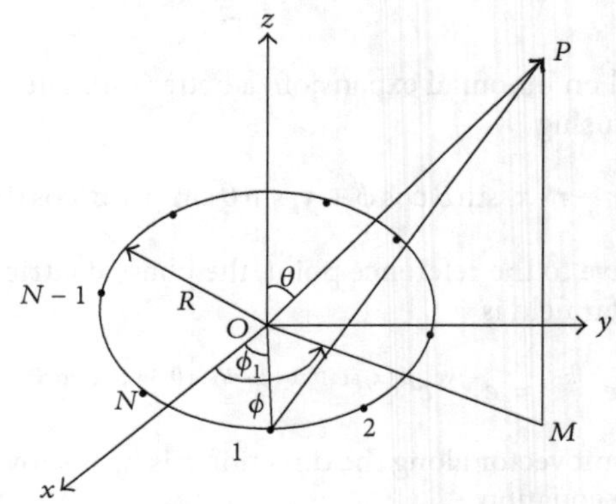

FIGURE 6: Illustration of circular array.

where

$$\alpha_i = \frac{2\pi r}{\lambda}\sin\theta\cos(\phi-\varphi_i) = \frac{1}{2} \cdot \pi\left[\cos\left(\frac{\pi}{2}-\theta+\phi-\varphi_i\right) + \cos\left(\frac{\pi}{2}-\theta-\phi+\varphi_i\right)\right],$$
$$\varphi_i = \frac{2(i-1)\pi}{N}, \quad i = 1,2,\ldots,N. \tag{15}$$

Through the conversion above, it can be seen that the complex weighting operation for each channel can be implemented by using three CORDIC algorithm modules: two cosine operations and one rotating module. Figure 3 depicts a single channel complex weighting module schematic.

In order to track target, we need to do parallel addition operation for I and Q signals, get total output $X'_{\Sigma} = x'_{I\Sigma} + jx'_{Q\Sigma}$, and then use the vector mode of CORDIC algorithm to obtain $|X'_{\Sigma}|$ as shown in Figure 4.

The operations above can be implemented by configuring CORDIC IP core in FPGA.

3.3. IF Digital Receiver in FPGA. Based on the realization scheme for DBF on top of CORDIC algorithm, we can construct the IF digital array radar receiver by combining it with A/D conversion and digital down conversion (DDC) [20]. A simplified block diagram of IF digital array radar receiver is shown in Figure 5.

4. Simulation for 16-Element Circular Array DBF

As shown in Figure 6, in the x-o-y plane, N isotropic radiation elements are evenly distributed on the circle with radius of R, $N = 16$.

Here

$$\phi_n = \frac{2\pi}{N}n, \quad n = 1,2,\ldots,16. \tag{16}$$

The coordinates of nth elements are

$$x_n = R\cos\phi_n,$$
$$y_n = R\sin\phi_n, \tag{17}$$
$$z_n = 0.$$

So,

$$\begin{aligned} k(r_n \cdot r_0) &= \psi_n \\ &= k(R\sin\theta\cos\phi\cos\phi_n + R\sin\theta\sin\phi\sin\phi_n + 0) \\ &= Rk\sin\theta(\cos\phi\cos\phi_n + \sin\phi\sin\phi_n) \\ &= Rk\sin\theta\cos(\phi_n - \phi). \end{aligned} \tag{18}$$

According to (10),

$$E(\theta,\phi) = \sum_{n=1}^{N} I_n \exp\left\{j\frac{2\pi}{\lambda}R\sin\theta\cos(\phi_n-\phi)\right\}, \tag{19}$$

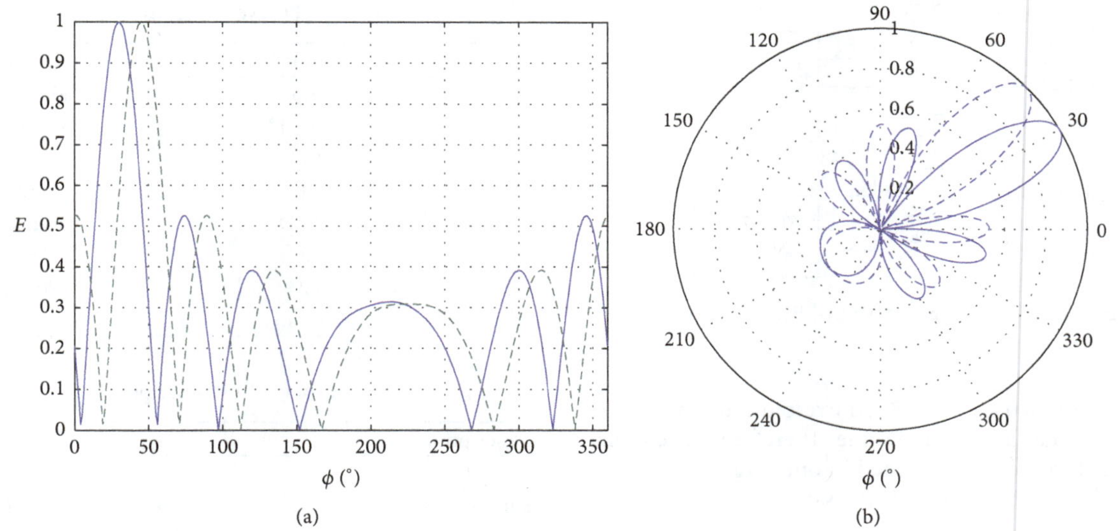

FIGURE 7: Pattern of 16-element circular array. (a) Normalized Cartesian coordinates. (b) Normalized polar diagram. ($\theta = 45°$, $\theta_0 = 60°$, $R/\lambda = 1:1$, and $\phi_0 = 30°$(—)/45°(- - -)).

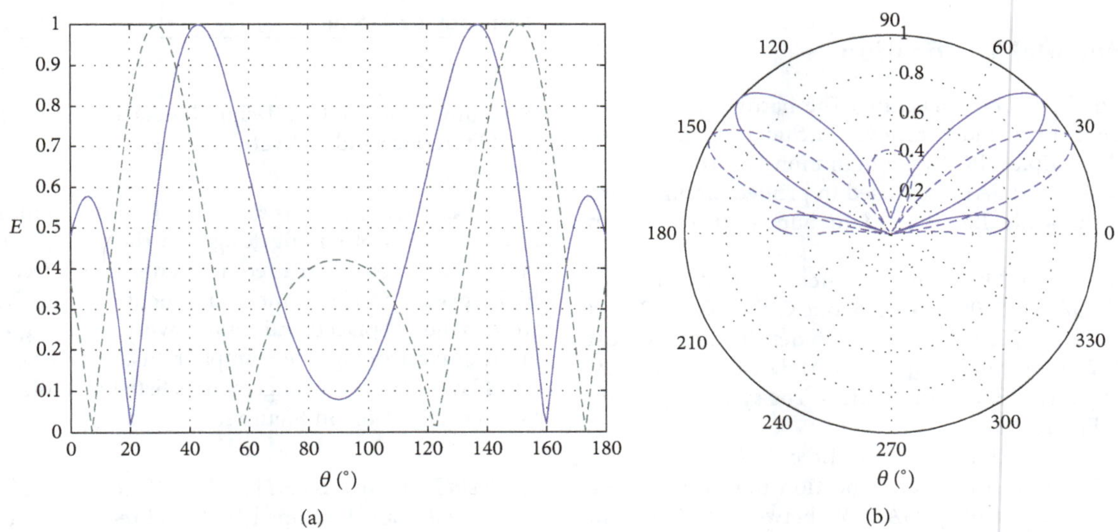

FIGURE 8: Pattern of 16-element circular array. (a) Normalized Cartesian coordinates. (b) Normalized polar diagram. ($\phi = 45°$, $\phi_0 = 30°$, $R/\lambda = 1:1$, and $\theta_0 = 30°$(- - -)/45°(—)).

where I_n is the complex current of nth element. For equal amplitude excitation, we have $I_n = 1$. When the array main beam points at (θ_0, ϕ_0), (19) can be expressed as follows:

$$E(\theta,\phi) = \sum_{n=1}^{N} I_n \exp\left\{ j\frac{2\pi}{\lambda} \cdot R\left[\sin\theta\cos(\phi_n - \phi) - \sin\theta_0\cos(\phi_n - \phi_0)\right]\right\}. \quad (20)$$

According to (20), circular pattern can scan in both θ and ϕ directions. For convenient observation and analysis, we make θ and ϕ constant separately. We first let θ be a constant; $\theta = 45°$, $\theta_0 = 60°$, and $R/\lambda = 1:1$; the simulation results are shown in Figure 7. The solid lines represent the pattern of $\phi_0 = 30°$, and the dashed lines represent the pattern of $\phi_0 = 45°$.

Then we let ϕ be a constant; $\phi = 45°$, $\phi_0 = 30°$, and $R/\lambda = 1:1$. Figure 8 shows the array pattern from our simulation.

The following can be observed from (20) and our simulation results:

(1) The beam of circular array can point at the specified direction in two dimensions, which can be used to distinguish the direction of a target, that is, the circular array has resolution with two dimensions, which can also distinguish the elevation and azimuth.

(2) Since θ is a constant in Figure 7, with the increase of ϕ, the beam width remains unchanged. That is to say, in

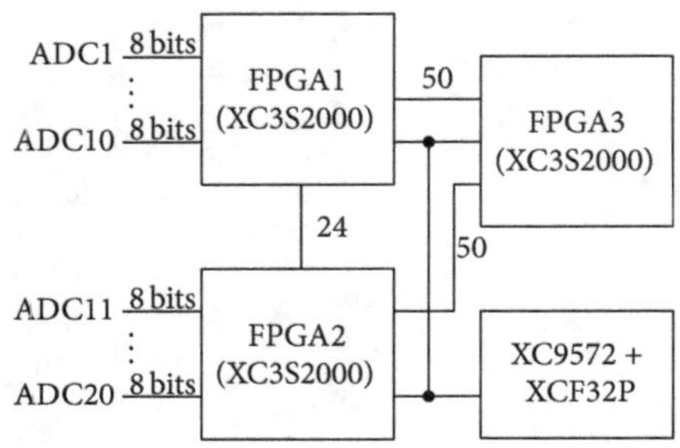

FIGURE 9: The diagram of Multi-FPGA system.

TABLE 1: The layer assignment of PCB.

	Symbol	Description
(1)	Top	Element layer
(2)	GND/CLK	Analog ground/clock
(3)	S2	Signal layer
(4)	P1	Power layer
(5)	SGND	Digital ground
(6)	S3	Signal layer
(7)	P2	Power layer
(8)	Bottom	Element layer

TABLE 2: Test description.

Description	Value
Array form	Circular array
Sum of array elements	16
Array element	Monopole antenna
RF frequency	1032 MHz
IF frequency	70 MHzSinglecarrier
Sampling frequency	66 MHz
Beamforming direction	ϕ

the range of $0^\circ \leq \phi \leq 360^\circ$, the resolution capability of the circular array is the same. Therefore, the circular array beam can be scanned in one direction, and the amplitude of the signal is the same.

(3) From Figure 8 where ϕ is a constant, the larger the value of θ is, the wider the width of the beam gets. This means that the ability for the circular array to distinguish a target will be reduced.

5. Experimental Verification

5.1. Multi-FPGA System Structure. The hardware design of the IF digital array radar receiver includes circuit board design and multiple FPGAs configuration control program design. OrCAD and PADS are used in printed circuit board (PCB) design, while ISE is used in our control program design.

Our developed Multi-FPGA system consists of three Spartan 3 FPGAXC3S2000FGG456, one CPLD XC9572, and one Prom XCF32p. XC3S2000 is a high density and low cost FPGA with 2 M logic gates, up to 326 MHz clock frequency, embedded by forty 18×18 multipliers. The system structure is shown in Figure 9.

The interconnection between three FPGAs is 8-way topology [16], where the interconnection pin numbers are 24 between FPGA1 and FPGA2, 50 between FPGA1 and FPGA3, and 50 between FPGA2 and FPGA3. There are 20 receiving IF branches, among which 16 are used to receive 16 channels sampling signal and the remaining 4 channels are spare. FPGA1 receives the first to the eighth channel IF sampling digital signals, while FPGA2 receives the ninth to the sixteenth channel IF sampling digital signals. The eight IF signals' DDC is realized in FPGA1 in parallel and the digital baseband signal is sent to FPGA3. Similarly, FPGA2 completes another 8 parallel IF signals' digital down conversion and the digital baseband signal is also sent to FPGA3. The operation of DBF is completed in FPGA3, which also operates display driver program and handles DBF result display on screen. The XC9572 plus XCF32p architecture is to complete the three FPGAs configuration. XCF32p is a 32 M flash memory, which can be divided into four 8 M segments for configuring four FPGAs in order to reduce the number of used flash memory chips. The chip configuration control program is stored in CPLD XC9572. All of the FPGA, flash memory, and CPLD chips are products from XILINX, which can simplify our system design thanks to the use of the same JTAG interface and software.

5.2. Layer Division Scheme of PCB. In order to minimize the adverse effects of circuit design on the performance of the system and to ensure the requirements of high frequency and high speed, the electronic components in our PCB design were manually placed and routed, with a thorough consideration of electromagnetic compatibility. Our developed PCB has 8 layers, whose arrangement is listed in Table 1. The PCB board is illustrated in Figure 10.

5.3. Field Experiments and Results. We conducted field experiments with our developed IF digital receiver. The receiving antenna array was a circular array, which included 16 array elements. The beacon was a monopole antenna, which spread signals. The antenna array searched the target, determined the target's range, and formed a strong beam in the direction of the target. One terminal screen displayed the beam to thereby complete the target searching and locating.

Test conditions and parameters are listed in Table 2. As a good trade-off between experimental effects and prototyping cost, the receiver antenna array was determined to be a 16-element circular array with the array element of monopole antenna.

The received radio frequency (RF) for antenna array was 1032 MHz, which is in L band. IF signal frequency after RF-analog conversion was 70 MHz, and the sampling frequency of IF signal was 66 MHz. The size of the verification site in our field experiment was about 200 meters $\times$ 200 meters.

Antenna array, analog conversion channel, and the receiving circuit board were placed in the middle of the experiment site, which is shown in Figure 11(a). The beacon

FIGURE 10: The snapshot of our developed PCB.

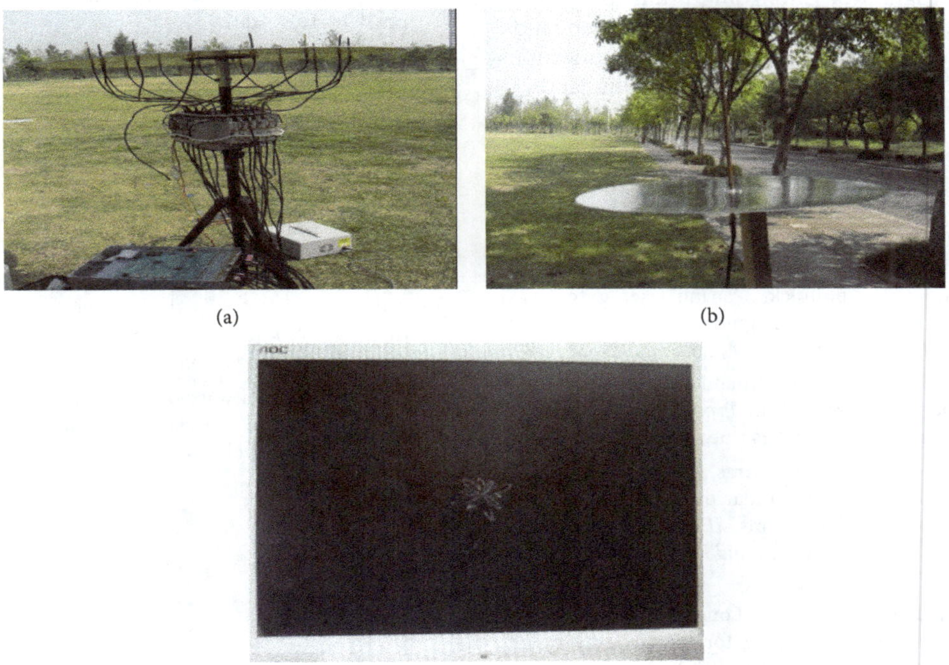

(a) (b)

(c)

FIGURE 11: Experiment results. (a) Verification platform. (b) The beacon. (c) The formed beam.

was placed at the edge of the site, which is shown in Figure 11(b). Due to the high position of the antenna array and the fixed beacon, θ is a fixed value. The radio frequency signal was transmitted by the beacon, and the receiver received the signal, formed the DBF beam, and pointed to the direction of the beacon, which is shown in Figure 11(c). When moving the position of the beacon during our experiments, which is equivalent to changing ϕ, the direction of the DBF beam was also moving along with the beacon. These field experiments have confirmed a great success for our developed algorithmic scheme and the implemented prototype.

In this work, we are only focused on the circular array digital receiver scheme. If the antenna is arranged as a linear array, the digital receiver DBF waveform shall point to the target in a specific direction with a certain θ or ϕ value. But the entire design would become much simpler.

6. Conclusions

In the field of aerospace measurement and control, the ground station signal receiving equipment based on analog technology is usually expensive, large in size, and inflexible. In this paper we proposed a scheme based on Multi-FPGA system to realize intermediate frequency digital receivers. The digital beam forming (DBF) was achieved by CORDIC algorithm and an experimental prototype was developed. Our field experiments show that the developed prototype worked perfectly. The power consumption of the IF receiver circuit was only 6 watts; the consuming capacity of FPGAs was about 60%. IF sampling and digital down conversion for 16 channels are developed in parallel, and process of programs in FPGA is hard wired circuit, so, our proposed scheme can provide a great convenience to the design of IF digital receivers, which provide a valuable reference for real-time application with strong features of low power consumption, high density, and compact utilization of FPGA logic gates.

Competing Interests

The authors declare that they have no competing interests.

Acknowledgments

This work was supported in part by Scientific and Technological Research Program of Chongqing Municipal Education Commission (KJ1500407), Youth Science Foundation of Chongqing University of Posts and Telecommunications (A2014-107), and Memorial University of Newfoundland.

References

[1] H. Sorg, *Advances in Inertial Navigation Systems and Components*, Advisory Group for Aerospace Research and Development, Neuilly-sur-Seine, France, 1981.

[2] R. J. Wright and J. V. Sponnick, "A ring laser gyro based navigator for space launch vehicle guidance," *IEEE Aerospace and Electronic Systems Magazine*, vol. 4, no. 3, pp. 29–38, 1989.

[3] L. Qian and X. Wang, "A new Wideband Digital Array Radar (WB-DAR) experiment system," in *Proceedings of the IEEE Region 10 Symposium (TENSYMP '14)*, pp. 440–445, April 2014.

[4] J. Del Castillo, S. Sanchez, R. De Porras, A. Pedreira, and J. R. Larranaga, "L-band digital array radar demonstrator for next generation multichannel SAR systems," *IEEE Journal of Selected Topics in Applied Earth Observations and Remote Sensing*, vol. 8, no. 11, pp. 5007–5014, 2015.

[5] F. Wang, J. Li, J. Liu, X. Chen, and W. Long, "System realization of broadband digital beam forming for digital array radar," *Journal of Radars*, vol. 2, no. 3, pp. 314–318, 2013.

[6] M. Woh, Y. Lin, S. Seo, S. Mahlke, and T. Mudge, "Analyzing the next generation software defined radio for future architectures," *Journal of Signal Processing Systems*, vol. 63, no. 1, pp. 83–94, 2011.

[7] A. L. G. Reis, A. F. Barros, K. G. Lenzi, L. G. P. Meloni, and S. E. Barbin, "Introduction to the software-defined radio approach," *IEEE Latin America Transactions*, vol. 10, no. 1, pp. 1156–1161, 2012.

[8] J. Mar and Y.-R. Lin, "Implementation of SDR digital beamformer for microsatellite SAR," *IEEE Geoscience and Remote Sensing Letters*, vol. 6, no. 1, pp. 92–96, 2009.

[9] G. Q. Wang, X. Z. Wei, and H. Z. Lu, "Double-IF quadrature demodulation of super-heterodyne radar receiver," in *Proceedings of the 9th International Conference on Signal Processing (ICSP '08)*, pp. 2505–2508, Beijing, China, October 2008.

[10] B. M. Albaker and N. A. Rahim, "Signal acquisition and parameter estimation of radio frequency pulse radar using novel method," *IETE Journal of Research*, vol. 55, no. 3, pp. 128–134, 2009.

[11] A. Kale, R. Thirumuru, and V. S. R. Pasupureddi, "Wideband channelized sub-sampling transceiver for digital RF memory based electronic attack system," *Aerospace Science and Technology*, vol. 51, pp. 34–41, 2016.

[12] K. George and C.-I. H. Chen, "A hybrid computing platform digital wideband receiver design and performance measurement," *IEEE Transactions on Instrumentation and Measurement*, vol. 60, no. 12, pp. 3956–3958, 2011.

[13] H. Wang, Y. Deng, B. Dong, and D. Su, "Analysis of the spectrum transform for narrow-band signal passing through nonlinear section of a digital radar receiver," *Eurasip Journal on Wireless Communications and Networking*, vol. 2015, article 265, 2015.

[14] R. Martinek and J. Zidek, "The real implementation of ANFIS channel equalizer on the system of software-defined radio," *IETE Journal of Research*, vol. 60, no. 2, pp. 183–193, 2014.

[15] P. P. Vaidyanathan, "Generalizations of the sampling theorem: seven decades after Nyquist," *IEEE Transactions on Circuits and Systems. I. Fundamental Theory and Applications*, vol. 48, no. 9, pp. 1094–1109, 2001.

[16] S. Hauck, *Multi-FPGA systems [Doctor of Philosophy]*, University of Washington, 1995.

[17] K. Kim, W. Seong, K. Lee, S. Kim, and T. Shim, "Range-dependent geoacoustic inversion of vertical line array data using matched beam processing," *The Journal of the Acoustical Society of America*, vol. 125, no. 2, pp. 735–745, 2009.

[18] C. Chen, J. Shao, S. Meng, G. Fang, and H. Yin, "4-element planar array antenna for UWB application," *Journal of Electronics*, vol. 31, no. 3, pp. 175–179, 2014.

[19] P. K. Meher and S. Y. Park, "CORDIC designs for fixed angle of rotation," *IEEE Transactions on Very Large Scale Integration*, vol. 21, no. 2, pp. 217–228, 2013.

[20] R. S. Aditya, "Software defined radio theoretical analysis and design approach," *International Journal of Engineering Trends and Technology*, vol. 4, no. 5, pp. 1788–1791, 2013.

3

Subpixel Mapping Method of Hyperspectral Images based on Modified Binary Quantum Particle Swarm Optimization

Shuhan Chen,[1] Xiaorun Li,[1] and Liaoying Zhao[2]

[1] *College of Electrical Engineering, Zhejiang University, Hangzhou 310027, China*
[2] *Institute of Computer Application Technology, Hangzhou Dianzi University, Hangzhou 310018, China*

Correspondence should be addressed to Xiaorun Li; lxrly@zju.edu.cn

Academic Editor: Qunming Wang

Subpixel mapping technology can determine the specific location of different objects in the mixed pixel and effectively solve the uncertainty of the ground features spatial distribution in traditional classification technology. Existing methods based on linear optimization encounter the premature and local convergence of the optimization algorithm. This paper proposes a subpixel mapping method based on modified binary quantum particle swarm optimization (MBQPSO) to solve the above issues. The initial subpixel mapping imagery is obtained according to spectral unmixing results. We focus mainly on the discretization of QPSO, which is implemented by modifying the discrete update process of particle location, to minimize the objective function, which is formulated based on different connected regional perimeter calculating methods. To reduce time complexity, a target optimization strategy of global iteration combined with local iteration is performed. The MBQPSO is tested on standard test functions and results show that MBQPSO has the best performance on global optimization and convergent rate. Then, we analyze the proposed algorithm qualitatively and quantitatively by simulated and real experiment; results show that the method combined with MBQPSO and objective function, which is formulated based on the gap length between region and background, has the best performance in accuracy and efficiency.

1. Introduction

Hyperspectral images are composed of hundreds of bands with a very high spectral resolution, generally from the visible to the infrared region. For every recorded pixel, rich spectral information provides a complete spectral description and a better characterization of the observed surface, which results in a very powerful tool for materials discrimination and earth observation. However, a common drawback of hyperspectral sensors is the relatively low spatial resolution, which leads to the problem of mixed pixels (e.g., pixels containing mixture of different materials). Ground feature spatial detail in mixed pixels is extremely important for land cover mapping, coast-line extraction, change detection, and landscape index estimating. Mixed pixels cannot be correctly addressed by traditional hard classification methods. Subpixel mapping is an effective measure to solve this problem.

Subpixel mapping and pixels spatial dependence theory were initially proposed by Atkinson [1]. Based on this theory, several subpixel mapping techniques have been proposed. Existing subpixel mapping methods contain the following typical algorithms: the neural network method, the geostatistics method, the Markov random field method, and the linear optimization method.

The first category is the neural network method. Among them, Hopfield's neural network (HNN) model combined with the constrained energy minimization principle of output neuron has been used in the subpixel mapping. In the HNN model, each pixel is a neuron that constructs an energy function with the correlation between subpixels and the neighboring subpixel, using the mixed pixels abundance as the constraint [2, 3]. The improved HNN method for small scale targets and small scale coexisting with large scale target had been proposed in [4]. A Back Propagation (BP) neural network is trained to learn the appropriate location of the subpixels, which belongs to the different classes inside the pixel [5]. At the same time, wavelet transform and genetic algorithms have been used to solve subpixel mapping

[6, 7]. The ARTMAP neural network [8] has been utilized to realize the learning model in [6]. Based on the subpixel shifted remote sensing images (SSRSI), component information obtained from SSRSI was put into the HNN model to increase ratio constraints and reduce uncertainty [9]. In order to eliminate isolated pixels, the results of the BP neural network model were postprocessed [10]. Multiple low spatial resolution subpixel shifted images were used to modify the BP neural network to reduce uncertainty and error in the BPNN model [11]. The general regression neural network (GRNN) has been proposed to achieve improved accuracy in subpixel mapping [12].

The second category is the geostatistics method. Existing approaches for subpixel mapping have been placed within an inverse problem framework; a geostatistical method was proposed for generating alternative synthetic land cover maps at the fine (target) spatial resolution [13, 14]. In [15], a model sequentially produced with local indicator variogram (SLIV) was proposed, and indicator variograms extracted from target-resolution classification were produced from a representative local area to develop an effective method.

The third category is the Markov random field method. The Markov random field (MRF) has been used in subpixel mapping [16]. Subsequently, scholars have done in-depth study on the MRF model [17, 18]. Considering the idea of the MRF, which is able to consider spatial and spectral information simultaneously, a novel sFCM-based (supervised fuzzy *c*-means) subpixel mapping model that incorporates the sFCM criterion for unmixing pixels into the objective function was proposed [19]. In terms of the energy minimization for MRF, one of the most commonly used traditional methods is simulated annealing (SA), but SA is very time-consuming. To overcome this limitation, graph cut is used in [20].

The fourth category is the linear optimization method. Through the decomposition of mixed pixels or fuzzy classification technique, the proportion of features was determined and then a mathematical model was defined for describing the spatial correlation to construct the objective function, converting the subpixel mapping into a linear optimization problem. Among linear optimization method, Verhoeye proposed the maximum spatial correlation between neighborhood pixels and solved the problem using simplex linear optimization technology [21]. Immune clonal selection algorithms and differential evolution algorithms were used to solve the mathematical model using Verhoeye's data [22, 23]. A new subpixel mapping method based on evolution agent was proposed [24]. Since subpixel/pixel spatial attraction model (SPSAM), which assumes that similar features have physical characteristics of mutual attraction, has effectively verified spatial correlation theory [25, 26], SPSAM ignores the correlation between subpixels. Wang put forward a modified SPSAM, which considers the correlation between and within the pixel simultaneously [27, 28]. By incorporating auxiliary datasets, a subpixel mapping framework based on a maximum a posteriori (MAP) model was proposed to utilize the complementary information of multiple shifted images [29]. Since the previous MAP-based subpixel mapping algorithm was obtained by downsampling a classification image without spectral unmixing errors, an adaptive subpixel mapping method based on a MAP model and a winner-take-all class determination strategy (AMCDSM) was proposed [30]. The existing spectral-spatial based SPM algorithms only use the maximal spatial dependence principle as the spatial term to describe the local spatial distribution of different land cover features; a novel spectral-spatial based SPM algorithm with multiscale spatial dependence was proposed [31]. In [32], the spectral unmixing predictions (i.e., coarse land cover proportions used as input for SPM) were considered a convolution of not only subpixels within the coarse pixel, but also subpixels from neighboring coarse pixels. A new SPM method based on optimization is developed which recognizes the optimal solution as the one that, when convolved with the PSF, is the same as the input coarse land cover proportion.

The computation complexity of those above objective functions is relatively highly complex. In order to intuitively reflect the spatial correlation, based on the principle that similar features attract each other, Villa utilized various types of regional perimeter minimum as a target after subpixel mapping, using the simulated annealing (SA) and pixel swapping algorithm (PSA) to achieve iterative optimization [33, 34]. Compared with SA and other similar evolutionary techniques, PSO has some attractive characteristics and has proven to have superior computational efficiency. This is because PSO is not applicable to problems where the position of the particle should be discrete [35]. Ertürk used binary particle swarm optimization (BPSO) to optimize the objective function that had been proposed by Villa, and the optimization algorithm has been fit for parallel implementation [35]. The results of the BPSO algorithm are better than SA. Compared with BPSO algorithm, Sun proposed the binary quantum particle swarm optimization algorithm (QPSO) [36], which demonstrates many advantages such as a simple evolution equation, fewer control parameters, fast convergence rate, and simple operation. Therefore, BQPSO algorithm was proposed for subpixel mapping [37], but the discrete update process of particle location may lead to algorithm premature convergence and local convergence. The modified version of BQPSO was proposed in this article.

In this study, a subpixel mapping method of hyperspectral images based on MBQPSO is proposed. The initial subpixel mapping imagery is obtained according to the results of spectral unmixing. Using regional perimeter minimizing to depict spatial correlation is intuitive and requires low computational complexity. The objective function is formulated by different connected regional perimeter calculating methods, such as the gap length between region and background (f_gap), the length of chain code (f_chain), and the sum of border point number (f_point). We focus mainly on the discretization of QPSO, which is implemented by modifying the discrete update process of particle location, to minimize the above three objective functions. In order to reduce time complexity, objective optimization strategy of global iteration combined with local iteration is performed.

The rest of this paper is organized as follows. Section 2 presents the basic methodology of the proposed subpixel mapping method based on MBQPSO. Section 3 provides the experimental results and analyses with simulated and real hyperspectral images. The conclusion is drawn in Section 4.

2. Subpixel Mapping Method Based on MBQPSO

The proposed method provides an accurate subpixel mapping method based on MBQPSO and local optimization strategy. In this paper, known endmembers and abundance are the premise of the subpixel mapping method. Specifically, the number of endmembers is estimated by VD [38] and extracted by VCA. Then, the abundance is obtained by FCLS in each pixel for each endmember. A pixel can be seen as a pure pixel only if the minimum difference value between the maximum abundance and the others in the pixel is greater than a certain threshold [33–35]. The initial subpixel mapping imagery is obtained according to spectral unmixing results.

Then the follow-up process of the proposed subpixel mapping algorithm based on MBQPSO consists of the following: (1) the objective function is formulated by different connected regional perimeter calculating methods; (2) the proposed MBQPSO is used to minimize the objective function; (3) optimization strategy of global iteration combined with local iteration is performed to reduce time complexity. The framework of the proposed subpixel mapping method is organized as follows: The objective function and adaptability analysis are presented in Section 2.1. The MBQPSO algorithm is introduced in Section 2.2. Two key problems of objective function optimization based on MBQPSO are described in Section 2.3 followed by the optimization strategy in Section 2.4.

2.1. Modified Binary Quantum Particle Swarm Optimization Algorithm. Before presenting the proposed MBQPSO, we first introduce QPSO method, containing the basic concept and evolution equation. The discrete update process of evolution equation is given in the form of pseudocode, and the specific details of the discrete update process of X_{ij}^{t+1} are given in the form of textual description.

2.1.1. QPSO Method. The QPSO algorithm [36] is theoretical a global convergence guaranteed algorithm, which has fewer parameters and better global searching ability than PSO. In QPSO algorithm, there are only the concepts of particle position and the distance between particles, while PSO includes the concept of speed and trajectory. In a QPSO, each particle only has position information $X_i^t = (X_{i1}^t, X_{i2}^t, \ldots, X_{id}^t)$ in a d-dimensional searching space. t is the current time and i is the ith particle in M particles. While time is t, the optimal position of the ith particle is $P_i^t = (P_{i1}^t, P_{i2}^t, \ldots, P_{id}^t)$ and the global optimal position of particle swarm is $P_g^t = (P_{g1}^t, P_{g2}^t, \ldots, P_{gd}^t)$. The evolution equations are as follows:

$$mbest^t = \sum_{i=1}^{M} \frac{P_i^t}{M} \tag{1}$$

$$p_{ij}^t = \emptyset_j^t \cdot P_{ij}^t + \left[1 - \emptyset_j^t\right] \cdot P_{gj}^t \tag{2}$$

$$X_{ij}^{t+1} = p_{ij}^t \pm \beta \cdot \left|mbest_j^t - X_{ij}^t\right| \cdot \ln\left(\frac{1}{u_{ij}^t}\right), \tag{3}$$

```
Input: P_ij^t, P_gj^t
Output: p_ij^t
r = rand()
p_ij^t = r P_ij^t + (1 - r) P_gj^t
  If p_ij^t > 0.5 then p_ij^t = 1
  else p_ij^t = 0
  end
```

ALGORITHM 1: Get_P: p_i^t discretization.

```
Input: P_i^t (1 ≤ i ≤ n)
Output: mbest^t
P̄ = (1/n) Σ_{i=1}^{n} P_i^t
for each bit in P̄
if P̄(i) > 0.5 then mbest^t(i) = 1
else if P̄(i) < 0.5 then mbest^t(i) = 0
else if rand() < 0.5 then mbest^t(i) = 0
else mbest^t(i) = 1
end
```

ALGORITHM 2: Get_mbest: $mbest^t$ discretization.

where j is a positive integer between 1 and d; u_{ij}^t and $\emptyset_j$ are random positive numbers between 0 and 1. If $u_{ij}^t \geq 0.5$, then the plus/minus sign in (3) takes a plus sign, or else it takes a minus [36]. p_i^t is the ith particle attractor, represented by a random vector that is defined by (2). $mbest^t$ is the average optimal position; β is contraction-expansion coefficient and determines the particle convergence speed. The expression is as follows:

$$\beta = (\alpha_0 - \alpha_1) \times \frac{\text{maxiteration} - \text{curiteration}}{\text{maxiteration}} + \alpha_0, \tag{4}$$

where curiteration is the current iteration number and maxiteration is the maximum iteration number. α_0 and α_1 are the initial and final value of the control parameter, respectively. In addition, α_0 is greater than α_1, and β is a linear decreasing function, controlled by the parameters.

However, like PSO and other evolutionary algorithms, QPSO readily relapses into local optimum when solving high-dimensional complex optimization problems. In order to adapt the QPSO algorithm to the practical problems in discrete search space, we introduce the concept of binary encoding to QPSO, which is implemented by modifying the discrete update process of particle location.

2.1.2. MBQPSO Method. In BQPSO [37], particle position is represented as a binary string, and continuous evolution equations (1)–(3) are all discretized. The discrete update process of $mbest^t$ and p_i^t is given in Algorithms 1 and 2.

```
Input: X_ij^t, p_ij^t, mbest^t
Output: X_ij^{t+1}
    Calculate l_ij, b, p_r
    X_ij^{t+1} = p_ij^t
    if b ≠ 0
    begin
      For each bit of X_ij^{t+1}, denote as X_ij^{t+1}(k),   k > 1
        if X_ij^{t+1} > 0
            if p_r < rand( ) then X_ij^{t+1}(k) = 1
            else X_ij^{t+1}(k) = 0
        else
            if p_r < rand( ) then X_ij^{t+1}(k) = 0
            else X_ij^{t+1}(k) = 1
        end
    end
```

Algorithm 3: Trans_X: X_{ij}^{t+1} discretization.

In order to discretize the particle updated location, deformation of (3) is as follows:

$$d_H\left(X_{ij}^{t+1}, p_{ij}^t\right) = [b]$$
$$b = \beta \cdot d_H\left(mbest_j^t, X_{ij}^t\right) \cdot \ln\left(\frac{1}{u_{ij}^t}\right), \tag{5}$$

where $d_H(\cdot)$ represents the function of Hamming distance and $[\cdot]$ represents integers. It should be noted that $b \geq 0$ due to u_{ij}^t is random positive number between 0 and 1 and $d_H(\cdot) \geq 0$

Substituting b and p_{ij}^t into (5), X_{ij}^{t+1} can be obtained in reverse. The time complexity of this procedure is $O(l_{ij}b)$. In order to reduce calculation amount, each bit of p_i^t is mutated with mutation probability p_r to obtain a new X_i^{t+1}:

$$p_r = \begin{cases} \dfrac{b}{l_{ij}}; \\ 1, & \text{if } \dfrac{b}{l_{ij}} > 1, \end{cases} \tag{6}$$

where l_{ij} is the length of X_{ij}^t.

It should be noted that "±" in (3) is not considered in (5), which may result in premature convergence. In order to overcome this problem, the discretization procedure of X_{ij}^{t+1} is modified and called Trans_X, which is shown in Algorithm 3. The time complexity of Trans_X is $O(l_{ij})$. The modified algorithm is named MBQPSO to distinguish it from the BQPSO algorithm.

The algorithm's design work is described as follows.

First, set $X_{ij}^{t+1} = p_{ij}^t$; then use mutation probability p_r to update X_{ij}^{t+1}. For X_{ij}^{t+1}, let $n10$ be the number of bits, which changes from 1 to 0 and $n01$ be the number of bits, which changes from 0 to 1. The detailed update process is divided into three cases.

(1) if $b = 0$, then $X_{ij}^{t+1} = p_{ij}^t$.

(2) According to the selection of plus or minus sign in (3), we set the following rule:

(2.1) If $u_{ij}^t \geq 0.5$, then $X_{ij}^{t+1} = p_{ij}^t + [b]$; this may result in $X_{ij}^{t+1} > p_{ij}^t$ due to $b > 0$. It is obvious that the bigger b, the bigger the difference of $n01$ and $n10$.

(a) For the case of $X_{ij}^{t+1} > 0$, in order to obtain updated $X_{ij}^{t+1} > p_{ij}^t$, the result of mutation should satisfy $n01 > n10$. According to the definition of p_r, the bigger b, the smaller p_r and the more the chance for $p_r <$ rand( ); that is, $n01 > n10$ will be satisfied if set $X_{ij}^{t+1}(k)$=1 when $p_r <$ rand( ) and otherwise set $X_{ij}^{t+1}(k) = 0$.

(b) For the case of $X_{ij}^{t+1} < 0$, in order to obtain a new $X_{ij}^{t+1} > p_{ij}^t$, the result of mutation should satisfy $n10 > n01$, which will be satisfied if set $X_{ij}^{t+1}(k) = 0$ when $p_r <$ rand( ) and otherwise set $X_{ij}^{t+1}(k) = 1$.

(2.2) If $u_{ij}^t < 0.5$, then $X_{ij}^{t+1} = p_{ij}^t - [b]$; this may result in $X_{ij}^{t+1} < p_{ij}^t$ due to $b > 0$. In the same way, the same conclusion can be obtained:

(a) For the case of $X_{ij}^{t+1} > 0$, set $X_{ij}^{t+1}(k) = 0$ when $p_r <$ rand( ) and else set $X_{ij}^{t+1}(k) = 1$.

(b) For the case of $X_{ij}^{t+1} < 0$, set $X_{ij}^{t+1}(k) = 1$ when $p_r <$ rand( ) and otherwise set $X_{ij}^{t+1}(k) = 0$.

One final comment is noteworthy. If X_{ij}^{t+1} is unsigned, a new binary string $[0, X_{ij}^{t+1}]$ will be constructed to complete the update of X_{ij}^{t+1} using the function Trans_X.

Using function analysis theory, the convergence of the BQPSO algorithm is proven [37]. Compared to BQPSO, the proposed MBQPSO algorithm simply modifies the discretization process of particle position. Therefore, it is easy to prove the convergence of the modified algorithm using the same method.

2.2. Objective Function and Adaptability Analysis

2.2.1. Objective Function. Relying on the spatial correlation tendency of features, we assume that each endmember within a pixel should be spatially close to the same endmembers in the surrounding pixels. Therefore, the cost function C to be minimized is chosen as the perimeter of the areas belonging to the same class:

$$C = \sum_{n=1}^{C_M} \sum_{m=1}^{C_n} P_i, \tag{7}$$

where C_M is the number of the classes, C_i is the number of connected components of the class n, and P_{ij} is the perimeter of the connected component j.

Table 1: The perimeter of Figure 1.

Method	(a)	(b)	(c)	(d)	(e)	(f)	(g)	(h)
f_gap	42	86	48	68	58	108	56	64
f_chain (4-connected)	26	17.31	30.73	24.38	40.24	33.07	40.49	39.90
f_chain (8-connected)	26	45.6	30.73	24.38	40.24	67.01	40.49	39.90
f_point	26	27	25	25	34	36	34	34

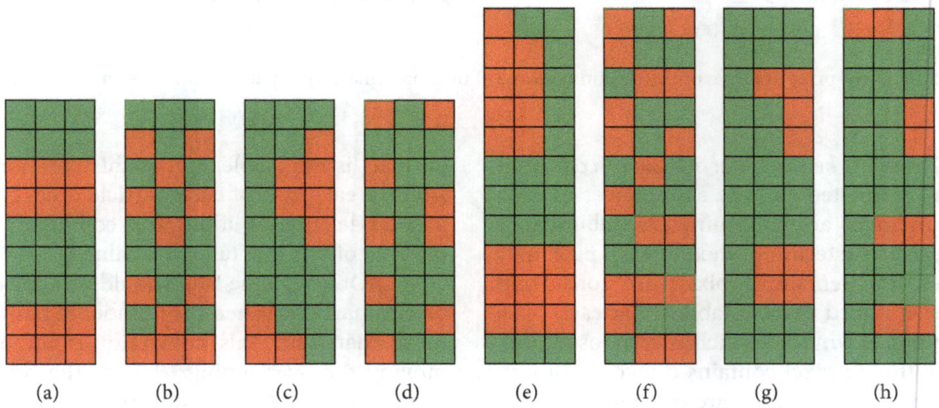

Figure 1: Special cases subpixel distribution.

In general, the perimeter calculation of the connected component has three methods: the gap length between region and background, the length of chain code (4-connected and 8-connected), and the sum of border points number [39]. For convenience, the objective functions corresponding to these three methods are named f_gap, f_chain, and f_point.

2.2.2. Adaptability Analysis of Perimeter Calculation Methods. Adaptability analysis of the three perimeter calculation methods (f_gap, f_chain, and f_point) are shown in Figure 1. Table 1 gives the perimeters that are calculated by the different methods in Figure 1. The real features' subpixel distribution maps are Figures 1(a), 1(c), 1(e), and 1(g), respectively. Based on the same subpixel abundance, Figures 1(b), 1(d), 1(f), and 1(h) are randomly selected subpixel mapping results with special case.

Special Case 1. There may be only one subpixel in a connected area (isolated single point). The perimeter of an isolated single point is 0. The total perimeter, including an isolated point and a certain region, is bound to be bigger than the region's perimeter, as shown in Figures 1(b) and 1(d).

Special Case 2. There may be only two subpixels in a connected area (isolated two points), as shown in Figures 1(f) and 1(h).

From Figure 1, it can be seen that the feature original pixel distributions (a), (c), (e), and (g) are better than the results of the special case distributions (b), (d), (f), and (h).

Combination of Figure 1 and Table 1 shows the following:

(1) f_gap method calculates connected components perimeter; isolated points or isolated two points will increase the perimeter length. This method satisfies the requirement that the smaller the value of objective function, the better the subpixel mapping result.

(2) For cases (b) and (f), the number of boundary points is increased with the f_point method, but for (d) and (h), the mapping results could not acquire (c) and (g), even with the same objective function values.

(3) The f_chain (8-connected) method has an uncertain trend of changing calculation results. For cases (b) and (f), the smaller the target function is, the better the result is. Because cases (d) and (h) have isolated single points, for which 8-connected chain code perimeter is 0, this method cannot obtain mapping results of (c) and (g).

(4) For Figure 1, the f_chain (4-connected) method cannot acquire the optimal results while the value of the objective function gets smaller.

Therefore, according to the analysis of these special cases, f_gap is the best, f_point takes second place, and f_chain is the worst. While using chain code method to calculate perimeter, 4-connected segmentation leads to the generation of isolated single points. However, 8-connected segmentation may not necessarily generate an isolated point. So the latter's performance is better than the former. Therefore, f_chain (8-connected) is used in the following experiments.

2.3. Objective Function Optimization Based on MBQPSO. In this paper, we use different connected regional perimeter to formulate the objective function and the proposed method is used to minimize the aforementioned objective function. Based on the MBQPSO algorithm, there are two problems that should be solved to optimize objective function (7). They are as follows.

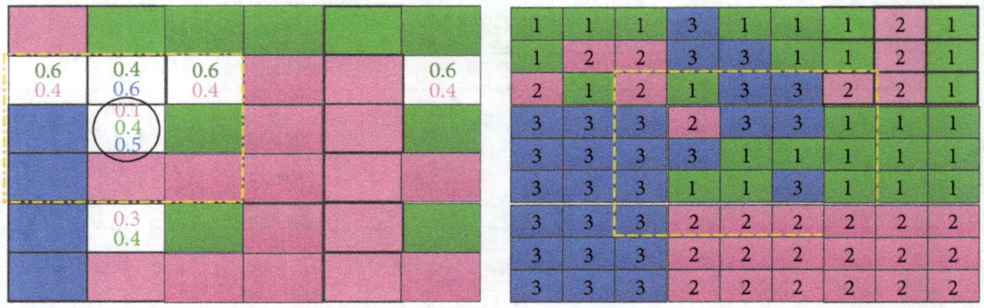

(a) (b)

FIGURE 2: Pixel-by-pixel iterative optimization region: (a) original image; (b) spatial resolution enhancement image.

(a) Corresponding Relations between Particle and Pixel. After the endmembers are extracted and the abundances in each pixel for each endmember are determined, an abundance threshold value is used to determine whether each pixel may be considered pure. The remaining pixels are considered mixed. Then, for each mixed pixel, establish that each pixel corresponds to a subpixel window, which is composed of r^2 subpixel. Supposing the ith pixel contains d pieces features, the abundance values $\{s_{i1}, s_{i2}, \ldots, s_{id}\}$ are ordered from low to high in accordance with $|s_{ik} \times r^2 - \text{round}(s_{ik} \times r^2)|$. The number of subpixels in the corresponding subpixel window is

$$n_{ik} = \text{round}\left(s_{ik} \times r^2\right), \quad k = 1, 2, \ldots, d-1$$
$$n_{id} = r^2 - \sum_{k=1}^{d-1} n_{ik}, \tag{8}$$

where d is the number of feature classes, i is the ith pixel, and k is the kth class feature. n_{ik} is the number of subpixels assigned to the kth endmember in the ith pixel and s_{ik} is the abundance of subpixels assigned to the kth endmember in the ith pixel. r is the rate of spatial resolution enhancement.

After the above step, the number of subpixel labels for each mixed pixel is known, whereas the subpixel position is not. In the subpixel mapping process of objective function optimization based on MBQPSO, a number of particles (particle population M) are assigned for each mixed pixel. Each of these particles is a bit series and contains $d-1$ rows and r^2 columns, where d is the number of endmembers that the pixel contains, and each column denotes a subpixel location for that pixel. These binary bit series initialize each particle X_i^t in particle swarm and set $P_i^t = X_i^t$.

The dth feature's distribution can be determined as long as the first $d-1$ kinds have been determined. The particle location of ith pixel can be expressed by $(d-1) \times r^2$ dimensional binary matrix. The position matrix of each pixel X_i must satisfy the needs of two constraints:

(a) The kth row only has n_k elements that are 1.

(b) Each column cannot have more than one element that is 1.

(b) Abundance Full Additive Constraint While Particle Updates. (a) The number of 1-bits in each row of a particle must be n_{ik}. However, the number of bits, taking 1 in each particle, is not stable. Under this condition, the number of 1-bits in each row of each particle is checked, and randomly located deletion or addition is conducted to ensure that the number of bits that takes the value of 1 is equal to n_{ik}.

(b) Only a single 1-bit should be allowed in each column of each particle. When the number of bits takes 1 in a column more than one, this constraint is imposed by randomly moving the overlapping 1-bit for the row with the smaller abundance rate to a column that is empty for all rows [35]. Note that this constraint is enforced under the first constraint. Furthermore, an overlapping 1-bit means that there exists an extra empty column.

2.4. Optimization Strategy. After construction of objective function and presentation of modified optimization algorithm, the target optimization strategy of subpixel mapping is the process of global iteration combined with local iteration. Local iteration is a pixel-by-pixel iteration, whereas traversing over all pixels can be considered a global iteration. In order to reduce the time complexity, the optimization strategy is as follows [39].

Pixel-by-Pixel Iteration. to construct cost function for each iteration, only calculate the current pixel's corresponding subpixel window and the surrounding structure's local area perimeter.

Figure 2 shows the pixel-by-pixel iterative optimization area diagram. Figure 2(b) is a 3×3 resolution enhancement image of the dotted area in Figure 2(a). Denote the pixel with a circle in Figure 2(a) and denote the dotted area in Figure 2(b) as region, which indicate the subpixel window and the surrounding structures local region for the pixel x. In each iteration, the subpixel mapping for pixel x is optimized by calculating the perimeter of region Rx as cost function.

3. Experimental Results and Discussion

Experiments are conducted to test the performance of the proposed modified BQPSO algorithm with standard test function, one simulated image, and one real image.

3.1. The Standard Test Function Experiment. In standard test function experiments, we use the same standard test function as done in [37]. Each test function implements three sets of

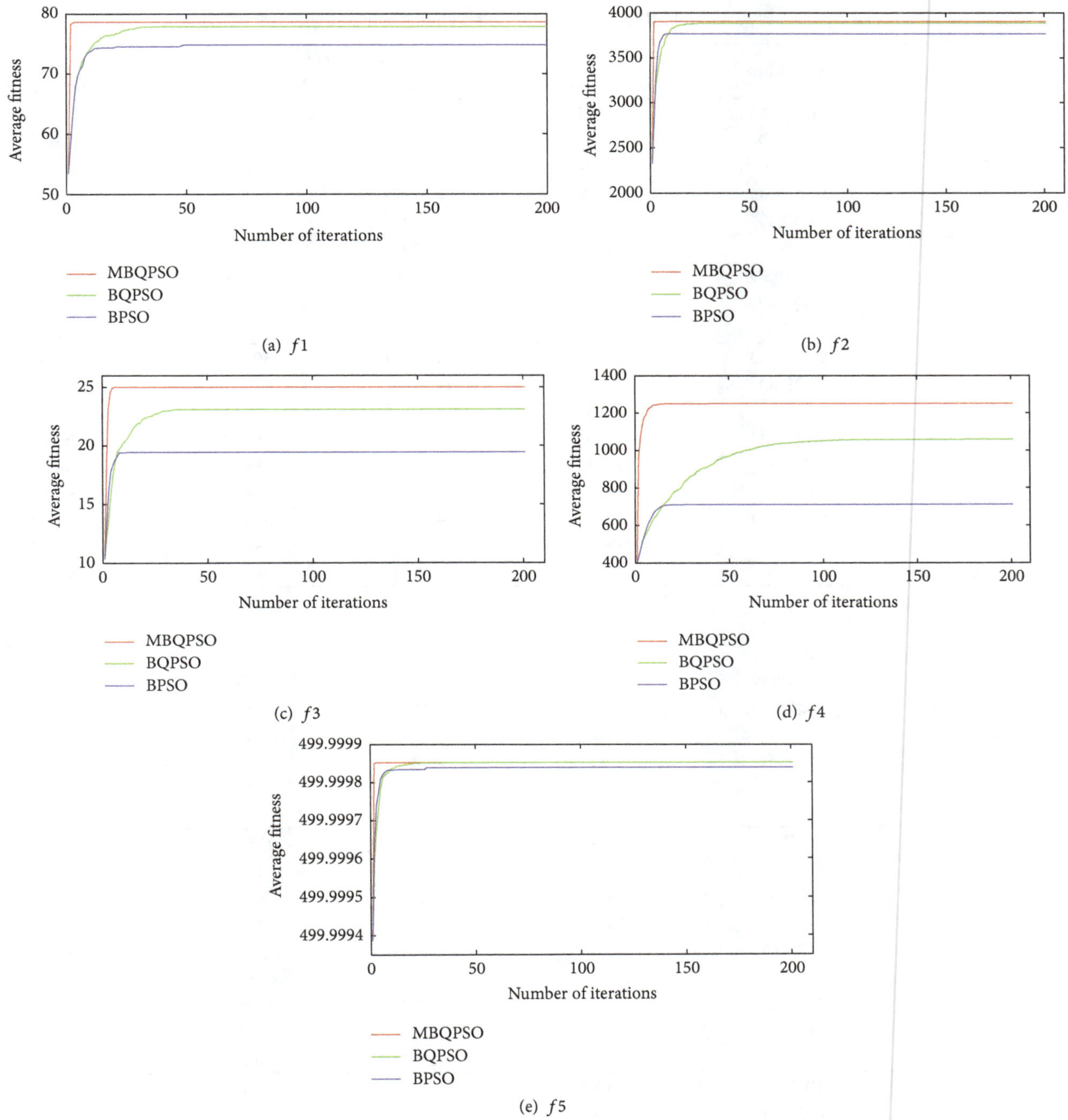

FIGURE 3: Comparison of the convergence curves of objective function values for different test functions with three optimization algorithms.

simulation test and the number of groups is 20, 40, and 80. The BPSO, BQPSO, and MBQPSO algorithm run 20 times [40] and the maximum iteration number is 200. Table 2 records the average optimal value of the three algorithms, the maximum value, the minimum value, and variance of the algorithm in the 20 times simulation.

From Table 2, we can see that the test results of the MBQPSO algorithm are better than those of the BQPSO and BPSO algorithm. In a low dimensional test function, the variance of the test results is greatly reduced. The average optimal value is significantly improved. The convergence curve from Figure 3 shows that the MBQPSO has the best performance on global optimization ability and convergent rate. The BQPSO algorithm's performance takes the second place and is better than the BPSO in simulation. The curve's adjustment time is shorter and the final result is optimized.

3.2. The Subpixel Mapping Experiment of Different Algorithms. In subpixel mapping experiments, consisting of simulated and real image, we compare the proposed six subpixel

Table 2: BPSO, BQPSO, and MBQPSO algorithm simulation results with five test functions.

Function	Number	Algorithm	Index parameter			
			Average optimal value	Variance	Maximum value	Minimum value
f_1	20	MBQPSO	**78.6**	**1.5e − 14**	**78.6**	**78.6**
		BQPSO	78.3	0.47	78.6	76.8
		BPSO	74.5	3.5	78.6	66.6
	40	MBQPSO	**78.6**	**1.5e − 14**	**78.6**	**78.6**
		BQPSO	78.6	0.11	78.6	78.2
		BPSO	76.9	2.6	77.7	67
	80	MBQPSO	**78.6**	**1.5e − 14**	**78.6**	**78.6**
		BQPSO	78.6	$3.5e-2$	78.6	78.5
		BPSO	78.4	0.4	78.6	77
f_2	20	MBQPSO	**3905.9**	**4.7e − 13**	**3905.9**	**3905.9**
		BQPSO	3885.3	37.2	3905.9	3735.2
		BPSO	3736.6	296.4	3900	2603.5
	40	MBQPSO	**3905.9**	**4.7e − 13**	**3905.9**	**3905.9**
		BQPSO	3902.1	11	3905.9	3860.1
		BPSO	3871.2	58.9	3900	3639.5
	80	MBQPSO	**3905.9**	**4.7e − 13**	**3905.9**	**3905.9**
		BQPSO	3904.8	4.56	3905.9	3885.5
		BPSO	3893.6	20.2	3905.9	3821.5
f_3	20	MBQPSO	**25**	**0**	**25**	**25**
		BQPSO	23.1	1.119	25	21
		BPSO	19.6	1.932	22	15
	40	MBQPSO	**25**	**0**	**25**	**25**
		BQPSO	24	0.999	25	22
		BPSO	21.5	1.317	23	19
	80	MBQPSO	**25**	**0**	**25**	**25**
		BQPSO	24.8	0.44	25	24
		BPSO	22.3	1.129	25	21
f_4	20	MBQPSO	**1251.8**	**0.2**	**1252.3**	**1251.6**
		BQPSO	1046.7	36.7	1119.3	968.9
		BPSO	734.45	43.131	818.7	661.39
	40	MBQPSO	**1251.8**	**0.2**	**1252.3**	**1251.5**
		BQPSO	1114.4	32.6	1157.5	1039.5
		BPSO	804.1	35.847	881.3	722.36
	80	MBQPSO	**1252**	**0.3**	**1252.7**	**1251.7**
		BQPSO	1149.2	31	1190.9	1082.3
		BPSO	882.8	42.979	961.7	809.75
f_5	20	MBQPSO	**499.9**	**0**	**499.9**	**499.9**
		BQPSO	499.9	$1e-6$	499.9	499.9
		BPSO	499.9	$0.8e-5$	499.9	499.9
	40	MBQPSO	**499.9**	**0**	**499.9**	**499.9**
		BQPSO	499.9	0	499.9	499.9
		BPSO	499.9	$0.5e-5$	499.9	499.9
	80	MBQPSO	**499.9**	**0**	**499.9**	**499.9**
		BQPSO	499.9	0	499.9	499.9
		BPSO	499.9	$0.1e-5$	499.9	499.9

(a) (b)

FIGURE 4: Simulated hyperspectral imagery (the thirtieth band): (a) image 1 (48 × 48); (b) image 2 (20 × 25).

mapping algorithms which are combined by three objective functions (f_point, f_chain, and f_gap) and two optimization algorithms (BQPSO and MBQPSO), named as PBQPSO, CBQPSO, GBQPSO, PMBQPO, CMBQPSO, and GMBQPSO. In addition, we also compare the six proposed methods with pixel swapping algorithm (PSA) in simulated and real experiment. The results of the six proposed subpixel mappings are compared under the conditions of different objective functions and different optimization strategies. Since the subpixel mapping result is the same as the classification result, the Kappa coefficient and recognition rate are used to evaluate the performance of the different subpixel mapping algorithms.

In order to construct an accurate abundance map, based on the real features distribution of the Indiana AVIRIS Indian pines hyperspectral data (size is: 145 × 145), the ground truth gives 17 kinds of different features (including the background), and then some of these classes are merged into nine categories. The simulated hyperspectral image is generated by nine kinds of spectral feature on the basis of category labels, which are selected from the USGS spectral library.

Two hyperspectral image pictures are constructed for the later experiments:

(1) Figure 4(a) is captured from the original image's [1 : 144, 1 : 144] region, which is then filtered by the 3 × 3 mean filter and narrowly shown in image 1.

(2) Figure 4(b) is captured from the original image's [1 : 60, 70 : 144] region, which is then filtered by the 3 × 3 mean filter and narrowly shown in image 2. So image 2 is in fact the subfigure of image 1.

This part has five experiments, and only the fourth experiment's spatial resolution is 5 × 5; the others are 3 × 3. In order to avoid the influence of the initial value sensitivity of the particle swarm algorithm, the same initial particle swarm is used in each experiment and the results run for five times.

We compare the results of the above seven algorithms for Figure 4(b). The initial recognition rate (the ratio of the number of correct classification result label to the total number of category labels) of the seven experiments is 93.89%. For BQPSO-based and MBQPSO-based method, the particle number is 50; the number of maximum iteration steps is 30.

Table 3 gives the results of the subpixel location result, combining the different objective functions and the different optimization algorithms. As can be seen from Table 3, we have the following:

TABLE 3: Comparison of subpixel mapping results for Figure 4(b) with six kinds of algorithms.

Algorithm	recognition rates%	Kappa coefficient	time (s)
PSA	94.1	0.84	3.9
PMBQPSO	96.5	0.91	13
CMBQPSO	96.1	0.90	238
GMBQPSO	98.7	0.96	23
PBQPSO	96.3	0.90	4
CBQPSO	95	0.87	178
GBQPSO	97.9	0.94	4

(1) Under the same objective function, recognition rate and Kappa coefficient of MBQPSO are greatly improved.

(2) Comparing the results of objective function calculation and subpixel mapping, we learn that the f_gap method is more suitable for constructing the target function of the subpixel position, which has the advantages of the shortest time and the best mapping results.

(3) Comparing the results of PSA with the six methods, based on BQPSO and MBQPSO, the proposed six methods can obtain higher recognition rates and Kappa coefficient.

As can be seen from Figure 5, MBQPSO had fewer isolated points than BQPSO and the result is better. From the iterative curves of Figure 6, all the results of BQPSO are only three times iterative convergence. The number of iteration steps is increased after optimization and the corresponding results are better. Compared with the above six, PSA had more isolated points and the recognition rates and Kappa coefficient are lower.

3.3. The Influence of Parameters

3.3.1. The Number of Particles. To investigate the influence of the number of particles, the following simulation parameters are used: set the maximum number of iterations to 30 and set the number of particles at 30, 40, 50, 60, and 100. The test results are shown in Figure 7.

The following information can be seen from Figure 7.

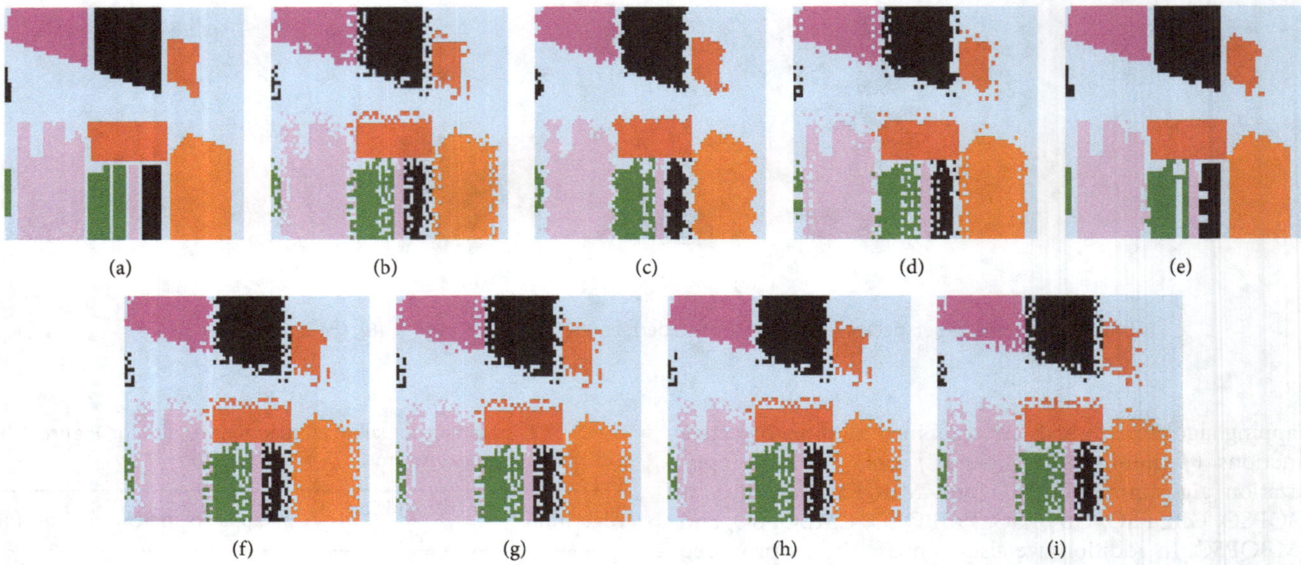

Figure 5: Subpixel mapping results with the six kinds of algorithms for Figure 4(b): (a) true distribution, (b) random distribution, (c) PMBQPSO, (d) CMBQPSO, (e) GMBQPSO, (f) PBQPSO, (g) CBQPSO, (h) GBQPSO, and (i) PSA.

(1) Under the same number of particles, the recognition rate and the Kappa coefficient are the best using MBQPSO combined with f_gap method to calculate perimeter.

(2) With the increase of the particles number, running time generally has an upward trend. Among them, the calculation results of the f_point method and MBQPSO present a rising trend on the whole, with an increasing number of particles. But at the middle of the function it slightly declines, which may be caused by the uncertainty of the two constraints during the iterative process.

(3) The running time of the f_gap method becomes longer with the increase of the number of particles. The recognition rate and Kappa coefficient decrease at first and then increase. The two parameters of the f_point and f_chain methods are increased at first and then decreased. The recognition rate of GMBQPSO algorithm is the highest in 50 particles.

3.3.2. Adaptability of Image Size. This section verifies the applicability of image size with the modified algorithm. Based on the real feature distribution of the whole Indian pines image, we design a simulation experiment, as shown in Figure 8(a). Using image 1 in Figure 4, the results of the above six algorithms are compared. The number of particles is 50 and the maximum iteration number is 30. Figure 8 depicts the result of subpixel mapping and the evaluation index of the mapping results is given as follows.

Table 4 and Figure 8 show the improvement of the modified algorithm. The results of Table 4 quantitatively confirm that the GMBQPSO algorithm has the best applicability and that the Kappa coefficient and the recognition rate are the highest, but the operation time is longer. From Figure 8 it can be seen that there are many isolated points among the results of the BQPSO algorithm. Most boundaries between categories of MBQPSO running results are more obvious. Comparing Tables 4 and 3, it can be seen that the image size has no effect on subpixel mapping accuracy. The bigger the image is, the larger the mixed pixels' number is. After statistical analysis in Figure 4, the number of mixed pixels in image 1 is 5.5 times the number of pixels in image 2. Comparing the operating time of Table 4 with Table 3, considering the two constraints, the operating time is proportional to the number of mixed pixels in the image. In summary, the GMBQPSO is superior to other algorithms in both visual perception and quantitative evaluation results (the Kappa coefficient and the recognition rate).

Table 4: Comparison of whole image's subpixel mapping results of six algorithms.

Algorithm	Recognition rates%	Kappa coefficient	Time (s)
PMBQPSO	97.3	0.92	96
CMBQPSO	96.6	0.90	2267
GMBQPSO	98.9	0.97	181
PBQPSO	97	0.92	20
CBQPSO	95.5	0.87	796
GBQPSO	98	0.95	21

3.3.3. Influence of Spatial Resolution Enhancement Ratio. In order to verify the subpixel mapping practicality of the optimization objective function on the image spatial resolution enhancement ratio, which is greater than 3×3, the original image is filtered by 5×5 filtering and subpixel mapping is implemented by a spatial enhancement ratio 5×5.

Because the 5×5 filtered image contained fewer pure pixels, it is necessary to adjust the abundance threshold. A pixel can be seen as a pure pixel only if the minimum

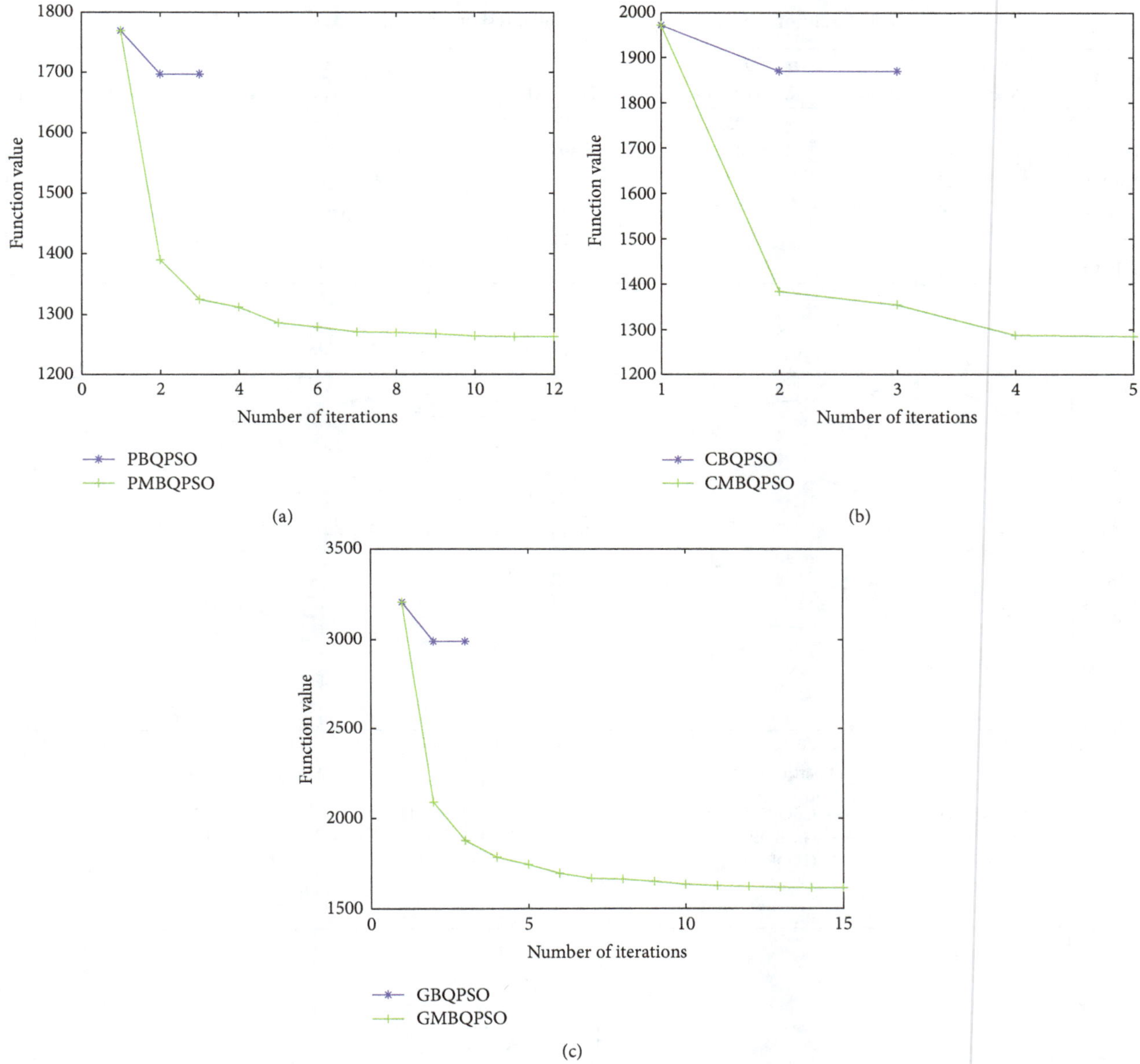

FIGURE 6: The iterations change curves with the space resolution enhancement ratio 3 × 3: (a) f_point method, (b) f_chain method, and (c) f_gap method.

difference value between the maximum abundance and the others in the pixel is greater than 0.9.

As can be seen from Table 5, we have the following:

(1) The improved MBQPSO results are obviously better than those of the BQPSO algorithm.

(2) In the case of 100 particles, the recognition rate and the Kappa coefficient are superior to the 50 particles, but the running time is longer. Considering time and accuracy, the choice of 50 particles is reasonable.

In Figure 9, the results show that MBQPSO combined with three kinds of perimeter calculation methods had fewer isolated points and the GMBQPSO results are optimal. But there still exist other isolated cases in the original 5 × 5 filtering image. In practical application, we should take further measures to achieve better results.

3.4. The Real Image Subpixel Mapping Experiment. The experiments on real images are conducted by considering two different hyperspectral pieces of data. The first experimental data uses a ROSIS hyperspectral remote sensing image, which is located in the north of Italy, Pavia University. The size of the original image is 610 × 340 pixels and it contains nine classes of pixel number ranging from 947 to 18649 pixels. The band number is 103 and the geometric resolution is

TABLE 5: Results of subpixel mapping with spatial resolution enhancement ratio 5 × 5.

Algorithm	$n = 50$			$n = 100$		
	Recognition rates%	Kappa coefficient	Time (s)	Recognition rates%	Kappa coefficient	Time (s)
PMBQPSO	92.9	0.82	27	93.7	0.83	59
CMBQPSO	91.4	0.78	394	91.6	0.78	446
GMBQPSO	93.4	0.82	44	93.8	0.85	123
PBQPSO	91.2	0.78	3	92.7	0.82	17
CBQPSO	90.8	0.77	88	91.2	0.78	343
GBQPSO	91.5	0.79	3	93.3	0.83	22

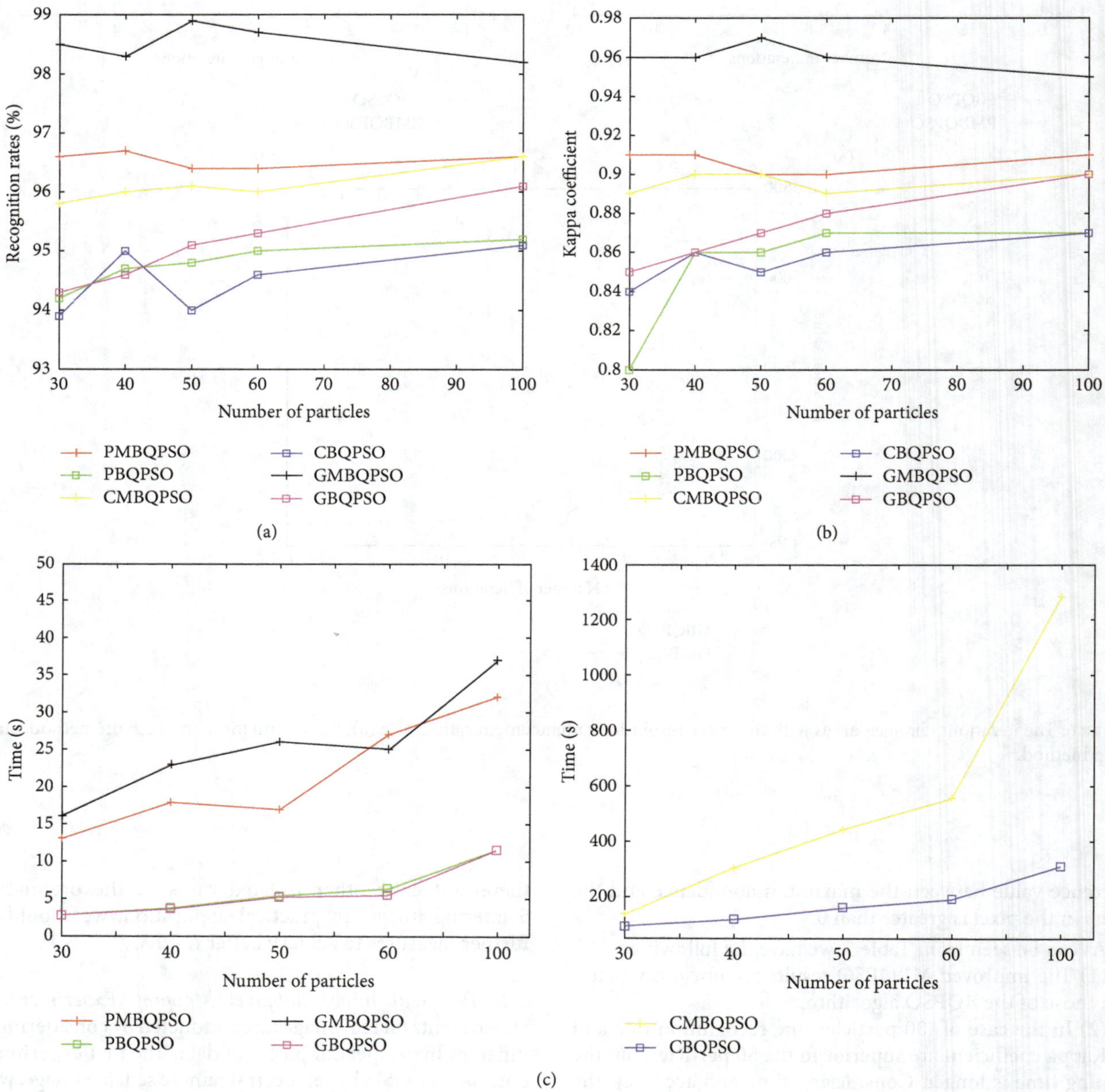

FIGURE 7: Comparison of different particle number: (a) recognition rates curve with particle number, (b) Kappa coefficient curve with particle number, and (c) iteration time curve with particle number.

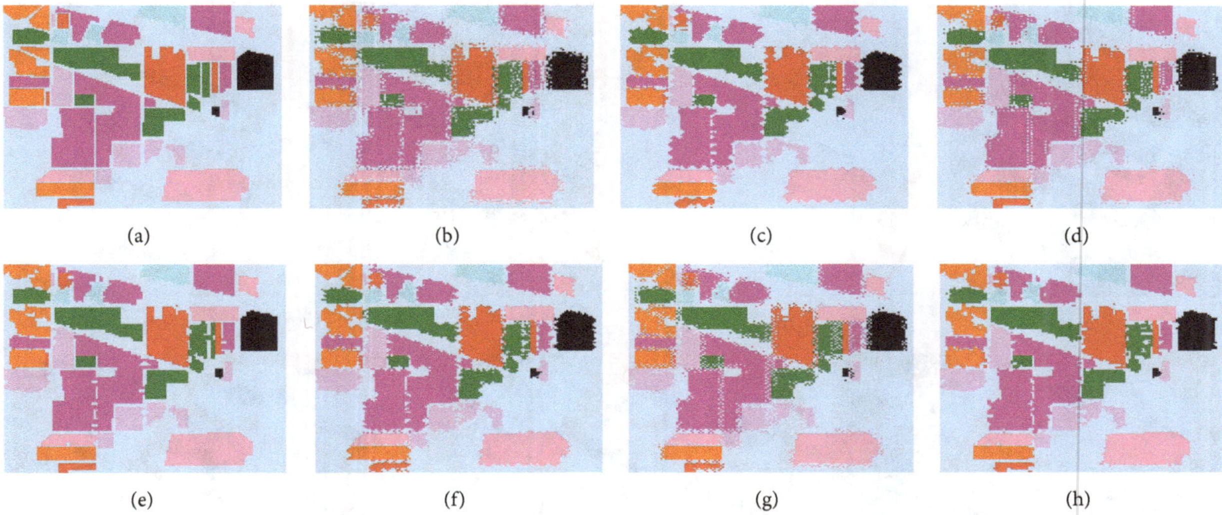

FIGURE 8: Subpixel positioning results of six algorithms for Figure 4(a): (a) true distribution, (b) random distribution, (c) PMBQPSO, (d) CMBQPSO, (e) GMBQPSO, (f) PBQPSO, (g) CBQPSO, and (h) GBQPSO.

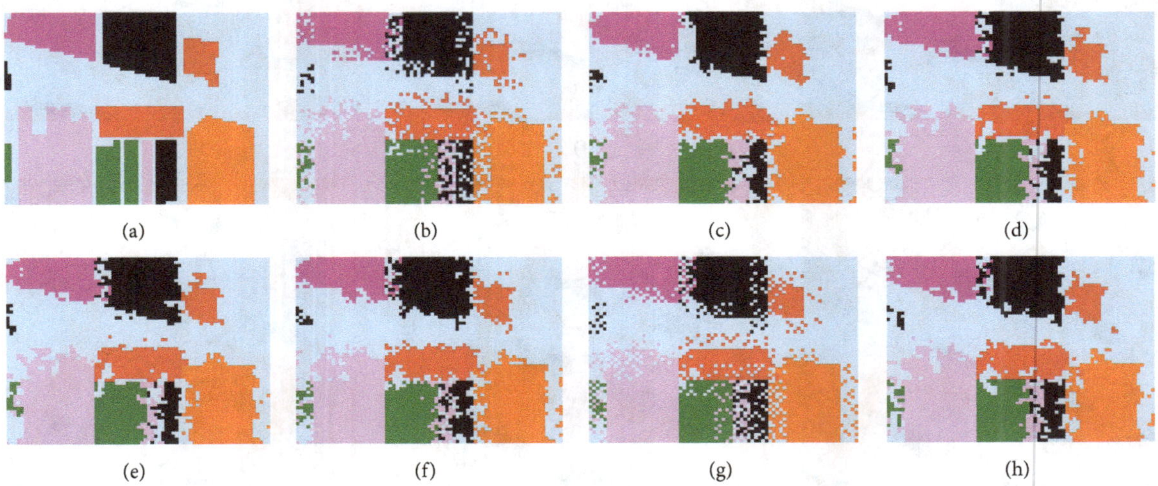

FIGURE 9: The subpixel mapping results of the 5 × 5 filter image with the six algorithms: (a) true distribution, (b) random distribution, (c) PMBQPSO, (d) CMBQPSO, (e) GMBQPSO, (f) PBQPSO, (g) CBQPSO, and (h) GBQPSO.

1.3 meters. We sample a 117 × 75 region where five kinds of features and the real surface features' distribution are shown in Figure 10(b). The 3 × 3 mean filter is developed for the original image to obtain the 60th band gray image Figure 10(a). The number of endmembers is estimated as 5 by VD [38]. A pixel can be seen as a pure pixel only if the minimum difference value between the maximum abundance and the others in the pixel is greater than a certain threshold [33–35]. Because a metal area is obvious, we only perform a statistical analysis of the recognition rate on a metal area. Figure 10(c) shows the random mapping results. Figures 10(d)–10(i) represent the subpixel mapping results with different optimization algorithm, respectively. Table 6 gives a comparison of recognition rate.

The second image analyzed in our experiments is a Salinas dataset. The size of the original image is 510 × 216 pixels and it contains 204 bands and contained 17 classes. The 3 × 3 mean filter is developed for the original image to obtain the 170 × 72 pixels downsampled image and we select three bands to obtain the pseudocolor image Figure 11(a). We combine class 1 and class 2 in one class, combine class 8 and class 15 in one class, and let class 16 be named class 2. The number of endmembers is estimated as 15 by VD and endmembers are extracted by calculating the mean of each category. A pixel can be seen as a pure pixel only if the minimum difference value between the maximum abundance and the others in the pixel is greater than a certain threshold. The background image is shown in

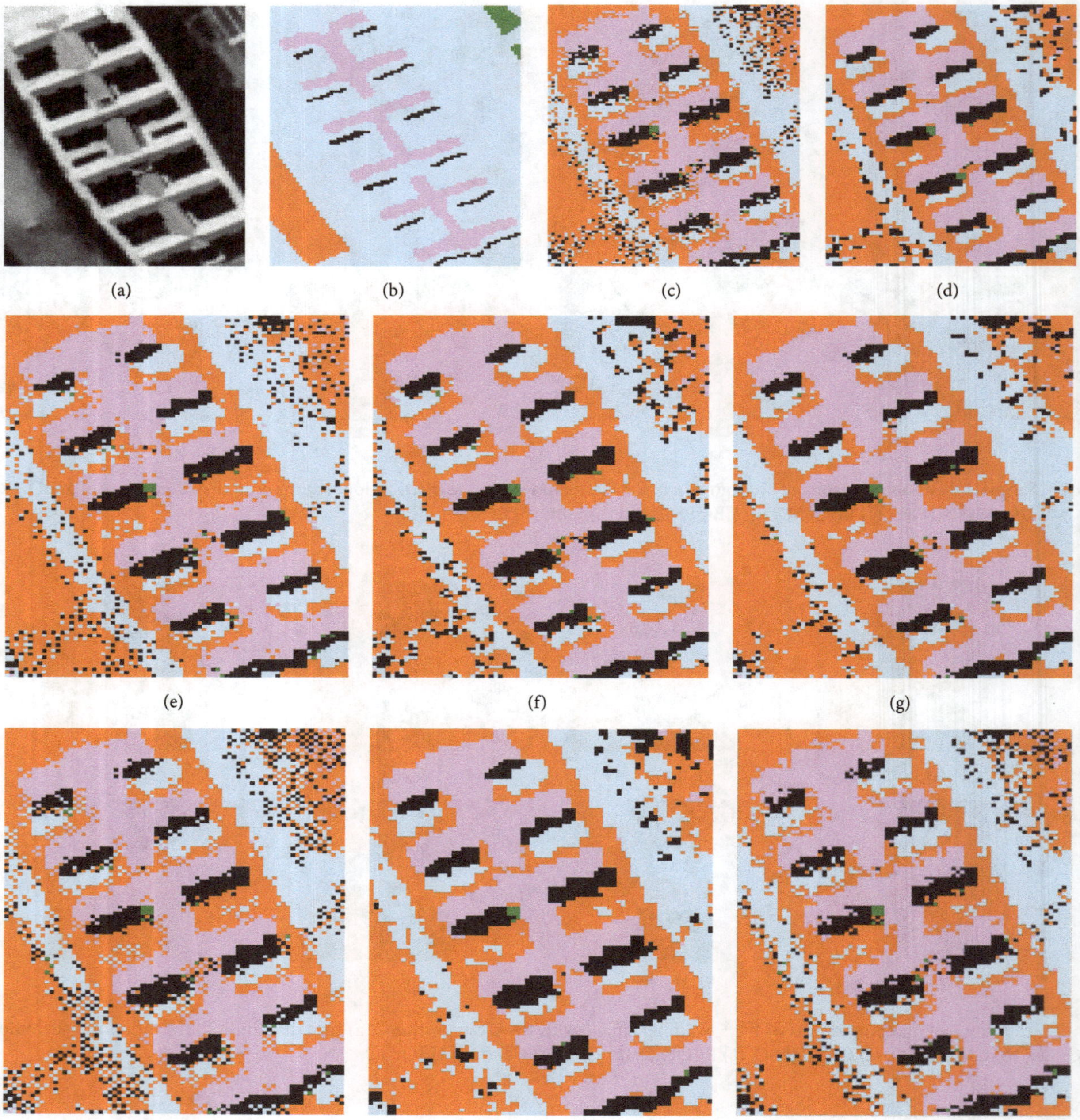

(a) (b) (c) (d) (e) (f) (g) (h) (i) (j)

FIGURE 10: The subpixel mapping results of the PaviaU original subimages: (a) 60th band gray image, (b) true distribution, (c) random distribution, (d) PMBQPSO, (e) CMBQPSO, (f) GMBQPSO, (g) PBQPSO, (h) CBQPSO, (i) GBQPSO, and (j) PSA.

Figure 11(b). In this part, we only perform a statistical analysis of the recognition rate on the other 14 categories without background. Figure 11(c) shows the random mapping results. Figures 11(d)–11(f) represent the subpixel mapping results with different optimization algorithm, respectively. Table 7 gives a comparison of recognition rate.

Table 6 gives a comparison of the recognition rate and the running time of various algorithms with the abundance threshold at 0.4 and 0.5, respectively. Figure 10 gives the results of the algorithm with the abundance threshold at 0.4 and 50 particles.

From Table 6, we have the following:

(1) With the same number of particles, the recognition rate of the MBQPSO is better than that of the BQPSO, and the recognition rate of GMBQPSO is the highest.

(2) The results of the abundance threshold at 0.4 are better than those at 0.5, and the recognition rate and Kappa coefficient are all improved.

TABLE 6: PaviaU subimage subpixel mapping results.

Algorithm	Particle number	0.5		0.4	
		Recognition rates%	Time (s)	Recognition rates%	Time (s)
PSA	—	92.7	7	94.3	6
PMBQPSO	50	97.4	88	97.7	98
	100	97.6	163	97.8	148
PBQPSO	50	97.1	12	97.4	11
	100	97.2	25	97.5	19
CMBQPSO	50	94.8	1036	96.2	400
	100	95.7	1089	95.9	793
CBQPSO	50	94.9	315	95.6	242
	100	95	625	95.5	473
GMBQPSO	50	97.7	199	98.2	115
	100	97.9	283	98.4	249
GBQPSO	50	97.5	13	97.4	11
	100	97.9	25	97.7	21

FIGURE 11: The subpixel mapping results of the PaviaU original subimages: (a) pseudocolor image, (b) true distribution, (c) random distribution, (d) GMBQPSO, (e) GBQPSO, and (f) PSA.

(3) From the point of view of the particle number, the Kappa coefficient and the recognition rate of the total particle number 100 can be improved, but time is longer.

(4) Compared with the above six, PSA had more isolated point, and the recognition rates and Kappa coefficient are lower.

As analysis in the first real experimental result, we present the comparison of the recognition rate and the running time of various algorithms with the abundance threshold at 0.4 in Table 7. Figure 11 gives the results of the algorithm with the abundance threshold at 0.4 and 50 particles. Based on the subpixel mapping results in the Table 6, we only present the subpixel mapping results image of GMBQPSO, GBQPSO, and PSA in Figure 11. We can see that the marginal area of MBQPSO's mapping results is better than others. The recognition rate of the MBQPSO is better than that of the corresponding BQPSO, and the recognition rate of GMBQPSO is the highest. The running time of the corresponding BQPSO is shorter than MBQPSO, owing to premature convergence of BQPSO-based method. In addition, the running time is

Table 7: Salinas subpixel mapping results.

Algorithm	0.4	
	Recognition rates%	Time (s)
PSA	99.14	94
PMBQPSO	99.56	1089
PBQPSO	99.52	104
CMBQPSO	99.39	20670
CBQPSO	99.33	9870
GMBQPSO	99.68	1127
GBQPSO	99.54	137

larger than the experiment result of ROSIS dataset because the size of the Salinas dataset is about 20 times larger than ROSIS dataset.

4. Conclusion

In this paper, a modified method of subpixel mapping based on binary quantum particle swarm optimization (MBQPSO) algorithm is proposed. In order to improve the adaptability of objective function and lower time complexity of optimization algorithm, we use three kinds of regional perimeter minimum to describe spatial correlation. At the same time, the discretization process of particle position updating is corrected to overcome premature convergence in the original BQPSO algorithm iteration process. Simulated and real image experiments show that the proposed method in this paper is superior to the original method in the Kappa coefficient, recognition rate, and running time. It can improve the mapping accuracy of the algorithm. The f_gap method obtains the optimal mapping result and the fastest speed, which is suitable for subpixel mapping. The method proposed in this paper can not only reflect the spatial detail information of the ground objects, but also reconstruct the shape feature information in a small area. Compared with the original linear optimization method, visual effect and practical classification accuracy are all improved greatly. The assumption of the spatial correlation between pixels is suitable in most cases, but because of the complexity of actual situations, this assumption will have some errors in some cases. Therefore, it is necessary to conduct further research to obtain a better expression of the pixel spatial distribution rules to improve accuracy of subpixel mapping.

Conflicts of Interest

The authors declare that there are no conflicts of interest regarding the publication of this paper.

Acknowledgments

This work was supported by the National Natural Science Foundation of China (no. 61671408), Shanghai Aerospace Science and Technology Innovation Fund (no. SAST2015033), and the Joint Funds of the Ministry of Education of China (no. 6141A02022314).

References

[1] P. M. Atkinson, "Mapping sub-pixel boundaries from remotely sensed images," in *Innovations in Gis Iv.taylor & Francis*, Z. Kemp, Ed., vol. 4, 1997.

[2] A. J. Tatem, H. G. Lewis, P. M. Atkinson, and M. S. Nixon, "Super-resolution target identification from remotely sensed images using a Hopfield neural network," *IEEE Transactions on Geoscience and Remote Sensing*, vol. 39, no. 4, pp. 781–796, 2001.

[3] A. J. Tatem, H. G. Lewis, P. M. Atkinson, and M. S. Nixon, "Super-resolution mapping of multiple-scale land cover features using a Hopfield neural network," in *Proceedings of theInternational Geoscience and Remote Sensing Symposium*, vol. 7, pp. 3200–3202, July 2001.

[4] A. J. Tatem, H. G. Lewis, P. M. Atkinson, and M. S. Nixon, "Super-resolution land cover pattern prediction using a Hopfield neural network," *Remote Sensing of Environment*, vol. 79, no. 1, pp. 1–14, 2002.

[5] K. Mertens, B. De Baets, L. Verbeke, and R. De Wulf, "Subpixel mapping with neural networks: real-world spatial configurations learned from artificial shapes," in *Proceedings of the 4th International symposium on Remote Sensing and Urban Areas; WG VII/4 Workshop*, pp. 3046–3049, Anchorage, AK, USA, September 2004.

[6] K. C. Mertens, L. P. C. Verbeke, T. Westra, and R. R. De Wulf, "Sub-pixel mapping and sub-pixel sharpening using neural network predicted wavelet coefficients," *Remote Sensing of Environment*, vol. 91, no. 2, pp. 225–236, 2004.

[7] K. C. Mertens, L. P. C. Verbeke, E. I. Ducheyne, and R. R. De Wulf, "Using genetic algorithms in sub-pixel mapping," *International Journal of Remote Sensing*, vol. 24, no. 21, pp. 4241–4247, 2003.

[8] K. Wu, R. Niu, P. Li, and L. Zhang, "Sub-pixel mapping of remote sensing images based on fuzzy ARTMAP neural network model," *Geomatics and Information Science of Wuhan University*, vol. 34, no. 3, pp. 297–300, 2009.

[9] F. Ling, Y. Du, F. Xiao, H. Xue, and S. Wu, "Super-resolution land-cover mapping using multiple sub-pixel shifted remotely sensed images," *International Journal of Remote Sensing*, vol. 31, no. 19, pp. 5023–5040, 2010.

[10] X. U. Xiong, Y. Zhong, and L. Zhang, "A sub-pixel mapping algorithm based on BP neural network with spatial autocorrelation function for remote sensing imagery," *Acta Geodaetica Et Cartographica Sinica*, vol. 40.3, pp. 307–311, 2011.

[11] W. Z. Shi, Y. L. Zhao, and Q. M. Wang, "Sub-pixel mapping based on BP neural network with multiple shifted remote sensing images," *Journal of Infrared & Millimeter Waves*, vol. 33, no. 5, pp. 527–532, 2014.

[12] L. Li, T. Xu, and Y. Chen, "Improved urban flooding mapping from remote sensing images using generalized regression neural network-based super-resolution algorithm," *Remote Sensing*, vol. 8, no. 8, article no. 625, 2016.

[13] A. Boucher and P. C. Kyriakidis, "Super-resolution land cover mapping with indicator geostatistics," *Remote Sensing of Environment*, vol. 104, no. 3, pp. 264–282, 2006.

[14] A. Boucher and P. C. Kyriakidis, "Integrating fine scale information in super-resolution land-cover mapping," *Photogrammetric Engineering and Remote Sensing*, vol. 73, no. 8, pp. 913–921, 2007.

[15] H. R. Jin, G. Mountrakis, and P. Li, "A super-resolution mapping method using local indicator variograms," *International Journal of Remote Sensing*, vol. 33, no. 24, pp. 7747–7773, 2012.

[16] T. Kasetkasem, M. K. Arora, and P. K. Varshney, "Super-resolution land cover mapping using a Markov random field based approach," *Remote Sensing of Environment*, vol. 96, no. 3-4, pp. 302–314, 2005.

[17] V. A. Tolpekin and A. Stein, "Quantification of the effects of land-cover-class spectral separability on the accuracy of markov-random-field-based superresolution mapping," *IEEE Transactions on Geoscience and Remote Sensing*, vol. 47, no. 9, pp. 3283–3297, 2009.

[18] J. P. Ardila, V. A. Tolpekin, W. Bijker, and A. Stein, "Markov-random-field-based super-resolution mapping for identification of urban trees in VHR images," *ISPRS Journal of Photogrammetry and Remote Sensing*, vol. 66, no. 6, pp. 762–775, 2011.

[19] X. Li, F. Ling, and Y. Du, "Super-resolution mapping based on the supervised fuzzy *c*-means approach," *Remote Sensing Letters*, vol. 3, no. 6, pp. 501–510, 2014.

[20] Q. Liu and J. Trinder, "Sub-pixel mapping of multispectral images using markov random field with graph cut optimization," *IEEE Geoscience & Remote Sensing Letters*, vol. 13, pp. 1–5, 2016.

[21] J. Verhoeye and R. De Wulf, "Land cover mapping at sub-pixel scales using linear optimization techniques," *Remote Sensing of Environment*, vol. 79, no. 1, pp. 96–104, 2002.

[22] Y. Zhong and L. Zhang, "Sub-pixel mapping algorithm based on adaptive differential evolution for remote sensing imagery," in *Proceedings of the 2011 IEEE International Geoscience and Remote Sensing Symposium, IGARSS 2011*, pp. 1724–1727, July 2011.

[23] Y. Zhong, L. Zhang, P. Li, and H. Shen, "A sub-pixel mapping algorithm based on artificial immune systems for remote sensing imagery," in *Proceedings of the 2009 IEEE International Geoscience and Remote Sensing Symposium, IGARSS 2009*, pp. 1007–1010, July 2009.

[24] K. Wu, P. X. Li, and L. P. Zhang, "Sub-pixel mapping of remote sensed images based on evolutionary agent algorithm," *Journal of Remote Sensing*, vol. 13, no. 1, pp. 60–66, 2009.

[25] K. C. Mertens, B. de Baets, L. P. C. Verbeke, and R. R. de Wulf, "A sub-pixel mapping algorithm based on sub-pixel/pixel spatial attraction models," *International Journal of Remote Sensing*, vol. 27, no. 15, pp. 3293–3310, 2006.

[26] F. Ling, "Sub-pixel mapping of remote sensing images based on cellular automata model," *Journal of Image & Graphics*, vol. 10, no. 7, pp. 916–921, 2005.

[27] Q. M. Wang, *Research on Sub-Pixel Mapping And Its Related Techniques for Remote Sensing Imagery, A Dissertation for degree of Master Engineering [M.S. thesis]*, Harbin Engineering University, 2012.

[28] Q. Wang, L. Wang, and D. Liu, "Particle swarm optimization-based sub-pixel mapping for remote-sensing imagery," *International Journal of Remote Sensing*, vol. 33, no. 20, pp. 6480–6496, 2012.

[29] X. Xu, Y. Zhong, L. Zhang, and H. Zhang, "Sub-pixel mapping based on a MAP model with multiple shifted hyperspectral imagery," *IEEE Journal of Selected Topics in Applied Earth Observations & Remote Sensing*, vol. 6, no. 2, pp. 580–593, 2013.

[30] Y. Zhong, Y. Wu, X. Xu, and L. Zhang, "An adaptive sub-pixel mapping method based on MAP model and class determination strategy for hyperspectral remote sensing imagery," *IEEE Transactions on Geoscience & Remote Sensing*, vol. 53, no. 3, pp. 1411–1426, 2015.

[31] Z. Shen and Z. Yu, "Sub-pixel mapping with partheno-genetic algorithm and sub-pixel/pixel spatial attraction model," *Remote Sensing Technology and Application*, vol. 30, no. 1, pp. 129–134, 2015.

[32] Q. Wang and P. M. Atkinson, "The effect of the point spread function on sub-pixel mapping," *Remote Sensing of Environment*, vol. 193, pp. 127–137, 2017.

[33] A. Villa, J. Chanussot, J. A. Benediktsson, C. Jutten, and R. Dambreville, "Unsupervised methods for the classification of hyperspectral images with low spatial resolution," *Pattern Recognition*, vol. 46, no. 6, pp. 1556–1568, 2013.

[34] M. W. Thornton, P. M. Atkinson, and D. A. Holland, "Sub-pixel mapping of rural land cover objects from fine spatial resolution satellite sensor imagery using super-resolution pixel-swapping," *International Journal of Remote Sensing*, vol. 27, no. 3, pp. 473–491, 2006.

[35] A. Ertürk, "Spatial resolution enhancement of hyperspectral images using un-mixing and binary particle swarm optimization," *IEEE Geoscience & Remote Sensing Letters*, vol. 11, no. 12, pp. 2100–2104, 2014.

[36] J. Sun, B. Feng, and W. Xu, *Particle Swarm Optimization with Particles Having Quantum Behavior, A Dissertation for Degree of Doctor Engineering [M.S. thesis]*, Jiangnan University, 2009.

[37] M. L. Xi, J. Sun, and Y. Wu, "Quantum-behaved particle swarm optimization with binary encoding," *Control and Decision*, vol. 25, no. 1, pp. 376–385, 2010.

[38] C. Chang and Q. Du, "Estimation of number of spectrally distinct signal sources in hyperspectral imagery," *IEEE Transactions on Geoscience & Remote Sensing*, vol. 42, no. 3, pp. 608–619, 2004.

[39] L. Y. Zhao, M. Y. Fan, and X. R. Li, "Sub-pixel mapping of hyperspectral imagery based on object optimization," *Journal of Image and Graphics*, vol. 21, no. 6, pp. 823–833, 2016.

[40] J. Kennedy and R. C. Eberhart, "A discrete binary version of the particle swarm algorithm," in *Proceedings of the IEEE International Conference on Systems, Man, and Cybernetics, Computational Cybernetics and Simulation*, vol. 5, pp. 4104–4108, Orlando, Fla, USA, October 1997.

4

Image Edge Detection based on Gaussian Mixture Model in Nonsubsampled Contourlet Domain

Li Yang,[1] Chang Xia,[2] and Chang Juan[3]

[1] *School of Computer Science and Engineering, Beifang University of Nationalities, Yinchuan 750021, China*
[2] *The Institute of Information and System Science, School of Mathematics and Information Science, Beifang University of Nationalities, Yinchuan 750021, China*
[3] *Graduate School, Ningxia University, Yinchuan 750021, China*

Correspondence should be addressed to Chang Xia; changxia0104@163.com

Academic Editor: Panajotis Agathoklis

In order to get accurate location and continuous edges, Gaussian mixture model and local direction modulus nonmaxima suppression are used in high frequency subbands of nonsubsampled Contourlet transform. The distribution of NSCT high frequency subbands coefficients has the "high spikes, long tail" non-Gaussian statistical characteristic. Gaussian mixture model (GMM) is used to distinguish the linear singular signal and the nonlinear singular signal on the high frequency subbands. Local direction modulus nonmaxima suppression is used to refine the linear singular signal. An appropriate threshold is used to distinguish edge pixels and nonedge pixels to get binary image. The experimental results demonstrate that the proposed method can capture more continuous edges in multiple directions and has accurate edge location. And the edges are with great convenience for the image recognition.

1. Introduction

Image edge detection plays a very important role in image processing [1]. The results of image edge detection will directly affect the performance of image understanding, analysis, and recognition. The classical edge detectors based on gradient are concise and easy to understand, such as Sobel and Prewitt operators. Because of doing the average operation in these algorithms, they will lose some detailed information and make the edges blur. So the accuracy of location is not high. Canny edge detection usually gives good performance. Canny algorithm uses two different thresholds to detect strong edges and weak edges, respectively. Weak edges will be detected when weak edges are connected to strong edges. But weak edges will be ignored when weak edges are not connected to strong edges. Due to using Gaussian filter in the Canny edge detection method, some weak edges will be smoothed. And the detected edges obtained are more miscellaneous, so it is difficult to identify the main contours.

As the multiscale analysis is introduced, the image processing methods based on wavelet have obtained good results. Wavelet transform is famous for its time frequency localization, multiscale, and multiresolution. It is a very effective tool for image edge detection in image processing [2], such as the wavelet modulus maxima edge detection method. But the basis functions of wavelet transform are isotropic, so it cannot capture more directional information in images [3]. In the recent years, multiscale geometric analysis has grown constantly, and new methods based on the new multiscale transforms are proposed, such as the curvelet and Contourlet transform [4]. The Contourlet transform was proposed to address the lack of geometrical structure in the separable two-dimensional wavelet transform. The research results show that the performance of the image edge detection based on the new multiscale transforms is better than those based on the wavelet transform. The Contourlet transform is a real 2D image representation using a directional filter and a Laplacian pyramid, which can effectively capture contours in an image

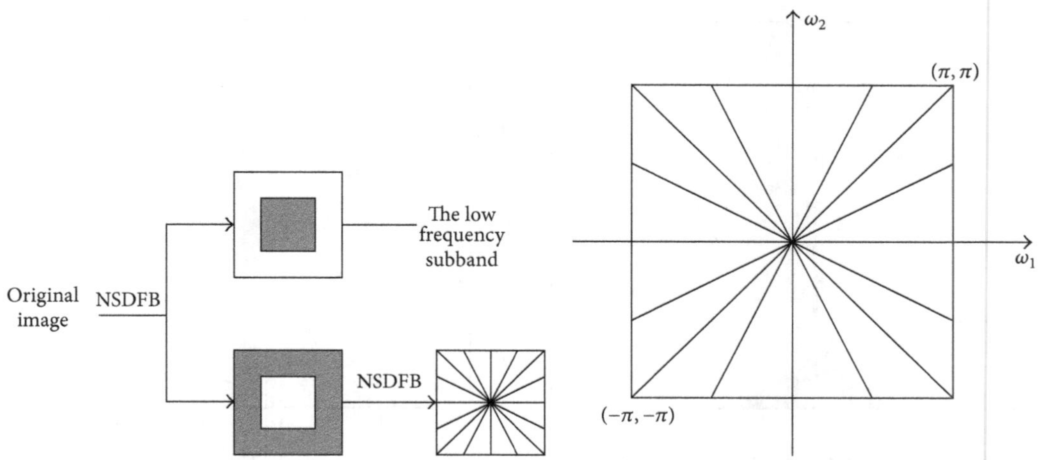

(a) Nonsubsampled directional filter bank (b) Frequency resolution

FIGURE 1: The structure of nonsubsampled Contourlet transform.

and can achieve better expression of image than the wavelet transform. It has the anisotropic characteristics. Because of the process of downsampling and upsampling, the Contourlet transform is shift-variance and always Gibbs phenomena around singularities. In 2006, the nonsubsampled Contourlet transform (NSCT) [5] was proposed by Cunha et al. The NSCT has better performance of the NSCT in image denoising and enhancement applications. NSCT is composed of the nonsubsampled pyramid filter and the nonsubsampled direction filter, so it is shift-invariance [6].

In this paper, a method of edge detection based on nonsubsampled Contourlet transform and Gaussian mixture model is proposed. The probability and statistics analysis is applied in the NSCT coefficients. The Gaussian mixture model is used to distinguish the linear singular signal and nonlinear singular signal on the high frequency subbands automatically. Local direction modulus nonmaxima suppression is used to refine the linear singular signal. An appropriate threshold is used to distinguish edge pixels and nonedge pixels to get binary image. We compare the proposed method with the classical Canny operator, Sobel operator, the wavelet method [2], and the method based on cellular neural network (CNN) [7]. The proposed method can capture more continuous edges in multiple directions and has accurate edge location.

2. Modeling NSCT High Frequency Coefficients

2.1. Nonsubsampled Contourlet Transform. NSCT consisted of the nonsubsampled pyramid filter banks (NSPFB) and the nonsubsampled directional filter banks (NSDFB) [6]. The two independent parts are shift-invariant. NSPFB is used for multiscale decomposition. NSDFB uses singular points of the same direction to synthesize the NSCT coefficients. NSDFB is shown in Figure 1(a). Frequency resolution is shown in Figure 1(b).

The distribution of NSCT high frequency subbands coefficients has the "high spikes, long tail" non-Gaussian statistical characteristic, as shown in Figure 2(a). High frequency subband coefficients consist of a few "big" coefficients and most "small" coefficients. The "small" coefficients contain less information and the "big" coefficients contain main information. Other high frequency directional subbands also have the similar distribution. Based on these characteristics to distinguish the linear singular signal and the nonlinear singular signal, Gaussian mixture model is used to model subbands, as shown in Figure 2(b); we use one Gaussian density function with big variance and one Gaussian density function with small variance to approximate the distribution. The "big" coefficients represent the linear singular signal. The linear singular signal is corresponding to the image edges.

2.2. Parameter Estimation of Gaussian Mixture Model. We model each high frequency directional subband through the Gaussian mixture model. $x^{v,w}$ is the coefficient which is in the v scale and the w high frequency directional subband; N is the number of all coefficients in $x^{v,w}$. Let $x^{v,w}_{m,n}$ be in $x^{v,w}$ matrix, where m is the column and n is the row. Assume that each sample in $x^{v,w}_{m,n}$ is independent of each other and obeys the K Gaussian distribution [8], $K = 2$. $x^{v,w}_{m,n}$ will be divided into two classes: one class is "big" state; the other is "small" state.

Density function of $x^{v,w}_{m,n}$ is $f^{v,w}(x) = \sum_{k=0}^{K} \pi_k^{v,w} N^{v,w}(x^{v,w}_{m,n} \mid \mu_k^{v,w}, \Sigma_k^{v,w})$. There, $\sum_{k=0}^{K} \pi_k^{v,w} = 1$, where $k = 0$ or 1: 0 means "big" state and 1 means "small" state; $\mu_k^{v,w}$ is the k class means; $\Sigma_k^{v,w}$ is the k class covariance, where v is the decomposition scale and w is the filter direction.

Expectation maximization algorithm can be used to calculate the optimal parameters of GMM by E-Step and M-Step.

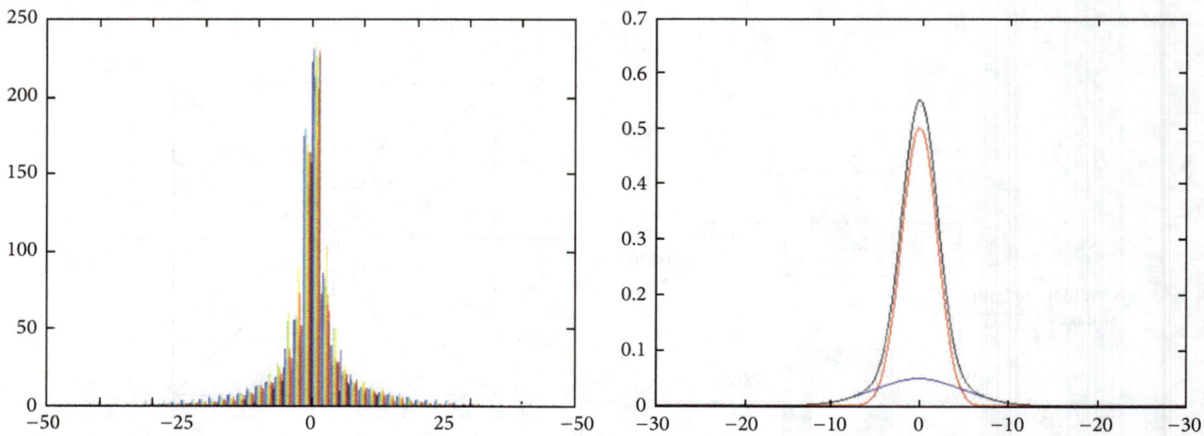

(a) "High spikes, long tail" non-Gaussian statistical characteristic

(b) The two classes of the coefficients

FIGURE 2: Statistics of the image NSCT coefficients.

Input: high-frequency directional sub-bands $x_{m,n}^{v,w}$
(1) Employ (1)~(3) to calculate the probability of $x_{m,n}^{v,w}$ generated by each component
(2) Employ (4)~(7) to calculate the $u_k^{v,w}$, $\pi_k^{v,w}$, $N_k^{v,w}$
(3) Repeated iteration, until the values of the likelihood function achieves convergence.
(4) Calculate the probability of x_i generated by each component according to $x_{m,n}^{v,w}$, $u_k^{v,w}$, $\pi_k^{v,w}$, $N_k^{v,w}$. Let pro0 be the probability of $x_{m,n}^{v,w}$ generated by the "big" variance of Gaussian model; let prol be the probability of $x_{m,n}^{v,w}$ generated by the "small" variance of Gaussian model.
(5) If pro0 < prol, set 0 to $x_{m,n}^{v,w}$
Output: $x_{m,n}^{v,w}$ (new high-frequency directional sub-bands $x_{m,n}^{v,w}$)

ALGORITHM 1: Data screening algorithm by Gaussian mixture model.

E-Step: calculating the probability of $x_{m,n}^{v,w}$ generated by each component

$$\gamma^{v,w}(i,k) = \frac{\pi_k^{v,w} N\left(x_{m,n}^{v,w} \mid \mu_k^{v,w}, \Sigma_k^{v,w}\right)}{\sum_{j=0}^{K} \pi_j^{v,w} N\left(x_{m,n}^{v,w} \mid \mu_j^{v,w}, \Sigma_j^{v,w}\right)}. \tag{1}$$

In formula (1), $\gamma^{v,w}(i,k)$ is the probability of $x_{m,n}^{v,w}$ and $\pi_k^{v,w}$ is the weighting factor.

M-Step: calculating $N_k^{v,w}$, $u_k^{v,w}$, and $\pi_k^{v,w}$, iteratively

$$N_k^{v,w} = \sum_{i=1}^{N} \gamma^{v,w}(i,k), \tag{2}$$

$$u_k^{v,w} = \frac{1}{N_k^{v,w}} \sum_{i=1}^{N} \gamma^{v,w}(i,k)\, x_{m,n}^{v,w}, \tag{3}$$

$$\pi_k^{v,w} = \frac{N_k^{v,w}}{N}. \tag{4}$$

2.3. Algorithm 1: Data Screening Algorithm by Gaussian Mixture Model. See Algorithm 1.

2.4. Local Direction Modulus Nonmaxima Suppression. After executing Algorithm 1, we use local direction modulus nonmaxima suppression to refine the linear signal. High frequency directional subbands contain the directional information, so there is no need to calculate the direction of the modulus; just compare the modulus of pixel and the modulus of the two gradient directional pixels of this pixel [9]. Determine whether it is a local maximum modulus and retain if it is.

For example, there are eight directions in some new high frequency directional subbands. So there are eight sectors, which are $\pi/16, 3\pi/16, 5\pi/16, 7\pi/16, 9\pi/16, 11\pi/16, 13\pi/16$, and $15\pi/16$. Mod $(x_{m,n}^{v,w})$ is the modulus of $x_{m,n}^{v,w}$. $\arg^{v,w}$ is the equivalent gradient of $x^{v,w}$. Figure 3 is the directional subband equivalent gradient.

2.5. Algorithm 2: Local Mold Maxima Suppression Algorithm. See Algorithm 2.

2.6. Our Proposed Edge Detection Algorithm. The proposed edge detection algorithm is summarized as follows:

Input: original image.

Step 1. Decompose the original image into high frequency directional subbands $x_{m,n}^{v,w}$ and low frequency subband by NSCT.

Input: new high-frequency directional sub-bands $x^{v,w}_{m,n}$
If $3\pi/16 < \arg^{v,w} < 5\pi/16 \parallel 11\pi/16 < \arg^{v,w} < 13\pi/16$
  If $\operatorname{mod}(x^{v,w}_{m,n}) < \operatorname{mod}(x^{v,w}_{m,n-1}) \parallel \operatorname{mod}(x^{v,w}_{m,n}) < \operatorname{mod}(x^{v,w}_{m,n+1})$
    $x^{v,w}_{m,n} = 0$;
  End
End
ElseIf $\pi/16 < \arg^{v,w} < 3\pi/16 \parallel 9\pi/16 < \arg^{v,w} < 11\pi/16$
  If $\operatorname{mod}(x^{v,w}_{m,n}) < \operatorname{mod}(x^{v,w}_{m-1,n+1}) \parallel \operatorname{mod}(x^{v,w}_{m,n}) < \operatorname{mod}(x^{v,w}_{m+1,n-1})$
    $x^{v,w}_{m,n} = 0$;
  End
End
ElseIf $5\pi/16 < \arg^{v,w} < 7\pi/16 \parallel 13\pi/16 < \arg^{v,w} < 15\pi/16$
  If $\operatorname{mod}(x^{v,w}_{m,n}) < \operatorname{mod}(x^{v,w}_{m-1,n}) \parallel \operatorname{mod}(x^{v,w}_{m,n}) < \operatorname{mod}(x^{v,w}_{m+1,n})$
    $x^{v,w}_{m,n} = 0$;
  End
End
Else
  If $\operatorname{mod}(x^{v,w}_{m,n}) < \operatorname{mod}(x^{v,w}_{m-1,n}) \parallel \operatorname{mod}(x^{v,w}_{m,n}) < \operatorname{mod}(x^{v,w}_{m+1,n})$
    $x^{v,w}_{m,n} = 0$;
  End
End
Output: $x^{v,w}_{m,n}$ (new high-frequency directional sub-bands $x^{v,w}_{m,n}$)

ALGORITHM 2: Local mold maxima suppression algorithm.

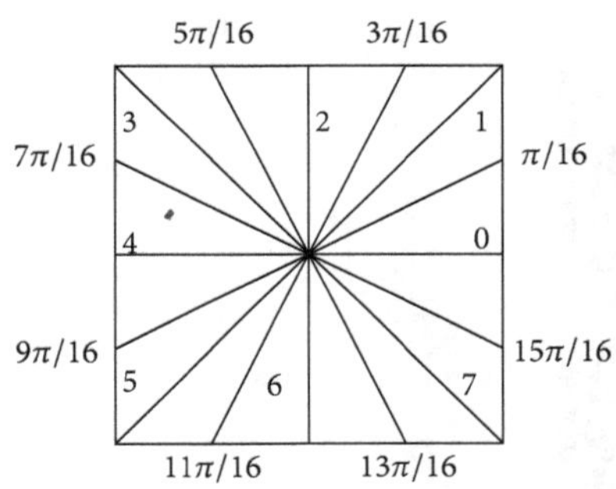

FIGURE 3: Directional subband equivalent gradient.

Step 2. Let low frequency subband coefficients be 0.

Step 3. Employ Algorithm 1 to process $x^{v,w}_{m,n}$, we get new $x^{v,w}_{m,n}$, and new $x^{v,w}_{m,n}$ are the linear singular signal here.

Step 4. Employ Algorithm 2 to process $x^{v,w}_{m,n}$, which is got by Step 3, we get new $x^{v,w}_{m,n}$, and new $x^{v,w}_{m,n}$ are the linear singular signal refined here.

Step 5. Reconstruct image using high frequency coefficients $x^{v,w}_{m,n}$ processed by Step 3 and low frequency coefficients processed by Step 1.

Step 6. After NSCT inverse transform, there will be a data matrix. There will be two types of data in that matrix, the modulus of one type is big, and the modulus of the other type is small. An appropriate threshold can be easily set to separate them. Set threshold to distinguish edge pixels and nonedge pixels to get binary image.

Output: edge detection result.

3. Experimental Results

In order to quantify the performance of the proposed method, FOM (Pratt's Figure of Merit) [10] is used to compare the performance of edge detection methods, because the edges of the standard image must be obtained while using FOM. Because of this, we can just use the synthetic images. The FOM is computed as follows:

$$\mathrm{FOM} = \frac{1}{\max\{N_I, N_A\}} \sum_{i=1}^{N_A} \frac{1}{1 + \alpha d_i}, \tag{5}$$

where N_I and N_A are the number of ideal and detected edge pixels, respectively, the parameter α is a constant, and d_i is the vertical distance between an actual edge pixel and the nearest ideal edge pixel. The FOM measures the accuracy of localization. The more the FOM is, the more the quality of the edge detection is. In order to objectively discuss the accuracy of the localization, the false edge pixels are not covered while computing the FOM. A synthetic image is used to calculate FOM.

The proposed method, CNN, Canny, Sobel operators, NSCT, and the wavelet method are used to detect the edge of the synthetic image. The NSCT method is used to detect edges without using GMM. Figure 4 shows the results detected by the different methods. Table 1 shows the FOM calculations by

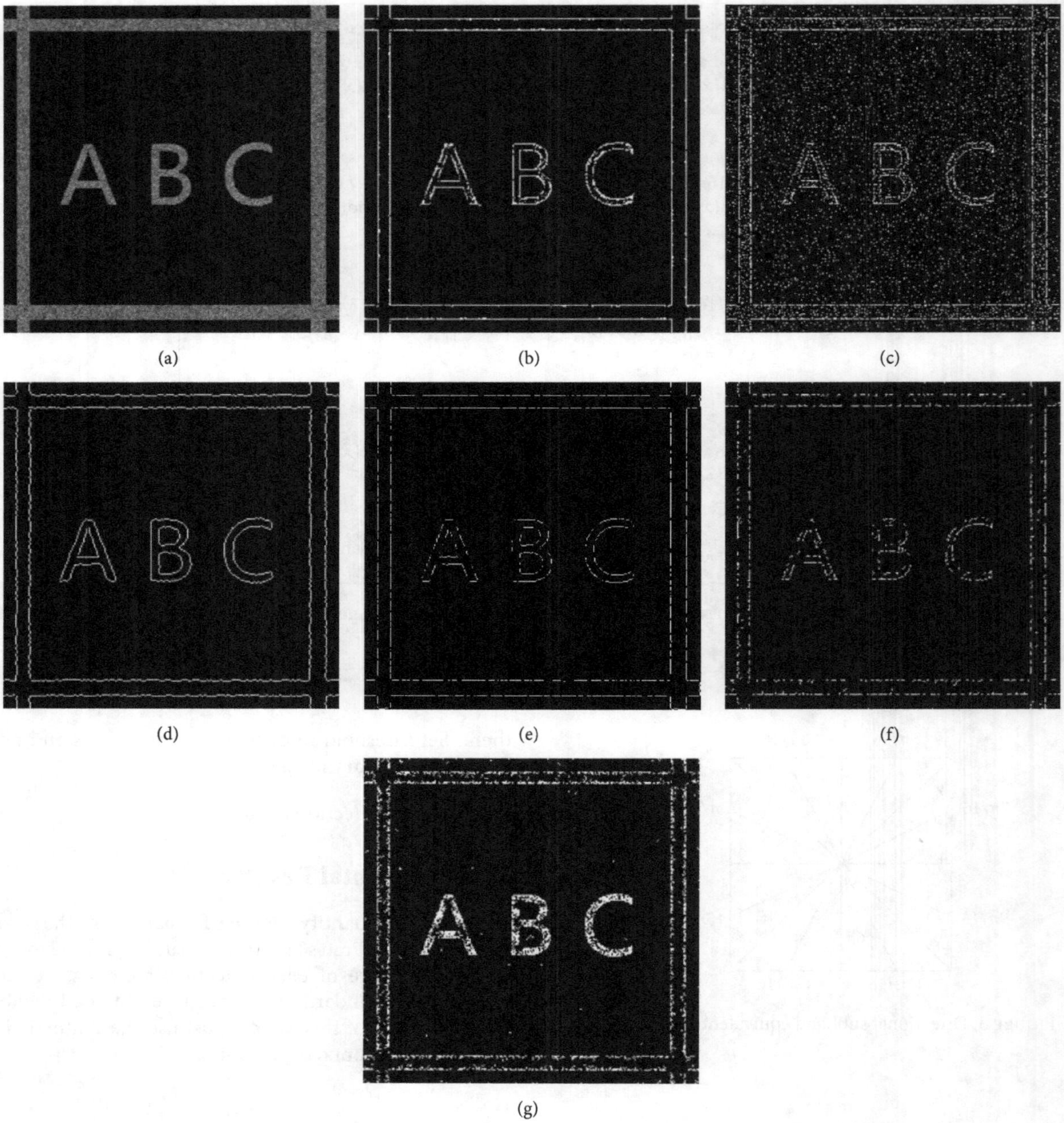

FIGURE 4: The results of the edge detection: (a) the synthetic image, (b) the proposed method, (c) the method based on CNN, (d) Canny operator, (e) Sobel operator, (f) wavelet method, and (g) NSCT method.

TABLE 1: The FOM calculations with different methods.

Figure 4	The proposed method	The method based on CNN	Canny operator	Sobel operator	NSCT	Wavelet method
FOM	0.9631	0.8934	0.8256	0.7937	0.6432	0.7231

different methods. The results demonstrate that the accuracy of the localization of the proposed method is more accurate than Canny operator, Sobel operator, wavelet method, NSCT method, and the method based on CNN. The accuracy of the localization of the proposed method is the best. Because of not using GMM to distinguish the linear singular signal and the nonlinear singular signal on the high frequency subbands, the edges detected by the NSCT method are interrupted by the nonlinear singular signal. So the accuracy of the localization of the NSCT method is very low. The Canny,

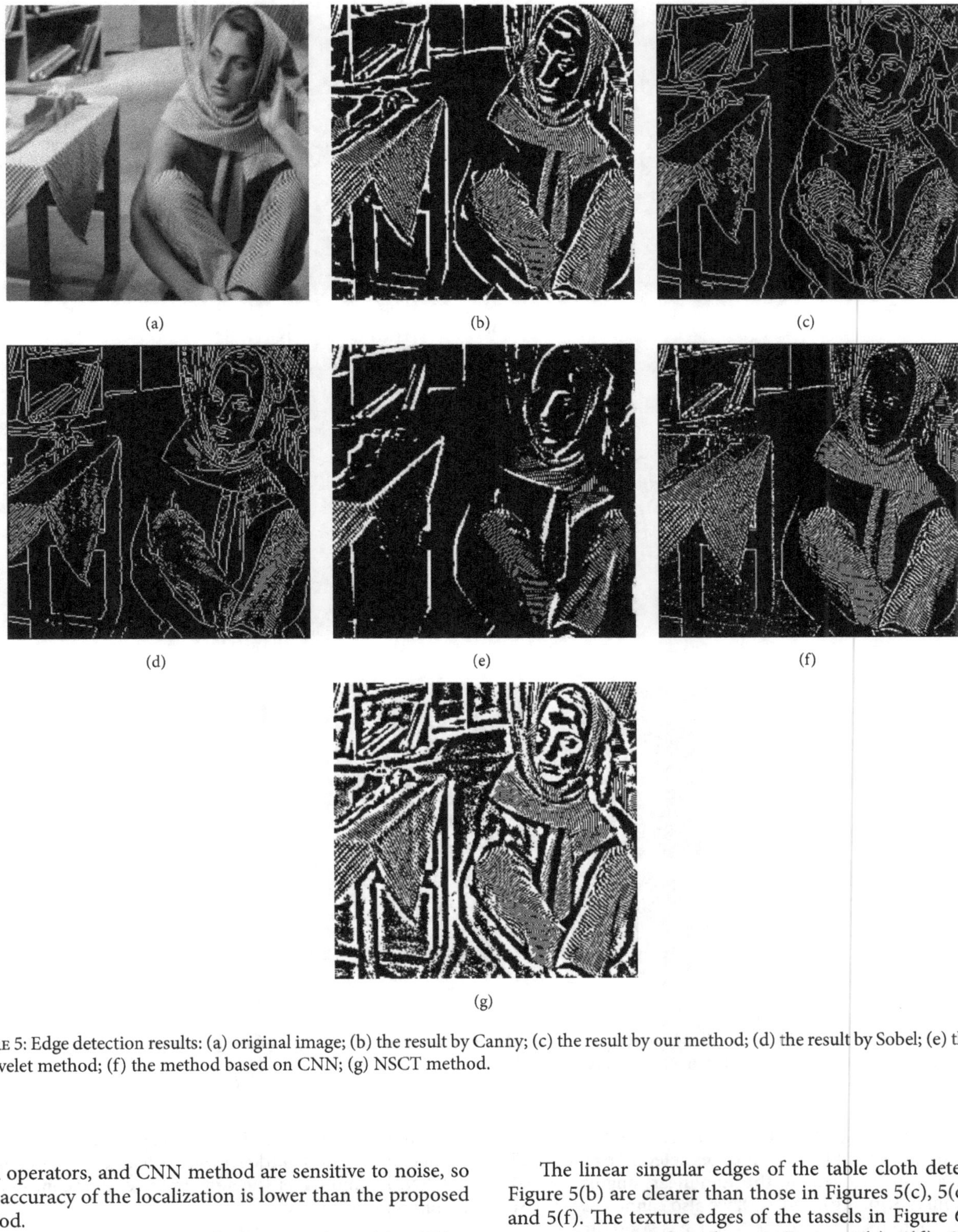

FIGURE 5: Edge detection results: (a) original image; (b) the result by Canny; (c) the result by our method; (d) the result by Sobel; (e) the result by wavelet method; (f) the method based on CNN; (g) NSCT method.

Sobel operators, and CNN method are sensitive to noise, so their accuracy of the localization is lower than the proposed method.

Three standard images, which are Barbara (size 256 ∗ 256), Lenna (size 256 ∗ 256), and boat (size 512 ∗ 512), are used to test in the experiments. We select "9-7" pyramid decomposition and "pkva" directional filter bank for NSCT decomposition. Decomposition level is 3 and each level has eight high frequency directions. Experimental results are shown in Figures 5, 6, and 7.

The linear singular edges of the table cloth detected in Figure 5(b) are clearer than those in Figures 5(c), 5(d), 5(e), and 5(f). The texture edges of the tassels in Figure 6(b) are more continuous than those in Figures 6(c), 6(d), 6(e), and 6(f). Contour edges detected of the hull in Figure 7(b) are more complete than those in Figures 7(c), 7(d), and 7(f). The edges are too miscellaneous to distinguish the main contours in Figures 5(c), 6(c), 7(c), 5(d), 6(d), and 7(d). Because the wavelet just has two directions, it cannot capture more directional details. Because of not using GMM to distinguish

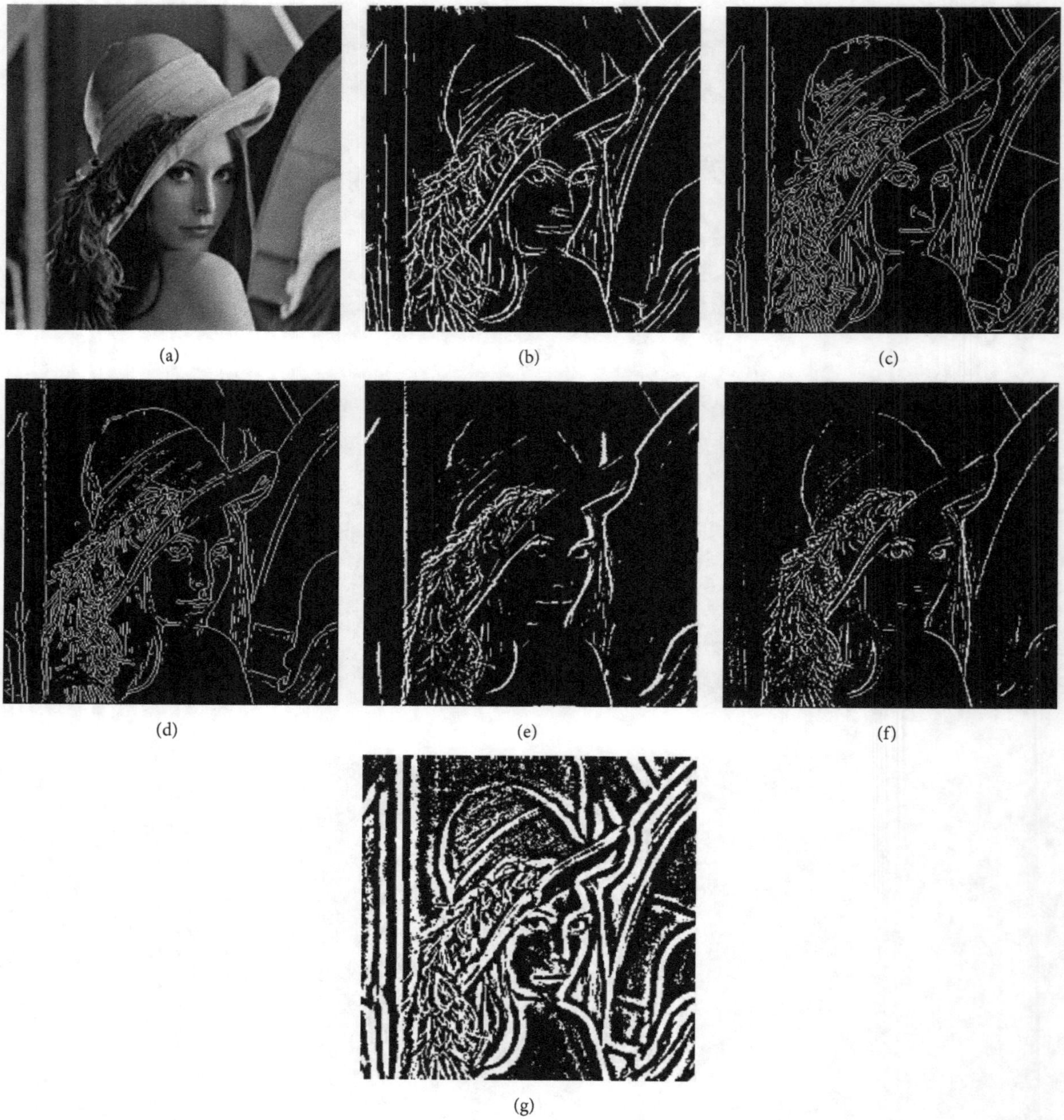

FIGURE 6: Edge detection results: (a) original image; (b) the result by Canny; (c) the result by our method; (d) the result by Sobel; (e) the result by wavelet method; (f) the method based on CNN; (g) NSCT method.

the linear singular signal and the nonlinear singular signal on the high frequency subbands, the edges detected by the NSCT method are interrupted by the nonlinear singular signal. From Figures 5(g), 6(g), and 7(g), we also cannot distinguish the main contours. Because the NSCT has good directionality, anisotropy, and decorrelation, the proposed method can capture more directional details. And the proposed method can capture the main contours because of using the GMM. From Figures 5(b), 6(b), and 7(b), we can clearly see the main contours in the images.

In order to evaluate the algorithm, the quantitative analysis is adopted [11]. In Table 2, CEN denotes the number of continuous edge pixels in edge image. TEN represents the number of all pixels in edge image. R denotes the ratio of TEN to CEN. The more R is, the more image edge is continuous. R is computed as follows:

$$R = \frac{\text{TEN}}{\text{CEN}}. \tag{6}$$

Table 2 shows that the edges obtained by the proposed method are more continuous than other methods. Because the wavelet just has two directions, it cannot capture more directional details. Though the NSCT method can capture more edges than the Canny and Sobel operators, the pixels

FIGURE 7: Edge detection results: (a) original image; (b) the result by Canny; (c) the result by our method; (d) the result by Sobel; (e) the result by wavelet method; (f) the method based on CNN; (g) NSCT method.

detected by the NSCT method contain so much the nonlinear singular signal. The proposed method can capture more details than other methods.

MSE (Mean Squared Error) is a standard to measure the accuracy of localization. Generally, when MSE is greater, the accuracy of localization is less. When MSE is less, the accuracy of localization is higher. The MSE is computed as follows:

$$\mathrm{MSE} = \frac{1}{mn}\sum_{i=0}^{m-1}\sum_{j=0}^{n-1}\left|I(i,j) - K(i,j)\right|^2. \tag{7}$$

Table 3 shows the MSE results of different methods. We can see that the proposed method has a lower MSE than other methods. It means that the proposed method has high accuracy of localization. Because of not using GMM to distinguish the linear singular signal and the nonlinear singular signal on the high frequency subbands, the edges detected by the NSCT method are interrupted by the nonlinear singular signal. So the accuracy of localization is low in the NSCT method.

The results show that, compared with traditional edge detection methods, the proposed method possessed better edge locating ability and keeps more edge details.

TABLE 2: The comparison of the different detection methods.

Image	Detection method	CEN	TEN	R
Barbara	Canny	9896	8713	0.8805
	Our method	16113	14984	0.9299
	Sobel	8646	7163	0.82848
	Wavelet	7556	6388	0.8454
	CNN	12703	10793	0.8496
	NSCT	21822	19790	0.9069
Lenna	Canny	7935	7184	0.9041
	Our method	8462	7924	0.9364
	Sobel	6731	5831	0.86629
	Wavelet	5470	4822	0.8815
	CNN	4544	3627	0.7982
	NSCT	23823	22043	0.9253
Boat	Canny	32011	29030	0.9069
	Our method	56781	54791	0.965
	Sobel	34536	30172	0.87364
	Wavelet	25920	23479	0.9058
	CNN	18805	13647	0.7257
	NSCT	80251	73809	0.9197

TABLE 3: MSE of the different detection methods.

Image	Sobel MSE	Canny MSE	CNN MSE	NSCT	Wavelet MSE	Proposed method MSE
Barbara	0.64891	0.4351	0.2574	0.7859	0.5874	0.1989
Boat	0.5945	0.3946	0.2415	0.6845	0.6842	0.1826
Lenna	0.84657	0.5874	0.4059	0.7548	0.3897	0.2834

4. Conclusions

We have presented an edge detection method based on nonsubsampled Contourlet transform and Gaussian mixture model. The proposed method is based on nonsubsampled Contourlet transform and Gaussian mixture model. It can capture more directional information and continuous edges. It overcomes the shortage of the limited directions of wavelet transform. The main contours in the image can be clearly expressed. This method can effectively locate the edge. The thick edges detected by the proposed method express the main edges in the images, which are with great convenience for the image recognition. The proposed method runs slowly; it needs to be improved in the future.

Competing Interests

The authors declare that they have no competing interests.

Acknowledgments

This work is supported by the National Natural Science Foundation of China (Grant no. 61440044, no. 61561001, no. 61102008, no. 61163017, no. 61261043, and no. 61462002), the Ningxia Science Foundation of China (Grant no. NZ13097), and the Foundations of Research Projects of State Ethnic Affairs Commission of P. R. China (Grant no. 14BFZ003).

References

[1] S. Bhattacharya, "Edge detection in range images using nonlinear laplacian operators," *IETE Journal of Research*, vol. 48, no. 3-4, pp. 205–209, 2002.

[2] H. Zhang, L. Luo, K. Yang, L. Wang, and X. Gao, "Improved multi-scale wavelet in pantograph slide edge detection," *Optik*, vol. 125, no. 19, pp. 5681–5683, 2014.

[3] M. T. Alonso, C. López-Martínez, J. J. Mallorquí, and P. Salembier, "Edge enhancement algorithm based on the wavelet transform for automatic edge detection in SAR images," *IEEE Transactions on Geoscience & Remote Sensing*, vol. 49, no. 1, pp. 222–235, 2011.

[4] M. N. Do and M. Vetterli, "The contourlet transform: an efficient directional multiresolution image representation," *IEEE Transactions on Image Processing*, vol. 14, no. 12, pp. 2091–2106, 2005.

[5] A. L. D. Cunha, J. Zhou, and M. N. Do, "Nonsubsampled contourlet transform: theory, design, and applications," *IEEE Transactions on Image Processing*, vol. 15, no. 10, pp. 3089–3101, 2006.

[6] L. Xu, J. Du, and Q. Li, "Image fusion based on nonsubsampled contourlet transform and saliency-motivated pulse coupled neural networks," *Mathematical Problems in Engineering*, vol. 2013, Article ID 135182, 10 pages, 2013.

[7] S. Deng, Y. Tian, X. Hu, P. Wei, and M. Qin, "Application of new advanced CNN structure with adaptive thresholds to color edge detection," *Communications in Nonlinear Science and Numerical Simulation*, vol. 17, no. 4, pp. 1637–1648, 2012.

[8] X. Chang, L. Jiao, F. Liu, and Y. Sha, "SAR image despeckling using scale mixtures of gaussians in the nonsubsampled contourlet domain," *Chinese Journal of Electronics*, vol. 24, no. 1, pp. 205–211, 2015.

[9] T. Y. Lo, K. S. Sim, C. P. Tso, and M. E. Nia, "Improvement to the scanning electron microscope image adaptive canny optimization colorization by pseudo-mapping," *Scanning*, vol. 36, no. 5, pp. 530–539, 2014.

[10] I. E. Abdou and W. K. Pratt, "Quantitative design and evaluation of enhancement/thresholding edge detectors," *Proceedings of the IEEE*, vol. 67, no. 5, pp. 753–763, 1979.

[11] G.-Y. Zhang, G.-Z. Gong, and W.-L. Zhu, "A novel method of color image edge detection based on cloning algorithm," *Acta Electronica Sinica*, vol. 34, no. 4, pp. 702–707, 2006.

5

Face Spoof Attack Recognition using Discriminative Image Patches

Zahid Akhtar and Gian Luca Foresti

Department of Mathematics and Computer Science, University of Udine, Via delle Scienze 206, 33100 Udine, Italy

Correspondence should be addressed to Zahid Akhtar; zahid.akhtar@uniud.it

Academic Editor: Kwok-Wai Cheung

Face recognition systems are now being used in many applications such as border crossings, banks, and mobile payments. The wide scale deployment of facial recognition systems has attracted intensive attention to the reliability of face biometrics against spoof attacks, where a photo, a video, or a 3D mask of a genuine user's face can be used to gain illegitimate access to facilities or services. Though several face antispoofing or liveness detection methods (which determine at the time of capture whether a face is live or spoof) have been proposed, the issue is still unsolved due to difficulty in finding discriminative and computationally inexpensive features and methods for spoof attacks. In addition, existing techniques use whole face image or complete video for liveness detection. However, often certain face regions (video frames) are redundant or correspond to the clutter in the image (video), thus leading generally to low performances. Therefore, we propose seven novel methods to find discriminative image patches, which we define as regions that are salient, instrumental, and class-specific. Four well-known classifiers, namely, support vector machine (SVM), Naive-Bayes, Quadratic Discriminant Analysis (QDA), and Ensemble, are then used to distinguish between genuine and spoof faces using a voting based scheme. Experimental analysis on two publicly available databases (Idiap REPLAY-ATTACK and CASIA-FASD) shows promising results compared to existing works.

1. Introduction

In the last years, face recognition systems have gained interest due to face's rich features that offer a strong biometric cue to recognize individuals for a wide variety of applications in both law and nonlaw enforcements [1]. In fact, facial recognition systems are already in operation worldwide, including USVISIT, which is a US Customs and Border Protection (CBP) management system, UIDAI that provides identity to all persons resident in India, and Microsoft Kinect which uses face recognition to access dashboard and automatic sign-in to Xbox Live profile. Similarly, face biometrics is also nowadays being used ubiquitously as an alternative to passwords on mobile devices such as Android KitKat mobile OS, Lenovo VeriFace, Asus SmartLogon, and Toshiba SmartFace.

Despite the great deal of progress in facial recognition systems, vulnerabilities to face spoof attacks are mainly overlooked [2]. Facial spoof attack is a process in which a fraudulent user can subvert or attack a face recognition system by masquerading as registered user and thereby gaining illegitimate access and advantages [1, 3–5]. Face spoofing attack is a major issue for companies selling face biometric-based identity management solutions [6]. For instance, at New York City, nonwhite robbers disguised themselves as white cops, using life-like latex masks, and were caught robbing a cash-checking store in 2014 (see Figure 1, also for other recent face spoof attacks).

Recent study reported in [1] suggests that the success rate of face spoof attacks could be up to 70%, even when a state-of-the-art Commercial Off-The-Shelf (COTS) face recognition system is used. Therefore, we could infer that even COTS face recognition systems are mainly not devised to effectively distinguish spoof faces from genuine live faces. As a matter of fact, this vulnerability of face spoofing to face recognition systems is now enlisted in the National Vulnerability Database of the National Institute of Standards and Technology (NIST) in the US.

Typical countermeasure to face spoof attacks is liveness detection method, which aims at disambiguating human live face samples from spoof artifacts [2, 7]. There exist

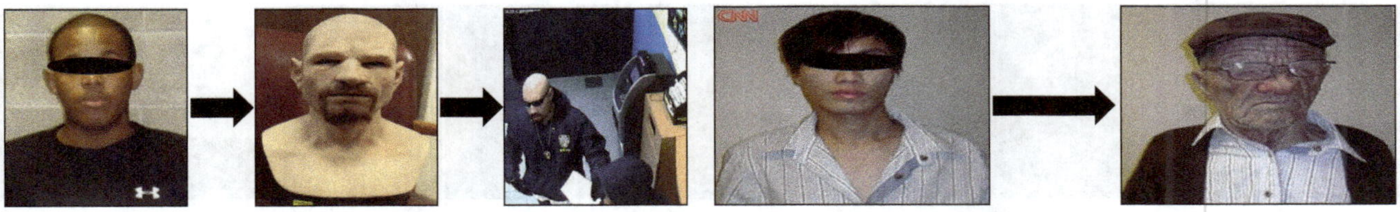

(a) (b)

FIGURE 1: (a) In New York, a robber disguised as white cop using latex masks robbing a cash-checking store. (b) In 2010, a passenger boarded a plane in Hong Kong with an old man mask and arrived in Canada to claim asylum.

several face antispoofing or liveness detection techniques [7–15]. However, face spoofing attacks remain a problem due to difficulties in finding discriminative and computationally inexpensive features and techniques for spoof recognition. Moreover, published methods are limited in their scope since they mainly use whole face image or complete video for liveness detection. Nevertheless, often certain face image regions (video frames) are redundant or correspond to the clutter in the image (video), leading thus generally to low performances.

It is thus essential to develop robust, efficient, and compact face antispoofing (or liveness detection) methods, which are capable of generalizing well to discriminative, class-specific information and imaging conditions. To this aim, in this paper, we propose a simple and effective solution based on discriminative image patches. In particular, we propose seven novel fully automated algorithms to highlight regions of interest in face images. We define these regions (or image patches) to be discriminative (i.e., specific to a particular class: live or spoof), consistent (i.e., reliably appearing in different face images or video frames), salient (i.e., conspicuous regions), and repetitive (i.e., frequently appearing in the image set of specific class). The basic notion is "interesting patches are those that are specific to a face image (or video frame) and should contain features that give assistance to discriminate a given live face image from spoofed one." Based on this definition, two of the seven proposed image patch selection methods (i.e., MAXDIST and DEND-CLUSTER) do not employ any training or prior learning. However, the remaining techniques use simple clustering (i.e., CP and CS), image intensity (i.e., IPI), image quality (i.e., IQA), or diversity filter (i.e., DF) to obtain discriminative patches. For final classification, we exploited four well-known classifiers, namely, support vector machine (SVM), Naive-Bayes, Quadratic Discriminant Analysis (QDA), and Ensemble, using voting based scheme. Experimental analysis on two publicly available databases (Idiap Replay-Attack and CASIA-FASD) shows good results compared to existing works.

The added advantages of the proposed framework are (i) being cheap, (ii) very low complexity, (iii) needing one face image (i.e., the same face sample acquired for face recognition) to detect whether it is genuine or spoof attack, (iv) being nonintrusive, (v) being user-friendly, and (vi) being easy to embed in already functional face recognition systems plus no requirement of new piece of hardware.

The reminder of the paper is organized as follows. Exiting literature works on face liveness detection are discussed in Section 2. The proposed approaches to determine the discriminative image patches and spoof classification schemes are described in Section 3. Experimental datasets, protocols, and results are presented in Section 4. A conclusion is drawn in Section 5.

2. Related Work

Despite great deal of advancements in face recognition systems, face spoofing still poses a serious threat. Most of the existing academic and commercial facial recognition systems may be spoofed by (see Figure 2) (i) a photo of a genuine user; (ii) a video of a genuine user; (iii) a 3D face model (mask) of a genuine user; (iv) a reverse-engineered face image from the template of a genuine user; (v) a sketch of a genuine user; (vi) an impostor wearing specific make-up to look like a genuine user; (vii) an impostor who underwent plastic surgery to look like a genuine user. The most easiest, cheapest, and common face spoofing attack is to submit a photograph of a legitimate user to the face recognition systems, which is also known as "photo attack."

Typical countermeasure (i.e., face liveness detection or antispoofing) techniques can be coarsely classified in three categories based on clues used for spoof attack detection: (i) *motion analysis based methods*, (ii) *texture analysis based methods*, and (iii) *hardware-based methods*. In what follows, we provide a brief literature overview of published face spoof recognition techniques along with their pros and cons.

(i) Motion Analysis Based Methods. These methods broadly try to detect spontaneous movement clues generated when two dimensional counterfeits are presented to the camera of the system, such as photographs or videos. Therefore, Pan et al. [7] exploited the fact that human eye-blink occurs once every 2–4 seconds and proposed eye-blink based liveness detection for photo-spoofing using (spontaneous) eye-blinks. This method uses an undirected conditional random field framework to model the eye-blinking, which relaxes the independence assumption of generative modelling and states dependence limitations from hidden Markov modelling. It is evident that real human faces (which are 3D objects) will move significantly differently from planer objects, and such deformation patterns can be employed for liveness detection. For example, Tan et al. [8] considered Lambertian

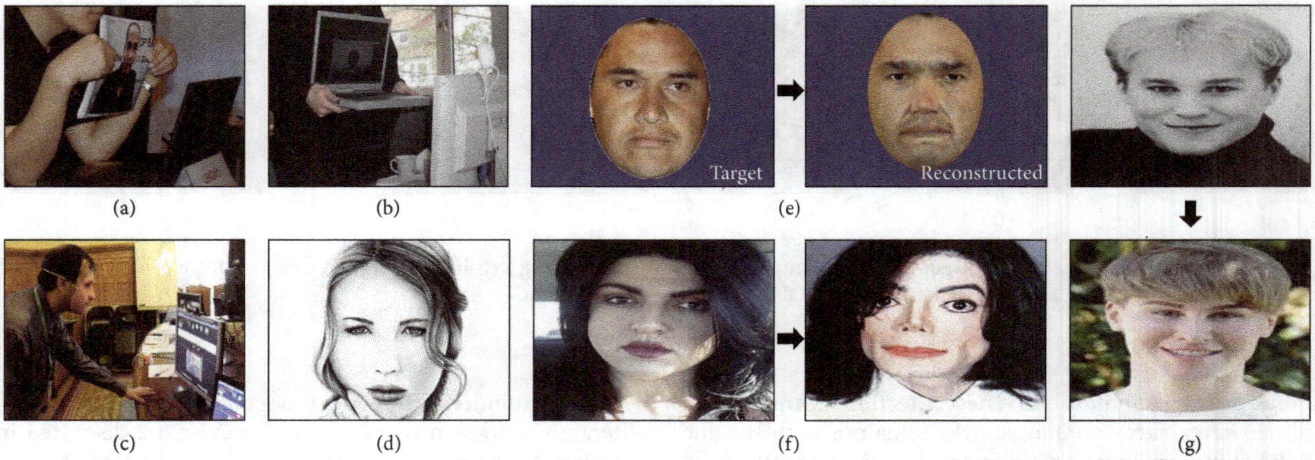

FIGURE 2: Examples of face spoofing using (a) photograph, (b) video, (c) 3D mask, (d) sketch, (e) reverse-engineered face image, (f) make-up (skillful application of make-up to look like Michel Jackson), and (g) plastic surgery (this boy underwent excessive plastic surgery to look like Justin Bieber).

reflectance model with difference-of-Gaussians (DoG) to derive differences of motion deformation patterns between 2D face photos presented during spoofing attacks and 3D live faces. It does so by extracting the features in both scenarios using a variational retinex-based method and difference-of-Gaussians (DoG) [9] based approach. Then, the features are used for live or spoof classification. Reported experiments showed promising results on a dataset consisting of real accesses and spoofing attacks to 15 clients using photo-quality and laser-quality prints. Kollreider et al. [10] proposed a liveness detection approach based on a short sequence of images using a binary detector, which captures and tracts the subtle movements of different selected facial parts using a simplified optical flow analysis followed by a heuristic classifier. The same authors also presented a method to combine scores from different experts systems, which concurrently observe the 3D face motion approach introduced in the former work as liveness attributes like eye-blinks or mouth movements. In the similar fashion, Bao et al. [11] also used optical flow to estimate motion for detecting attacks produced with planar media such as prints or screens.

Since the frequency of facial motion is restricted by the human physiological rhythm, thus motion based methods take a relatively long time (usually > 3 s) to accumulate stable vitality features for face spoof detection. Moreover, they may be circumvented or confused by other motions, for example, background motion in the video attacks.

(ii) Texture Analysis Based Methods. This kind of methods examines the skin properties, such as skin texture and skin reflectance, under the assumption that surface properties of real faces and prints, for example, pigments, are different. Examples of detectable texture patterns due to artifacts are printing failures or blurring. Li et al. [13] described a method for print-attack face spoofing by exploiting differences in the 2D Fourier spectra of live and spoof images. The method assumes that photographs are normally smaller in size and contain fewer high frequency components compared to real faces. The method only works well for downsampled photos of the attacked identity but likely fails for higher-quality samples. In [14, 16, 17], authors developed microtexture analysis based methods to detect printed photo attacks. One limitation of presented methods is the requirement of reasonably sharp input image. Recently, Galbally et al. [3] designed a face spoof detection scheme based on 25 different image quality measures: 21 full-reference measures and 4 nonreference measures. However, all 25 image quality measures are required to get good results and no face-specific information has been considered in designing informative features for face spoof detection.

Compared to other techniques, texture analysis based algorithms are generally faster to classify a spoof attack. Nevertheless, they could be easily overfitted to one particular illumination and imagery condition and hence do not generalize well to different spoofing conditions.

(iii) Hardware-Based Methods. Few interesting hardware-based face antispoofing techniques have been proposed so far based on imaging technology outside the visual spectrum, such as 3D depth [18], complementary infrared (CIR), or near infrared (NIR) images [15] by comparing the reflectance information of real faces and spoof materials using a specific set-up of LEDs and photodiodes at two different wavelengths. Preliminary efforts on thermal imaging for face liveness detection have also been exploited, including the acquisition of large database of thermal face images for real and spoofed access attempts [19]. Besides, numbers of researchers have explored multimodality as antispoofing techniques for face spoofing attacks. They have mainly considered the combination of face and voice by utilizing the correlation between the lips movement and the speech being produced [20], where a microphone and a speech analyzer were required. Similarly, challenge-response strategy considering voluntary eye-blinking and mouth movement following a request from the system has been studied in [10]. Though hardware-based methods provide better results and performances, they

TABLE 1: Summary of different face spoof detection techniques.

Method	Main features used	Pros and cons
Motion analysis based methods	Motion detection [10] Eye-blink detection using conditional random fields (CRF) [7] Face motion detection using Optical Flow Lines (OFL) [10] Context-based using correlation between face motion and background motion [17]	(i) Good generalization capability (ii) High computational cost (iii) Easily circumvented by fake motions
Texture analysis based methods	Face texture using Lambertian model [8] Face texture using LBPs [17] Texture + shape combining LBPs + Gabor Wavelets + HOG [5]	(i) Fast response (ii) Low computational cost (iii) Poor generalization capability
Hardware-based methods	Multimodality: face and voice [20] Thermal images [19] Reflectance in 3D [21]	(i) Better generalization capability (ii) Extra hardware requirement (iii) High cost of the system

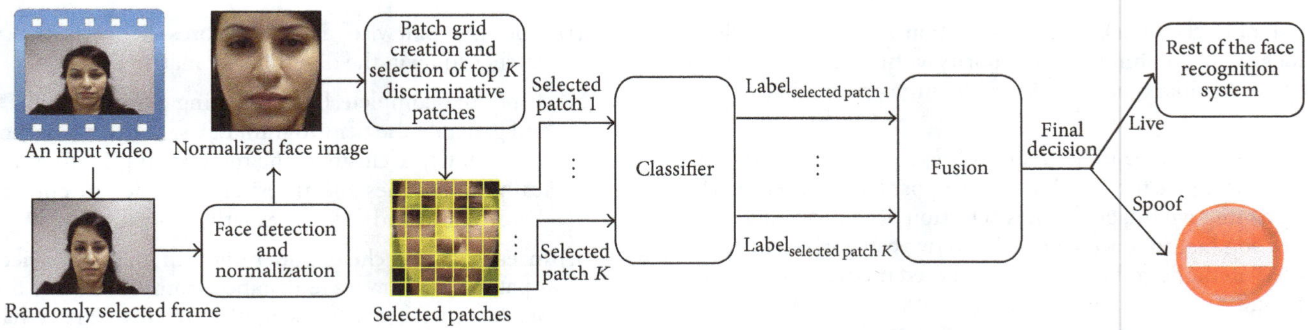

FIGURE 3: The proposed face spoof recognition algorithm based on discriminative image patches.

require extra piece of hardware which increases the cost of the system. A summary with relevant features of the most representative works in face antispoofing is presented in Table 1.

Though there exist several face antispoofing or liveness detection techniques, face spoof attacks remain an issue because of difficulty in finding discriminative and computationally inexpensive features and mechanisms for spoof recognition. Reported methods are limited in their scope since they mainly use full image or complete video for liveness detection. In particular, there is a lack of investigation on how specific image patches rather than full image perform in spoof detection. As often image regions (video frames) are redundant or correspond to the clutter in the image (video), leading thus generally to low performances and high computational cost, towards this direction, we propose seven novel methods to find discriminative image patches, which we define as regions that are salient, instrumental, and class-specific. Experimental results show that the proposed methods obtain comparable performances to existing techniques.

3. The Proposed Method for Face Spoof Recognition

Face spoof detection can be seen as a two-class classification problem, where the input face image (or video) has to be flagged as either live or spoof. The keynote of the process is attaining a discriminant feature set together with an appropriate classification scheme that gives the probability of the image (or video) realism. Practical face spoof detection requires that a decision be made based on single image or a limited number of frames in the video-based system. In this work, thus we aim to design simple but effective solution based on discriminative image patches using a single face frame/image. We define these image patches to be discriminative, consistent, salient, and repetitive. The notion is that the interesting patches are those that are specific to a face image (or video frame) and should contain features that help discriminate a given live face image from spoofed one.

Figure 3 shows the schematic diagram of the proposed face spoof recognition algorithm based on discriminative image patches. The proposed framework first randomly selects a single frame from a given face video (in case of image-based system, the given single image is used). Then, face is detected using Local SMQT Features and Split-Up Snow Classifier [30]. Subsequently, the detected face is densely divided into a grid of nonoverlapping local patches. These patches are ranked based on their discriminative power. The top K patches are selected using specific discriminative image patch selection method among the proposed techniques (explained below in detail). For each selected image patch, features are extracted that are then fed into particular classifier (i.e., SVM, Naive-Bayes, QDA, or

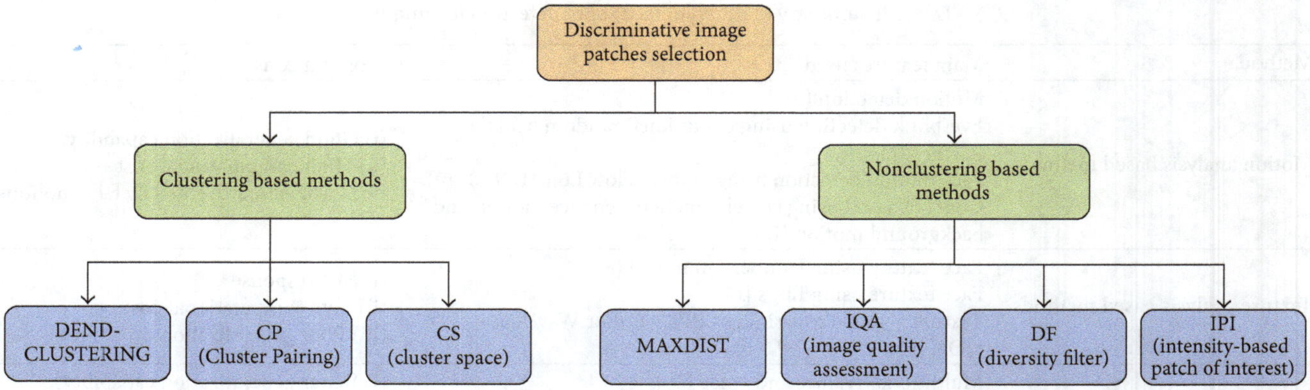

FIGURE 4: Classification of the seven discriminative image patches selection methods proposed in this work.

Ensemble classifier). The classification results of individual patches are combined by a majority-voting based scheme to obtain the final binary decision: genuine or spoof face.

3.1. Discriminative Image Patches Selection Methods. In what follows, we give the details of the proposed seven methods for discriminative image patches selection. The proposed patch selection methods are grouped into two categories: clustering based methods and nonclustering based methods. For clarity, in Figure 4 we show a diagram with the patch selection methods classification followed in this section.

3.1.1. Clustering Based Methods. The patch selection methods in this category rely on a clustering algorithm at any specific stage of the procedure. Three clustering based methods proposed in this work are as follows.

(1) DEND-CLUSTERING. In this discriminative image patches selection technique, the n patches in the given face image are grouped into $\mathbf{T}$ clusters, such that patches within a cluster are more similar to each other than patches belonging to different clusters. Then, for each cluster a prototype (representative) patch that typifies the members of that cluster is chosen, resulting in $\mathbf{T}$ discriminative patches. Since this method uses the dendrogram [31] to choose the discriminative patches, thus we call it DEND-CLUSTERING.

For each normalized face image ($\mathbf{I}$), first a set of dense patches $\{\mathbf{P}_i(\mathbf{I}) \in \mathbb{R}^{M\times N}\}_{i=1}^{n}$ is taken, where n is the total number of dense patches. Computation of the dissimilarity scores between patches is needed to perform clustering. Therefore, first the dissimilarity between successive patches is computed by comparing the respective features. Then, hierarchical clustering [31] is exploited, because the representation of the n patches is in the form of an $n \times n$ dissimilarity matrix instead of an $n \times p$ pattern matrix, where p is the dimension of feature vector. In particular, an agglomerative complete link clustering algorithm [31] is used in this work. The outcome of this algorithm is a dendrogram (a binary tree), where each terminal node corresponds to a patch, and the intermediate nodes indicate the formation of clusters. The discriminative K patches are selected as follows:

(i) Find the pairwise distance scores between the n patches to form the dissimilarity matrix $\mathbf{D}$.

(ii) Apply the complete link clustering algorithm on $\mathbf{D}$, and generate the dendrogram $\mathbf{L}$. Use the dendrogram $\mathbf{L}$ to identify $\mathbf{T}$ clusters. The method in [31] automatically determines the threshold distance to cut the dendrogram and identify exactly $\mathbf{T}$ clusters.

(iii) In each of the clusters identified in step (ii), select a patch whose average distance from the rest of the patches in the cluster is minimum. If a cluster has only 2 patches, choose any one of the two patches at random.

(iv) The patches selected in step (iii) are arranged in descending order on the basis of their ideal selection measure (ISM) value, which is computed as

$$\mathrm{ISM}(\mathbf{P}) = \sum_{x=1}^{P-2}\sum_{y=1}^{Q-2} G(x,y), \quad (1)$$

where $\mathbf{P}$ is a patch of size $P \times Q$ and $G(x, y)$ is the image gradient at location (x, y).

(v) The top K patches are selected as discriminative patches.

It is worth mentioning that steps (i)–(iii) in DEND-CLUSTERING method have close resemblance with the technique in [31] for fingerprint template selection. Here, we extended the technique by proposing step (iv) to be utilized for ranking and selection of discriminative patches.

(2) CP (Cluster Pairing). Apparently, the discrimination power of patches (features) decides maximum possible classification accuracy, and thus prior knowledge of "how cluttered the features (patches) may be" and "their contribution to classes separability in the feature space" can help to design and accomplish better classification scheme and accuracy. To this aim, in this method, first two independent sets of clusters are generated using genuine and spoof attack samples, respectively. Since overlapping of interclass clusters has great effect on classification accuracy, therefore pairs

of overlapped corresponding clusters of two independent sets are formed using minimum distance between them. Finally, patches which do not belong to both clusters of a given interclass highly overlapped cluster pair are selected as discriminative patches. In other words, if a patch belongs to both clusters of a given interclass cluster pair, it means that its features cause overlapping different classes in the feature space, which might thus lead to lower classification accuracy. The steps followed to obtain top K discriminative patches using this method are as follows:

(i) Two independent sets of clusters are generated using live and spoof attack training samples, respectively. Each class is represented by X number of clusters scattered in the feature space. K-means clustering technique is exploited in this work for cluster creation.

(ii) All possible pairs of corresponding clusters of two independent sets are formed using

$$D_{\text{mean}}\left(C_i, C_j\right) = \left\|m_i - m_j\right\|_{j=1}^{X} \leq \eta, \tag{2}$$

where η is threshold, m_x is center of C_x, C_i is a given cluster from live class clusters set, and C_j is a given cluster from spoof class clusters set. D_{mean} is appropriate for detecting spherical and compact cluster pairs, since each cluster is represented only by its center point.

(iii) For a given face image, K patches are chosen as discriminative patches, which do not belong to both elements of the interclass clusters.

(3) CS (Cluster Space). Principally, information classes cannot be described efficaciously by a single well-defined grouping in a spectral space. Thus, it is better to represent them by a group of spectral classes (clusters), which is prime inference of this method. It is worth noting that this method is identical to the above-mentioned Cluster Pairing (CP) method. In this method, X number of clusters are generated using both live and fake labeled training patches together; we name resulting spectral space as cluster space. For each cluster, an individual classifier (IC) is trained, hence resulting in X number of individual classifiers. Given an image patch, its discriminative value (DV) is computed as an average of the probabilities given by all ICs. Later, the patches are sorted based on their DV with respect to other patches. Finally, patches corresponding to K largest DV values are selected. This method (cluster space + IC) provides a means of optimizing the variance and correlation present in all classes and samples. Following are the steps executed to designate top discriminative patches.

(i) Using training dataset's labeled patches, X number of clusters are generated using both live and fake samples together. K-means clustering algorithm is exploited in this work for cluster creation.

(ii) For each cluster, an individual classifier (IC) (in this work, SVM) is trained using ground truth.

(iii) The patches of a given face image are arranged in descending order on the basis of their respective DV:

$$\text{DV} = \exp\left(\frac{1}{X}\sum_{i=1}^{X} P_i(\mathbf{P})\right), \tag{3}$$

where P_i is the probability given by ith classifier trained on ith cluster and $\mathbf{P}$ is the candidate patch.

(iv) The top K patches are selected as discriminative patches.

3.1.2. Nonclustering Based Methods. Unlike the clustering based methods, techniques in this category do not require any clustering algorithm. Following are the four nonclustering based discriminative patch selection methods.

(1) MAXDIST. This method of discriminative patches selection is based on the assumption that candidate discriminative patches are maximally dissimilar from the other patches in the given face image. Therefore, first the dissimilarity between successive patches is computed. The method then sorts the patches based on their average dissimilarity score with respect to other patches and selects those patches (i.e., discriminative patches) that correspond to the K largest average dissimilarity scores. We refer to this method as MAXDIST since discriminative patches are selected using a maximum dissimilarity criterion.

Following steps are followed to select top discriminative patches.

(i) An $n \times n$ dissimilarity matrix ($\mathbf{D}$) is generated, where each element $\mathbf{D}(i, j)$, $i, j \in 1, 2, \ldots, n$ is the distance score between features of patches i and j.

(ii) For the jth patch, the average dissimilarity score (d_j) with respect to the remaining ($n - 1$) patches is computed by finding the average of the elements in jth row of $\mathbf{D}$.

(iii) The average values obtained in step (ii) are ordered in descending order and the top K patches that have the largest average dissimilarity scores are selected as discriminative patches, since they are the most "dissimilar" patches in the image and hence they are representing typical data measurements.

For classification performance point of view, smaller K values might not be able to sufficiently seize the inter- and intraclass variability, which may lead to inferior performance. However, larger K values, on the other hand, would be computationally demanding. Thus, a rational value of K, by taking above-mentioned factors into account, has to be specified. A similar method has been proposed in [32] for peak frame selection in a given facial expression video. Unlike [32], in this work, we employ the technique to select discriminative patches in an image/frame.

(2) IQA (Image Quality Assessment). This algorithm uses image quality measures to select discriminative patches, thus named as image quality assessment (IQA). The method

assumes that the discriminative patches will have different quality from nondiscriminative patches. The expected quality differences between discriminative and nondiscriminative patches may be local artifacts, color levels, luminance levels, degree of sharpness, entropy, structural distortions, or natural appearance.

This framework exploits 4 general reference-based image quality measures, thus having a very low degree of complexity. First, four distinct label-sets for all patches are obtained using four different image quality measures. The labels are "discriminative" and "nondiscriminative." The method selects only those patches, which are flagged as discriminative by all four-image quality assessment techniques.

In particular, reference-based IQA methods are employed in this scheme that rely on the availability of a clean undistorted reference image to estimate the quality of the test image. However, in typical spoof recognition such a reference image is unknown, because only the input sample is available. Therefore, to circumvent this limitation, the same technique (filtering the image with a low-pass Gaussian kernel) that is successfully being used for image manipulation detection [33] and for steganalysis [34] is utilized. Following steps are executed to attain top K discriminative patches:

(i) The normalized face image ($\mathbf{I}$) is filtered with a low-pass Gaussian kernel in order to generate a smoothed version $\hat{\mathbf{I}}$.

(ii) Two corresponding sets of dense patches $\{\mathbf{P}_i(\mathbf{I}) \in \mathbb{R}^{M\times N}\}_{i=1}^{n}$ and $\{\mathbf{P}_j(\hat{\mathbf{I}}) \in \mathbb{R}^{M\times N}\}_{j=1}^{n}$ are taken, where n is the total number of patches.

(iii) Four label matrices ($\mathbf{L}_{\mathrm{PSNR}}$, $\mathbf{L}_{\mathrm{NCC}}$, $\mathbf{L}_{\mathrm{TED}}$, and $\mathbf{L}_{\mathrm{GMSD}}$) using following four-image quality measures (IQM) are generated. The patches are flagged as "discriminative" if their IMQ is greater than or equal to the threshold. The image quality measures are as follows:

(a) *Peak Signal to Noise Ratio (PSNR).* It computes the distortion between two corresponding patches (of size $P\times Q$) on the basis of their pixel-wise differences as follows:

$$\mathrm{PSNR}\left(\mathbf{P},\hat{\mathbf{P}}\right) = 10\log\left[\frac{\max\left(\mathbf{P}^2\right)}{\mathrm{MSE}\left(\mathbf{P},\hat{\mathbf{P}}\right)}\right], \tag{4}$$

where

$$\mathrm{MSE}\ (\text{Mean Squared Error}) = \frac{1}{PQ}\sum_{x=1}^{P}\sum_{y=1}^{Q}\left(\mathbf{P}_{x,y}-\hat{\mathbf{P}}_{x,y}\right)^2. \tag{5}$$

(b) *Normalized Cross-Correlation (NCC).* The correlation function can also be used to quantify the similarity between two digital image patches [3]. Here, a variant of correlation based measure is obtained by considering the statistics of the angles between the pixel vectors of the original and distorted patches as

$$\mathrm{NCC}\left(\mathbf{P},\hat{\mathbf{P}}\right) = \frac{\sum_{x=1}^{P}\sum_{y=1}^{Q}\left(\mathbf{P}_{x,y}\cdot\hat{\mathbf{P}}_{x,y}\right)}{\sum_{x=1}^{P}\sum_{y=1}^{Q}\left(\mathbf{P}_{x,y}\right)^2}. \tag{6}$$

(c) *Total Edge Difference (TED).* Edge features are generally considered as one of the most informative datasets in computer vision algorithms. Thus, we considered edge-related quality measure since the structural distortion of an image is deeply linked with its edge degradation. The TED measure is calculated as follows:

$$\mathrm{TED}\left(\mathbf{P},\hat{\mathbf{P}}\right) = \frac{1}{PQ}\sum_{x=1}^{P}\sum_{y=1}^{Q}\left|\mathbf{P}_{\mathbf{E}_{x,y}}-\hat{\mathbf{P}}_{\mathbf{E}_{x,y}}\right|. \tag{7}$$

In this work, we use Sobel operator to build the binary edge maps $\mathbf{P}_{\mathbf{E}}$ and $\hat{\mathbf{P}}_{\mathbf{E}}$.

(d) *Gradient Magnitude Similarity Deviation (GMSD).* GMSD finds the pixel-wise Gradient Magnitude Similarity (GMS) between the reference and distorted patches. It uses a pooling strategy based on standard deviation of the GMS map to predict accurately perceptual image quality. The GMSD is obtained as follows:

$$\mathrm{GMSD}\left(\mathbf{P},\hat{\mathbf{P}}\right) = \sqrt{\frac{1}{U}\sum_{i=1}^{U}\left(\mathrm{GMS}\left(i\right)-\mathrm{GMSM}\right)^2}, \tag{8}$$

where U is the total number of pixels in the patch, GMS is Gradient Magnitude Similarity map, and GMSM is Gradient Magnitude Similarity Mean calculated by applying average pooling to the GMS map. The GMS is calculated as

$$\mathrm{GMS}\left(i\right) = \frac{2\mathbf{m}_r\left(i\right)\mathbf{m}_d\left(i\right)+c}{\mathbf{m}_r^2\left(i\right)+\mathbf{m}_d^2\left(i\right)+c}, \tag{9}$$

where c is a positive constant that supplies numerical stability, while $\mathbf{m}_r$ and $\mathbf{m}_d$ are gradient magnitude images obtained by convolution using Prewitt filters along horizontal and vertical directions, respectively. The GMSM is obtained as

$$\mathrm{GMSM} = \frac{1}{U}\sum_{i=1}^{U}\mathrm{GMS}\left(i\right). \tag{10}$$

Clearly, a higher GMSM score means higher image quality. We refer reader to [35] for further details of GMSD technique.

(iv) The patches flagged as discriminative by all above four-image quality assessment techniques are selected:

$$\mathbf{L}_{\mathrm{PSNR}}\cap\mathbf{L}_{\mathrm{NCC}}\cap\mathbf{L}_{\mathrm{TED}}\cap\mathbf{L}_{\mathrm{GMSD}}. \tag{11}$$

(v) The patches selected in step (iv) are arranged in descending order on the basis of their average values of (4), (6), (7), and (8).

(vi) The top K patches are selected as discriminative patches.

(3) DF (Diversity Filter). This method considers that the discriminative patches are discernibly diverse from the other patches in the given face image. The discriminative patches are obtained using the combination of a trained classifier (we refer to such classifier as diversity filter) and a selection procedure that selects the patches based on their incremental contribution to the discriminative patch detection performance. This method is close to object detection methods [36, 37], where similar procedure is carried out to capture visually varied parts of objects at a fixed pose or viewpoint. Unlike the proposed method, techniques in [36, 37] use, before diversity filter, a preprocessing step, that is, creating various clusters corresponding to different parts appearances in images. Also distinct diversity filter is employed for each such cluster, while our method single diversity filter is used for all samples and image patches without any clustering. Moreover, the diversity tradeoff parameter in our method is computed dynamically for each image, whereas in [36, 37] a static value is utilized for all images.

In particular, we learn the diversity model of patches based on their properties that can be computed from the filter itself. The intuition is that across image categories good filter exhibits common traits such as low clutter and gradients that are spatially correlated. Therefore, we train a ranking function with the objective to produce the order of diversity quality of patches. The function uses a weight to control tradeoff between estimated rank of a patch and the diversity it adds, which discourages adding patches similar to the ones already selected, even if this patch is highly ranked. Following are the steps required to select K discriminative patches:

(i) Using training dataset, a classifier (diversity filter, here SVM) is trained.

(ii) The patches (n) of a given face image are arranged in descending order on the basis of following equation:

$$\underset{i}{\operatorname{argmax}} \left\{ \hat{y}_i - \alpha \max_{1,\dots,t} S_{i,j} \right\}, \tag{12}$$

where $\hat{y}_i$ is the probability given by diversity filter, $S_{i,j}$ denotes similarity between patches i and j, and $\alpha = \max(\text{eigenvalues}(\text{cov}(S_{i,j})))$ is the diversity tradeoff parameter. It is worth mentioning that in (12) during the first iteration $t = n$ (total number of patches), and then in each successive iteration t is reduced by 1, such that the patch selected in foregoing iteration is removed.

(iii) The top K patches are selected as discriminative patches.

(4) IPI (Intensity-Based Patch of Interest). Local intensity inhomogeneity can be exploited to find the regions, shapes, and edges of similar kind in an image [38]. However, our aim here is to disregard the image patches (regions) with similar features in order to avoid redundancy. Therefore, in this method to determine the discriminative patches, we apply an approach identical to standard statistical background-subtraction approach (which is most commonly used to accost intensity inhomogeneity) [39]. The proposed method does not use any preprocessing step, that is, foreground and background models based on recursive or nonrecursive techniques like in [39]. Following steps are executed to attain K discriminative patches:

(i) A set of dense patches $\{\mathbf{P}_i(\mathbf{I}) \in \mathbb{R}^{M\times N}\}_{i=1}^{n}$ are taken, where n is the total number of patches (of size $P \times Q$).

(ii) A label matrix ($\mathbf{F}_{\text{IPI}}$) is generated using a standard statistical background-subtraction approach:

$$\mathbf{F}_{\text{IPI}} = \begin{cases} \text{Discriminative}, & \text{if } \dfrac{\sum_{x=1}^{P}\sum_{y=1}^{Q}\left|\mathbf{P}_i(x,y) - \mathbf{P}_i\right|}{\sigma(\mathbf{P}_i)} > \eta \\ \text{Nondiscriminative}, & \text{otherwise}, \end{cases} \tag{13}$$

where η is threshold, which is estimated using similar procedure as explained above in IQA method.

(iii) The patches flagged as discriminative in step (ii) are arranged in descending order on the basis of their values using (13).

(iv) The top K patches are selected as final discriminative patches.

3.2. Classification Method. For final classification whether the face is genuine or spoof, we used majority-voting based scheme that exploits four well-known classifiers: support vector machine (SVM), Naive-Bayes (NB), Quadratic Discriminant Analysis (QDA), and Ensemble based on AdaBoost algorithm.

4. Experiments

In this section, we evaluate the proposed approach on two different publicly available databases: REPLAY-ATTACK [4] and CASIA-Face Antispoofing Database (FASD) [22].

4.1. Datasets

4.1.1. REPLAY-ATTACK [4]. This dataset is composed of short videos of both real-access and spoofing attempts of 50 different subjects, acquired with a 320 × 240 resolution camera. The datasets were collected under controlled (with a uniform background and artificial lighting) and adverse (with natural illumination and nonuniform background) conditions. The face spoof attacks were created by forging genuine verification attempts of the respective subjects via printed photos, displayed photos/videos on mobile phone's screen, and displayed photos/videos on HD screen.

TABLE 2: Summary of two databases used in this study.

Database	Number of subjects	Number of videos	Resolution	Attack type
REPLAY-ATTACK [4]	50	(i) 200 live (ii) 1000 spoof	320 × 240	(i) Printed photo (ii) Displayed photo (mobile/HD) (iii) Replayed video (mobile/HD)
CASIA-FASD [22]	50	(i) 150 live (ii) 450 spoof	640 × 480‡ 480 × 640⊕ 1280 × 720⋆	(i) Printed photo (ii) Cut photo (iii) Replayed video

‡, ⊕, and ⋆ indicate low-, normal-, and high-quality camera.

4.1.2. CASIA-FASD [22]. This database contains video recordings of real and fake faces for 50 different identities. Both real-access and spoof attacks were captured using three camera resolutions: low resolution, normal resolution, and high resolution. Three kinds of attack attempts were considered: warped photo attacks, cut photo attacks, and video attacks. The dataset is divided into two subsets for training and testing: 20 and 30 identities, respectively. Table 2 provides a summary of the above two databases.

4.2. Evaluation Protocols. For REPLAY-ATTACK dataset, we followed the same standard protocols specified in [4] for the experiments. The dataset contains three totally independent datasets in terms of users. The train and development sets are used for training and parameter tuning, respectively. The final results are computed on test. The performance of the proposed liveness detection system was evaluated, as in [3, 4, 17], using Half Total Error Rate (HTER) computed as HTER = (FAR + FRR)/2, where FRR and FAR stand for False Rejection Rate and False Acceptance Rate, respectively.

For CASIA-FASD database, we followed the benchmark protocols specified in [22]. The test protocol consists of seven scenarios. The first three scenarios are to study the effect of imaging quality: (1) low quality, (2) normal quality, and (3) high quality. The next three scenarios are (4) warped photo attacks, (5) cut photo attacks, and (6) video attacks. Finally, (7) is the overall scenario (here all data are combined together to give a general and overall evaluation). The classifier training and parameters tuning were performed on training set, while the results are reported in terms of Equal Error Rate (EER) on the test set.

In a given video frame, first the face is detected. The detected face image is then normalized to 128 × 128 pixels and densely divided into a grid of nonoverlapping local patches of size 16 × 16. Out of total n number of patches, only 40% peculiar patches are selected as discriminative patches. The LBP (Local Binary Patterns) technique is utilized to extract the features both for final classification and for discriminative patch selection methods (to obtain dis(similarity) score and clustering). Figure 5 shows examples from REPLAY-ATTACK and CASIA-FASD database of a given face image and corresponding selected discriminative patches using proposed seven image patches selection methods.

4.3. Experimental Results. The experimental results on REPLAY-ATTACK and CASIA-FASD databases are reported in Tables 3 and 4, respectively. We compared the performance of proposed method with most eminent techniques published in the literature such as methodologies in [4] (based on local binary pattern features with SVM classifier), [22] (grounded on multiple difference-of-Gaussian (DoG) filters to extract the high frequency information), [23] (using Multiscale Local Binary Patterns with SVM), [3] (which makes use of general full-reference and nonreference image quality measures), [16] (exploiting correlation between head motion and background that is estimated using optical flow), [28] (encoding information with a Histogram of Oriented Optical Flow (HOOF)), [23] (utilizing both texture and motion estimation along with preprocessing for motion magnification), and [1] (based on image distortion analysis features which is combination of specular reflection, blurriness, chromatic moment, and color diversity properties).

The results in both Tables 3 and 4 show that the proposed method in general achieves better accuracy than existing techniques under specific combination of discriminative patch selection method and classification scheme. For instance, in the experiment using the REPLAY-ATTACK database, it is easy to see in Table 3 that when the proposed framework is implemented using DEND-CLUSTERING-Ensemble or MAXDIST-Ensemble combinations, the Half Total Error Rate (HTER) is 5.00%, which is much lower than method in [25] (i.e., PCA + LBP + SVM (20.50%)). Similarly, in Table 4, we can see that the proposed system achieves error rate better than or similar to the state-of-the-art methods under overall scenario.

The MAXDIST patch selection method achieves better performances on average with the four classification techniques used in this study. Additionally, MAXDIST, CS, and DF patch selection algorithms demonstrate good generalization capability not only for disparate datasets but also for spoofing attacks with varying qualities and fabrication methods. Quite to the contrary, CP and IQA methods fail to attain proficient generalization aptitude. Beside patch selection algorithm, choice of feature classification scheme also plays vital role in accomplishing preferable performances. To this end, it is easy to see in Tables 3 and 4 that, among SVM, Naive-Bayes (NB), QDA, and Ensemble based on AdaBoost classifiers, Ensemble performs best under varying features, datasets, attack types, and amount of training and testing samples, owing to its ability of reducing the variances, averaging out the biases, and most unlikeliness of overfitting. The NB and QDA classifiers in this study are quite sensitive

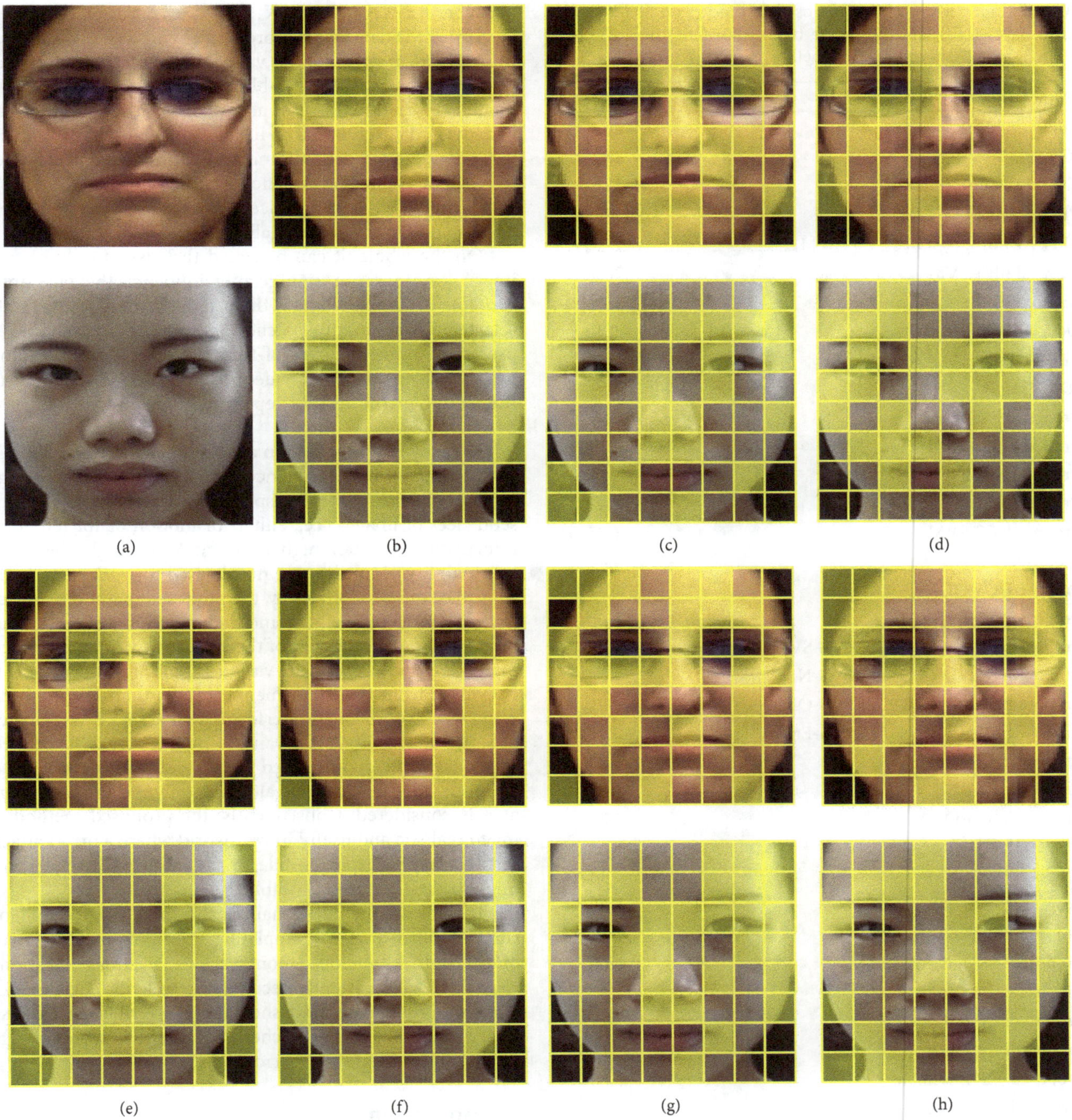

FIGURE 5: Examples of selected discriminative patches using proposed patches selection methods. Top row of (a)–(h), face from REPLAY-ATTACK database. Bottom row of (a)–(h), face from CASIA-FASD database. A normalized face image (a) of a subject and the discriminative selected patches using (b) CS, (c) DEND-CLUSTERING, (d) IPI, (e) IQA, (f) CP, (g) MAXDIST, and (h) DF methods.

to patch selection approaches. Specifically, though NB enjoys simplicity and computational efficiency, it substantially performs poorly under diverse attack conditions. This may be occurring due to its assumption that all attributes are independent (i.e., no correlation between variables), since it was pointed out in [20, 40] that correlation mapping is beneficial to procure better accuracy and generalization capability in biometric liveness detection. Moreover, NB also assumes that the samples follow Gaussian distribution. However, Gaussian distribution assumption is generally true for small biometric datasets. But, spoofing databases are heterogeneous that contain different spoof attack types and sizes, and thereby NB either gets overfitted or fails to address the problem of concept-drift.

TABLE 3: Comparison of the proposed method (with SVM, QDA, Naive-Bayes (NB), and Ensemble based classifiers) on REPLAY-ATTACK database with existing methods.

Method	HTER (%)
Multi-LBP [23]	20.25
IQA [3]	15.20
GLCM (Unicamp) [24]	15.62
$LBP^{u2}_{8,1}$ [4]	16.10
$LBP^{u2}_{8,2} + LBP^{u2}_{16,2} + LBP^{u2}_{8,1}$ + SVM [4]	13.87
PCA + LBP + SVM [25]	20.50
Motion [16]	11.70
DoG-LBP + SVM [1]	11.10
LBP-TOP [26]	8.51
IDA [1]	7.41
Proposed: DF-SVM	6.87
Proposed: DF-NB	8.01
Proposed: DF-QDA	7.30
Proposed: DF-Ensemble	6.23
Proposed: CS-SVM	6.25
Proposed: CS-NB	7.44
Proposed: CS-QDA	6.87
Proposed: CS-Ensemble	6.00
Proposed: DEND-CLUSTERING-SVM	5.98
Proposed: DEND-CLUSTERING-NB	8.87
Proposed: DEND-CLUSTERING-QDA	6.11
Proposed: DEND-CLUSTERING-Ensemble	**5.00**
Proposed: IQA-SVM	6.23
Proposed: IQA-NB	11.05
Proposed: IQA-QDA	7.75
Proposed: IQA-Ensemble	5.62
Proposed: IPI-SVM	7.50
Proposed: IPI-NB	8.30
Proposed: IPI-QDA	6.19
Proposed: IPI-Ensemble	6.00
Proposed: CP-SVM	8.37
Proposed: CP-NB	9.18
Proposed: CP-QDA	7.12
Proposed: CP-Ensemble	6.80
Proposed: MAXDIST-SVM	5.87
Proposed: MAXDIST-NB	8.01
Proposed: MAXDIST-QDA	6.12
Proposed: MAXDIST-Ensemble	**5.00**

By metaknowledge analysis, it was found that spoof attacks with higher resolution/quality are comparatively harder to be recognized, as also pointed out in [3, 22], for instance, high-quality eye cut-off attacks in which the cropped portions are filled by real eyes of the attackers leading thus to the high quality spoofing attacks that are having a combination of real and spoofed face features. Furthermore, between REPLAY-ATTACK and CASIA-FASD databases, CASIA-FASD database is more challenging as it incorporates more practical states such as variant of spoof attacks (e.g., cut photo attack simulating eye-blinking) and samples with high quality (resolutions). All in all, results also suggest that, for systems or datasets based on low- or normal-quality samples, it is advisable to adopt CS method with Ensemble classifier to reach desirable performance accuracies, while MAXDIST with Ensemble is better choice for systems/datasets based on high-quality samples, especially videos.

On the whole, it can be stated that use of only certain image locations can significantly influence the face anti-spoofing accuracy. Namely, the proposed method uses only selected discriminative patches and attains higher-ranking precision, unlike the state-of-the-art methods which exploit whole face image/frame/video, leading hence generally to the clutter in the feature representations and to their low performances.

In many face recognition applications, there is no access to the video or image sequences of the user. However, a large number of existing face antispoofing solutions need video or sequences of images (i.e., either for motion or for temporal information) to attain high accuracy. Accordingly, they have less usability, since they are not devised to work on a single static face image. Conversely, the proposed method is single-image algorithm (i.e., the method requires just one input image and not a sequence of them). Therefore, the proposed method is more useful in various applications. Further, it is evident from the experimental results that the proposed framework is robust and performs well across diverse types of spoof attacks, materials, and techniques (such as printed image attack, video-replayed attack, cut photo attack, and image or video resolutions), although only specific face image area is considered. Consequently, the proposed method is much realistic and useful in real-world where a priori attack (artifact) types (paper, mobile, or resolution), which attacker might utilize, are unpredictable.

To sum up, the performances shown by proposed algorithm confirm that contemplating the inherent differences of discriminant abilities among various face image locations is useful for consistently recognizing well the facial spoof attacks. In other words, we show that it seems feasible to use only certain face image patches instead of whole face image to reduce significantly the error rates.

5. Conclusion

The vulnerability of face recognition systems to spoofing attacks is a largely accepted reality, which has led to great advances in face antispoofing (especially face liveness detection) technologies. Despite the remarkable advances, counteracting face spoof attacks has yet proven to be a challenging task. Moreover, existing face liveness detection methods use whole face image or complete video for liveness detection. However, often image regions (video frames) are redundant or correspond to the clutter in the image (video), thus leading generally to low performances. Therefore, in this paper, we propose using just discriminative image patches for face liveness detection. In particular, we present seven

TABLE 4: Comparison of the proposed method (with SVM, QDA, Naive-Bayes (NB), and Ensemble based classifiers) on CASIA-FASD database with existing methods in terms of EER (%).

Method	Low quality	Normal quality	High quality	Warped photo attack	Cut photo attack	Video attack	Overall
DoG [22]	13.00	13.00	26.00	16.00	6.00	24.00	17.00
IQA [27]	31.70	22.20	5.69	26.10	18.31	34.41	32.45
LBP + SVM baseline	16.50	17.20	23.00	24.70	16.70	27.00	24.80
Multi-LBP baseline	12.77	16.66	26.66	15.55	25.55	17.77	17.77
HOOF [28]	16.66	30.00	26.11	15.55	17.77	38.88	21.11
Mag-HOOF	17.22	33.33	22.77	12.22	20.00	36.60	22.22
HOOF + Multi-LBP	9.44	20.55	16.66	10.00	16.66	24.44	15.55
Motion-MLBP [23]	7.22	13.33	29.44	14.44	22.22	13.33	15.74
Motion magnification [23]	6.11	23.33	13.88	10.00	14.44	20.00	14.44
Color texture [29]	7.80	10.10	6.40	7.50	**5.40**	8.10	6.20
Proposed: DF-SVM	7.53	6.65	6.28	6.94	8.21	7.97	6.71
Proposed: DF-NB	7.77	7.79	6.66	7.00	7.66	8.16	9.00
Proposed: DF-QDA	5.78	7.01	5.65	6.97	7.88	7.15	7.81
Proposed: DF-Ensemble	**4.65**	5.99	6.57	5.94	6.49	6.00	6.11
Proposed: CS-SVM	6.46	6.43	5.97	6.56	8.72	7.27	8.54
Proposed: CS-NB	6.69	6.30	5.99	7.49	8.01	7.68	8.69
Proposed: CS-QDA	7.33	8.89	6.09	7.43	8.21	7.68	7.97
Proposed: CS-Ensemble	6.51	**5.59**	**5.30**	**5.31**	6.13	7.12	7.59
Proposed: DEND-CLUSTERING-SVM	7.39	7.09	5.93	7.35	8.22	8.42	8.07
Proposed: DEND-CLUSTERING-NB	6.98	6.72	7.40	7.65	8.23	8.00	8.45
Proposed: DEND-CLUSTERING-QDA	7.11	6.89	7.90	7.03	6.99	8.89	8.67
Proposed: DEND-CLUSTERING-Ensemble	5.89	6.06	5.58	5.33	5.42	6.02	5.16
Proposed: IQA-SVM	6.32	7.00	5.97	7.73	7.71	7.69	7.33
Proposed: IQA-NB	6.66	7.66	6.32	7.33	8.45	8.00	8.22
Proposed: IQA-QDA	6.35	7.78	8.31	8.78	7.87	8.06	8.69
Proposed: IQA-Ensemble	6.02	6.57	5.76	7.27	6.87	7.27	7.27
Proposed: IPI-SVM	8.34	7.99	8.72	7.89	8.56	8.74	8.36
Proposed: IPI-NB	8.88	7.45	8.35	7.22	8.90	8.12	8.83
Proposed: IPI-QDA	6.01	7.33	6.88	7.33	8.78	9.14	8.57
Proposed: IPI-Ensemble	6.00	6.78	6.10	6.27	7.24	8.00	7.22
Proposed: CP-SVM	8.87	8.23	9.08	7.64	9.37	8.90	9.50
Proposed: CP-NB	9.10	9.07	8.39	9.74	9.85	9.45	9.34
Proposed: CP-QDA	8.67	8.34	9.00	8.78	8.51	8.34	8.59
Proposed: CP-Ensemble	8.01	7.27	7.34	6.82	7.13	8.23	7.60
Proposed: MAXDIST-SVM	7.25	5.76	6.68	9.28	8.28	8.42	8.56
Proposed: MAXDIST-NB	7.37	6.98	7.33	7.33	8.00	8.31	8.00
Proposed: MAXDIST-QDA	7.11	6.78	7.32	8.84	8.02	8.42	8.40
Proposed: MAXDIST-Ensemble	5.26	6.00	**5.30**	5.78	5.49	**5.02**	**5.07**

novel methods to obtain discriminative patches in a face image (or randomly selected lone video frame). The features of selected discriminative image patches are fed to a specific classifier (i.e., SVM, Naive-Bayes, QDA, or Ensemble). The classification results of these patches are combined by a majority-voting based scheme for the final classification of genuine and spoof faces. Experimental results on two publicly available databases show comparative performances compared to the existing works. The future works include devising more novel techniques for attaining discriminative image patches and inclusion of temporal information in the proposed method for higher security applications.

Competing Interests

The authors declare that they have no competing interests.

References

[1] D. Wen, H. Han, and A. K. Jain, "Face spoof detection with image distortion analysis," *IEEE Transactions on Information Forensics and Security*, vol. 10, no. 4, pp. 746–761, 2015.

[2] Z. Akhtar, C. Micheloni, and G. L. Foresti, "Biometric liveness detection: challenges and research opportunities," *IEEE Security & Privacy*, vol. 13, no. 5, pp. 63–72, 2015.

[3] J. Galbally, S. Marcel, and J. Fierrez, "Image quality assessment for fake biometric detection: application to iris, fingerprint, and face recognition," *IEEE Transactions on Image Processing*, vol. 23, no. 2, pp. 710–724, 2014.

[4] I. Chingovska, A. Anjos, and S. Marcel, "On the effectiveness of local binary patterns in face anti-spoofing," in *Proceedings of the International Conference of the Biometrics Special Interest Group (BIOSIG '12)*, pp. 1–7, Darmstadt, Germany, September 2012.

[5] J. Määttä, A. Hadid, and M. Pietikäinen, "Face spoofing detection from single images using texture and local shape analysis," *IET Biometrics*, vol. 1, no. 1, pp. 3–10, 2012.

[6] Z. Akhtar, C. Micheloni, C. Piciarelli, and G. L. Foresti, "MoBio_LivDet: mobile biometric liveness detection," in *Proceedings of the 11th IEEE International Conference on Advanced Video and Signal Based Surveillance (AVSS '14)*, pp. 187–192, Seoul, Republic of Korea, August 2014.

[7] G. Pan, L. Sun, Z. Wu, and S. Lao, "Eyeblink-based anti-spoofing in face recognition from a generic webcamera," in *Proceedings of the IEEE 11th International Conference on Computer Vision (ICCV '07)*, pp. 1–8, Rio de Janeiro, Brazil, October 2007.

[8] X. Tan, Y. Li, J. Liu, and L. Jiang, "Face liveness detection from a single image with sparse low rank bilinear discriminative model," in *Proceedings of the 11th European Conference on Computer Vision (ECCV '10)*, K. Daniilidis, P. Maragos, and N. Paragios, Eds., vol. 6316 of *Lecture Notes in Computer Science*, pp. 504–517, Crete, Greece, September 2010.

[9] Y. Li and X. Tan, "An anti-photo spoof method in face recognition based on the analysis of fourier spectra with sparse logistic regression," in *Proceedings of the Chinese Conference on Pattern Recognition (CCPR '09)*, Nanjing, China, November 2009.

[10] K. Kollreider, H. Fronthaler, and J. Bigun, "Non-intrusive liveness detection by face images," *Image and Vision Computing*, vol. 27, no. 3, pp. 233–244, 2009.

[11] W. Bao, H. Li, N. Li, and W. Jiang, "A liveness detection method for face recognition based on optical flow field," in *Proceedings of the International Conference on Image Analysis and Signal Processing (IASP '09)*, pp. 233–236, IEEE, Taizhou, China, April 2009.

[12] G. Pan, L. Sun, Z. Wu, and Y. Wang, "Monocular camera-based face liveness detection by combining eyeblink and scene context," *Telecommunication Systems*, vol. 47, no. 3, pp. 215–225, 2011.

[13] J. Li, Y. Wang, T. Tan, and A. K. Jain, "Live face detection based on the analysis of Fourier spectra," in *Proceedings of the Biometric Technology for Human Identification*, vol. 5404 of *Proceedings of SPIE*, pp. 296–303, Orlando, Fla, USA, April 2004.

[14] J. Bai, T. Ng, X. Gao, and Y. Shi, "Is physics-based liveness detection truly possible with a single image?" in *Proceedings of IEEE International Symposium on Circuits and Systems*, pp. 3425–3428, Paris, France, May-June 2010.

[15] Z. Zhang, D. Yi, Z. Lei, and S. Z. Li, "Face liveness detection by learning multispectral reflectance distributions," in *Proceedings of the IEEE International Conference on Automatic Face and Gesture Recognition and Workshops (FG '11)*, pp. 436–441, Santa Barbara, Calif, USA, March 2011.

[16] J. Komulainen, A. Hadid, M. Pietikäinen, A. Anjos, and S. Marcel, "Complementary countermeasures for detecting scenic face spoofing attacks," in *Proceedings of the International Conference on Biometrics (ICB '13)*, pp. 1–7, Madrid, Spain, June 2013.

[17] A. Anjos and S. Marcel, "Counter-measures to photo attacks in face recognition: a public database and a baseline," in *Proceedings of the International Joint Conference on Biometrics (IJCB '11)*, pp. 1–7, Washington, DC, USA, October 2011.

[18] T. Wang and S. Z. Li, "Face liveness detection using 3d structure recovered from a single camera," in *Proceedings of the International Conference on Biometrics (ICB '13)*, pp. 1–6, IEEE, Madrid, Spain, 2013.

[19] T. I. Dhamecha, A. Nigam, R. Singh, and M. Vatsa, "Disguise detection and face recognition in visible and thermal spectrums," in *Proceedings of the 6th IAPR International Conference on Biometrics (ICB '13)*, pp. 1–6, Madrid, Spain, June 2013.

[20] G. Chetty and M. Wagner, "Liveness detection using cross-modal correlations in face-voice person authentication," in *Proceedings of the 9th European Conference on Speech Communication and Technology (Interspeech '05)*, pp. 2181–2184, Lisbon, Portugal, September 2005.

[21] N. Kose and J.-L. Dugelay, "Reflectance analysis based countermeasure technique to detect face mask attacks," in *Proceedings of the 18th International Conference on Digital Signal Processing (DSP '13)*, pp. 1–6, IEEE, Fira, Greece, July 2013.

[22] Z. Zhang, J. Yan, S. Liu, Z. Lei, D. Yi, and S. Z. Li, "A face antispoofing database with diverse attacks," in *Proceedings of the 5th IAPR International Conference on Biometrics (ICB '12)*, pp. 26–31, IEEE, New Delhi, India, April 2012.

[23] S. Bharadwaj, T. I. Dhamecha, M. Vatsa, and R. Singh, "Face anti-spoofing via motion magnification and multifeature videolet aggregation," Tech. Rep. IIITD-TR-2014-002, 2014.

[24] I. Chingovska, J. Yang, Z. Lei, and D. Yi, "The 2nd competition on counter measures to 2D face spoofing attacks," in *Proceedings of the IEEE International Conference on Biometrics (ICB '13)*, pp. 1–6, Madrid, Spain, June 2013.

[25] S. Tirunagari, N. Poh, D. Windridge, A. Iorliam, N. Suki, and A. T. S. Ho, "Detection of face spoofing using visual dynamics," *IEEE Transactions on Information Forensics and Security*, vol. 10, no. 4, pp. 762–777, 2015.

[26] T. de Freitas Pereira, A. Anjos, J. M. De Martino, and S. Marcel, "Can face anti-spoofing countermeasures work in a real world scenario?" in *Proceedings of the 6th IAPR International Conference on Biometrics (ICB '13)*, pp. 1–8, Madrid, Spain, June 2013.

[27] J. Galbally and S. Marcel, "Face anti-spoofing based on general image quality assessment," in *Proceedings of the 22nd International Conference on Pattern Recognition (ICPR '14)*, pp. 1173–1178, Stockholm, Sweden, August 2014.

[28] R. Chaudhry, A. Ravichandran, G. Hager, and R. Vidal, "Histograms of oriented optical flow and Binet-Cauchy kernels on nonlinear dynamical systems for the recognition of human actions," in *Proceedings of the IEEE Conference on Computer Vision and Pattern Recognition (CVPR '09)*, pp. 1932–1939, Miami, Fla, USA, June 2009.

[29] Z. Boulkenafet, J. Komulainen, and A. Hadid, "Face anti-spoofing based on color texture analysis," in *Proceedings of the IEEE International Conference on Image Processing (ICIP '15)*, pp. 2636–2640, Quebec City, Canada, September 2015.

[30] M. Nilsson, J. Nordberg, and I. Claesson, "Face detection using local SMQT features and split up snow classifier," in *Proceedings of the IEEE International Conference on Acoustics, Speech and Signal Processing (ICASSP '07)*, pp. II-589–II-592, Honolulu, Hawaii, USA, April 2007.

[31] U. Uludag, A. Ross, and A. K. Jain, "Biometric template selection and update: a case study in fingerprints," *Pattern Recognition*, vol. 37, no. 7, pp. 1533–1542, 2004.

[32] S. Zhalehpour, Z. Akhtar, and C. Eroglu Erdem, "Multimodal emotion recognition based on peak frame selection from video," *Signal, Image and Video Processing*, 2015.

[33] S. Bayram, I. Avcibaş, B. Sankur, and N. Memon, "Image manipulation detection," *Journal of Electronic Imaging*, vol. 15, no. 4, Article ID 041102, 17 pages, 2006.

[34] I. Avcibas, N. Memon, and B. Sankur, "Steganalysis using image quality metrics," *IEEE Transactions on Image Processing*, vol. 12, no. 2, pp. 221–229, 2003.

[35] W. Xue, L. Zhang, X. Mou, and A. C. Bovik, "Gradient magnitude similarity deviation: a highly efficient perceptual image quality index," *IEEE Transactions on Image Processing*, vol. 23, no. 2, pp. 684–695, 2014.

[36] L. Bourdev and J. Malik, "Poselets: body part detectors trained using 3D human pose annotations," in *Proceedings of the IEEE 12th International Conference on Computer Vision (ICCV '09)*, pp. 1365–1372, Kyoto, Japan, September 2009.

[37] P. F. Felzenszwalb, R. B. Girshick, D. McAllester, and D. Ramanan, "Object detection with discriminatively trained part-based models," *IEEE Transactions on Pattern Analysis and Machine Intelligence*, vol. 32, no. 9, pp. 1627–1645, 2010.

[38] U. Vovk, F. Pernus, and B. Likar, "A review of methods for correction of intensity inhomogeneity in MRI," *IEEE Transactions on Medical Imaging*, vol. 26, no. 3, pp. 405–421, 2007.

[39] S. Cheung and C. Kamath, "Robust techniques for background subtraction in urban traffic video," in *Proceedings of the IEEE Conference on Visual Communications and Image Processing (VCIP '07)*, pp. 1–12, 2007.

[40] Z. Akhtar, C. Micheloni, and G. L. Foresti, "Correlation based fingerprint liveness detection," in *Proceedings of the International Conference on Biometrics (ICB '15)*, pp. 305–310, Phuket City, Thailand, May 2015.

6

Fetal Heart Rate Monitoring from Phonocardiograph Signal using Repetition Frequency of Heart Sounds

Hong Tang,[1] Ting Li,[2] Tianshuang Qiu,[1] and Yongwan Park[3]

[1]*Department of Biomedical Engineering, Dalian University of Technology, Dalian, China*
[2]*College of Information and Communication Engineering, Dalian Minzu University, Dalian, China*
[3]*Department of Information and Communication Engineering, Yeungnam University, Gyeongsan, Republic of Korea*

Correspondence should be addressed to Hong Tang; tanghong@dlut.edu.cn

Academic Editor: Jit S. Mandeep

As a passive, harmless, and low-cost diagnosis tool, fetal heart rate (FHR) monitoring based on fetal phonocardiography (fPCG) signal is alternative to ultrasonographic cardiotocography. Previous fPCG-based methods commonly relied on the time difference of detected heart sound bursts. However, the performance is unavoidable to degrade due to missed heart sounds in very low signal-to-noise ratio environments. This paper proposes a FHR monitoring method using repetition frequency of heart sounds. The proposed method can track time-varying heart rate without both heart sound burst identification and denoising. The average accuracy rate comparison to benchmark is 88.3% as the SNR ranges from −4.4 dB to −26.7 dB.

1. Introduction

As a well-being policy, pregnant women in many countries are periodically demanded to monitor the variations in fetal heart rate (FHR). This monitoring of FHR and its variability provide up-to-date information about the fetus to prevent intrauterine death or permanent damage to the fetus [1]. Many techniques can monitor fetal heart rate previously. Fetal magnetocardiography (fMCG) using superconducting quantum interference device allows accurate assessment of beat-to-beat fetal heart rate variability [2]. However, the equipment of fMCG is too expensive to be widely accessed in primary hospital. Cardiotocography (CTG) is continuous monitoring of the fetal heart rate using an ultrasound transducer placed on the mother's abdomen [3–5]. Physicians evaluate specific clinical CTG parameters by means of visual inspection. Hence, the accuracy of CTG monitoring is generally depending on observer's expertise. Fetal electrocardiogram (fECG) signal taken from the abdominal electrocardiogram of pregnant woman is another diagnostic tool for evaluating FHR [6]. However, the fECG signal is often masked by maternal ECG, power line interference, maternal electromyogram, and so forth. Very complex algorithms are needed to separate the fECG signal from mixed signal before further analysis. As a totally passive methodology, fetal phonocardiography (fPCG) signal collected from mother's abdomen using a microphone transducer was proposed about ten years ago to monitor FHR [7]. This new technique makes long-term and frequent measurements of FHR possible. They detected the burst shaped heart sounds from signal envelope and estimated FHR via the time difference of heart sound occurrence [8–16]. Consequently, the accuracy of FHR estimation relies on heart sound detection rate. The performance degrades greatly if heart sounds are missed and/or wrong heart sounds are detected. Denoising technique is usually used as preprocessing due to the low energy of fetal heart sounds and multiple interferences [17].

It is known that fetal heart sounds are originated from interaction between fetal heart and blood flow therein. The sounds are repetitive from one cardiac cycle to another. FHR monitoring is thus equal to tracking the local repetition frequency of heart sound in adjacent cardiac cycles. Hence, heart beat identification and denoising are then not necessary.

2. Methodology

2.1. Repetition Frequency Estimated by Cyclic Frequency Spectrum. In the field of signal processing theory, repetition frequency is usually characterized by cycle frequency. It means occurrence rate of a repetitive event in a period of time. For example, the cycle frequency of fetal heart sounds is 2.5 Hz if they repeat every 0.4 seconds. It is assumed that $x(t)$ is a virtual digital sequence of an fPCG signal with an exact cycle period T. The objective of FHR estimation is to extract the cycle frequency of the fetal heart sound. The time-varying autocorrelation is written as

$$R_x(t,\tau) \triangleq \lim_{N\to\infty} \frac{1}{2N+1} \cdot \sum_{n=-N}^{N} x\left(t+\frac{\tau}{2}+nT\right) x^*\left(t-\frac{\tau}{2}+nT\right), \quad (1)$$

where n is cycle index, T is a real number denoting cycle period, τ is time delay, and N determines the number of cycles involved in analysis. $R_x(t,\tau)$ must be periodic because fetal heart sounds are exactly cyclic. It can be expanded into Fourier series:

$$R_x(t,\tau) = \sum_{\alpha=-\infty}^{+\infty} R_x(\alpha,\tau)\, e^{j2\pi\alpha t}, \quad (2)$$

where α is a real number called cycle frequency. The coefficient of the Fourier series can be calculated by the following:

$$R_x(\alpha,\tau) = \left\langle x\left(t+\frac{\tau}{2}\right) x\left(t-\frac{\tau}{2}\right) e^{j2\pi\alpha t} \right\rangle_t. \quad (3)$$

The operator $\langle\cdot\rangle_t$ means the time average. $R_x(\alpha,\tau)$ is called cyclic correlation function. It can be seen that it reduces to the traditional correlation if the cycle frequency α is fixed to zero. $R_x(\alpha,\tau)$ can be transformed into frequency domain via Fourier transform:

$$S_x(\alpha,f) = \int_{-\infty}^{\infty} R_x(\alpha,\tau)\, e^{-j2\pi f\tau} d\tau. \quad (4)$$

$S_x(\alpha,f)$ is called cyclic spectral density. The physical meaning of $S_x(\alpha,f)$ indicates that $S_x(\alpha,f) \neq 0$ if the signal $x(t)$ has any cyclic component at cycle frequency α. In the point of view of fPCG signal, $S_x(\alpha,f)$ must have a peak at heart rate because the heart sounds are dominant cyclic components at the heart rate in adjacent fetal cardiac cycles. Therefore, detecting FHR is equal to detecting the basic cycle frequency of the digital sequence. The frequency spectrum is not of interest in this paper. An integral is performed to $S_x(\alpha,f)$ to remove the frequency variable f for the purpose of simplification:

$$\gamma_x(\alpha) = \int_{\infty}^{\infty} |S_x(\alpha,f)|\, df. \quad (5)$$

This is called cyclic frequency spectrum (CFS). Therefore, the FHR is indicated by the dominant peak of $\gamma_x(\alpha)$:

$$\text{HR} = \arg\{\max(\gamma_x(\alpha))\} \times 60, \quad (6)$$

where arg{·} is the argument that meets the condition in the brace and max(·) is the maximum operator. A sharp and outstanding peak means a high degree of repetition of heart sounds in the signal. The normal FHR is commonly in [110160] bpm. That is, the cycle frequency in consideration is in [1.82.7] Hz. For extreme value, the signals in this paper are analyzed in a greater range, [1.33.5] Hz. These extreme heart rates are 80 bpm and 210 bpm, which correspond to bradycardia and tachycardia, respectively.

To test the performance of CFS, the well-known simulated fPCG signals proposed by Cesarelli et al. [18, 19] are used. These simulated signals considered different fetal physiological and pathological conditions and recording situations by simulation technique. The noises considered consist of vibrations created by maternal body organs, fetal movements, and unwanted sounds from surrounding environment. These simulated signals have been openly published in PhysioNet for freely public access and have been widely accepted for algorithm evaluation. For example, two segments of a simulated fPCG signal both have 6 s with sampling frequency of 1 KHz, where the true real-time heart rate is between 140 bpm and 155 bpm, seen in Figure 1. There are noises which saturate the signal due to maternal fast body movements, acoustic sensor displacements, and/or high magnitude external noises. CFS analysis is applied to the signal. Figure 1(a) shows the low noise part of the fPCG signal in time domain. Figure 1(b) shows the CFS of the signal. It can be found that the dominant peak occurs at the cycle frequency of 2.35 Hz; that is, the FHR of this segment is $2.35 * 60 = 141$ bpm. Figure 1(c) shows the heavy noise part of the signal. The signal is so heavily contaminated by noises that the heart beats cannot be identified by human visual inspection. Figure 1(d) shows that the CFS still has an outstanding prominent peak as if the noise has no effect on the peak. The peak is located at 2.51 Hz (151 bpm). The example illustrates that CFS can indicate the repetition frequency of the heart sounds in this segment. CFS analysis takes the segment as an integral input. Hence, heart sound detection that is essential in previous methods becomes not necessary in the proposed method. The noise is generally random and does not have any repetition feature. So the dominant peak of CFS will not be affected even the noise is so heavy to saturate the amplitude, as the example given in Figure 1(c).

It should be noted that the heart sounds are not perfectly periodic due to the heart rate variability (HRV). So, the heart sounds are quasi-periodic. The degree of the periodic property can be reflected by the sharpness of the peak. That is, the more periodic the heart sounds become, the sharper the dominant peak will be. In the extreme case, if the heart sounds having ultimate recording time are perfectly periodic, the CFS will have only one nonzero magnitude at the cycle frequency. On the contrary, if the heart rate varies much, the degree of periodic property of the heart sounds reduces greatly; then the peak of the CFS will become flat. In another extreme case, the heart sounds appear randomly; then the CFS will not have any dominant peak. In this sense, the repetition frequency of heart sounds reflects the average frequency of heart rate. So, the repetition frequency can be used as a metric of fetal heart rate.

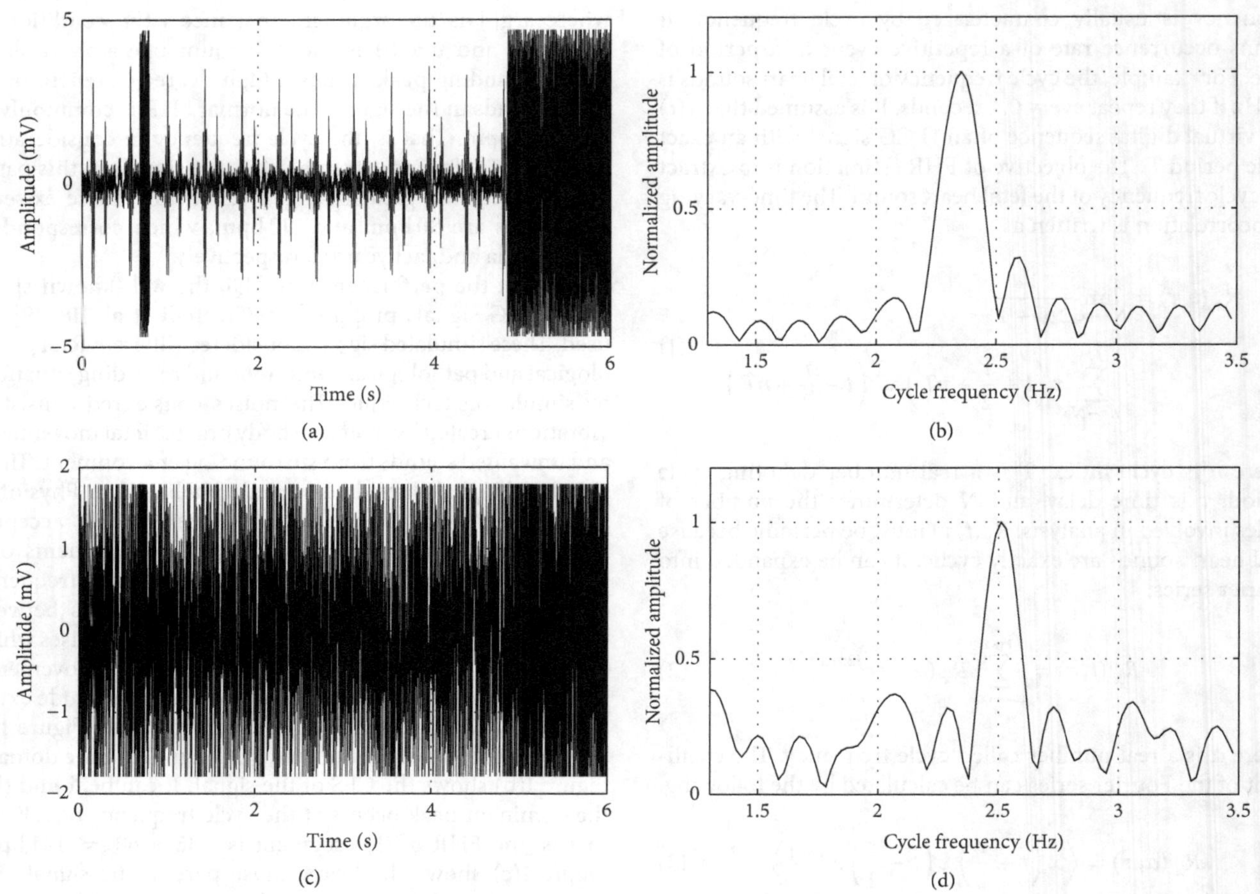

FIGURE 1: Fetal heart rate detection for an fPCG signal. (a) A low noise fPCG signal; (b) CFS analysis. Cycle frequency indicated by the dominant peak is 2.35 Hz (141 bpm). (c) A heavy noise fPCG signal; (d) CFS analysis. Cycle frequency indicated by the dominant peak is 2.51 Hz (151 bpm).

TABLE 1: Summary of the tests.

Record number	SNR (dB)	Heart rate variation (bpm)	Accuracy rate
1	−4.4	108–159	92.3%
2	−6.6	108–159	91.9%
3	−8.1	108–156	90.1%
4	−10.2	108–159	89.1%
5	−14.8	108–168	89.7%
6	−17.2	108–156	90.1%
7	−20.1	108–177	88.5%
8	−23.8	111–153	88.8%
9	−25.6	111–168	82.0%
10	−26.7	108–168	81.0%

Note: the accuracy rate is defined in (9).

2.2. Time-Varying Estimation Using Sliding Window. To track time-varying FHR, a sliding window covering the fPCG signal is used. Therefore, (5) becomes time-dependent:

$$\gamma_x(\alpha,t) = \int_{\infty}^{\infty} \left|S_x(\alpha,f,t)\right| df, \tag{7}$$

where $S_x(\alpha, f, t)$ is a cyclic spectral density over a time window $[t - \zeta, t + \zeta]$. So, $\gamma_x(\alpha, t)$ can be called time-varying cyclic spectrum. The width of the sliding window is 2ζ. From the repetition feature of heart sound, it can be concluded that the width of the sliding window must be greater than two cardiac cycles to ensure that the fetal heart beats at least

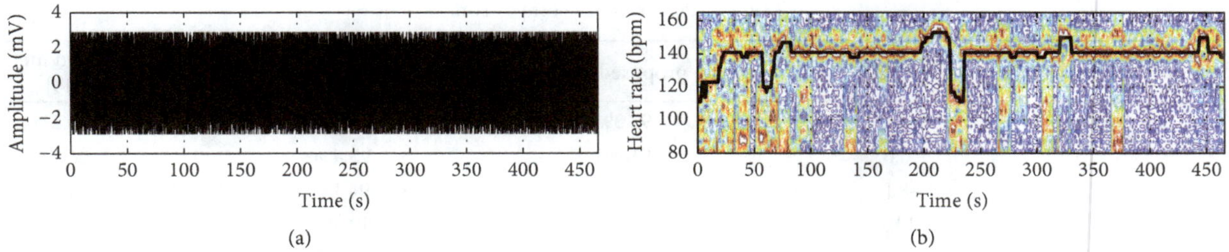

FIGURE 2: Time-varying fetal heart rate monitoring. (a) An fPCG signal with saturated noise. (b) Detected time-varying fetal heart rate.

two times in the window. Otherwise, there is no repetition feature in the windowed signal. The time-varying FHR can be tracked by searching the peak in $\gamma_x(\alpha, t)$:

$$\mathrm{Hr}(t) = \arg\{\max(\gamma_x(\alpha, t))\} \times 60. \quad (8)$$

So, the instantaneous FHR can be detected by the peak location of the time-varying cyclic frequency spectrum.

An fPCG signal is 478 s in time length, as seen in Figure 2(a). The signal is contaminated by saturated noise. The estimated SNR is −15.7 dB. A sliding window with width of 8 s and overlap of 0.1 s is applied to the signal. The detected time-varying FHR is shown by the black solid line in Figure 2(b). The tracking trajectory accurately reflects the FHR variations. Accelerations and decelerations are clearly observed in the trajectory.

3. Experiments

3.1. Time Resolution and Cycle Frequency Resolution. The primary purpose of the sliding window is to limit the extent of the subsequence to be analyzed, so that the cyclic characteristics are approximately constant over the duration of the window. The more rapidly the repetition frequency changes, the shorter the window should be. So, it is reasonable to conclude that as the window length becomes longer, the peak will become sharper; that is, the repetition frequency resolution increases. On the other hand, as the window length decreases, the ability to resolve changes with time increases. Consequently, the choice of window length becomes a trade-off between repetition frequency resolution and time resolution. With the preknowledge of the change rate of FHR, the window length can be empirically selected as 5–10 s to obtain an acceptable trade-off.

3.2. Performance of Fetal Heart Rate Monitoring Using the Repetition Frequency. To evaluate the proposed method, it is necessary to test the performance under controlled conditions. This is achieved by the simulated fPCG signals [18, 19]. Each signal lasts 478 s. These signals considered different fetal physiological and pathological conditions and recording situations by simulation technique. The noise considered consists of vibrations created by maternal body organs, fetal movements, unwanted sounds from surrounding environment, and so forth. The SNR ranges from −4.4 dB to −26.7 dB. Visual check by human eyes shows that fetal heart sounds cannot be recognized as the SNR is lower than −15 dB. The FHR in each signal is tracked by the proposed method. If the difference between the detected heart rate and the benchmark is within ±5 bpm, the detected heart rate is considered accurate. The accuracy rate is defined as

$$\text{accuracy rate} = \frac{\text{number of beats within difference} \leqslant 5\,\text{bpm}}{\text{total number of beats}}. \quad (9)$$

A summary of the tests is presented in Table 1. It can be found that the proposed method can accurately track the fetal heart rate even the SNR is lowered to −26.7 dB, where the heart sounds are totally covered by the noise. The key point why the proposed method is so robust to noise is because it detects the repetition frequency of heart sounds instead of detecting the time difference of heart sounds.

3.3. Performance Comparison to the Previous Methods. Two typical previous methods are selected to be compared to the proposed method in this subsection. One is a rule-based method proposed by Kovâcs et al. [16], where the FHR was evaluated by searching S1-S2 pairs. The other method is an advanced method proposed by Várady et al. [14], where the FHR was estimated by the periodicity of heart sound bursts from cross-correlation of signal envelope. Both methods need detection of the fetal heart sound burst. Hence, the performance of the two methods is heavily dependent on the detection accuracy of fetal heart sound burst. Ten simulated fPCG signals with SNR varying from −4.4 dB to −26.7 dB are used to evaluate the three methods and the results are given in Table 2. It is found that the rule-based method and the advanced method outperform the proposed method in low noise environments. However, both the rule-based and the advanced methods degrade greatly due to the fast increasing of missing rate for sound burst detection with decreasing SNR. The proposed method does not need sound burst detection. It is to estimate the repetition frequency of heart sound, which is less affected by random noise. Hence, the FHR estimation is robust to noise even if the sound burst is destroyed by heavy noise because the repetition feature still remains.

4. Conclusions

fPCG is a promising technique to monitor fetal heart rate. However, the performance of previous methods generally

TABLE 2: Performance comparison.

Record number	SNR (dB)	The proposed method	Rule-based method [16]	Advanced method [14]
1	−4.4	92.3%	95.4%	96.2%
2	−6.6	91.9%	95.4%	96.1%
3	−8.1	90.1%	94.1%	96.3%
4	−10.2	89.1%	90.3%	94.5%
5	−14.8	89.7%	82.6%	90.4%
6	−17.2	90.1%	60.8%	86.7%
7	−20.1	88.5%	—	82.4%
8	−23.8	88.8%	—	63.6%
9	−25.6	82.0%	—	—
10	−26.7	81.0%	—	—

Note: (a) the accuracy rate is defined in (9). (b) "—" means that so many heart sound burst pairs were missed that the accuracy rate was very low.

depends on the detection rate of heart sound bursts from an fPCG recording. The authors find that the fetal heart rate can be evaluated by the repetition frequency of heart sounds, which can be extracted from peaks in the cyclic frequency spectrum without sound burst detection and denoising. The feature of repetition can remain even if the SNR is lowered to −26.7 dB. As the SNR is lower than −15 dB, the proposed method outperforms the two typical previous methods.

Competing Interests

The authors declare that there are no competing interests regarding the publication of this paper.

Acknowledgments

This work was supported in part by the National Natural Science Foundation of China (Grant nos. 61471081 and 61601081) and Fundamental Research Funds for the Central Universities (Grant nos. DUT15QY60, DUT16QY13, DC201501056, and DCPY2016008).

References

[1] M. Romano, L. Iuppariello, A. M. Ponsiglione, G. Improta, P. Bifulco, and M. Cesarelli, "Frequency and time domain analysis of foetal heart rate variability with traditional indexes: a critical survey," *Computational and Mathematical Methods in Medicine*, vol. 2016, Article ID 9585431, 12 pages, 2016.

[2] R. T. Wakai, "Assessment of fetal neurodevelopment via fetal magnetocardiography," *Experimental Neurology*, vol. 190, S1, pp. 65–71, 2004.

[3] R. M. Grivell, Z. Alfirevic, G. M. Gyte, and D. Devane, "Antenatal cardiotocography for fetal assessment," *Cochrane Database of Systematic Reviews*, no. 1, Article ID CD007863, 2010.

[4] H. Cao, D. E. Lake, J. E. Ferguson, C. A. Chisholm, M. P. Griffin, and J. R. Moorman, "Toward quantitative fetal heart rate monitoring," *IEEE Transactions on Biomedical Engineering*, vol. 53, no. 1, pp. 111–118, 2006.

[5] M. Romano, M. Bracale, M. Cesarelli et al., "Antepartum cardiotocography: a study of fetal reactivity in frequency domain," *Computers in Biology and Medicine*, vol. 36, no. 6, pp. 619–633, 2006.

[6] M. Anisha, S. S. Kumar, and M. Benisha, "Methodological survey on fetal ECG extraction," *Journal of Health & Medical Informatics*, vol. 5, article 469, 2014.

[7] F. Kovács, C. Horváth, Á. T. Balogh, and G. Hosszú, "Fetal phonocardiography—past and future possibilities," *Computer Methods and Programs in Biomedicine*, vol. 104, no. 1, pp. 19–25, 2011.

[8] F. Kovács, N. Kersner, K. Kádár, and G. Hosszú, "Computer method for perinatal screening of cardiac murmur using fetal phonocardiography," *Computers in Biology and Medicine*, vol. 39, no. 12, pp. 1130–1136, 2009.

[9] F. Kovács, C. Horváth, Á. T. Balogh, and G. Hosszú, "Extended noninvasive fetal monitoring by detailed analysis of data measured with phonocardiography," *IEEE Transactions on Biomedical Engineering*, vol. 58, no. 1, pp. 64–70, 2011.

[10] V. S. Chourasia, A. K. Tiwari, and R. Gangopadhyay, "Time-frequency characterization of fetal phonocardiographic signals using wavelet scalogram," *Journal of Mechanics in Medicine and Biology*, vol. 11, no. 2, pp. 391–406, 2011.

[11] M. Moghavvemi, B. H. Tan, and S. Y. Tan, "A non-invasive PC-based measurement of fetal phonocardiography," *Sensors and Actuators A: Physical*, vol. 107, no. 1, pp. 96–103, 2003.

[12] V. S. Chourasia, A. K. Tiwari, and R. Gangopadhyay, "A novel approach for phonocardiographic signals processing to make possible fetal heart rate evaluations," *Digital Signal Processing: A Review Journal*, vol. 30, pp. 165–183, 2014.

[13] M. Ruffo, M. Cesarelli, M. Romano, P. Bifulco, and A. Fratini, "An algorithm for FHR estimation from foetal phonocardiographic signals," *Biomedical Signal Processing and Control*, vol. 5, no. 2, pp. 131–141, 2010.

[14] P. Várady, L. Wildt, Z. Benyó, and A. Hein, "An advanced method in fetal phonocardiography," *Computer Methods and Programs in Biomedicine*, vol. 71, no. 3, pp. 283–296, 2003.

[15] J. Chen, K. Phua, Y. Song, and L. Shue, "A portable phonocardiographic fetal heart rate monitor," in *Proceedings of the IEEE International Symposium on Circuits and Systems (ISCAS '06)*, May 2006.

[16] F. Kovâcs, M. Török, and I. Habermajer, "A rule-based phonocardiographic method for long-term fetal heart rate monitoring," *IEEE Transactions on Biomedical Engineering*, vol. 47, no. 1, pp. 124–130, 2000.

[17] S. Vaisman, S. Yaniv Salem, G. Holcberg, and A. B. Geva, "Passive fetal monitoring by adaptive wavelet denoising method," *Computers in Biology and Medicine*, vol. 42, no. 2, pp. 171–179, 2012.

[18] M. Cesarelli, M. Ruffo, M. Romano, and P. Bifulco, "Simulation of foetal phonocardiographic recordings for testing of FHR extraction algorithms," *Computer Methods and Programs in Biomedicine*, vol. 107, no. 3, pp. 513–523, 2012.

[19] Simulated fetal PCGs, July 2015, http://www.physionet.org/physiobank/database/simfpcgdb/.

7

Improving Robustness of Biometric Identity Determination with Digital Watermarking

Juha Partala, Angelos Fylakis, Anu Pramila, Anja Keskinarkaus, and Tapio Seppänen

Physiological Signal Analysis Team, Center for Machine Vision and Signal Analysis, University of Oulu, Oulu, Finland

Correspondence should be addressed to Juha Partala; juha.partala@ee.oulu.fi

Academic Editor: Isao Echizen

The determination of an identity from noisy biometric measurements is a continuing challenge. In many applications, such as identity-based encryption, the identity needs to be known with virtually 100% certainty. The determination of identities with such precision from face images taken under a wide range of natural situations is still an unsolved problem. We propose a digital watermarking based method to aid face recognizers to tackle this problem in applications. In particular, we suggest embedding multiple face dependent watermarks into an image to serve as expert knowledge on the corresponding identities to identity-based schemes. This knowledge could originate, for example, from the tagging of those people on a social network. In our proposal, a single payload consists of a correction vector that can be added to the extracted biometric template to compile a nearly noiseless identity. It also supports the removal of a person from the image. If a particular face is censored, the corresponding identity is also removed. Based on our experiments, our method is robust against JPEG compression, image filtering, and occlusion and enables a reliable determination of an identity without side information.

1. Introduction

An identity is naturally captured using the physical characteristics of a person. Those characteristics can be measured using different biometric traits such as face, iris, voice, or fingerprint that have been successfully used in many security oriented applications such as identification, authentication, and access control. Biometrics has enabled the development of user-friendly security mechanisms and, for certain applications, has enabled us to get rid of cryptographic keys and identifiers and to exchange those with biometric ones. Fuzzy identity-based encryption [1] enables the application of public biometric identifiers for public key encryption without the hassle of certificates. In contrast to traditional public key encryption, where a certificate is needed to ensure the identity of the recipient, identity-based encryption [2] derives public keys directly from the public identity. In fuzzy identity-based encryption, biometric templates can be used with a certain threshold of error tolerance without any secrecy requirements on the biometric features.

In order to be conveniently applicable for fuzzy identity-based encryption, the biometric modality has to be public and easy to acquire by others. Face is a natural public biometric trait that we all use to identify individuals in everyday life. It is also one of the most studied traits regarding biometric identification and authentication. Therefore, it is a natural biometric modality to derive identities for identity-based encryption and other identity-based schemes. However, to achieve good accuracy, face recognition techniques often assume that the faces are aligned and photometrically and geometrically normalized. In addition, occlusion and non-frontal facial poses create significant challenges even to state-of-the-art face recognition algorithms [3]. These changes in the biometric acquisition process cause intrasubject variations that are common for images taken under natural conditions and posted to social networks or public photo-sharing services such as Facebook or Instagram. These variations lead to a large number of errors in the extracted biometric features and render their use in many applications impossible.

In this paper, we suggest a digital watermarking based solution to this problem. While our approach works for any biometric modality, we concentrate on the determination of biometric identities from an image taken under natural conditions. We boost the performance of *image independent* biometrics, the determination of identity in any condition, with *image dependent* biometrics, features computed from a single image, combined with digital watermarking. Image dependent biometrics are subjective to variations caused by image processing such as filtering and compression. However, since the pose, lighting, and occlusion are fixed for a fixed image, image dependent biometrics are spared from a large class of intrasubject variations. Therefore, we are able to improve the performance of applications that need robust biometrics such as fuzzy identity-based encryption and other privacy related applications.

Our solution is based on the embedding of supplementary information about the identities in an image to the image itself. We apply a print-scan robust embedding algorithm and tie its payload directly to the biometric features of the people in the image. Thus, the biometric features of a person in that image work as a key to the supplementary information. If the amount of errors in biometric feature extraction is greater than the threshold of the application, the watermark can be used in correcting those errors. Note that the biometric features for the key are computed from *that particular image* which means that biometric feature extraction is extremely robust compared to unconstrained situations. In addition, since the key is based on biometrics, the removal of a person from an image also removes his or her identity.

The paper is organized as follows. In Section 2, we present the background on face biometrics and digital watermarking relevant to the paper. Section 3 is devoted to the details of our scheme. The experiments are described in Section 4. Finally, Section 5 provides the discussion and the conclusion.

2. Background

2.1. Face Biometrics. Biometrics refers to the automatic recognition of individuals based on their characteristics [3]. They are often used in verifying the claimed identity of the subject (one-to-one matching) or to identify individuals (one-to-many matching). In a biometric system, raw biometric data is first acquired using a sensor. This raw data can be in the form of an image, audio, or a physiological signal. *Feature extraction* is applied on the raw data to extract an identifying set of features into a *biometric template* that should be unique to each individual. For certain applications, such as fuzzy identity-based encryption and fuzzy extractor [4] based schemes, we are not interested in classification but directly apply the templates.

Automatic computation of 2D face biometrics consists of several subtasks such as face detection, alignment, normalization, and description. Face detection methods attempt to detect and indicate face regions in arbitrary images. Face alignment refers to the geometric normalization of the face region and is often based on eye locations. In addition to alignment, the image is photometrically normalized to remove lighting variations. Finally, face description refers to the process of feature extraction and can be based on principal component analysis [5], linear discriminant analysis [6], local binary patterns (LBP) [7], or deep learning [8].

Acquisition of biometrics (with different sensors for instance) induces variations into the extracted features. These are called intrasubject variations that can be caused, among other things, by differences in sensors, pose and expression changes, illumination, and occlusion [3]. Intrasubject variations lead to errors in the extracted biometric template. These errors in turn affect the true and false acceptance rates (TAR and FAR). According to a test conducted by NIST in 2012, state-of-the-art face recognition algorithms reach TAR of approximately 96% at FAR 0.1% on frontal images in controlled conditions. On a challenging dataset with a larger intraclass variation, such as the People In Photo Albums set [9], the performance is much worse. For example, the deep learning based DeepFace [8] has an overall accuracy of 46.66%. Performance in uncontrolled conditions is too low for applications such as fuzzy identity-based encryption where the identity of the person has to be certain.

2.2. Watermarking. Watermarking is often used in biometric systems to add another layer of security, generally either by directly embedding a biometric template into the host data or by protecting the biometric data with a watermark [10, 11]. Hämmerle-Uhl et al. [11] discuss different applications for biometric watermarking. These include steganographic approaches, multibiometric recognition, two-factor authentication, sample replay prevention, and sensor and sample authentication.

In steganography, biometric data is hidden in arbitrary data for transmitting so that the attacker is unaware that data is being transferred. In multibiometric recognition, biometric data is embedded into the biometric sample. The advantage of such method is that two different modalities can be transmitted at the same time and/or recognition performance is increased. Somewhat similar application is two-factor authentication in which authentication data is put, for example, on a smart card. The smart card can contain, for example, a fingerprint of the person [12] and the fingerprint image is watermarked with the face of the person therefore enabling a second layer of authentication. The idea of sample replay prevention is to prevent the use of sniffed sample data to fool the sensor. The sensor embeds a watermark to the sample image before transmitting it for feature extraction. An attacker tries to remove the watermark in order to be able to use sniffed data for replay attacks or as fake traits and consequently the watermark must be robust. Sensor and sample authentication are very similar except that the attacker aims at inserting a watermark in order to mimic correctly acquired sensor data. This can be prevented with semifragile watermarking [11].

Using the biometrics as a key, however, is a more recent concept. Dutta et al. [13] proposed a method for applying iris biometrics as a key in audio to prove the ownership of a piece of music. They argued that a random pseudorandom sequence is not enough for proving ownership unless the sequence is uniquely mapped to an entity that is logically or physically owned by the claimant. They proceeded to extract features from an iris image and used these features as seed of the audio watermark.

An image can have multiple faces and therefore our method should support multiple watermarking. Sheppard et al. [14] divided the multiple watermarking methods into rewatermarking, segmented watermarking, and composite watermarking. We are the most interested in a special case of segmented watermarking in which the multiple watermarks are embedded by interleaving the separate watermarks instead of embedding watermarks on top of each other.

3. Suggested Scheme

Our goal is to embed supplementary information into an image to aid in automatic face recognition and identity determination. Since embedding capacity is a precious resource, we want to minimize the size of this supplementary information. Therefore, instead of real valued feature vectors, we apply binary biometric templates. In this section, we describe our proposal in detail. We first give an overview and then proceed to explain our face feature extraction and the generation of the binary biometric templates. Finally, we give the details on embedding and extraction.

3.1. Overview

3.1.1. Embedding. Given an image $\mathcal{J}$ our method first detects the faces inside it. Let the correct identities for those faces and their corresponding binary biometric template vectors $B_1, B_2, \ldots, B_m$ be given. These correct identities are given as expert knowledge, for example, by humans tagging people in those images on a social network. For that particular image $\mathcal{J}$, our scheme computes the face features of the detected faces and converts them into binary biometric template vectors $B'_{\mathcal{J},1}, B'_{\mathcal{J},2}, \ldots, B'_{\mathcal{J},m}$ which we call the *image dependent templates*. Note that these templates might contain a significant amount of noise compared to the true identity templates B_i due to intrasubject variations such as pose changes, make up, or lighting. To aid in any identity determination task that later uses this image $\mathcal{J}$, our scheme computes supplementary information $I_{\mathcal{J},i} = B_i \oplus B'_{\mathcal{J},i}$ for all $i \in \{1, 2, \ldots, m\}$, where $\oplus$ is bitwise addition, and embeds $I_{\mathcal{J},1}, I_{\mathcal{J},2}, \ldots, I_{\mathcal{J},m}$ as a payload into $\mathcal{J}$ and thus creates a watermarked image $\mathcal{J}_w$. Note that $B'_{\mathcal{J},i}$ is effectively used as a key which can be later used to derive the correct template B_i from the supplementary information $I_{\mathcal{J},i}$. In contrast to direct embedding of identities into the image, in our scheme the removal of the face of person i from the image renders the computation of $B'_{\mathcal{J},i}$ impossible and thus prevents the determination of B_i and the identification of that person. The embedding process has been depicted in Figure 1.

3.1.2. Identity Determination. To determine the identities of people in a watermarked image $\mathcal{J}_w$, faces are first detected. Image dependent templates $\overline{B'_{\mathcal{J},i}}$ are then computed for those faces and the supplementary information $\overline{I_{\mathcal{J},i}}$ extracted for every $i \in \{1, 2, \ldots, m\}$. We have denoted by overlining that errors might have been introduced to both the image dependent features and the supplementary information due to image processing such as compression. That is, $\overline{B'_{\mathcal{J},i}} = B'_{\mathcal{J},i} \oplus e_i$ and $\overline{I_{\mathcal{J},i}} = I_{\mathcal{J},i} \oplus e'_i$ for every $i \in \{1, 2, \ldots, m\}$, where e_i and e'_i are some error vectors. The final template

$$\begin{aligned} \overline{B_i} &= \overline{I_{\mathcal{J},i}} \oplus \overline{B'_{\mathcal{J},i}} = B_i \oplus B'_{\mathcal{J},i} \oplus e'_i \oplus B'_{\mathcal{J},i} \oplus e_i \\ &= B_i \oplus e_i \oplus e'_i \end{aligned} \tag{1}$$

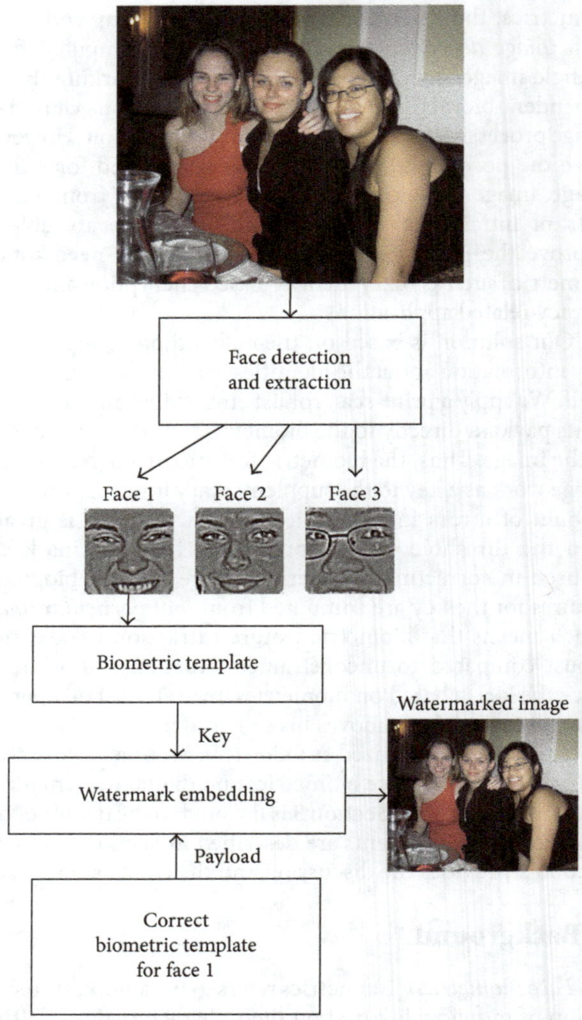

FIGURE 1: General overview of the embedding process of our proposed scheme. The same process is repeated for all of the faces in the image.

is computed for every face in the image. Compared to the original template B_i, the computed template $\overline{B_i}$ may contain errors e_i due to the extraction of the image dependent template and errors e'_i due to the extraction of the supplementary information. However, being image dependent, $B'_{\mathcal{J},i}$ and $\overline{B'_{\mathcal{J},i}}$ differ only by the amount of errors introduced by image processing variations such as color changes and scaling. Compared to nonrestricted face biometrics, the image dependent features of a person are robust yielding a significant boost to identity determination.

3.2. Face Feature Extraction. For face biometrics, we apply local binary patterns (LBP). LBP based descriptors are computationally efficient compared with deep learning based methods and suitable even for real-time applications and constrained devices. For details on LBP, its variants, and applications, see, for example, [15]. Face recognition based on LBP was pioneered by Ahonen et al. [7] and we follow the same methodology. In their work, the (extended) LBP operator computes a label for each pixel by thresholding P sampling points in an evenly spaced circle of radius R. The label is considered as a binary number which means that each pixel is mapped to a bit vector of length P. The pattern of this bit vector is called uniform if there are at most two transitions from 0 to 1 or from 1 to 0 when considering the vector circular. Even in face images, most of the patterns are uniform [7]. Given a face image, it is partitioned into $n \times n$ blocks and a histogram of uniform patterns is computed for each block. This spatially enhanced histogram is used as a face descriptor and classification is based on the Chi squared histogram dissimilarity measure. For details, see [7].

Following the general subtask structure of face recognition, our method consists of face detection, alignment, normalization, and face description. For face detection, we apply the method of Zhu and Ramanan [16] that applies a unified approach to face detection, pose estimation, and landmark point extraction. It reliably estimates head pose and facial landmarks such as eye locations in unconstrained situations and is robust to background clutter. We refer to [16] for details on the algorithm. To extract the face region, the image is first converted into grayscale and rotated based on the eye locations. Next, the image is scaled and a 128×128 pixels' face region is extracted with eyes fixed at coordinates $(27, 37), (101, 37)$. The resulting face region is smoothed with median filtering using 3×3 neighborhood and the lighting is normalized using the method of Tan and Triggs [17].

For face description, we follow Ahonen et al. [7]. The extracted face region is divided into 7×7 rectangular regions and the extended LBP operator is applied on each of those regions separately. Following [7], we chose $R = 2$ and $P = 8$ with uniform patterns giving a total of 49 histograms of length 59 for a single face. These histograms are combined into a single spatially enhanced histogram X (subsequently referred to as a histogram). The process has been depicted in Figure 2.

(a)

(b)

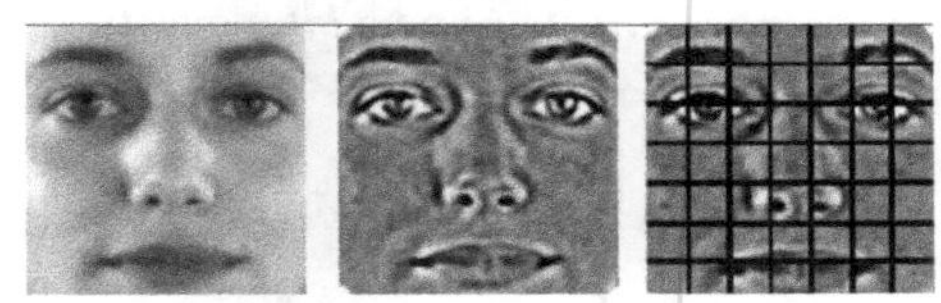

(c)

Figure 2: Example image from the BioID face database [18]. The original image (a) is first rotated based on the eye locations (b). The resulting image is scaled, eye locations are fixed, and a 128×128 pixel face region is extracted ((c) left). The lighting is normalized ((c) middle) and the face is divided into 7×7 rectangular regions ((c) right). LBP histograms are computed for each region separately and combined into a spatially enhanced histogram.

3.3. Biometric Template Generation. Based on the spatially enhanced histogram, we generate a binary biometric template for each person using a dictionary of training images. For the training set, we apply the FaceScrub database containing unconstrained face images of 530 celebrities (265 male and 265 female) [19]. A single facial image from each person in the dataset is used. The set of $k = 530$ training images is processed with the feature extraction described above to form a dictionary D of spatially enhanced histograms $D = \{X_1, X_2, \ldots, X_k\}$.

For each pair $X, Y \in D$, weighted Chi square distance

$$\chi^2_W(X,Y) = \sum_{j,i} w_j \frac{\left(x_{i,j} - y_{i,j}\right)^2}{x_{i,j} - y_{i,j}} \tag{2}$$

is computed, where i runs over the histogram bins and the weight terms w_j are determined by the region j in the 7×7 grid. Based on the work of Ahonen et al. [7], the weight terms given by the matrix

$$W = \begin{pmatrix} 2 & 1 & 1 & 1 & 1 & 1 & 2 \\ 2 & 4 & 4 & 1 & 4 & 4 & 2 \\ 1 & 1 & 1 & 0 & 1 & 1 & 1 \\ 0 & 1 & 1 & 0 & 1 & 1 & 0 \\ 0 & 1 & 1 & 1 & 1 & 1 & 0 \\ 0 & 1 & 1 & 2 & 1 & 1 & 0 \\ 0 & 1 & 1 & 1 & 1 & 1 & 0 \end{pmatrix} \tag{3}$$

are chosen. Those terms emphasize the eyebrows, eyes, mouth, and temples. Cheeks and the nose are given zero weight. For every $X_i \in D$ a bound c_i is computed by taking the median of the set

$$\left\{\chi^2_W(X_i, Y) : Y \in D, Y \neq X_i\right\} \quad (4)$$

for every $i \in \{1, 2, \ldots, n\}$.

To compute an n-bit binary biometric template $B = (b_1, b_2, \ldots, b_n)$ for a given face, we extract its spatially enhanced histogram Z, its distance $\chi^2_W(Z, X_i)$ to a selected set $\{X_1, X_2, \ldots, X_n\} \subseteq D$ of dictionary histograms, and set

$$b_i = \begin{cases} 0, & \chi^2_W(Z, X_i) < c_i, \\ 1, & \chi^2_W(Z, X_i) \geq c_i \end{cases} \quad (5)$$

for every $i \in \{1, 2, \ldots, n\}$. We prune the histogram dictionary for those histograms that produce the most consistent templates for frontal face images. As a second set, we apply the BioID face database (https://www.bioid.com/About/BioID-Face-Database) that consists of 1521 grayscale images of 23 persons captured in realistic settings [18].

To select n histograms, we first generate templates of length k for every image of every person in the test set. For a single person j, each of these templates should be identical. However, due to variations some of the bits have flipped. Let $z_{j,i}$ denote the amount of 0's divided by the amount of 1's for bit position i in the templates of person j. A *consistency vector* $P_j = (p_{j,1}, p_{j,2}, \ldots, p_{j,k})$ and a *bit majority vector* $Q_j = (q_{j,1}, q_{j,2}, \ldots, q_{j,k})$ are computed where

$$p_{j,i} = \max\left(z_{j,i}, 1 - z_{j,i}\right),$$
$$q_{j,i} = \begin{cases} 0, & z_{j,i} \geq 1 - z_{j,i}, \\ 1, & z_{j,i} < 1 - z_{j,i} \end{cases} \quad (6)$$

for every $i \in \{1, 2, \ldots, k\}$. The value $p_{j,i}$ measures the probability of bit flipping on position i for person j. The value $q_{j,i}$ represents the most probable bit in this position.

Let z'_i denote the amount of 0's divided by the amount of 1's in the set $\{q_{j,i} : j \in \{1, 2, \ldots, T\}\}$, where T is the number of persons (for our test set $T = 23$). To pick histograms into the pruned dictionary D', we first compute a *differentiation vector* $R = (r_1, r_2, \ldots, r_k)$, where

$$r_i = \max\left(z'_i, 1 - z'_i\right) \quad (7)$$

that measures the amount of variation in bits for position i across persons. In addition, a *total consistency vector* $P = (p_1, p_2, \ldots, p_k)$ is computed by summing the consistency values on bit position i across persons: $p_i = \sum_{j=1}^{T} p_{j,i}/T$. Based on the difference values $R = (r_1, r_2, \ldots, r_k)$, bit indices $i \in \{1, 2, \ldots, k\}$ are sorted into descending order. A total of n dictionary histograms X_i are selected into a pruned dictionary D' based on this ordering provided that the total consistency value p_i is over a predefined bound δ. For our experiments, we chose $\delta = 0.85$. The pruned dictionary D' is used for template generation.

The template length n needs to be chosen based on the capacity of the watermarking scheme. There should be a unique biometric template for every person from the world population so theoretically the threshold of 33 bits has to be exceeded. However, a larger template size provides better resiliency for misidentification. Based on the capacity of our watermarking scheme detailed in Section 3.4, we chose $n = 40$. It gives us enough capacity to host the supplementary information for at least one biometric template even in the smallest images (close to 500 × 500 pixels) found in social networks.

3.4. Watermarking. In our use-case scenario, watermarking is used as a second layer of the biometric feature acquisition process. This layer will provide the supplementary information $I_{\mathcal{F},i}$ embedded into the structure of the image itself. This allows the detection and correction of errors in identity-based schemes. The first requirement of this use-case is to have enough capacity. Specifically, each face in the image corresponds to a payload of 40 bits because of the biometric template size. The second watermarking requirement includes robustness to certain transformations which are common on images that are usually uploaded to social networks. Such transformations include JPEG compression with different compression ratios, color filtering, and possibly removal or modifications of some of the faces. Furthermore, high fidelity is required so that there will be no easily distinguishable artifacts on the images under normal view on a computer monitor. For this requirement, peak signal-to-noise ratio (PSNR) values over 40 dB are generally considered to be acceptable.

The watermarking method used in this scheme is a modification of the multibit watermarking technique proposed by Keskinarkaus et al. [20]. It was designed to be robust to print-scan attacks; that is, the host image can be printed and scanned and the watermark should still be reliably extractable. In the current scheme's case, this satisfies the robustness requirements with some margin so that the strength of the watermark can be lowered to meet the fidelity requirements.

The principal idea is that a message sequence, that is, the payload, is mapped to a directional angle of periodic patterns which are scattered and embedded in using triangular masks placed in permuted locations. The permutations are pseudo-randomly generated. The permuted triangles are grouped in polygon sets. Next, the polygons are grouped in segments which are 40 bit bins, the size of the biometric templates. Each one of the bins is capable of hosting the supplementary biometric information for one face.

Note that the first polygon of permuted triangles is always reserved to host the information about the number of initial faces detected in the image, unlike the original watermarking method where this polygon was reserved for synchronization purposes. This feature enables the proper extraction of information even when some faces have been removed from the image. This is because the number of biometric templates that have to be extracted is always known. Further modification on this method is the adaptability to size. The number of the triangular sequences is proportional to the image size. Thus, bigger images can host a larger number

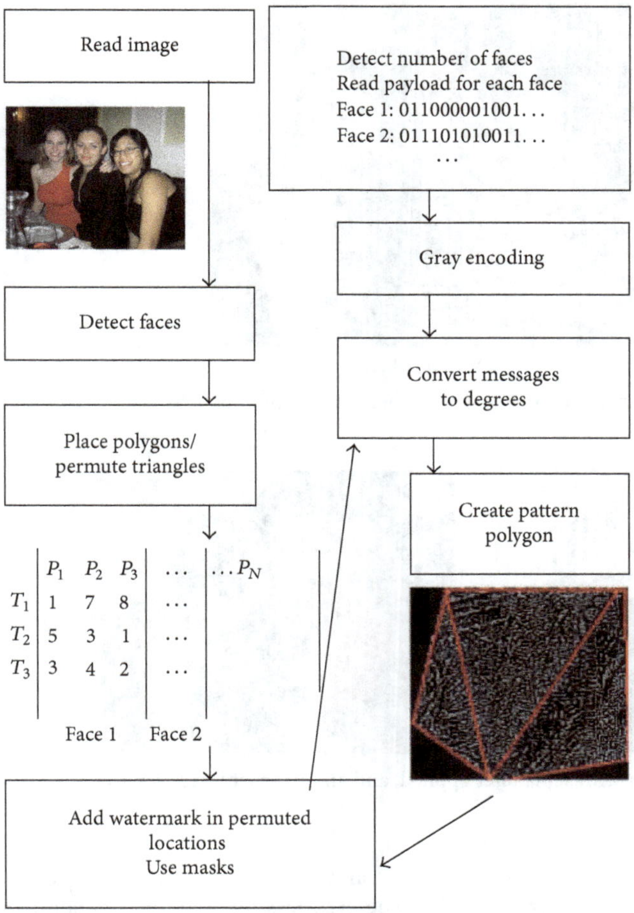

FIGURE 3: Overview of the watermarking process. T_1–T_3 refer to the three triangles acquired using the Delaunay triangulation. P_1–P_N refer to the new polygons being sets of three scattered triangles.

of faces. Compared with the original method, lower watermarking strength settings have been tested to improve fidelity while still maintaining high enough levels of robustness to withstand transformations such as JPEG compression and filter applications which are very common for images that are uploaded on social networks. Also the feature point detection scheme has been removed from the original method [20]. Instead faces can be considered as feature points in order to align an image in its proper position before extraction.

The watermarking technique based on [20] including the modifications to fit this use-case follows the steps below. The process is also depicted in Figure 3.

Embedding Steps

(1) Apply Zhu and Ramanan's face detection method and calculate the number of faces N_F as well as the locations of their feature points in the coordinate system.

(2) Determine the number of convex polygons k that fit to tile the input image $\mathcal{I}$. More information follows in the capacity subsection.

(3) Use Delaunay triangulation to divide each polygon into triangle areas.

(4) Produce a fixed pseudorandom permutation of the triangles that form the polygons. Thus, we create a vector of polygons which are actually sets of 3 scattered triangles. Segment the set vector so that it forms 40 bit bins. That will be 8 polygons. Every face is assigned to one bin. Note that the first polygon is reserved to store the number of detected faces.

(5) Embed the watermarks using the directional patterns in each polygon for each polygon that is now a set of scattered triangles. Use masks to restrict the marked region. Return the watermarked image $\mathcal{I}_w$.

Extracting Steps

(1) Detect faces in input image $\mathcal{I}_w$ using the same method. If the image has been rotated face features can be used in rotating it back to its original angle.

(2) Determine the number of convex polygons k that fit to tile the image.

(3) In the same way divide each polygon into triangle areas.

(4) Use the same seed number to permute the triangles and assign the faces to them.

(5) By detecting the angles of the periodic patterns extract the number of faces from the first polygon and continue by extracting the supplementary biometric information from the 40 bit bins, each one being a set of polygons.

The periodic directional patterns are generated in the same way as described in the original paper [20]. Changes focus on the adaptability of the number of polygons to the image size, the removal of reference points, and the use of number of faces and face locations to allow the embedding of multiple independent watermarks. Last the scaling factor λ is multiplied by a strength varying from 0.4 to 0.6 to reduce the watermarking strength.

Capacity and thus the maximum number of faces of which the information can be embedded in an image is dynamic as it is determined by the image size. In our case since we used a fixed polygon that is circumscribed in a 162×164 rectangle, an image of size $M \times N$ is tiled using $k = \lfloor N/162 \rfloor \times \lfloor M/164 \rfloor$ convex polygons. Because of the fact that each polygon is able to host 5 bits of data and one block is used for storing the initial number of faces, the total capacity is $C = (\lfloor N/162 \rfloor \times \lfloor M/164 \rfloor - 1) \times 5$ bits of data. Each face required $C_F = 40$ bits of space. In a use-case scenario where there are N_F faces there is $C_r = C - N_F C_F$ capacity left which is available to host extra information, for example, error correcting codes or a checksum. To give an example a typical image of size 1280×960 has a capacity of 170 bits and thus it is able to host the biometric templates of 4 faces.

4. Experiments

Both the biometrics and the watermarking algorithm were implemented in Matlab (trademark: The Mathworks Inc.).

Original Watermarked

Figure 4: Comparison of an unwatermarked and a watermarked image with embedding strength of 60%.

Figure 5: The three tested scenarios. An original image, an image with sepia filter applied, and an image with face removed.

Table 1: Mean bit error rate of image dependent template under JPEG compression.

	JPEG 80	JPEG 70	JPEG 60	JPEG 50
Mean BER	0.08	0.08	0.14	0.20

4.1. Test Dataset. For the experiments, we chose 10 images from the Annotated Face in-the-Wild (AFW) dataset [16]. The dataset contains 205 images with a total of 468 faces with large variations in face orientation, appearance, and background clutter. Resolutions range from 1024 × 768 up to 2606 × 1733. To capture a wide range of situations, we chose images containing one to four faces in different viewpoints. No annotations were used. Face landmark points were determined by the algorithm of Zhu and Ramanan [16] as a part of our scheme.

4.2. Image Dependent Templates under JPEG Compression. First, the image dependent biometric template performance was measured under JPEG compression. Four JPEG quality factors were evaluated: 80, 70, 60, and 50. First, a reference template was computed for each face in the test set. Then, the image was compressed and the template was recomputed and compared to the reference template. The mean bit error rate (BER) under compression was computed. The experimental results have been collected into Table 1.

4.3. Watermarking Quality Assessment and Performance. To evaluate the combination of biometrics and watermarking, for each individual j, a random identity template B_j of 40 bits representing the true errorless template was generated. To evaluate robustness against JPEG compression, we tested quality factors 100, 90, and 80. Three watermark embedding strengths were evaluated: 40%, 50%, and 60%. An example is depicted in Figure 4. For a single image, all combinations of these experiments were performed resulting in a total of 252 tests. Based on these tests, the mean PSNR and the mean BER of the extracted templates were evaluated.

To measure the robustness to image filtering, as demonstrated in Figure 5, we tested the effect of sepia filtering which is a common image effect. In addition, to measure the integrity of multiple identities in the image, we tested the case of face removal. In this test, one of the faces was selected and the face region determined by the method of Zhu and Ramanan [16] was rendered completely black. The above experiments were repeated for both of these modified images resulting in two additional sets of mean PSNR and mean BER measurements.

The experimental results have been collected into Table 2, where Orig stands for the original image, Sepia stands for the sepia filtered image, and FR stands for the image where a single face has been removed. Finally the parentheses (p%) denote the embedding strength of the watermark.

5. Discussion and Conclusion

The exact determination of an identity is hard in challenging conditions even if we have millions of training samples available [3]. Our suggestion provides a method of improving

Table 2: The mean bit error rate of the final template and the mean peak signal-to-noise ratio [dB] of the watermarked image for different JPEG quality levels.

	JPEG 100		JPEG 90		JPEG 80	
	BER	PSNR	BER	PSNR	BER	PSNR
Orig (40%)	0.08	45.6	0.06	45.6	0.06	45.6
Sepia (40%)	0.16	45.6	0.13	45.6	0.19	45.6
FR (40%)	0.10	45.6	0.11	45.6	0.26	45.0
Orig (50%)	0.06	43.5	0.07	43.5	0.07	43.5
Sepia (50%)	0.13	43.5	0.12	43.5	0.12	43.5
FR (50%)	0.06	43.4	0.10	43.4	0.11	43.4
Orig (60%)	0.08	42.2	0.07	42.2	0.07	42.2
Sepia (60%)	0.17	42.2	0.11	42.2	0.11	42.2
FR (60%)	0.06	42.0	0.08	42.0	0.11	42.0

the performance of identity-based schemes especially in challenging conditions. Our method provides this resilience without side information; merely the watermarked raw data suffices. Furthermore, our method supports manual removal of a person from an image. If the face of a particular person is censored from the image, his or her identity is also removed. Based on local binary patterns, our face biometrics scheme is efficient and suitable for constrained devices. However, our method is not restricted to LBP. Our methodology can be applied with any face recognizer. We believe that the biometric performance can be greatly increased with a state-of-the-art face descriptor such as DeepFace [8] in exchange of computational performance. The classic LBP is not particularly robust to noise [21] which is also seen in our experiments (Table 1). The periodic pattern induced by the watermarking scheme also increases the BER of image dependent templates. Performance could be increased, for example, by masking faces in the watermarking algorithm to prevent the periodic pattern from affecting those regions. There are also LBP variants more robust to such noise [22]. However, based on our experiments, even with the classic LBP, the performance is acceptable for high quality JPEG and embedding strength $\leq 75\%$.

Two datasets were used in the computation of the dictionary for template generation: FaceScrub [19] and BioID [18]. Both sets contain variations in lighting, expression, and gender. In addition, FaceScrub contains a lot of variations in size, compression, and noise. We have not explicitly tested the performance of template generation with regard to individual characteristics such as gender. However, the dictionary generation we applied maximizes the consistency of a template across all of the variations present in the data. Naturally, there are limits to these datasets. They are not able to capture all of the variations encountered in the wild. For instance, we are unsure how the template generation works, for example, for elder people or children since they are largely missing from the training data. To the best of our knowledge, there are no biometric face databases containing in-the-wild variations. Such considerations are left for future work.

By our experiments, our proposal is robust against JPEG compression down to quality level 80. Mean BER of 0.08, 0.07, and 0.07 were reached for JPEG quality levels 100, 90, and 80, respectively, for the original image with embedding strength 60%. Sepia filtering causes an increase in the mean BER. Since this value is not decreased when embedding strength is increased, it seems that the LBP based face descriptor is sensitive to sepia filtering. For robustness against sepia filtering and face removal, embedding strength less than 50% should not be used. In general, face removal does not affect the mean BER for the remaining faces provided that an adequately high embedding strength is used. For all of our tests, the PSNR is high meaning that watermarks were imperceptible.

For extremely short template lengths, the achieved BER may lead to false identification. However, even under sepia filtering the BER ≤ 0.2 is within the applicable range of existing fuzzy identity-based schemes provided that the template length is adequately large. To increase the template length from 40, we suggest choosing a less robust watermarking scheme and to cut down the number of attacks the scheme needs to withstand. For certain applications, zero BER is required. In such a case, we suggest applying fuzzy randomness extractors such as the fuzzy extractor [4] or related schemes [23–25]. These methods apply error correcting codes to bring the BER down to zero in exchange of the final template length. Such errorless determination of the template enables new applications that are not possible when the possibility of false identification is present. For example, traditional errorless identity-based encryption could be used if the template can be extracted without errors.

The payload in our method is derived from the correct identity in the image combined with image dependent features. It is possible for an adversary to exchange the payload of a particular person in the image with his or her chosen payload. To counter such manipulation attacks, the extractor needs to check that the extracted identity corresponds to the one in the image. One possibility is to use the extracted biometric template as a link to a frontal face image of that person taken in good conditions. Tampering of the payload is easy to detect provided that the system shows such a high quality image to the user whenever the identity is requested. However, we note that our proposal does not guarantee cryptographic security against such tampering or cryptographic protection for an identity when a face is removed. Such considerations are left for future work.

The watermarking method we applied is highly robust. It has been shown to perform well even in the challenging print-scan scenario with correctly chosen parameters [20]. The main drawback is the relatively low capacity. In particular, we would want to increase the biometric template size to increase its accuracy and robustness in applications. Capacity increase is naturally possible by sacrificing robustness.

Competing Interests

The authors declare that they have no competing interests.

Acknowledgments

This research was supported by Tekes, the Finnish Funding Agency for Technology and Innovation in VitalSens and INKA projects. This work was also supported in part by Infotech Oulu.

References

[1] A. Sahai and B. Waters, "Fuzzy identity-based encryption," in *Advances in Cryptology—EUROCRYPT 2005*, R. Cramer, Ed., vol. 3494 of *Lecture Notes in Computer Science*, pp. 457–473, Springer, Berlin, Germany, 2005.

[2] A. Shamir, "Identity-based cryptosystems and signature schemes," in *Advances in Cryptology—Proceedings of CRYPTO 84*, G. R. Blakley and D. Chaum, Eds., vol. 196 of *Lecture Notes in Computer Science*, pp. 47–53, Springer, Berlin, Germany, 1985.

[3] A. K. Jain, K. Nandakumar, and A. Ross, "50 years of biometric research: accomplishments, challenges, and opportunities," *Pattern Recognition Letters*, vol. 79, pp. 80–105, 2016.

[4] Y. Dodis, L. Reyzin, and A. Smith, "Fuzzy extractors: how to generate strong keys from biometrics and other noisy data," in *Advances in Cryptology—EUROCRYPT 2004*, C. Cachin and J. L. Camenisch, Eds., vol. 3027 of *Lecture Notes in Computer Science*, pp. 523–540, Springer, Berlin, Germany, 2004.

[5] M. Turk and A. Pentland, "Eigenfaces for recognition," *Journal of Cognitive Neuroscience*, vol. 3, no. 1, pp. 71–86, 1991.

[6] P. N. Belhumeur, J. P. Hespanha, and D. J. Kriegman, "Eigenfaces vs. fisherfaces: recognition using class specific linear projection," *IEEE Transactions on Pattern Analysis and Machine Intelligence*, vol. 19, no. 7, pp. 711–720, 1997.

[7] T. Ahonen, A. Hadid, and M. Pietikäinen, "Face description with local binary patterns: application to face recognition," *IEEE Transactions on Pattern Analysis and Machine Intelligence*, vol. 28, no. 12, pp. 2037–2041, 2006.

[8] Y. Taigman, M. Yang, M. Ranzato, and L. Wolf, "DeepFace: closing the gap to human-level performance in face verification," in *Proceedings of the 27th IEEE Conference on Computer Vision and Pattern Recognition (CVPR '14)*, pp. 1701–1708, June 2014.

[9] N. Zhang, M. Paluri, Y. Taigman, R. Fergus, and L. Bourdev, "Beyond frontal faces: improving person recognition using multiple cues," in *Proceedings of the IEEE Conference on Computer Vision and Pattern Recognition (CVPR '15)*, pp. 4804–4813, IEEE, Boston, Mass, USA, June 2015.

[10] J. Dong and T. Tan, "Effects of watermarking on iris recognition performance," in *Proceedings of the 10th International Conference on Control, Automation, Robotics and Vision (ICARCV '08)*, pp. 1156–1161, December 2008.

[11] J. Hämmerle-Uhl, K. Raab, and A. Uhl, "Watermarking as a means to enhance biometric systems: a critical survey," in *Information Hiding*, pp. 238–254, Springer, 2011.

[12] A. K. Jain and U. Uludag, "Hiding biometric data," *IEEE Transactions on Pattern Analysis and Machine Intelligence*, vol. 25, no. 11, pp. 1494–1498, 2003.

[13] M. K. Dutta, P. Gupta, and V. K. Pathak, "Biometric based unique key generation for authentic audio watermarking," in *Pattern Recognition and Machine Intelligence*, vol. 5909 of *Lecture Notes in Computer Science*, pp. 458–463, Springer, Berlin, Germany, 2009.

[14] N. P. Sheppard, R. Safavi-Naini, and P. Ogunbona, "On multiple watermarking," in *Proceedings of the Workshop on Multimedia and Security: New Challenges*, pp. 3–6, ACM, October 2001.

[15] M. Pietikäinen and G. Zhao, "Two decades of local binary patterns: a survey," in *Advances in Independent Component Analysis and Learning Machines*, E. Bingham, S. Kaski, J. Laaksonen, and J. Lampinen, Eds., pp. 175–210, Elsevier, New York, NY, USA, 2015.

[16] X. Zhu and D. Ramanan, "Face detection, pose estimation, and landmark localization in the wild," in *Proceedings of the IEEE Conference on Computer Vision and Pattern Recognition (CVPR '12)*, pp. 2879–2886, Providence, RI, USA, June 2012.

[17] X. Tan and B. Triggs, "Enhanced local texture feature sets for face recognition under difficult lighting conditions," *IEEE Transactions on Image Processing*, vol. 19, no. 6, pp. 1635–1650, 2010.

[18] O. Jesorsky, K. J. Kirchberg, and R. W. Frischholz, "Robust face detection using the hausdorff distance," in *Proceedings of the International Conference on Audio-and Video-Based Biometric Person Authentication (AVBPA '01)*, pp. 90–95, Springer, Halmstad, Sweden, 2001.

[19] H. W. Ng and S. Winkler, "A data-driven approach to cleaning large face datasets," in *Proceedings of the IEEE International Conference on Image Processing (ICIP '14)*, pp. 343–347, 2014.

[20] A. Keskinarkaus, A. Pramila, and T. Seppänen, "Image watermarking with feature point based synchronization robust to print-scan attack," *Journal of Visual Communication and Image Representation*, vol. 23, no. 3, pp. 507–515, 2012.

[21] J. Chen, V. Kellokumpu, G. Zhao, and M. Pietikäinen, "Rlbp: robust local binary pattern," in *Proceedings of the British Machine Vision Conference (BMVC '13)*, pp. 1–10, Bristol, UK, 2013.

[22] G. Kylberg and I.-M. Sintorn, "Evaluation of noise robustness for local binary pattern descriptors in texture classification," *EURASIP Journal on Image and Video Processing*, vol. 2013, article 17, 2013.

[23] Y. Dodis and Y. Yu, "Overcoming weak expectations," in *Theory of Cryptography*, A. Sahai, Ed., vol. 7785 of *Lecture Notes in Computer Science*, pp. 1–22, Springer, Berlin, Germany, 2013.

[24] G. Cohen, R. Raz, and G. Segev, "Nonmalleable extractors with short seeds and applications to privacy amplification," *SIAM Journal on Computing*, vol. 43, no. 2, pp. 450–476, 2014.

[25] X. Li, "Non-malleable condensers for arbitrary min-entropy, and almost optimal protocols for privacy amplification," in *Theory of Cryptography*, vol. 9014 of *Lecture Notes in Computer Science*, pp. 502–531, Springer, 2015.

8

3D Curvelet-Based Segmentation and Quantification of Drusen in Optical Coherence Tomography Images

M. Esmaeili,[1,2] A. M. Dehnavi,[1] and H. Rabbani[1]

[1] *Department of Bioelectrics and Biomedical Engineering, Medical Image & Signal Processing Research Center, School of Advanced Technologies in Medicine, Isfahan University of Medical Sciences, Isfahan, Iran*
[2] *Department of Advanced Technologies in Medicine, Faculty of Medicine, Tabriz University of Medical Sciences, Tabriz, Iran*

Correspondence should be addressed to H. Rabbani; hoss_rab@yahoo.com

Academic Editor: Panajotis Agathoklis

Spectral-Domain Optical Coherence Tomography (SD-OCT) is a widely used interferometric diagnostic technique in ophthalmology that provides novel in vivo information of depth-resolved inner and outer retinal structures. This imaging modality can assist clinicians in monitoring the progression of Age-related Macular Degeneration (AMD) by providing high-resolution visualization of drusen. Quantitative tools for assessing drusen volume that are indicative of AMD progression may lead to appropriate metrics for selecting treatment protocols. To address this need, a fully automated algorithm was developed to segment drusen area and volume from SD-OCT images. The proposed algorithm consists of three parts: (1) preprocessing, which includes creating binary mask and removing possible highly reflective posterior hyaloid that is used in accurate detection of inner segment/outer segment (IS/OS) junction layer and Bruch's membrane (BM) retinal layers; (2) coarse segmentation, in which 3D curvelet transform and graph theory are employed to get the possible candidate drusenoid regions; (3) fine segmentation, in which morphological operators are used to remove falsely extracted elongated structures and get the refined segmentation results. The proposed method was evaluated in 20 publically available volumetric scans acquired by using Bioptigen spectral-domain ophthalmic imaging system. The average true positive and false positive volume fractions (TPVF and FPVF) for the segmentation of drusenoid regions were found to be 89.15% ± 3.76 and 0.17% ± .18%, respectively.

1. Introduction

Age-related Macular Degeneration (AMD) is the most common degenerative eye disease that affects the central portion of the retina, known as the macula, and is the leading cause of vision loss and blindness in patients who are aged 65 and over in the developing countries [1]. There are two types of AMD: dry (atrophic or nonneovascular) and wet (neovascular or exudative) [2]. Dry form of AMD affects approximately 80–90% of individuals with AMD [3] and in this type of macular degeneration, extracellular small white or yellowish deposits (made up of lipids, a type of fatty protein), called drusen, accumulated between the Retinal Pigment Epithelium (RPE) and the inner collagenous layer of Bruch's membrane (vitreous lamina). Drusen are produced when the photoreceptors of the eye drop off older parts of the cell and could cause a protrusion of RPE over Bruch's membrane which leads to deterioration or degeneration of the macula over time [4].

Drusen in the macula are a common early sign of AMD that can be defined as hard and soft. Hard drusen are small and distinct lesions with sharp borders (this type of drusen may not cause vision problems for a long time and is normal with aging, and most people over 40 have some hard drusen). In contrast, soft drusen are those with large lesions (clusters close together) with indistinct borders that have been recognized as intermediate to advanced form of AMD [5].

Most patients with macular degeneration have the dry form of the disease. However, the dry form of macular degeneration can lead to the more damaging wet form that usually leads to more serious vision loss [2]. The Age-Related Eye Disease Study (AREDS) reported that certain antioxidant vitamins and zinc delay the progression of advanced dry

TABLE 1: Four defined categories by AREDS depending on the stages of AMD.

Category 1 No AMD	Category 2 Early stage AMD	Category 3 Intermediate AMD	Category 4 Advanced AMD
No or a few small drusen	Several small sized drusen or a few medium-sized drusen in one or both eyes	Many medium-sized drusen or one or more large drusen in one or both eyes	Degeneration of the deepest cells of the retina (in one eye only) or abnormal and fragile blood vessels known as choroidal neovascularization under the retina (wet form)

AMD or vision loss by about 25 percent over a six-year period [6]. The early diagnosis of the patients at higher risk for advanced AMD allows earlier protective intervention and preventive treatment to reduce severe vision loss. Thus, it is more important to identify the main changes in drusen and the RPE morphology to inspect the progression of nonneovascular AMD [7]. As shown in Table 1, AREDS defines four categories depending on the stages of AMD [8].

The previous studies on the color fundus (CF) images have shown that there is strong correlation between the maximum drusen size and total drusen area, with the risk of progression of AMD to its advanced form [1, 8]. Many different approaches and algorithms for automatic soft and hard drusen grading and quantification from CF images have been developed [9–12]. However, detecting and locating the drusen in a color retinal image are difficult to measure because of their differences in size and shape and also irregular and blurred boundaries.

Spectral-Domain Optical Coherence Tomography (SD-OCT) is a widely used interferometric diagnostic technique in ophthalmology that provides in vivo novel information of depth-resolved inner and outer retinal structures. The high resolution, sensitivity, and speed of SD-OCT provides a valuable opportunity for detection and volume measurement of drusen from SD-OCT images compared to color fundus photographs [13] and can be used as a method of following the AMD development and monitoring retinal changes over the time [8, 14, 15]. However, manual segmentation of macular drusen area and volume is time consuming and labor intensive, which limits its use in large-scale population. So, the automatic segmentation of drusen becomes more appealing to process SD-OCT data more efficiently.

OCT can provide useful information about drusen and results in more accurate differentiation of drusenoid deposits, including cuticular drusen, soft drusen, and subretinal drusenoid deposits [16]. Soft drusen are yellow-white dome shaped elevations that their central portion appear lighter than its edge and are typically 63 to ≥1000 μm in diameter. Smaller soft drusen have a little effect on the RPE and IS/OS junction layer morphology. Cuticular drusen appear as a round and punctate, with saw tooth pattern in an SD-OCT and are typically 50 to 75 μm in diameter [16]. But, OCT subretinal drusenoid deposits are seen in the subretinal space, in similar size to soft drusen, above the RPE. Soft and cuticular drusen, in contrast, are under the RPE. They often have a more punctate appearance and are interconnected so that their sizes range from 25 to >1000 μm [16].

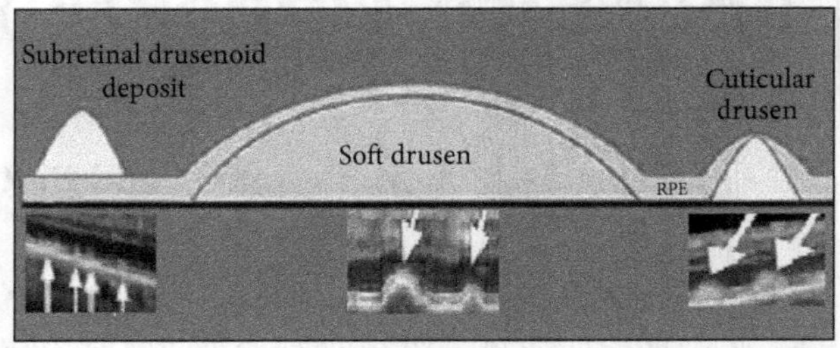

FIGURE 1: Different appearance of drusen in OCT image: soft, cuticular, and subretinal drusen.

As shown in Figure 1, small and medium-size drusen may cause subtle elevation of discreet areas of RPE with variable reflectivity, indicating the variable accumulations of the underlying material. Greater elevation of RPE may be seen as the result of larger confluent drusen or Drusenoid Pigment Epithelium Detachment (DPED), with a hyporeflective or medium-reflective deposit separating the RPE from the underlying Bruch membrane (BM) [17].

Although a number of fully/semiautomatic techniques have been proposed to detect drusen from OCT images (Table 2), only the accuracy of a few methods has been validated and they mainly only identify the absence or presence of drusen. The challenging issues associated with drusen detection from OCT images include the variation of texture and intensity distribution (hyporeflective or medium-reflective) of drusen which prevents accurate segmentation of the RPE and BM for quantification of drusen complex (the distance between the abnormal RPE and the normal estimated RPE floor), and methods which are based on accurate detection of RPE layer [18, 19] may fail due to difficulty in RPE layer segmentation even in normal patients, because the RPE layers are often contiguous with the inner segment/outer segment (IS/OS) retinal layer.

In this paper, we tackle the above-mentioned challenges and present a novel fully automated drusen segmentation method in SD-OCT images. For this purpose, in order to prevent inaccurate detection of retinal layers, we preprocess OCT images and remove highly reflective posterior hyaloid or other reflective layers beneath the BM layer, then we use curvelet transform and modify curvelet coefficients to enhance and fill the possible appeared vertically black shadowing effects due to blood vessels. These possible gaps due to large vessels may degrade graph-based segmentation

TABLE 2: Available image analysis techniques for AMD.

Diagnostic techniques for AMD			
Image analysis techniques for AMD	Color fundus image analysis	Texture-based segmentation	Gabor filter and wavelet analysis [33]
			Statistical structural information-based texture analysis [34]
			The Artificial Neural Network (ANN) based supervised classifiers [35]
			Expectation Maximization (EM) based unsupervised classifiers [36]
		Thresholding-based segmentation	Probability based thresholding [37]
			Otsu method for background selection and intensity thresholding [38]
			Histogram based adaptive local thresholding [39]
			Edge based thresholding [11]
		Clustering-based segmentation	Spatial histogram and similarity based classification [40]
			Distance based clustering [41]
			Wavelet based feature extraction & SVM based classification [42]
		Edge and template matching	Gaussian template matching [43]
			Gradient based segmentation [44]
	OCT image analysis techniques	Graph-based segmentation	Graph and dynamic programming [31]
			A graph-based multilayer segmentation approach [45]
		Active contour-based segmentation	Anisotropic noise suppression and deformable splines [46]
			Based on active contours and Markov random fields [47]
		Polynomial curve-fitting based technique	Local convexity condition and fitting second- or fourth-order polynomials [48, 49]
			Based on the distance between the abnormal RPE and the normal RPE floor [19, 50]

of IS/OS and BM layers. The areas located between the shifted extracted IS/OS junction layer and BM pixels by the use of graph theory from modified image are marked as drusen.

The paper is organized as follows. Section 2 provides an introduction to 3D digital curvelet transform (3DCUT). In Section 3 our proposed method is described and the results and performance evaluation are presented in Section 4. Finally, the conclusions are given in Section 5.

2. 3D Digital Curvelet Transform

The curvelet transform is a time-frequency analysis of images that gives a sparse representation of objects. The basis elements of this transform have good directional selectivity and are highly anisotropic [20]. The directional selectivity of curvelets and spatially localized property of each curvelet can be utilized to preserve the image features along certain directions in each subband. Following this reasoning, curvelets are appropriate basis elements (atoms) for representing d-D objects which are smooth apart from singularities along smooth manifolds of codimension 1 [21].

Although the direct analyzing of 3D data as a volume and considering the 3D geometrical nature of the data are computationally expensive, it has been shown that 3D analysis of 3D data outperforms 2D slice-by-slice analyzing of data [22]. The parabolic scaling, good directional selectivity, tightness and sparse representation properties of 3D curvelet singularities, provide new opportunities to analyze and study large datasets in medical image processing. 3D curvelet elements are plate-like shapes of $2^{-j/2}$ in two directions and width about 2^{-j} in the orthonormal direction which are smooth within the plate and oscillate along the normal direction of the plate.

The 3D discrete curvelet transform is an extension to the 2D curvelet transform proposed in [21]. In 2D, the curvelet family is generated by translation ($b \in \mathfrak{R}^2$) and rotation (θ) of the basic element $\varphi_{a,0,0}$:

$$\varphi_{a,b,\theta}(x) = \varphi_{a,0,0}\left(R_\theta (x-b)\right), \quad (1)$$

where $R_\theta = \begin{pmatrix} \cos\theta & -\sin\theta \\ \sin\theta & \cos\theta \end{pmatrix}$ is 2×2 rotation matrix with angle θ and $\varphi_{a,0,0} \in L^2(\mathfrak{R}^2)$ is given by its Fourier transform $\widehat{\varphi}_{a,0,0}(\xi) = U_a(\xi)$. The a-scaled window U_a is a polar wedge in frequency domain that is used for building curvelet functions [23]. Since

$$\widehat{\varphi}_{a,b,\theta}(\xi) = e^{-i\langle b,\xi\rangle}\widehat{\varphi}_{a,0,0}\left(R_\theta \xi\right) = e^{-i\langle b,\xi\rangle} U_a\left(R_\theta \xi\right) \quad (2)$$

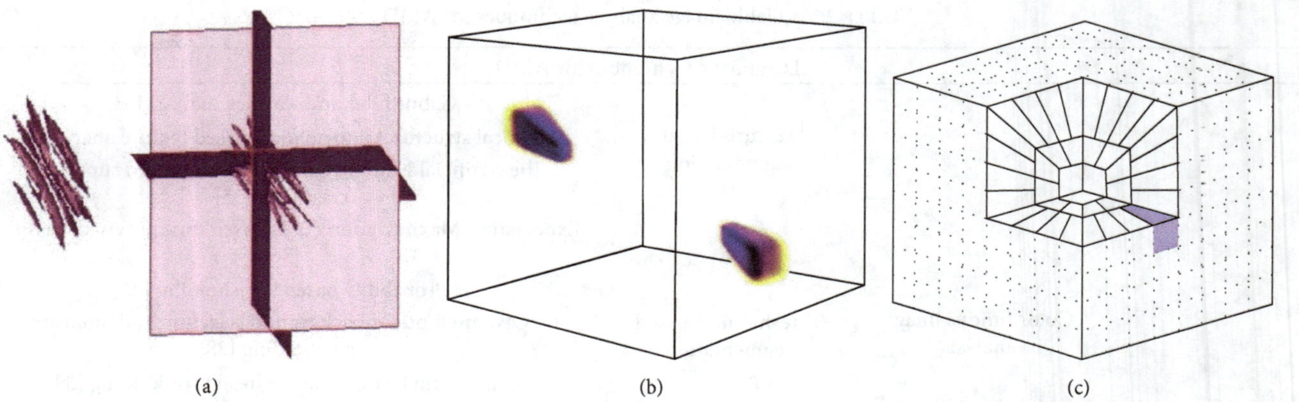

FIGURE 2: 3D rendering of curvelet atom in (a) space, (b) frequency, and (c) discrete frequency tiling. The shaded area separates 3D wedge.

using a suitable sampling at the range of scales $a_j = 2^{-j}$, $j \geq 0$, orientations $\theta_{j,l} = \pi l 2^{-\lceil j/2 \rceil}/2$, $l = 0, 1, \ldots, 4.2^{\lceil j/2 \rceil} - 1$ ($\lceil x \rceil / \lfloor x \rfloor$ denote the smallest integer being greater/smaller than or equal to x), and locations $b_k^{j,l} = b_{k1,k2}^{j,l} = R_{\theta_{j,l}}^{-1}(k_1/2^j, k_2/2^{j/2})^T$, $k_1, k_2 \in Z$, the curvelet coefficients are defined as

$$c_{j,k,l}(f) = \int_{R^2} \hat{f}(\xi) U_j \left(R_{\theta_{j,l}} \xi\right) \exp\left(i \left\langle b_k^{j,l}, \xi \right\rangle\right) d\xi. \quad (3)$$

To have Cartesian arrays instead of the polar tiling of the frequency plane, the Cartesian window $\widetilde{U}_j(\xi)$ is defined [23] that determines the frequencies in the trapezoid $\{(\xi_1, \xi_2): 2^{j-1} \leq \xi_1 \leq 2^{j+1}, -2^{-\lfloor j/2 \rfloor} \leq (3\xi_2)/(2\xi 1) \leq 2^{-\lfloor j/2 \rfloor}\}$. So, the Cartesian counterpart of the coefficients can be defined by

$$\tilde{c}_{j,k,l}(f) = \int_{R^2} \hat{f}(\xi) \widetilde{U}_j \left(S_{\theta_{j,l}}^{-1} \xi\right) \exp\left(i \left\langle \tilde{b}_k^{j,l}, \xi \right\rangle\right) d\xi, \quad (4)$$

where $k_j = (k_1 2^{-j}, k_2 2^{-\lfloor j/2 \rfloor})^T$, $(k_1, k_2)^T \in Z^2$, and $\tilde{b}_k^{j,l} = S_{\theta_{j,l}}^{-T}(k_1 2^{-j}, k_2 2^{-\lfloor j/2 \rfloor}) = S_{\theta_{j,l}}^{-T} k_j$ and $S_\theta = \begin{pmatrix} 1 & 0 \\ -\tan\theta & 1 \end{pmatrix}$.

By defining regular rectangular grid instead of tilted grid, Cartesian curvelets are defined as

$$\tilde{c}_{j,k,l}(f) = \int_{R^2} \hat{f}(\xi) \widetilde{U}_j \left(S_{\theta_{j,l}}^{-1} \xi\right) e^{i\langle k_j, \xi \rangle} d\xi. \quad (5)$$

$\tilde{b}_k^{j,l} = S_{\theta_{j,l}}^{-T} k_j$ of formula (4) has been replaced by $k_j = (k_1 2^{-j}, k_2 2^{-\lfloor j/2 \rfloor})^T$ that is being taken on values on a rectangular grid.

In 3D, similar to 2D case, the Cartesian window $\widetilde{U}_j(\xi)$ is defined by $\widetilde{U}_j(\xi) = \widetilde{U}_j(\xi_1, \xi_2, \xi_3)$ that isolates the frequencies in the truncated pyramid

$$\left\{ (\xi_1, \xi_2, \xi_3)^T : 2^{j-1} \leq \xi_1 \leq 2^{j+1}, \; -2^{-\lfloor j/2 \rfloor} \leq \frac{3\xi_2}{2\xi_1} \leq 2^{-\lfloor j/2 \rfloor}, \; -2^{-\lfloor j/2 \rfloor} \leq \frac{3\xi_3}{2\xi_1} \leq 2^{-\lfloor j/2 \rfloor} \right\}. \quad (6)$$

With the angles $\theta_{j,l}$ and $\upsilon_{j,m}$ the 3D shear matrix is defined as $S_{\theta_{j,l},\upsilon_{j,m}} = \begin{pmatrix} 1 & 0 & 0 \\ -\tan\theta_{j,l} & 1 & 0 \\ -\tan\upsilon_{j,m} & 0 & 1 \end{pmatrix}$, where

$$\begin{aligned} \tan\theta_{j,l} &= l 2^{-\lfloor j/2 \rfloor} \quad l = -2^{\lfloor j/2 \rfloor} + 1, \ldots, 2^{\lfloor j/2 \rfloor} + 1, \\ \tan\upsilon_{j,m} &= m 2^{-\lfloor j/2 \rfloor} \quad m = -2^{\lfloor j/2 \rfloor} + 1, \ldots, 2^{\lfloor j/2 \rfloor} + 1, \\ k_j &= \left(k_1 2^{-j}, k_2 2^{-\lfloor j/2 \rfloor}, k_3 2^{-\lfloor j/2 \rfloor}\right)^T, \\ &(k_1, k_2, k_3)^T \in Z^3. \end{aligned} \quad (7)$$

In 3D, every Cartesian corona has six components, one for each face of the unit cube. Each component is regularly partitioned into wedges with same volume as shown in Figure 2.

The curvelet functions in the cone $\{(\xi_1, \xi_2, \xi_3): 0 < \xi_1, -1 \leq \xi_2/\xi_1 < 1, -1 \leq \xi_3/\xi_1 < 1\}$ are given by

$$\tilde{\varphi}_{j,k,l,m} = \tilde{\varphi}_{j,0,0,0}\left(S_{\theta_{j,l},\upsilon_{j,m}}^T \left(x - \tilde{b}_k^{j,l,m}\right)\right). \quad (8)$$

So, its Fourier transform would be

$$\begin{aligned} \hat{\tilde{\varphi}}_{j,k,l,m} &= \exp\left(-i \left\langle \tilde{b}_k^{j,l,m}, \xi \right\rangle\right) \hat{\tilde{\varphi}}_{j,0,0,0}\left(S_{\theta_{j,l},\upsilon_{j,m}}^{-1} \xi\right) \\ &= \exp\left(-i \left\langle \tilde{b}_k^{j,l,m}, \xi \right\rangle\right) \widetilde{U}_j\left(S_{\theta_{j,l},\upsilon_{j,m}}^{-1} \xi\right), \end{aligned} \quad (9)$$

where $\hat{\tilde{\varphi}}_{j,0,0,0}(\xi) = \widetilde{U}_j(\xi)$. Analogously as in (5), the curvelet coefficients are given by

$$\begin{aligned} \widetilde{C}_{j,k,l,m}(f) &= \left\langle f, \tilde{\varphi}_{j,k,l,m} \right\rangle \\ &= \int_{R^3} \hat{f}(\xi) \widetilde{U}_j\left(S_{\theta_{j,l},\upsilon_{j,m}}^{-1} \xi\right) \exp\left(i \left\langle \tilde{b}_k^{j,l,m}, \xi \right\rangle\right) d\xi \\ &= \int_{R^3} \hat{f}\left(S_{\theta_{j,l},\upsilon_{j,m}} \xi\right) \widetilde{U}_j(\xi) \exp\left(i \left\langle k_j, \xi \right\rangle\right) d\xi. \end{aligned} \quad (10)$$

In this paper we use a new implementation of the 3D Fast Curvelet Transform (3DFCT) [24, 25] with strong directional

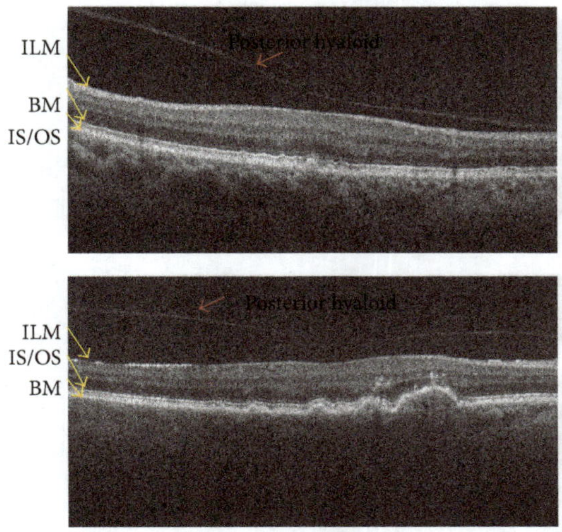

FIGURE 3: Presence of detached posterior hyaloid in B-scans. ILM: internal limiting membrane, IS/OS: inner segment/outer segment, and BM: Bruch's membrane.

selectivity property at the finest scale that has a reduced redundancy factor than the wrapping-based implementation proposed in CurveLab Toolbox [20, 21]. According to this implementation, 3D curvelet coefficients are obtained as follows:

(1) Cartesian coronization is performed that decomposes the object into dyadic coronae based on concentric cubes. Each corona is subdivided into trapezoidal regions conforming the usual parabolic scaling as shown in Figure 2.

(2) The 3D coefficients are obtained by applying an inverse 3D FFT to each wrapped wedge as shown in Figure 2, which appropriately fits into a 3D rectangular parallelepipeds of dimensions $\sim (2^j, 2^{j/2}, 2^{j/2})$ centered at the origin.

3. Proposed Method

3.1. Preprocessing. The posterior hyaloid membrane separates the vitreous from the retina. A Posterior Vitreous Detachment (PVD) is a condition of the eye, which is common in over 75% of patients who are age 65 and older, in which the vitreous membrane is separated from the retina [26]. As shown in Figure 3, the presence of detached posterior hyaloid (thin layer above retina) in some B-scans of 3D OCT volumetric data, creates an abnormal layer that may be seen as a thin hyperreflective layer above the internal limiting membrane (ILM). Some previous algorithms [27, 28] that assume ILM to be the first hyperreflective layer may determine this abnormal detached vitreous membrane as the retinal boundary, which results in an inaccurate segmentation of the remaining layers, also overestimation of retinal thickness.

To overcome this problem, each OCT image is thresholded with an optimal threshold selected efficiently for each image based on entropy-based thresholding algorithm [29], which takes into account the spatial distribution of gray levels in order to coarsely extract the Region of Interest (ROI) from the nonretinal layer background.

Some possible connected pixels due to highly reflective posterior hyaloid in thresholded image can be removed by applying length filtering and removing elongated short components. In the extracted ROI, possible holes near ILM are also removed by dilating and eroding (morphological operators) with the disk shaped structuring element of radius 5.

In some cases upper band of ROI is segmented with some kind of bumps (Figure 4(b), right). In order to segment this layer smoothly, we fit a third order polynomial to the upper pixels of ROI region. We use this boundary to limit the search region in segmentation process of IS/OS and BM and differentiate them from ILM (3 distinct hyperreflective layers with similar characteristics). Once the binary mask is generated, 10 pixels below the upper detected boundary in created binary image is used to separate inner and outer portion of retina and reduce the search range which results in an accurate and fine segmentation of BM and IS/OS junction layer by the use of graph theory.

Figure 4(d) shows the created binary mask in two sample B-scans contain posterior hyaloid and bumps in detected upper boundary of ROI.

3.2. Drusenoid Region Segmentation. The appeared vertically black shadowing effects due to blood vessels may degrade the continuity of IS/OS junction layer, and methods based on thresholding and graph theory may fail for accurate detection of these layers in presence of large vessels. So, in order to enhance IS/OS layer and fill these gaps we modify curvelet coefficients in OCT images. Since the curvelet coefficients have a sparse distribution and the bright objects get large coefficients in the curvelet domain in order to enhance the bright hyperreflective regions in OCT images, similar to one defined by Starck et al. [30], we take curvelet transform of OCT image and modify DCUT coefficients using function k_c:

$$k_c(x) = \begin{cases} 0.7x^3 & \text{if } x < N \\ 0.1x & \text{if } N \le x < 6N \\ 0.6x & \text{if } 6N \le x < 9N \\ x & \text{if } 9N \le x. \end{cases} \tag{11}$$

In this equation $N = 0.1M$, where M is the maximum curvelet coefficient of the relative band. Then, we reconstruct the enhanced image from the modified curvelet coefficients by applying inverse DCUT. Figure 5(b) shows the reconstructed image from modified curvelet coefficients.

To delineate the BM, we use modified reconstructed image with upper masked region using the extracted binary mask in previous section (Figure 5(c)). By using this image that the upper cropped region is set to zero, the BM layer would be the first light-to-dark boundary (vertically). Then, we apply the graph theory based segmentation method in [31]

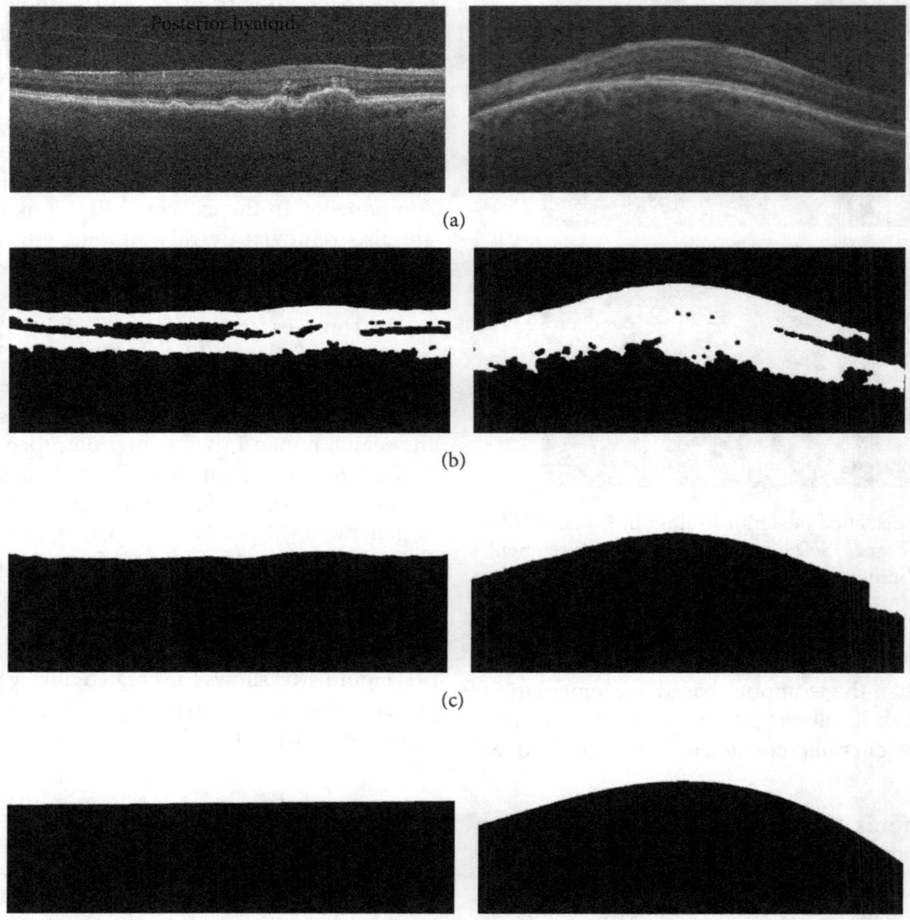

FIGURE 4: Created binary mask to accurate segmentation of BM and RPE. First and second columns, respectively, show the results for a sample B-scan with posterior hyaloid and a sample data with bumps in detected ROI region. (a) Original image. (b) Thresholded + length filtered + filled ROI image. (c) Segmented upper band pixels of ROI. (d) Created binary mask by fitting 3rd-order polynomial to upper band pixels of ROI.

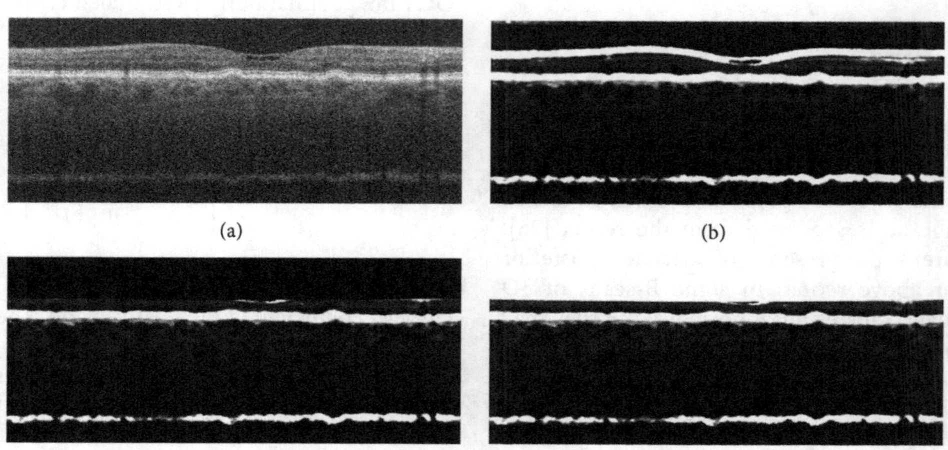

FIGURE 5: Reconstructed image from modified curvelet coefficients. (a) Original image, (b) modified image, (c) masked image used to delineate the BM, and (d) masked image used to delineate the IS/OS junction layer.

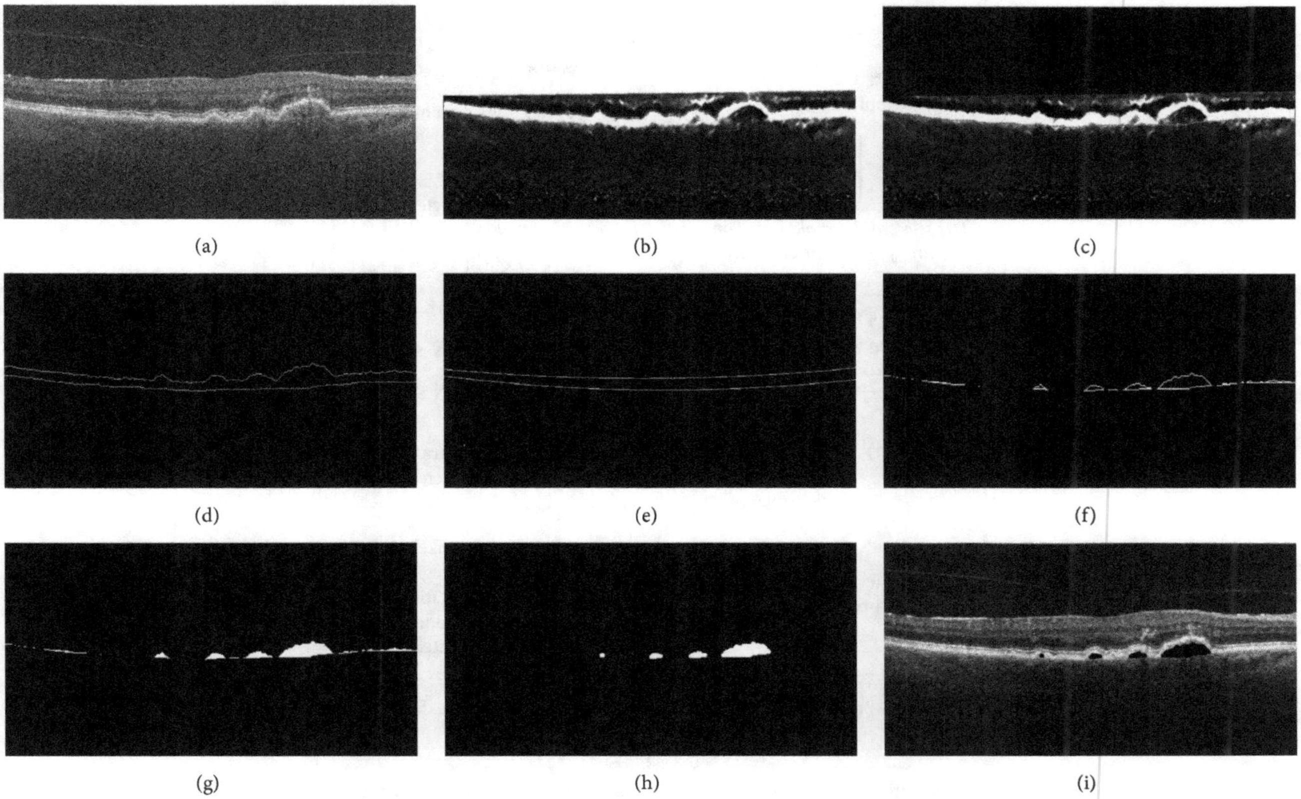

FIGURE 6: Drusenoid region segmentation results. (a) Original noisy image, (b) detected IS/OS junction boundary, (c) detected BM layer, (d) extracted BM and IS/OS boundaries, (e) 2nd-order polynomial fitted lines to extracted boundaries in order to calculate the mean difference of two boundaries, (f) created image by shifting the upper detected IS/OS junction pixels, (g) filled image, (h) reconstructed image by applying morphological erosion and then dilation operators, and (i) segmented drusenoid region.

with an adjacency matrix that its weights are calculated using the following equation:

$$w_{ab} = \text{norm}\left(-\left(g_a^{\text{LD}} + g_b^{\text{LD}}\right), 3, 7\right) + \text{norm}\left(-\left(g_a^{\text{DL}} + g_b^{\text{DL}}\right), 0, 0.5\right) + \text{norm}\left(-d_{ab}, 2, 4\right) + w_{\min}. \quad (12)$$

In this equation, w_{ab} is the weight assigned to the edge connecting nodes a and b, g_a^{DL} and g_b^{DL} are the vertical dark-to-light gradients of the image at nodes a and b, g_a^{LD} and g_b^{LD} are the vertical light-to-dark gradients of the image at nodes a and b and d_{ab} is the Euclidian distance from node a to node b, and $w_{\min} = 1 \times 10^{-5}$ is the minimum weight of the graph.

The norm notation norm(A, a, b) indicates the normalization of the value A to range from a to b.

For IS/OS junction boundary, we use modified reconstructed image after setting the upper cropped regions to maximum intensity of image (Figure 5(d)). By using this image, the IS/OS layer is the first vertical dark-to-light boundary. For this image, the adjacency matrix is constructed by calculating the weights as follows:

$$w_{ab} = \text{norm}\left(-\left(g_a^{\text{LD}} + g_b^{\text{LD}}\right), 0, 0.5\right) + \text{norm}\left(-\left(g_a^{\text{DL}} + g_b^{\text{DL}}\right), 2, 3\right) + \text{norm}\left(-d_{ab}, 0, 0.5\right) + w_{\min} \quad (13)$$

Figures 6(b) and 6(c) illustrate the results of applying the mentioned graph-based method on the masked images for a sample B-scan (Figure 6(a)).

In order to estimate RPEDC (RPE + drusen complex) thickness, at first we apply two 2nd order polynomial fitting functions to the detected IS/OS and BM (Figure 6(d)), then the mean difference of these two 2nd order polynomial curves is calculated. The calculated mean value is used to estimate RPEDC thickness (Figure 6(e)) and to shift upper detected IS/OS pixels toward detected BM (Figure 6(f)).

For extraction of drusenoid regions, we produce an image that in each column of this image if row location (in MATLAB coordinate system) of IS/OS pixel + mean difference is less than or equal to the row location of detected BM we set the row location of shifted IS/OS (row location of IS/OS pixel + mean difference) and row location of BM in that column to 1; otherwise this column is set to zero (Figure 6(f)).

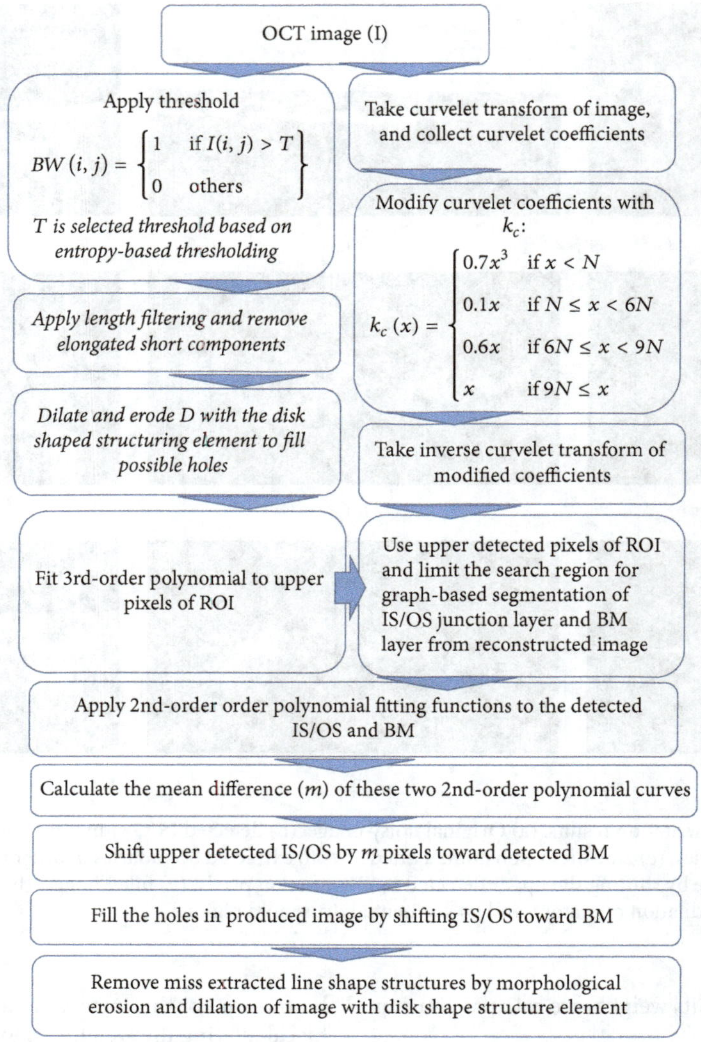

FIGURE 7: The block diagram of the proposed method.

The created holes between connected layers (shifted IS/OS and BM) are filled by performing a flood-fill morphological operation on the pixels of the binary image (Figure 6(g)). Then we remove miss extracted liny shape structures by morphological erosion and dilation of image with disk shape structure element (Figure 6(h)). Figure 6(i) shows the final results of our proposed method.

A sketch of the proposed algorithm after detecting IS/OS and BM layers with the use of graph theory is as in Algorithm 1.

Figure 7 illustrates the general block diagram of our proposed approach.

4. Results

For this study we use 20 volumetric scans acquired by using Bioptigen spectral-domain ophthalmic imaging system (the dataset is publicly available at http://people.duke.edu/~sf59/Chiu_IOVS_2011_dataset.htm) [31]. Two independent expert graders performed manual segmentation of the drusenoid regions using the 3D Slicer software (the software is freely available at https://www.slicer.org/). We then performed automatic segmentation using the algorithm described earlier, which was implemented in MATLAB.

To test the accuracy of proposed algorithm, automatic segmentation results for 220 B-scans across 20 patients were compared against the available results from two graders.

Segmentation results are quantitatively measured using Dice Coefficient (DC) and true positive and false positive volume fractions (TPVF and FPVF) [32] are defined as follows:

$$\text{Dice coefficient} = 2\frac{|V_{seg} \cap V_{ref}|}{|V_{seg}| + |V_{ref}|}, \quad (14)$$

where V_{seg} and V_{ref} are the extracted volume in the automatic segmented image and in the reference ground truth image. The maximum possible value for DC is 1 (indicating a perfect match between the result of the algorithm and ground truth)

```
Input: I% I is an image m × n, with extracted locations of IS/OS and BM (Figure 6(d))
Output: DR% Create the image DR, for extraction of drusenoid regions
DR = zeros(size(I));
For l = 1 : n
A = I(:, l);
[i, j] = Find(A == 1);
If min(i) + mean_diff-max(i)<=0% mean_diff is obtained from I (Figure 6(e))
DR(min(i) + mean_diff,l) = 1;
DR(max(i), l) = 1;
End
End
```

ALGORITHM 1: The proposed algorithm for drusenoid region segmentation.

TABLE 3: Comparison of automatic versus manual results of drusenoid region segmentation.

In comparison to	Dice coefficient (DC)	TPVF	FPVF
Grader 1	89.45% ± 3.46%	89.21% ± 4.61%	0.15% ± 0.21%
Grader 2	89.85% ± 4.34%	90.12% ± 4.43%	0.12% ± 0.26%
Graders 1 & 2	90.23% ± 4.12%	89.15% ± 3.76%	0.17% ± 0.18%

and DC is zero for a slice when no drusenoid region is detected or when there is no overlap between V_{seg} and V_{ref}.

The mean and standard deviation of DC for all the volumes with ground truth taken as Graders 1 and 2 and union of Graders 1 and 2 are mentioned in Table 3.

TPVF also indicates the fraction of the total amount of tissue inside the reference delineation, and FPVF denotes the amount of tissue falsely identified. They are defined as follows:

$$\mathrm{TPVF} = \frac{\left|V_{seg} \cap V_{ref}\right|}{\left|V_{ref}\right|}$$
$$\mathrm{FPVF} = \frac{\left|V_{seg}\right| - \left|V_{seg} \cap V_{ref}\right|}{\left|V_{t.v} - V_{ref}\right|}, \qquad (15)$$

where $V_{t.v}$ denotes the volume in the whole image between ILM and BM.

In order to show the effect of applying 3D curvelet transform in the proposed method in this paper, we can implement the different stages of proposed segmentation algorithm without applying 3D curvelet transform and compare the results. As an example, Figure 8 shows the effect of applying 3D curvelet transform in accurate detection of BM. In comparison to the proposed method in [18], which is based on assuming the RPE layer to be constant in thickness (20 μm) and using the threshold-based method to identify RPE layer, our proposed method can robustly identify RPE and IS/OS layer for drusen quantification. Figure 9 illustrates the final results of proposed drusenoid region segmentation method in this paper for different B-scans.

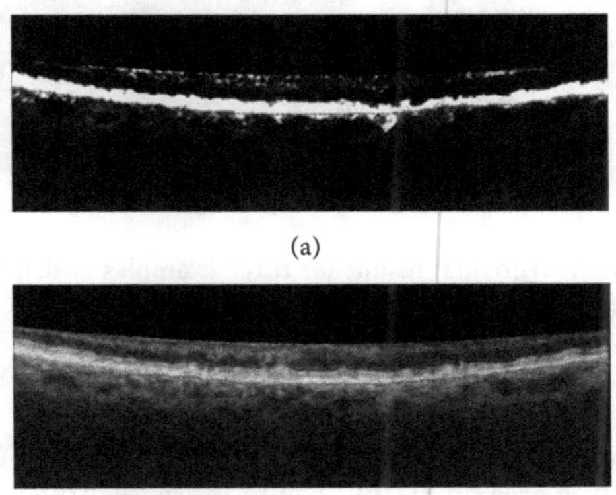

(a)

(b)

FIGURE 8: The results of detected BM by applying the graph-based segmentation method on (a) modified image by curvelet transform and (b) original image without modification.

5. Conclusion

SD-OCT provides high axial resolution and dense 3D imaging of the pathological changes in the eye, which makes it an appropriate device for detection of abnormal elevations of the RPE and IS/OS layer in various pathologies such as drusenoid pigment epithelial detachment. Previous methods that are based on difference between detected normal and abnormal RPE layer may fail due to difficulty in RPE layer segmentation even in normal patients because the RPE layer are often contiguous with the IS/OS retinal layer.

This paper presents a new fully automatic method to segment and quantify the drusen volume from OCT images, in order to provide a metric that can assist clinicians in monitoring the progression of AMD. The previous studies

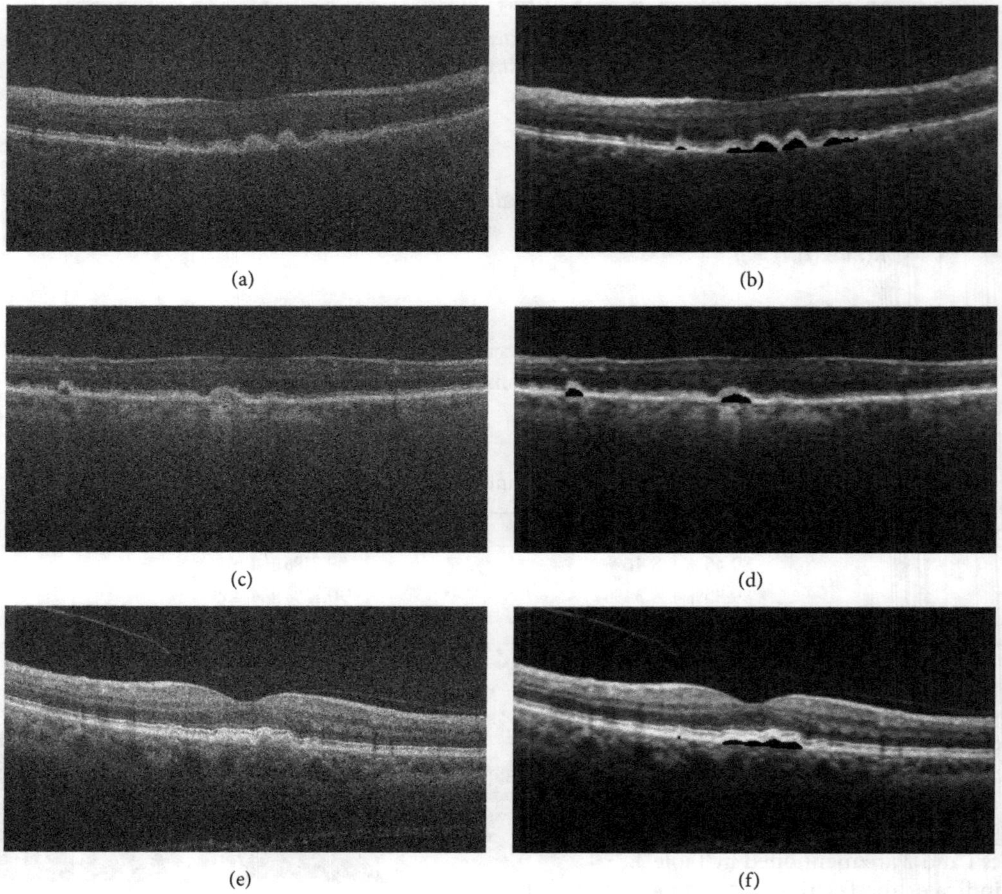

Figure 9: Experimental results for three examples of drusen segmentation. The first column shows the original image and the 2nd column shows the final fine segmentation results.

show that there are good agreement of drusen diameter and mean drusen area on SD-OCT with those identified on color fundus photographs. In this base, drusen detection using 3D SD-OCT may be a potentially useful alternative method to drusen assessment by human graders using color fundus photographs.

In order to prevent inaccurate detection of retinal layers by graph-based intraretinal layer segmentation methods, we preprocessed OCT images and removed highly reflective posterior hyaloid (or other reflective layers above RPE), and then we applied 3D curvelet transform and modified curvelet coefficients to enhance and fill the possible appeared vertically black shadowing effects due to blood vessels. These possible gaps due to large vessels may degrade graph-based segmentation of IS/OS and BM layers.

The shifted upper detected IS/OS pixels that are above BM are considered as a possible candidate region and dome shaped regions are considered as a drusenoid regions. The proposed method for drusen segmentation is not able to identify small size drusen (approximately less than 8 pixels). The information about the drusenoid area can be better assessed by creating OCT fundus images as an enface projection image from the SD-OCT dataset and registering these images to corresponding color fundus photographs and mapping the delineated drusen area on each color fundus photograph.

Competing Interests

The authors declare that they have no competing interests.

References

[1] R. Klein, B. E. K. Klein, S. C. Jensen, and S. M. Meuer, "The five-year incidence and progression of age-related maculopathy: the beaver dam eye study," *Ophthalmology*, vol. 104, no. 1, pp. 7–21, 1997.

[2] J. Z. Nowak, "Age-related macular degeneration (AMD): pathogenesis and therapy," *Pharmacological Reports*, vol. 58, no. 3, pp. 353–363, 2006.

[3] Y. Kanagasingam, A. Bhuiyan, M. D. Abràmoff, R. T. Smith, L. Goldschmidt, and T. Y. Wong, "Progress on retinal image analysis for age related macular degeneration," *Progress in Retinal and Eye Research*, vol. 38, pp. 20–42, 2014.

[4] I. Barbazetto and U. Schmidt-Erfurth, "Photodynamic therapy: a novel approach to the treatment of AMD," in *Age-Related Macular Degeneration*, pp. 143–158, Springer, Berlin, Germany, 2004.

[5] N. M. Bressler, S. B. Bressler, and S. L. Fine, "Age-related macular degeneration," *Survey of Ophthalmology*, vol. 32, no. 6, pp. 375–413, 1988.

[6] Age-Related Eye Disease Study Research Group, "A randomized, placebo-controlled, clinical trial of high-dose supplementation with vitamins C and E and beta carotene for age-related cataract and vision loss," *Archives of Ophthalmology*, vol. 119, no. 10, p. 1439, 2001.

[7] K. I. Hartmann, M. L. Gomez, D.-U. G. Bartsch, A. K. Schuster, and W. R. Freeman, "Effect of change in drusen evolution on photoreceptor inner segment/outer segment junction," *Retina*, vol. 32, no. 8, pp. 1492–1499, 2012.

[8] Age-Related Eye Disease Study Research Group, "The Age-Related Eye Disease Study severity scale for age-related macular degeneration: AREDS report no. 17," *Archives of Ophthalmology*, vol. 123, no. 11, pp. 1484–1498, 2005.

[9] M. Niemeijer, B. Van Ginneken, S. R. Russell, M. S. A. Suttorp-Schulten, and M. D. Abràmoff, "Automated detection and differentiation of drusen, exudates, and cotton-wool spots in digital color fundus photographs for diabetic retinopathy diagnosis," *Investigative Ophthalmology & Visual Science*, vol. 48, no. 5, pp. 2260–2267, 2007.

[10] A. Güven, "Automatic detection of age-related macular degeneration pathologies in retinal fundus images," *Computer Methods in Biomechanics and Biomedical Engineering*, vol. 16, no. 4, pp. 425–434, 2013.

[11] Z. Liang, D. W. K. Wong, J. Liu, K. L. Chan, and T. Y. Wong, "Towards automatic detection of age-related macular degeneration in retinal fundus images," in *Proceedings of the 32nd Annual International Conference of the IEEE Engineering in Medicine and Biology Society (EMBC '10)*, pp. 4100–4103, Buenos Aires, Argentina, September 2010.

[12] M. Esmaeili, H. Rabbani, A. M. Dehnavi, and A. Dehghani, "Automatic detection of exudates and optic disk in retinal images using curvelet transform," *IET Image Processing*, vol. 6, no. 7, pp. 1005–1013, 2012.

[13] Z. Yehoshua, G. Gregori, S. R. Sadda et al., "Comparison of drusen area detected by spectral domain optical coherence tomography and color fundus imaging," *Investigative Ophthalmology & Visual Science*, vol. 54, no. 4, pp. 2429–2434, 2013.

[14] T. E. Clemons, L.-Y. Lee, E. Y. Chew, R. C. Milton, S. B. Bressler, and R. Klein, "A simplified severity scale for age-related macular degeneration," *Archives of Ophthalmology*, vol. 123, no. 11, pp. 1570–1574, 2005.

[15] R. Klein, B. E. K. Klein, M. D. Knudtson, S. M. Meuer, M. Swift, and R. E. Gangnon, "Fifteen-year cumulative incidence of age-related macular degeneration: the Beaver Dam Eye Study," *Ophthalmology*, vol. 114, no. 2, pp. 253–262, 2007.

[16] R. F. Spaide and C. A. Curcio, "Drusen characterization with multimodal imaging," *Retina*, vol. 30, no. 9, pp. 1441–1454, 2010.

[17] P. A. Keane, P. J. Patel, S. Liakopoulos, F. M. Heussen, S. R. Sadda, and A. Tufail, "Evaluation of age-related macular degeneration with optical coherence tomography," *Survey of Ophthalmology*, vol. 57, no. 5, pp. 389–414, 2012.

[18] Q. Chen, T. Leng, L. Zheng et al., "Automated drusen segmentation and quantification in SD-OCT images," *Medical Image Analysis*, vol. 17, no. 8, pp. 1058–1072, 2013.

[19] G. Gregori, F. Wang, P. J. Rosenfeld et al., "Spectral domain optical coherence tomography imaging of drusen in nonexudative age-related macular degeneration," *Ophthalmology*, vol. 118, no. 7, pp. 1373–1379, 2011.

[20] L. Ying, L. Demanet, and E. Candes, "3D discrete curvelet transform," *Applied and Computational Mathematics*, vol. 50, 2005.

[21] E. Candès, L. Demanet, D. Donoho, and L. Ying, "Fast discrete curvelet transforms," *Multiscale Modeling & Simulation*, vol. 5, no. 3, pp. 861–899, 2006.

[22] H. Rabbani, M. Sonka, and M. D. Abramoff, "Optical coherence tomography noise reduction using anisotropic local bivariate gaussian mixture prior in 3D complex wavelet domain," *International Journal of Biomedical Imaging*, vol. 2013, Article ID 417491, 23 pages, 2013.

[23] J. Ma and G. Plonka, "A review of curvelets and recent applications," *IEEE Signal Processing gazine*, vol. 27, no. 2, pp. 118–133, 2010.

[24] A. Woiselle, J.-L. Starck, and J. Fadili, "3-D data denoising and inpainting with the low-redundancy fast curvelet transform," *Journal of Mathematical Imaging and Vision*, vol. 39, no. 2, pp. 121–139, 2011.

[25] A. Woiselle, J.-L. Starck, and J. Fadili, "3D curvelet transforms and astronomical data restoration," *Applied and Computational Harmonic Analysis*, vol. 28, no. 2, pp. 171–188, 2010.

[26] J. Yonemoto, Y. Noda, N. Masuhara, and S. Ohno, "Age of onset of posterior vitreous detachment," *Current Opinion in Ophthalmology*, vol. 7, no. 3, pp. 73–76, 1996.

[27] S. J. Chiu, X. T. Li, P. Nicholas, C. A. Toth, J. A. Izatt, and S. Farsiu, "Automatic segmentation of seven retinal layers in SDOCT images congruent with expert manual segmentation," *Optics Express*, vol. 18, no. 18, pp. 19413–19428, 2010.

[28] M. Salarian, R. Ansari, J. Wanek, and M. Shahidi, "Acuurate segmentation of retina nerve fiber layer in OCT images," in *Proceedings of the IEEE International Conference on Electro/Information Technology (EIT '15)*, pp. 653–656, IEEE, Dekalb, Ill, USA, May 2015.

[29] W.-T. Chen, C.-H. Wen, and C.-W. Yang, "A fast two-dimensional entropic thresholding algorithm," *Pattern Recognition*, vol. 27, no. 7, pp. 885–893, 1994.

[30] J.-L. Starck, F. Murtagh, E. J. Candes, and D. L. Donoho, "Gray and color image contrast enhancement by the curvelet transform," *IEEE Transactions on Image Processing*, vol. 12, no. 6, pp. 706–717, 2003.

[31] S. J. Chiu, J. A. Izatt, R. V. O'Connell, K. P. Winter, C. A. Toth, and S. Farsiu, "Validated automatic segmentation of AMD pathology including drusen and geographic atrophy in SD-OCT images," *Investigative Ophthalmology & Visual Science*, vol. 53, no. 1, pp. 53–61, 2012.

[32] J. K. Udupa, V. R. LaBlanc, H. Schmidt et al., "Methodology for evaluating image-segmentation algorithms," in *Medical Imaging 2002*, pp. 266–277, 2002.

[33] N. Lee, A. F. Laine, and T. R. Smith, "Learning non-homogenous textures and the unlearning problem with application to drusen detection in retinal images," in *Proceedings of the 5th IEEE International Symposium on Biomedical Imaging: From Nano to Macro (ISBI '08)*, pp. 1215–1218, Paris, France, May 2008.

[34] C. Köse, U. Şevik, O. Gençalioğlu, C. İkibaş, and T. Kayıkıçıoğlu, "A statistical segmentation method for measuring age-related macular degeneration in retinal fundus images," *Journal of Medical Systems*, vol. 34, no. 1, pp. 1–13, 2010.

[35] L. Brandon and A. Hoover, "Drusen detection in a retinal image using multi-level analysis," in *Medical Image Computing and Computer-Assisted Intervention-MICCAI 2003*, pp. 618–625, Springer, Berlin, Germany, 2003.

[36] D. E. Freund, N. Bressler, and P. Burlina, "Automated detection of drusen in the macula," in *Proceedings of the IEEE International Symposium on Biomedical Imaging: From Nano to Macro (ISBI '09)*, pp. 61–64, Boston, Mass, USA, July 2009.

[37] P. Soliz, M. P. Wilson, S. C. Nemeth, and P. Nguyen, "Computer-aided methods for quantitative assessment of longitudinal changes in retinal images presenting with maculopathy," in *Medical Imaging 2002*, pp. 159–170, 2002.

[38] R. T. Smith, J. K. Chan, T. Nagasaki et al., "Automated detection of macular drusen using geometric background leveling and threshold selection," *Archives of Ophthalmology*, vol. 123, no. 2, pp. 200–206, 2005.

[39] K. Rapantzikos and M. Zervakis, "Nonlinear enhancement and segmentation algorithm for the detection of Age-related Macular Degeneration (AMD) in human eye's retina," in *Proceedings of the IEEE International Conference on Image Processing (ICIP '01)*, pp. 1055–1058, Thessaloniki, Greece, October 2001.

[40] M. H. A. Hijazi, F. Coenen, and Y. Zheng, "Retinal image classification for the screening of age-related macular degeneration," in *Research and Development in Intelligent Systems XXVII: Incorporating Applications and Innovations in Intelligent Systems XVIII Proceedings of AI-2010, The Thirtieth SGAI International Conference on Innovative Techniques and Applications of Artificial Intelligence*, pp. 325–338, Springer, Berlin, Germany, 2011.

[41] G. Quellec, S. R. Russell, and M. D. Abramoff, "Optimal filter framework for automated, instantaneous detection of lesions in retinal images," *IEEE Transactions on Medical Imaging*, vol. 30, no. 2, pp. 523–533, 2011.

[42] S. Lahmiri, C. S. Gargour, and M. Gabrea, "Automated pathologies detection in retina digital images based on complex continuous wavelet transform phase angles," *Healthcare Technology Letters*, vol. 1, no. 4, pp. 104–108, 2014.

[43] B. Remeseiro, N. Barreira, D. Calvo, M. Ortega, and M. G. Penedo, "Automatic drusen detection from digital retinal images: Amd prevention," in *Computer Aided Systems Theory-EUROCAST 2009: 12th International Conference, Las Palmas de Gran Canaria, Spain, February 15–20, 2009, Revised Selected Papers*, vol. 5717 of *Lecture Notes in Computer Science*, pp. 187–194, Springer, Berlin, Germany, 2009.

[44] A. D. Mora, P. M. Vieira, A. Manivannan, and J. M. Fonseca, "Automated drusen detection in retinal images using analytical modelling algorithms," *Biomedical Engineering Online*, vol. 10, article 59, 2011.

[45] M. K. Garvin, M. D. Abràmoff, X. Wu, S. R. Russell, T. L. Burns, and M. Sonka, "Automated 3-D intraretinal layer segmentation of macular spectral-domain optical coherence tomography images," *IEEE Transactions on Medical Imaging*, vol. 28, no. 9, pp. 1436–1447, 2009.

[46] M. Mujat, R. C. Chan, B. Cense et al., "Retinal nerve fiber layer thickness map determined from optical coherence tomography images," *Optics Express*, vol. 13, no. 23, pp. 9480–9491, 2005.

[47] I. Ghorbel, F. Rossant, I. Bloch, S. Tick, and M. Paques, "Automated segmentation of macular layers in OCT images and quantitative evaluation of performances," *Pattern Recognition*, vol. 44, no. 8, pp. 1590–1603, 2011.

[48] S. Farsiu, S. J. Chiu, J. A. Izatt et al., "Fast detection and segmentation of drusen in retinal optical coherence tomography images," in *Proceedings of the Ophthalmic Technologies XVIII, 68440D*, vol. 6844 of *Proceedings of SPIE*, February 2008.

[49] C. Toth, S. Farsiu, S. Chiu, A. Khanifar, and J. Izatt, "Automatic Drusen segmentation and characterization in spectral domain optical coherence tomography (SDOCT) images of AMD eyes," *Investigative Ophthalmology & Visual Science*, vol. 49, p. 5394, 2008.

[50] D. Iwama, M. Hangai, S. Ooto et al., "Automated assessment of drusen using three-dimensional spectral-domain optical coherence tomography," *Investigative Ophthalmology & Visual Science*, vol. 53, no. 3, pp. 1576–1583, 2012.

9

Hand Depth Image Denoising and Superresolution via Noise-Aware Dictionaries

Huayang Li,[1] Dehui Kong,[1] Shaofan Wang,[1] and Baocai Yin[2]

[1] *Beijing Key Laboratory of Multimedia & Intelligent Software Technology, College of Metropolitan Transportation, Beijing University of Technology, Beijing 100124, China*

[2] *Faculty of Electronic Information and Electrical Engineering, Dalian University of Technology, Dalian 116024, China*

Correspondence should be addressed to Shaofan Wang; wangshaofan@bjut.edu.cn

Academic Editor: Jiri Jan

This paper proposes a two-stage method for hand depth image denoising and superresolution, using bilateral filters and learned dictionaries via noise-aware orthogonal matching pursuit (NAOMP) based K-SVD. The bilateral filtering phase recovers singular points and removes artifacts on silhouettes by averaging depth data using neighborhood pixels on which both depth difference and RGB similarity restrictions are imposed. The dictionary learning phase uses NAOMP for training dictionaries which separates faithful depth from noisy data. Compared with traditional OMP, NAOMP adds a residual reduction step which effectively weakens the noise term within the residual during the residual decomposition in terms of atoms. Experimental results demonstrate that the bilateral phase and the NAOMP-based learning dictionaries phase corporately denoise both virtual and real depth images effectively.

1. Introduction

With the development of 3D range imaging devices such as laser scanner, Kinect sensor, and Time-of-Flight (ToF) cameras, depth images are widely used in various research fields including computer vision, computer graphics, virtual reality, and human computer interaction. While laser scanners provide 3D measurements with precise accuracy, Kinect sensor and ToF cameras provide a convenient way to accomplish 3D range imaging in faithful time, which facilitate many applications with high requirement of efficiency and convenience.

However, depth images provided by Kinect sensor or ToF cameras, either in structured light principle or in ToF principle, suffer from lower quality and resolutions because of the deficiency of received light speckles and the noise incurred from ranging environment. Typically, depth images produce holes, missing regions, or unstable boundaries and nonzero-mean Gaussian noise (see Figure 1).

The research work devoted to enhancement of depth images, including superresolution of depth images and denoising depth images, can be roughly divided into three categories: filtering methods, probabilistic methods, and sparse representation methods. In general, filtering methods [1–6] perform depth enhancement by using filters, based on the assumption that faithful depth data and noise be separable in frequency domains; probabilistic methods [7–12] formulate the depth enhancement as the uncertainty problem of depth measurement and use probabilistic graphical models to resolve; sparse representation methods [13–18] model the depth enhancement problem as a sparse optimization by assuming that faithful depth data have an underlying sparse or low-rank structure.

Hand depth image denoising and superresolution are important for hand-based human-machine interaction. Although many research works have been proposed for RGB/depth image denoising and superresolution, conventional approaches do not work well for hand depth images. This is because the resolutions of depth images captured from Kinect sensor are 512×424, where the hand takes only a very small subregion (typically 170×150 in our experiments). Thus traditional approaches usually confuse depth with noise and are unsuitable for such small scale depth data.

This paper proposes hand depth image denoising and superresolution using bilateral filters and NAOMP-based dictionaries. The bilateral filtering phase recovers singular

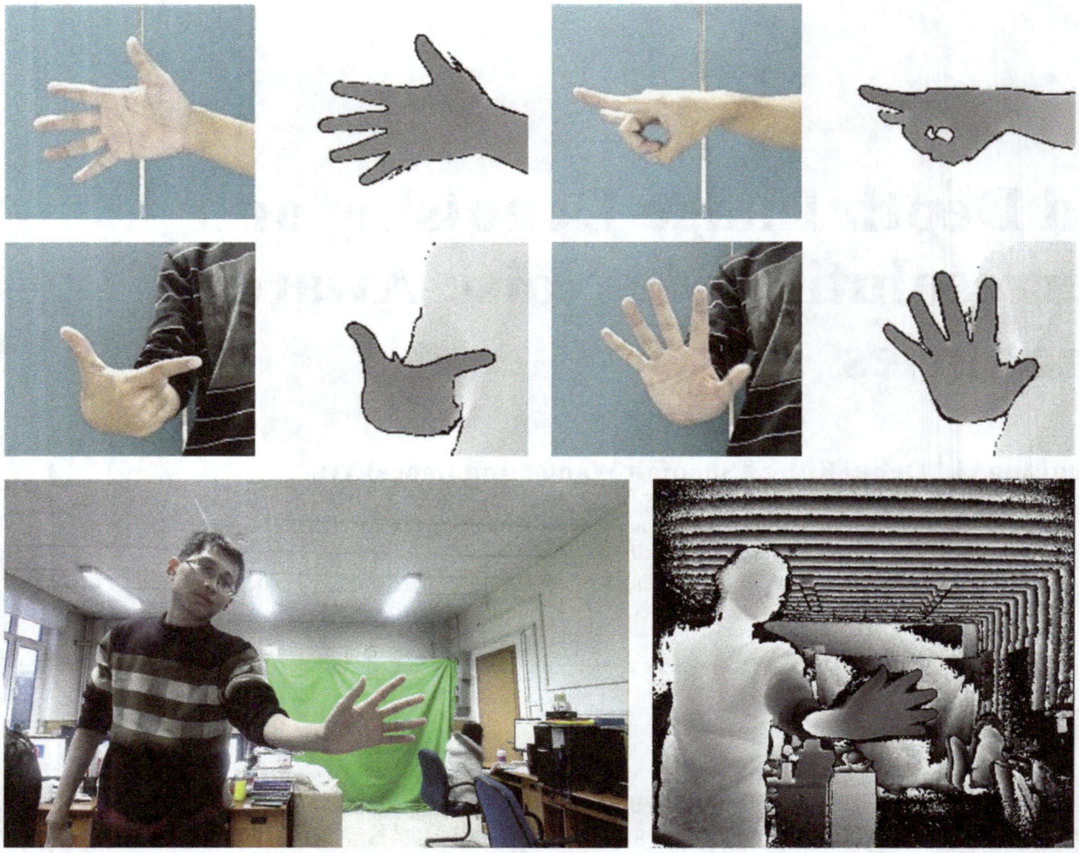

FIGURE 1: Color/depth images of human hands captured from Kinect sensor v2.

points and removes artifacts on silhouettes by averaging depth data with restrictions of both depth difference and RGB similarity. The dictionary learning phase uses the K-SVD method with the NAOMP for training dictionaries to separate faithful depth from noisy data. While traditional orthogonal matching pursuit (OMP) works well for training dictionaries to denoise RGB images, the performance of denoising depth images deteriorates badly as depth images involve nonzero-mean Gaussian noise. As a result, traditional dictionary learning algorithm (e.g., OMP-based K-SVD) cannot prevent noisy data from penetrating dictionaries which results in an unsatisfied denoising effect. To improve traditional OMP-based K-SVD for fitting nonzero-mean Gaussian noise which frequently appears in depth data, we propose NAOMP for replacing traditional OMP in K-SVD, where the noise term within residuals is weakened in each atom updating step. Such a modification gives dictionaries capable of representing faithful depth data in a more precise fashion. Experimental results show that the proposed bilateral filters and NAOMP-based dictionaries corporately give promising results for denoising both virtual and real depth data, compared with traditional bilateral filters and traditional OMP-based dictionaries.

This paper is organized as follows. Section 2 reviews previous work on enhancing depth images. Section 3 proposes the bilateral filtering phase using RGBD data, which first recovers singular points and then removes incorrect points over silhouettes and finally removes nonsingular points over nonsilhouette regions. Sections 4 and 5 propose depth image denoising and superresolution, respectively, both using learning dictionaries via NAOMP-based K-SVD. Section 6 shows experimental results with both virtual and real hand depth images.

2. Previous Work

Previous work on enhancing depth images is reviewed in this section, including superresolution of depth images and denoising depth images in the following three categories.

2.1. Filtering Methods. Yang et al. [1] construct a 3D volume of depth probability, impose a bilateral filter over the volume iteratively, and obtain high-resolution depth images by taking the winner-takes-all approach on the weighted volume and a subpixel refinement afterward. Huhle et al. [2] present a two-stage depth enhancement method, which first removes outliers from depth data and then performs smoothing via a nonlocal means filter which uses the similarity of both color and intrapatch of depth. Wasza et al. [3] propose a GPU-based depth image preprocessing, including a normalized convolution for restoring depth images, a bilateral temporal averaging for dynamic scenes, and a guided filter for edge-preserving denoising. Min et al. [4] propose a joint histogram of depth maps for measuring color similarity between reference and neighboring pixels and find a global mode solution via ℓ_1-norm minimization for depth video enhancement. Fu et al. [5] propose a spatial-temporal denoising algorithm which exploits both the intraframe spatial correlation and the interframe temporal correlation to fill the depth hole and suppress the depth noise. Camplani and Salgado [6] propose a joint-bilateral filtering framework for denoising depth images, by evaluating missing depth values with a filter to neighboring pixels involving both spatial and temporal information, with the filter weights selected as a function of a photometric similarity measure of the neighbor pixels.

2.2. Probabilistic Methods. Mac Aodha et al. [7] explore the height field of patches of low-resolution depth images and select high-resolution candidate depth patches by solving a Markov Random Field (MRF) labeling problem. Shen and Cheung [8] use depth layers to account for the differences between foreground objects and background scene, the missing depth value phenomenon, and the correlation between color and depth channels and consider the depth layer labeling as a maximum a posteriori estimation problem. Wang et al. [9] evaluate the confidence of depth map for adaptive weighting of MRF energy terms and introduce a guided depth recovering method in the framework of MRF optimization for handling large holes across multiple image regions. Li et al. [10] segment the input low-resolution depth image into several regions with different labels which correspond to a high-resolution counterpart on training images and formulate the depth superresolution as an MRF-based patchwork assembly problem. Hui and Ngan [11] propose a variational-based depth map enhancement by fusing the depth maps from the active sensor of a moving RGBD system and the depth cues from an induced optical flow. Yang et al. [12] propose a regression method for enhancing depth images using RGB-D data, which first fits the regression model for depth images and then designs pixel-wise regression predictors using the similarity of depth images and the accompanied color images.

2.3. Sparse Representation Methods. Schuon et al. [13] combine several low-resolution noisy depth images of a static scene from slightly displaced viewpoints and merge them into a high-resolution depth image with ToF calibration data. The depth superresolution problem is formulated as an optimization of a data reconstruction term plus a sparsity term of spatial gradient for separating noise from features. Li et al. [14] propose a novel joint example-based depth map superresolution method which reconstructs high-resolution depth images by learning a mapping function from a set of training samples of an image database. Kiechle et al. [15] propose a joint intensity and depth cosparse model for depth map superresolution, by assuming that the cosupports of corresponding intensity and depth image structures be aligned. Zheng et al. [16] propose constructing multiple dictionaries with different structures and different number of atoms for sparse representing each low-resolution patch of depth images. Xie et al. [17] learn a coupled dictionary with local coordinate constraints and incorporate an adaptively regularized shock filter to sharpen the edges and implement both depth superresolution and depth denoising. Lu et al. [18] assemble similar RGBD patches into a low-rank matrix in order to prevent the noise or weak correlation between color and depth.

3. RGBD-Based Bilateral Filters

This section proposes the bilateral filtering phase, which preprocesses depth images by using bilateral filters with both depth and RGB restrictions. The phase includes the following three steps. The depth of each singular point of depth images is corrected by averaging depths over the neighborhood according to a rule of both RGB comparison and depth histogram difference; the depth of each point over silhouettes is corrected by averaging depths over the neighborhood according to a rule of both RGB comparison and depth difference; the depth of other points is corrected using traditional bilateral filters, that is, by averaging depths over spatial neighborhood.

Let I_q be the intensity values/depth value at the pixel q of an intensity/depth image. Traditional filters at the pixel p with respect to its neighborhood Ω_p are given by

$$J_p = \frac{1}{k_p} \sum_{q \in \Omega_p} I_q f(p, q)\, g\left(\left\|I_p - I_q\right\|_2\right),$$
$$k_p = \sum_{q \in \Omega_p} f(p, q)\, g\left(\left\|I_p - I_q\right\|_2\right), \quad (1)$$

where $\Omega_p = \{q \in \mathbb{N}^2 : |p(1) - q(1)| \leq \epsilon,\ |p(2) - q(2)| \leq \epsilon\}$ denotes an $\epsilon \times \epsilon$ spatial neighborhood of p with $p(1)$, $p(2)$ representing the row index, the column index of p within images, respectively, k_p denotes the normalization term, $f(p, q)$ is the 2D Gaussian smoothing kernel (known as the domain term) which measures the closeness of the pixels, and $g(\|I_p - I_q\|_2)$ is the 1D Gaussian smoothing kernel (known as the range term) which measures the similarity of RGB/depth values of the pixels in RGB/depth images.

3.1. Recovering Singular Points. The singular points of a depth image are referred to as the points whose depth is undetected. For each singular point p of depth images, we first set an initial depth value D_p^{initial} at p as the average of depths which are no smaller than $(1/2)D_{\max}^{\Omega}$ in the neighborhood Ω_p of p (choosing such an initial value because experimental results indicate that the depths over the neighborhood of a singular point belong to two regions generally: an interval determined by the depths of the target hand and an interval determined by the depths of the background. Experimental results show that $(1/2)D_{\max}^{\Omega}$ separates two intervals well and hence gives an initial approximation of depth at p), where $D_{\max}^{\Omega}$ is the maximum depth over Ω_p, and then choose a suitable subset of Ω_p whose histogram of depth is much greater than the histogram of depth at p (choosing such a subregion of the neighborhood of p overcomes directly choosing a spatial

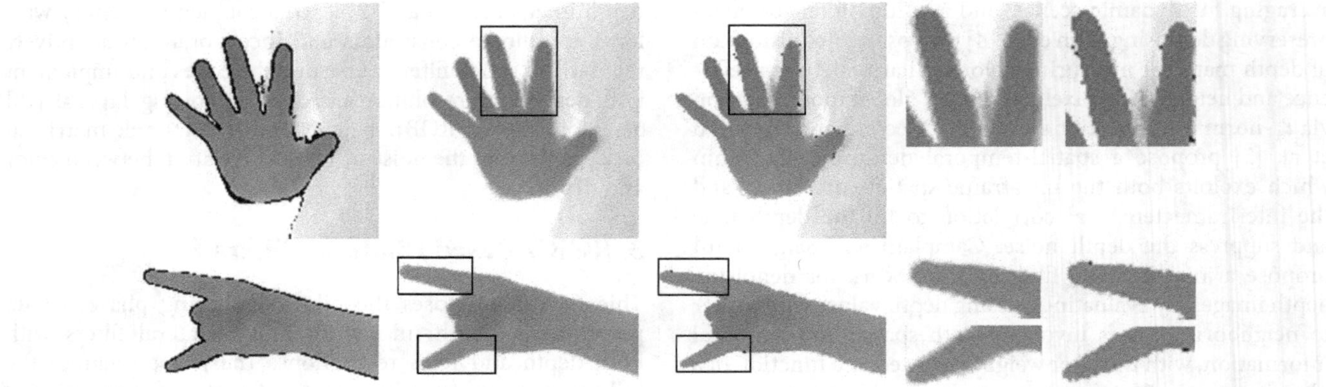

Figure 2: Left to right: noisy depth images, images preprocessed by traditional filters, images preprocessed by the proposed filters (2)–(4), and comparison details. One can check from details that the proposed filters (2)–(4) make artifacts on silhouettes either removed or remained with a small number while traditional filters turn artifacts into a shadow effect. Although those shadow artifacts give smaller depth values, the continuous region where the artifacts locate make them difficult to be removed in the dictionary learning phase.

neighborhood of p, because the RGB comparison improves the confidence of the pixels and because the histogram of the depth avoids producing shadows within the silhouette. We illustrate this improvement in Figure 2). Finally we select the filtered value of the depth at the pixel p by averaging the depth data with respect to both the domain term and the range term provided that the confidence pixels are enough; otherwise the singular points remain as the same value (such untreated singular points are much less than before and are dispersed within the depth image and can hence easily be treated in the second phase). That is,

$$J_p = \frac{1}{k_p} \sum_{q \in \widehat{\Omega}_p} D_q f(p,q)\, g\left(\left\|I_p - I_q\right\|_2\right) \quad \text{if } \left|\widehat{\Omega}_p\right| \geq \epsilon_1, \tag{2}$$

where I_q, D_q denote the RGB, depth values at the pixel q of an image, respectively, k_p denotes the normalization term, and $\widehat{\Omega}_p = \{q \in \Omega_p : \text{hist}_{\Omega_p}(D_q) \geq \text{hist}_{\Omega_p}(D_p^{\text{initial}}) + \epsilon_2, \|I_p - I_q\|_2 \leq \epsilon_3\}$ denotes the confidence subset of a square neighborhood Ω_p of p with both a depth difference restriction and an intensity similarity restriction, with $\text{hist}_{\Omega}(d) := |\{(i,j) \in \Omega : D_{i,j} = d\}|$ denoting the number of pixels within Ω whose depth value is equal to d.

3.2. Removing Incorrect Points on Silhouettes. For all nonsingular points p located the silhouettes of depth images, we modify its depth value at the pixel p by first choosing a suitable subset of Ω_p whose depth is much greater than the depth of p and whose RGB values are similar to p's and then averaging the depth data within the background domain (corresponding to the restriction of depth difference) with similar RGB values at p, provided that such suitable pixels are enough. That is,

$$J_p = \frac{1}{k_p} \sum_{q \in \widehat{\Omega}_p} D_q f(p,q)\, g\left(\left\|I_p - I_q\right\|_2\right) \quad \text{if } \left|\widehat{\Omega}_p\right| \geq \epsilon_4, \tag{3}$$

where I_q, D_q, k_p, $f(\cdot,\cdot)$, and $g(\cdot)$ have the same meanings as in (2) and $\widehat{\Omega}_p = \{q \in \Omega_p : D_q \geq D_p + \epsilon_5, \|I_p - I_q\|_2 \leq \epsilon_6\}$ denotes the confidence subset of a square neighborhood Ω_p of p with both a depth difference restriction and an intensity similarity restriction.

3.3. Removing Other Incorrect Points. For all other nonsingular points p of depth images, we treat them with a smoothing filter based on the similarity of both depth value and RGB values. That is,

$$J_p = \frac{1}{k_p} \sum_{q \in \Omega_p} D_q f(p,q)\, g\left(\left|D_p - D_q\right|\right) g\left(\left\|I_p - I_q\right\|_2\right), \tag{4}$$

where I_q, D_q, k_p, $f(\cdot,\cdot)$, and $g(\cdot)$ have the same meanings as in (2) and $\Omega_p = \{q \in \mathbb{N}^2 : |p(1) - q(1)| \leq \epsilon_7, |p(2) - q(2)| \leq \epsilon_7\}$ denotes an $\epsilon_7 \times \epsilon_7$ square neighborhood of p (different from subtle filtering (2) and (3) where both of the neighborhoods are determined by a depth difference restriction and an intensity similarity restriction, the filtering (4) simply selects a square neighborhood without additional restrictions. This is because the filtering (2) and (3) treats singular points and incorrect points on silhouettes which require careful treatment, whereas the filtering (4) performs a smoothing effect over the whole square neighborhood of the point which needs smoothing.).

Figures 2 and 3 show some comparison results of depth images using traditional filters [6] and the proposed filters (2)–(4). In a word, the proposed filters recover singular points using neighboring points with a quantity restriction (2) and correct artifacts using neighboring points with an intensity restriction (3), while traditional bilateral filters accomplish such tasks by directly averaging points within spatial neighborhood. Therefore, traditional filters always produce new artifacts while the proposed filters either correct the artifacts over silhouettes or suppress them in a small number without creating new ones, so that they can be easily treated in the next dictionary learning phase.

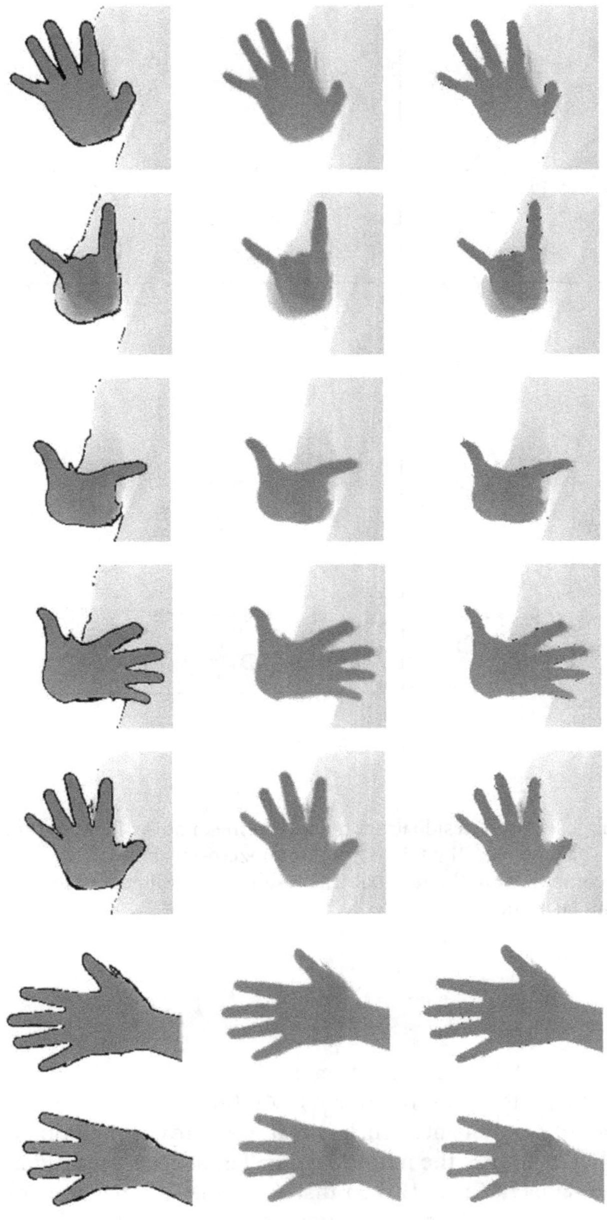

FIGURE 3: Bilateral filters on hand depth images. Left: noisy depth images. Middle: depth images with bilateral filters (1). Right: depth images with bilateral filters (2)–(4).

4. Hand Depth Denoising Using Noise-Aware Dictionaries

This section proposes the dictionary-based denoising phase, followed by the bilateral filtering phase given in Section 3. Recovering the original image from a degraded image with additive white Gaussian noise can be modelled by solving the ill-posed system $\mathbf{Y} = \mathbf{X} + \mathbf{v}$, where $\mathbf{Y}$ denotes the degraded image, $\mathbf{X}$ denotes the original image, and $\mathbf{v}$ denotes the noise term. Sparse representation is an important tool for solving such an ill-posed system. According to sparse representation, natural images can be represented by a linear combination of a series of overcomplete basis (known as a *dictionary*) with very few nonzero combinational coefficients. Therefore, by denoting $\mathbf{D} \in \mathbb{R}^{n \times K}$ to be such a dictionary whose columns are basis vectors (known as *atoms*) with $K \gg n$, the original image can be obtained by imposing an ℓ_0-norm constraint of the coefficients α over the system $\mathbf{Y} = \mathbf{D}\alpha + \mathbf{v}$. In particular, we select image patches of size $\sqrt{n} \times \sqrt{n}$ pixels randomly from $\mathbf{Y}$ as training data and order each patch lexicographically as column vectors $\mathbf{Y}_{ij} \in \mathbb{R}^n$. Then we obtain a trained dictionary $\mathbf{D}$ and sparse coefficients α_{ij} simultaneously via the following ℓ_0-minimization of all coefficients of training data:

$$\begin{aligned} \min_{\mathbf{D},\{\alpha_{ij}\}} \quad & \sum_{i,j} \left\|\alpha_{ij}\right\|_0 \\ \text{s.t.} \quad & \mathbf{D}\alpha_{ij} \approx \mathbf{Y}_{ij} \\ & \forall (i,j) \in \Omega_{\text{train}}, \end{aligned} \tag{5}$$

where Ω_{train} denotes the training set of indices of patches of $\mathbf{Y}$ we randomly select. The original image $\mathbf{X}$ is finally reconstructed by assembling all image patches $\mathbf{X}_{ij} = \mathbf{D}\alpha_{ij}$ with α_{ij} given by

$$\begin{aligned} \min_{\alpha_{ij} \in \mathbb{R}^K} \quad & \left\|\alpha_{ij}\right\|_0 \\ \text{s.t.} \quad & \mathbf{D}\alpha_{ij} \approx \mathbf{Y}_{ij} \\ & \forall (i,j). \end{aligned} \tag{6}$$

Because (5) and (6) are both nonconvex systems involving the ℓ_0-norm minimization, developing efficient algorithms for solving them is an important task. Elad and Aharon [19] propose the K-SVD algorithm to solve (5), by constructing a dictionary iteratively from training signals with a sparse coding phase and an atom update phase. Within the sparse coding phase of K-SVD and the optimization problem (6), OMP is a greedy algorithm frequently used for solving the ℓ_0-norm minimization.

While traditional K-SVD denoises traditional images well for zero-mean Gaussian additive noise, it is not well suited for denoising depth images with nonzero-mean noise because traditional OMP can hardly separate noise data from noiseless image when the amplitude of noise has an irregular distribution. To remove such noise from depth images in a more effective fashion, traditional OMP is improved by modifying the amplitude of entries of residual whenever such entries have large amplitude and a small quantity. Figure 4 helps readers understand how this idea works. The y-axis represents the value of each component of residuals while the x-axis represents the index of each component of residuals. Let r_{s-1} be the residual obtained in an iteration of OMP. The left subfigure of Figure 4 shows how traditional OMP represents residuals when the noise is zero-mean Gaussian. While r_{s-1} contains four components heavily contaminated by noise (denoted by blue cubes, i.e., the original training data $\mathbf{Y}$ has noise terms in the positions of those four components), the least square fitting, obtained when computing the sparse coefficients, approximates noiseless components

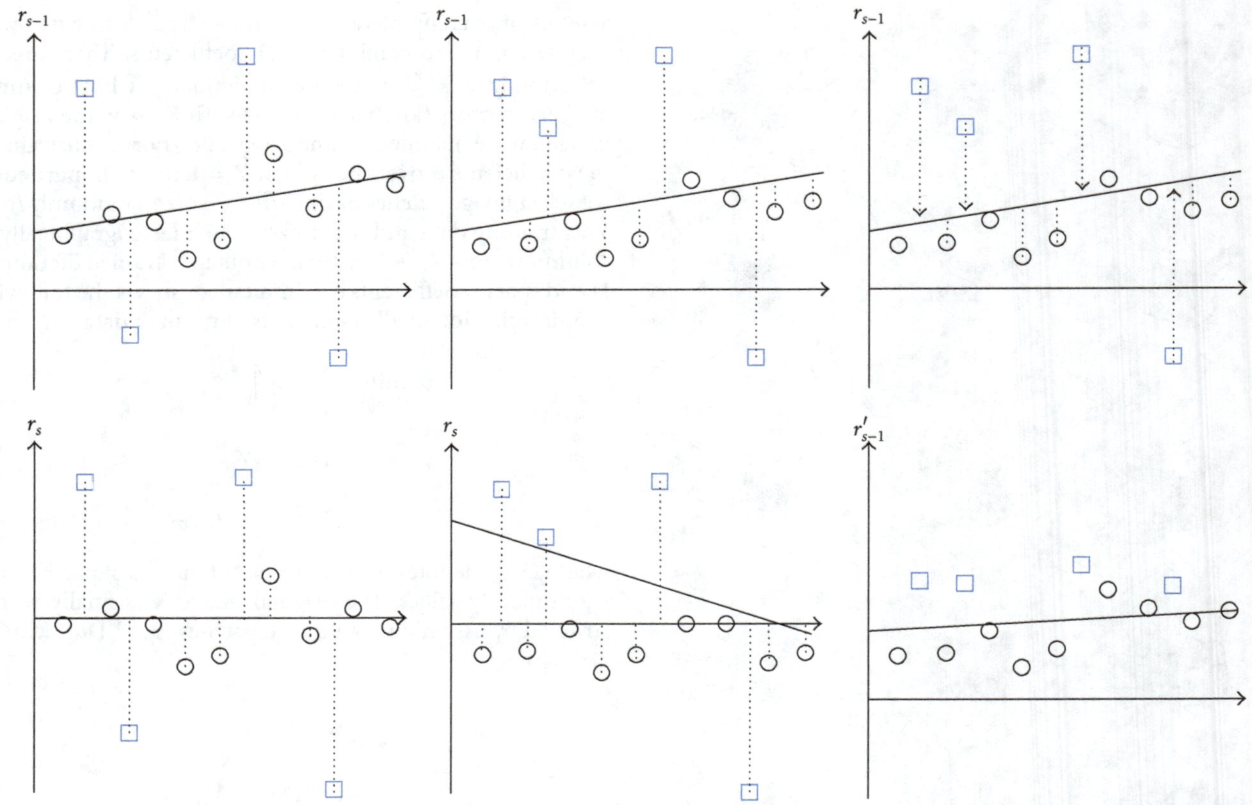

FIGURE 4: The difference between how OMP and NAOMP work on decompositions of residuals with nonzero-mean noise. Left: traditional OMP for zero-mean Gaussian noise. Middle: traditional OMP for nonzero-mean noise. Right: NAOMP for nonzero-mean noise. The x-axis: the index of each component of the residual; the y-axis: the value of each component of the residual. The residual in this figure is a vector of length sixteen. From top to bottom: a procedure for updating a residual and an atom.

well and hence makes most of noiseless components vanish at the next iteration before the decomposition of the noise terms with respect to current atoms, because the stopping criteria of the number of sparse coefficients are satisfied. The middle subfigure of Figure 4 shows how traditional OMP fails to remove noise from depth images when the noise is nonzero-mean. In this case, the least square fitting deviates from most noiseless terms and hence the noisy terms begin decomposition in the next iteration, because neither the criteria of the number of sparse coefficients nor the criteria of the residual amplitude are satisfied. From this step, the new atoms obtained shall be contaminated by noise. The right subfigure of Figure 4 shows how NAOMP improves this issue. When the current residual r_{s-1} contains greater-magnitude entries and those entries have a relatively small number, it is reasonably believed that those entries correspond to the noisy components of training data. In this case, each of those components of r_{s-1} is modified by reevaluating it as the amplitude of corresponding component within the fitting line. Then the updated residual is redecomposed over atoms. By doing so, the least square fitting approximates noiseless terms of residuals and weakens the effect of noisy terms. We illustrate the idea using numerical examples in Appendix.

The NAOMP algorithm is illustrated in Algorithm 3. After performing traditional OMP (lines (4)–(9)), the number of components which is relatively greater than others is checked. Once the number is smaller, the current residual is reevaluated (line (12)) so that the values of greater components approach those of other components. Then the atom and the residual are reupdated as in OMP (lines (13)–(17)). Detailed parameter setting shall be given in Section 6.

5. Hand Depth Image Superresolution Using Noise-Aware Dictionaries

We apply NAOMP to hand depth image superresolution, where we use NAOMP for joint dictionary training. The main idea is similar to the work of [20]; hence we only give the different part of our work (which is the dictionary training algorithm) without giving details. We randomly select patches of virtual hand depth images as a training set, each of which is stacked as a vector, denoted by X_i, $i \in \Omega_{\text{train}}$. Then we obtain its corresponding downsampled version Y_i and form pairwise training sets $\{(X_i, Y_i) : i \in \Omega_{\text{train}}\}$. We denote $\mathbf{D}_h$ and $\mathbf{D}_l$ to be the joint dictionary, which is the sparse representations for high-resolution and low-resolution image patches, respectively. The joint dictionary is given by the following optimization:

$$\min_{\mathbf{D}_h,\mathbf{D}_l,\{z_i\}} \sum_{i\in\Omega_{\text{train}}} \left(\frac{1}{N} \|X_i - \mathbf{D}_h z_i\|_2^2 + \frac{1}{M} \|Y_i - \mathbf{D}_l z_i\|_2^2 + \lambda \left(\frac{1}{N} + \frac{1}{M} \right) \|z_i\|_0 \right). \quad (7)$$

By rewriting the first and the second terms in a single term of 2-norm, we obtain the optimization model similar to the last section and solve it using the NAOMP-based K-SVD algorithm. To recover the high-resolution patch of an input image $\mathbf{Y}$, we find a sparse representation of each patch of $\mathbf{Y}$ with respect to $\mathbf{D}_l$ and then obtain the corresponding high-resolution patch by combining the high-resolution atoms $\mathbf{D}_h$ with the same sparse coefficients with respect to $\mathbf{D}_l$.

6. Experimental Results

The experimental results are given in this section. The experiments are run on Core™ 2 Quad Q6600 2.4 GHz machine with 2 GB RAM using Visual Studio 2010. The proposed filters are given by (2)–(4), and the NAOMP is given by Algorithm 3. All the parameters in traditional filters [6], the proposed filters (2)–(4), OMP-based K-SVD [19] (Algorithms 1 and 2), and NAOMP-based K-SVD (Algorithms 1 and 3) are given in Table 1, respectively. One can see from Table 1 that traditional OMP-based K-SVD has to select different values for the residual threshold ε according to the intensity of noise, while NAOMP-based K-SVD selects a single value. In fact, according to [19], the denoising effect heavily depends on the choice of the residual threshold. Such a threshold is difficult to determine for OMP-based K-SVD when no a priori information of noise is given, while this threshold is fixed and easily determined in NAOMP-based K-SVD.

6.1. Denoising Virtual Hand Depth Images. Artificial Gaussian noise is added on three virtual hand models and the images are denoised using OMP-based K-SVD and NAOMP-based K-SVD, both without filter preprocessing. Qualitative results are given in Figure 5 (with 0.5%, 2%, and 5% of noise) and quantitative results are given in Table 2. We see that traditional OMP fails to recover the wrist part of models while NAOMP treats them well. Moreover, within all cases, the proposed method provides higher PSNR than OMP-based K-SVD except for the first example with 0.5% noise.

6.2. Denoising Hand Depth Images Obtained from Kinect v2. We show qualitative results of denoising and superresolution of six hand depth images obtained from Kinect sensor v2 in Figure 6 and give the average running time of different approaches in Table 3. The comparison results include five approaches: traditional bilateral filters [6], OMP-based K-SVD [19], NAOMP-based K-SVD, the bilateral filters (2)–(4) plus OMP-based K-SVD, and the bilateral filters (2)–(4) plus NAOMP-based K-SVD.

In general, both bilateral filters preprocess depth images well in that singular points are removed (the different effect of two filters is given in Section 3 using Figures 2 and 3). Moreover, one can see from the fifth column and the sixth column that NAOMP-based K-SVD produces clearer silhouettes than OMP-based K-SVD, as the trained dictionaries from NAOMP remove noisy terms well.

6.3. Discussions. The two-stage depth image denoising and superresolution enhance hand depth images well mainly because of the following two reasons. For one thing, the proposed bilateral filter functions choose suitable neighborhood pixels with exquisite depth and RGB restrictions than traditional filters. For another, the NAOMP modifies the noisy terms of the residual in each atom updating step so that they are reevaluated with closed values to noiseless terms. This enables the residual to decompose in a more noiseless fashion and results in dictionaries which are less contaminated by noise.

It should be noted that the proposed method may fail in denoising and superresolution of depth images of objects other than human hands. This is because while the skin of human hands shares RGB data in a small range, other objects do not show such an advantage, which make the proposed bilateral filters fail to select suitable neighborhood for filter functions.

In future work, the proposed image denoising and superresolution framework shall be developed for enhancing depth images of more complex objects or scenes. More accurate RGB/depth restrictions can be designed for filter functions to preprocess depth images, so that the singular points and artifacts remain in a more dispersing fashion. Furthermore, the modification of residuals within NAOMP can be improved so that the updated atom can represent the noiseless data more precisely.

Appendix

A. OMP versus NAOMP: The First Example

Let

$$\begin{aligned} \mathbf{Y} &= \mathbf{X} + \mathbf{v} = (k+1, k, k)^{\top}, \\ \mathbf{X} &= (k, k, k)^{\top} \end{aligned} \quad (A.1)$$

be an input signal, a faithful signal, respectively, where $\mathbf{v} = (1, 0, 0)^{\top}$ denotes the noise term, $k \gg 1$, and let the atom number be one. Without loss of generality we assume that both traditional K-SVD and NAOMP-based K-SVD return the same DCT dictionary (the expression of the trained dictionary depends on the training set and residual

input: training data $\mathbf{Y}_{ij} \in \mathbb{R}^n, (i, j) \in \Omega_{\text{train}}$, residual threshold ε, and iteration number J
output: a dictionary $\mathbf{D} \in \mathbb{R}^{n\times K}$
(1) Initialize $\mathbf{D}^{(1)}$ as overcomplete DCT dictionary;
(2) **for** $m = 1, \ldots, J$ **do**
(3) **foreach** $(i, j) \in \Omega_{\text{train}}$ **do** /* *sparse coding phase* */
(4) Using the OMP (Algorithm 2 or Algorithm 3) to solve $\min_{\alpha_{ij}} \|\alpha_{ij}\|_0$ s.t. $\|\mathbf{Y}_{ij} - \mathbf{D}^{(m)}\alpha_{ij}\|_2 \le \varepsilon$;
(5) **end**
(6) **for** $k = 1, \ldots, K$ **do** /* *atom update phase* */
(7) Find the set of patches which use the kth atom $\omega_k \leftarrow \{(i, j) \in \Omega_{\text{train}} : \alpha_{ij}(k) \ne 0\}$;
(8) **foreach** $(i, j) \in \omega_k$ **do**
(9) $\mathbf{e}_{ij}^k \leftarrow \mathbf{Y}_{ij} - \sum_{l\ne k} \mathbf{d}_l \alpha_{ij}(l)$; /* $\mathbf{d}_l$ *denotes the* l*th column of* $\mathbf{D}^{(m)}$ */
(10) **end**
(11) Set $\mathbf{E}_k \in \mathbb{R}^{n\times|\omega_k|}$ as the matrix whose columns are $\{\mathbf{e}_{ij}^k\}_{(i,j)\in\omega_k}$;
(12) Apply SVD decomposition $\mathbf{E}_k = \mathbf{U}\Delta\mathbf{V}^\top$;
(13) Update the kth column of dictionary $\mathbf{D}^{(m)}$ by $\mathbf{d}_k \leftarrow$ the first column of $\mathbf{U}$;
(14) Update the coefficient values $\{\alpha_{ij}(k)\}_{(i,j)\in\omega_k}$ to be the first column of $\mathbf{V}$ multiplied by $\Delta(1, 1)$;
(15) **end**
(16) **end**

Algorithm 1: The K-SVD algorithm [19].

input: vector $\mathbf{Y} \in \mathbb{R}^n$, dictionary $\mathbf{D} \in \mathbb{R}^{n\times K}$, sparsity S
output: sparse coefficient $\alpha \in \mathbb{R}^K$ of $\mathbf{Y}$ with respect to $\mathbf{D}$
(1) Initialize residual $r_0 \leftarrow \mathbf{Y}$, index set $\Lambda_0 \leftarrow \varnothing$, $\mathbf{D}^{(0)} \leftarrow \varnothing$;
(2) **for** $s = 1, \ldots, S$ **do**
(3) Denote $\mathbf{d}_j, j = 1, \ldots, K$ to be all columns of $\mathbf{D}$;
(4) $\lambda_s \leftarrow \arg\max_{j=1,\ldots,K} |\langle r_{s-1}, \mathbf{d}_j\rangle|$;
(5) Update the index set $\Lambda_s \leftarrow \Lambda_{s-1} \cup \{\lambda_s\}$;
(6) Update the atom matrix $\mathbf{D}^{(s)} \leftarrow [\mathbf{D}^{(s-1)}, \mathbf{d}_{\lambda_s}]$;
(7) $\alpha \leftarrow \arg\min_x \|\mathbf{Y} - \mathbf{D}^{(s)}x\|_2$;
(8) Update residual $r_s \leftarrow \mathbf{Y} - \mathbf{D}^{(s)}\alpha$;
(9) **end**

Algorithm 2: The orthogonal matching pursuit.

threshold we choose. Here we make this assumption so that the dictionary can exactly represent the faithful signal):

$$\mathbf{D} = \begin{pmatrix} 1 \\ 1 \\ 1 \end{pmatrix}. \quad \text{(A.2)}$$

After the dictionary is obtained, we recover the signal using OMP and NAOMP, respectively. OMP obtains sparse coefficients and the reconstruction signal as follows:

$$\alpha = \arg\min_{x\in\mathbb{R}} \|r_0 - \mathbf{D}x\|_2 = \arg\min_{x\in\mathbb{R}} \|\mathbf{Y} - \mathbf{D}x\|_2 = k + \frac{1}{3}, \quad \text{(A.3)}$$

$$\mathbf{X}_{\text{OMP}} = \mathbf{D}\alpha = \left(k + \frac{1}{3}, k + \frac{1}{3}, k + \frac{1}{3}\right)^\top.$$

Nevertheless, NAOMP obtains the same sparse coefficients $\alpha = k + 1/3$ and finds that the first component of the residual $r_1 = (2/3, -1/3, -1/3)^\top$ is relatively greater than the second and the third ones. As a result, the residual is modified as $r_0'(1) \leftarrow r_0(1) - r_1(1)$ (noticing that $r_0 = \mathbf{Y}$ holds in the first step). Thus the sparse coefficients are recomputed by

$$\alpha = \arg\min_{x\in\mathbb{R}} \|r_0' - \mathbf{D}x\|_2 = k + \frac{1}{9}, \quad \text{(A.4)}$$

where $r_0' = (k + 1/3, k, k)^\top$ is the modified residual term, and the reconstruction signal is given by

$$\mathbf{X}_{\text{NAOMP}} = \mathbf{D}\alpha = \left(k + \frac{1}{9}, k + \frac{1}{9}, k + \frac{1}{9}\right)^\top \quad \text{(A.5)}$$

which approximates the faithful signal compared with $\mathbf{X}_{\text{omp}}$.

input: vector $\mathbf{Y} \in \mathbb{R}^n$, dictionary $\mathbf{D} \in \mathbb{R}^{n\times K}$, sparsity S, residual threshold $\widehat{r}$, noise index threshold $\widehat{L}$
output: sparse coefficient $\alpha \in \mathbb{R}^K$ of $\mathbf{Y}$ with respect to $\mathbf{D}$
Initialize residual $r_0 \leftarrow \mathbf{Y}$, index set $\Lambda_0 \leftarrow \varnothing$, $\mathbf{D}^{(0)} \leftarrow \varnothing$;
for $s = 1, \ldots, S$ **do**
  Denote $\mathbf{d}_j$, $j = 1, \ldots, K$ to be all columns of $\mathbf{D}$;
  $\lambda_s \leftarrow \arg\max_{j=1,\ldots,K} |\langle r_{s-1}, \mathbf{d}_j\rangle|$;
  Update the index set $\Lambda_s \leftarrow \Lambda_{s-1} \cup \{\lambda_s\}$;
  Update the atom matrix $\mathbf{D}^{(s)} \leftarrow [\mathbf{D}^{(s-1)}, \mathbf{d}_{\lambda_s}]$;
  $\alpha \leftarrow \arg\min_x \|\mathbf{Y} - \mathbf{D}^{(s)}x\|_2$;
  Update residual $r_s \leftarrow \mathbf{Y} - \mathbf{D}^{(s)}\alpha$;
  $L \leftarrow \{1 \le l \le n : |r_s(l)| \ge \widehat{r}\}$; /* *collect all indices of great components of residual* r_s */
  if $|L| \le \widehat{L}$ **then** /* *when* r_s *has only a small number of great components, we re-update the atom by removing those components from* r_{s-1} */
    foreach $l \in L$ **do** $r_{s-1}(l) \leftarrow r_{s-1}(l) - r_s(l)$;
    $\lambda'_s \leftarrow \arg\max_{j=1,\ldots,K} |\langle r_{s-1}, \mathbf{d}_j\rangle|$;
    Re-update the index set $\Lambda_s \leftarrow \Lambda_{s-1} \cup \{\lambda'_s\}$;
    Re-update the atom matrix $\mathbf{D}^{(s)} \leftarrow [\mathbf{D}^{(s-1)}, \mathbf{d}_{\lambda'_s}]$;
    $\alpha \leftarrow \arg\min_x \|\mathbf{Y} - \mathbf{D}^{(s)}x\|_2$;
    Re-update residual $r_s \leftarrow \mathbf{Y} - \mathbf{D}^{(s)}\alpha$;
  end
end

ALGORITHM 3: The noise-aware orthogonal matching pursuit.

B. OMP versus NAOMP: The Second Example

Let

$$\mathbf{Y} = \mathbf{X} + \mathbf{v} = \left((k+1)c^2 + c, c^2 + c, c^2, c^2\right)^{\top},$$
$$\mathbf{X} = \left(c^2 + c, c^2 + c, c^2, c^2\right)^{\top} = c^2\mathbf{d}_1 + c\mathbf{d}_3 \tag{B.1}$$

be an input signal, a faithful signal, respectively, where $\mathbf{v} = (kc^2, 0, 0, 0)^{\top}$ denotes the noise term and where $\mathbf{d}_1$, $\mathbf{d}_3$ are atoms of the following dictionary:

$$\mathbf{D} = (\mathbf{d}_1, \mathbf{d}_2, \mathbf{d}_3, \mathbf{d}_4) = \begin{pmatrix} 1 & 1 & 1 & 1 \\ 1 & 1 & 1 & 0 \\ 1 & 1 & 0 & 0 \\ 1 & 0 & 0 & 0 \end{pmatrix}. \tag{B.2}$$

According to the expression of $\mathbf{X}$, the best choice of atoms is $\{\mathbf{d}_1, \mathbf{d}_3\}$. Suppose that $2 < kc < 10$, $c > 0$. Let us recover the signal using OMP and NAOMP, respectively, both with two iteration steps. For OMP, in the first step, the atom which takes the maximum inner product with $r_0 = \mathbf{Y}$ is $\mathbf{d}_1$, and the residual is accordingly given by

$$r_1 = r_0 - \alpha_1\mathbf{d}_1 = \left(\frac{3}{4}kc^2 + \frac{1}{2}c, -\frac{1}{4}kc^2 + \frac{1}{2}c, -\frac{1}{4}kc^2 - \frac{1}{2}c, -\frac{1}{4}kc^2 - \frac{1}{2}c\right)^{\top}, \tag{B.3}$$

where $\alpha_1 = \arg\min_{\alpha\in\mathbb{R}} \|r_0 - \alpha\mathbf{d}_1\|_2$. In the second step, using simple calculation and the inequalities $kc > 2$, $c > 0$, the atom which takes the maximum inner product with r_1 is $\mathbf{d}_4$. Finally we recover the signal by

$$\mathbf{X}_{\mathrm{OMP}} = \left(c^2 + \frac{1}{3}c\right)\mathbf{d}_1 + \left(kc^2 + \frac{2}{3}c\right)\mathbf{d}_4 \tag{B.4}$$

with the coefficients given by

$$\left[c^2 + \frac{1}{3}c, kc^2 + \frac{2}{3}c\right] = \arg\min_{\alpha,\beta\in\mathbb{R}} \|\mathbf{Y} - \alpha\mathbf{d}_1 - \beta\mathbf{d}_4\|_2. \tag{B.5}$$

Thus the reconstruction error is given by

$$\|\mathbf{X}_{\mathrm{OMP}} - \mathbf{X}\|_2 = \sqrt{k^2c^4 + \frac{2}{3}c^2}. \tag{B.6}$$

For NAOMP, the first step finds the same atom $\mathbf{d}_1$ and the same residual r_1 as in OMP. Using $kc > 2$ we can see that the first component of r_1 is greater than the other three components and hence we modified the residual by $r'_0(1) \leftarrow r_0(1) - r_1(1)$, which gives

$$r'_0 = \left(\left(\frac{1}{4}k + 1\right)c^2 + \frac{1}{2}c, c^2 + c, c^2, c^2\right)^{\top}. \tag{B.7}$$

We reselect the atom taking the maximum inner product with r'_0 which is still $\mathbf{d}_1$ and the residual is given by

$$r_1 = \left(\frac{3}{16}kc^2 + \frac{1}{8}c, -\frac{1}{16}kc^2 + \frac{5}{8}c, -\frac{1}{16}kc^2 - \frac{3}{8}c, -\frac{1}{16}kc^2 - \frac{3}{8}c\right)^{\top}. \tag{B.8}$$

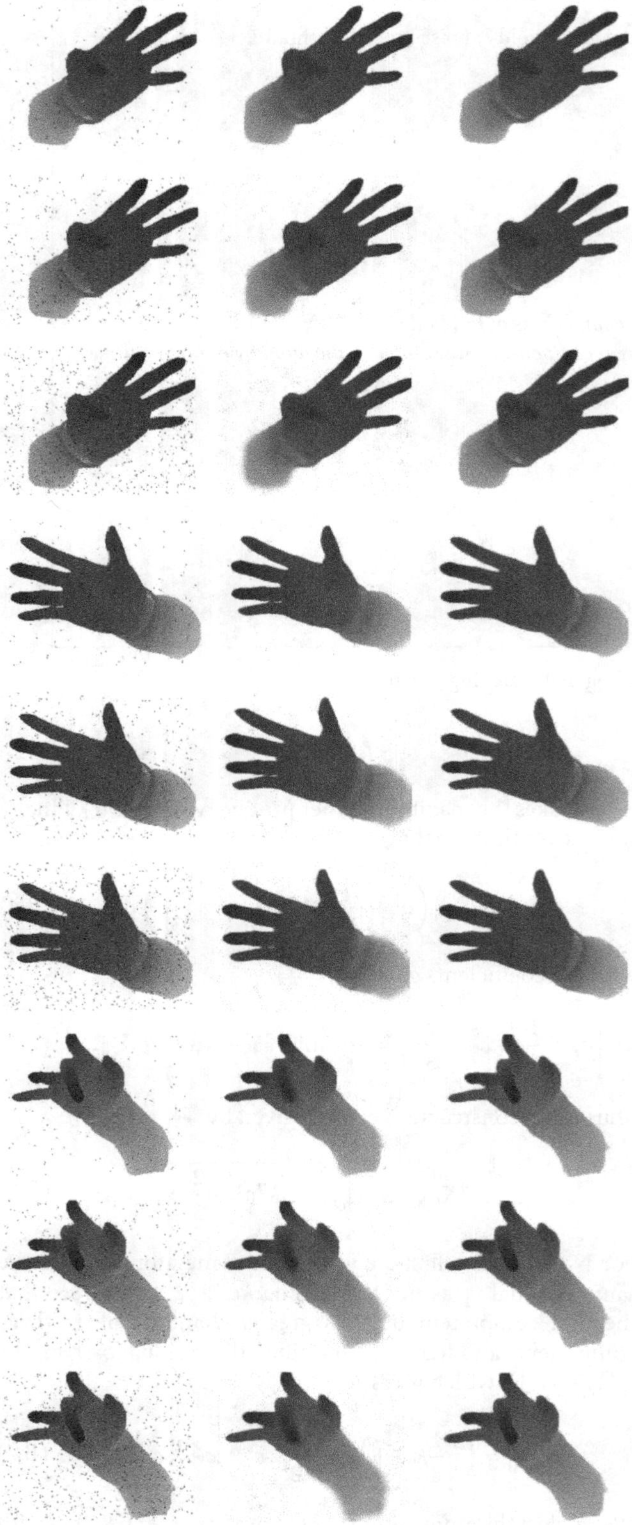

FIGURE 5: Denoising additive Gaussian noise using traditional OMP and NAOMP (both without the filtering phase). Left: hand depth images with artificial noise. Middle: denoised images using traditional OMP. Right: denoised images using NAOMP. Rows 1–3: 0.5%, 2%, and 5% of noise for *hand 1*; rows 4–6: 0.5%, 2%, and 5% of noise for *hand 2*; rows 7–9: 0.5%, 2%, and 5% of noise for *hand 3*.

TABLE 1: Parameter setting in filters [6], filters (2)–(4), OMP-based K-SVD, and NAOMP-based K-SVD.

Stages	Parameters	Values
Traditional bilateral filters [6]	σ for $f(\cdot,\cdot)$	4
	σ for $g(\cdot)$	15
	ϵ	3
Bilateral filters (2)–(4)	σ for $f(\cdot,\cdot)$	4
	σ for $g(\cdot)$	15
	ϵ_1	5
	ϵ_2	30
	ϵ_3	5
	ϵ_4	25
	ϵ_5	30
	ϵ_6	6
	ϵ_7	3
OMP-based K-SVD ([19], Algorithms 1 and 2)	n	64
	K	512
	J	10
	S	30
	ε for Figure 6	10^5
	ε for Figure 5 (0.5% noise)	$5 * 10^4$
	ε for Figure 5 (2% noise)	$1.2 * 10^5$
	ε for Figure 5 (5% noise)	$2 * 10^6$
NAOMP-based K-SVD (Algorithms 1 and 3)	n	64
	K	512
	J	10
	S	30
	ε	10^3
	$\hat{r}$	30
	$\hat{L}$	15
NAOMP-based superresolution (7)	N	25
	M	100
	λ	0.15

In the second step, using simple calculation and the relationship $kc < 10$, the atom taking the maximum inner product with r_1 is $\mathbf{d_3}$. Then we update the residual by

$$r_2 = \left(\frac{1}{2}kc^2, -\frac{1}{2}kc^2, 0, 0\right)^{\mathrm{T}}. \tag{B.9}$$

Because no single component has a relatively great magnitude, we do not implement the modification step and hence return the reconstruction signal by

$$\mathbf{X}_{\mathrm{NAOMP}} = c^2\mathbf{d_1} + \left(\frac{1}{2}kc^2 + c\right)\mathbf{d_3} \tag{B.10}$$

with the coefficients given by

$$\left[c^2, \frac{1}{2}kc^2 + c\right] = \arg\min_{\alpha,\beta\in\mathbb{R}} \left\|\mathbf{Y} - \alpha\mathbf{d_1} - \beta\mathbf{d_3}\right\|_2. \tag{B.11}$$

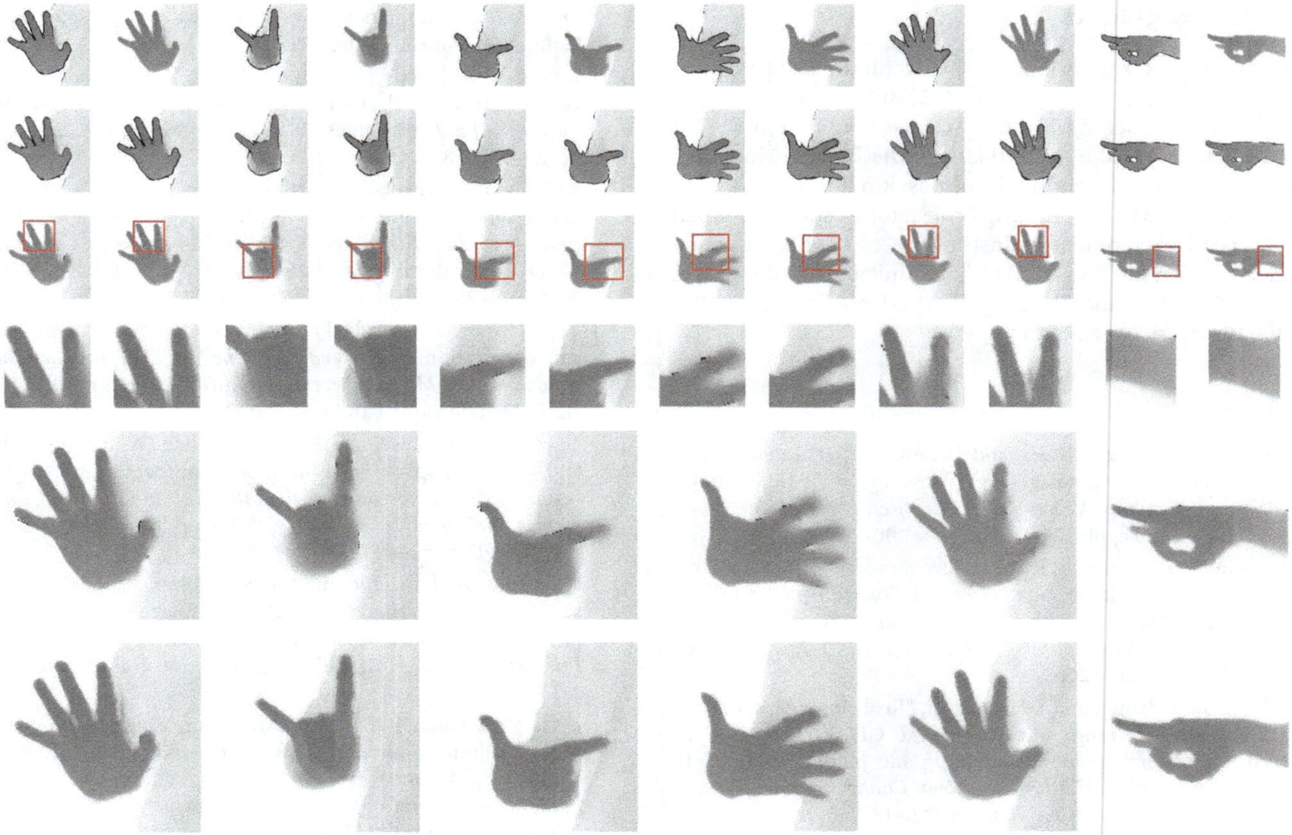

FIGURE 6: Denoising Gaussian noise from hand depth images captured by Kinect sensor v2. Columns 1–3 are three examples of hand depth images. Row 1 left: noisy depth images of resolutions 150 × 150. Row 1 right: denoised images using bilateral filters [6]. Row 2 left: denoised images using OMP-based K-SVD [19]. Row 2 right: denoised images using NAOMP-based K-SVD. Row 3 left: denoised images using bilateral filters (2)–(4) plus OMP-based K-SVD [19]. Row 3 right: denoised images using the proposed filters (2)–(4) plus NAOMP-based K-SVD. Row 4: amplified comparison details of images of row 3. Row 5: superresolution of denoised images of left subfigure of row three via OMP-based K-SVD. Row 6: superresolution of denoised images of right subfigure of row three via NAOMP-based K-SVD.

TABLE 2: PSNR (unit: dB) comparison of three models in Figure 5.

Noise percent	PSNR (*hand 1*)		PSNR (*hand 2*)		PSNR (*hand 3*)	
	OMP	NAOMP	OMP	NAOMP	OMP	NAOMP
0.5%	40.2939	39.8312	39.7280	39.7308	39.7076	41.4870
2%	37.6736	39.6812	37.5300	39.5911	37.0001	41.1889
5%	35.4332	39.4074	35.5310	39.3932	35.0824	40.9597

TABLE 3: Average running time comparison of six models in Figure 6.

Methods	Time (s)
Filters [6]	19.644
Filters (2)–(4)	9.444
K-SVD [19]	85.848
NAOMP K-SVD (Algorithms 1 and 3)	95.163
Filters [6] + K-SVD [19]	105.492
Filters (2)–(4) + NAOMP K-SVD	104.607

The reconstruction error is given by

$$\|\mathbf{X}_{\text{NAOMP}} - \mathbf{X}\|_2 = \frac{1}{\sqrt{2}}kc^2 \quad \text{(B.12)}$$

which is smaller than the error of OMP. Moreover, NAOMP selects the correct atoms $\{\mathbf{d}_1, \mathbf{d}_3\}$ while OMP selects the atoms $\{\mathbf{d}_1, \mathbf{d}_4\}$.

Competing Interests

The authors declare that there are no competing interests regarding the publication of this paper.

Acknowledgments

This work was supported by the Natural Science Foundation of China (Grant nos. 61227004, 61402024, and 61300065), the Beijing Natural Science Foundation (Grant nos. 4162009, 4152009, and 4142010), the Beijing Municipal Commission of Education (Grant nos. km201410005013 and km201610005033), the Funding Project for Academic Human Resources Development in Institutions of Higher Learning under the Jurisdiction of Beijing Municipality (Grant no. IDHT20150504), and the Jing-Hua Talents Project of the Beijing University of Technology.

References

[1] Q. Yang, R. Yang, J. Davis, and D. Nistér, "Spatial-depth super resolution for range images," in *Proceedings of the IEEE Conference on Computer Vision and Pattern Recognition (CVPR '07)*, pp. 1–8, Minneapolis, MN, USA, June 2007.

[2] B. Huhle, T. Schairer, P. Jenke, and W. Straßer, "Robust nonlocal denoising of colored depth data," in *Proceedings of the IEEE Computer Society Conference on Computer Vision and Pattern Recognition Workshops (CVPRW '08)*, pp. 1–7, Anchorage, Alaska, USA, June 2008.

[3] J. Wasza, S. Bauer, and J. Hornegger, "Real-time preprocessing for dense 3-D range imaging on the GPU: defect interpolation, bilateral temporal averaging and guided filtering," in *Proceedings of the IEEE International Conference on Computer Vision Workshops (ICCV '11)*, pp. 1221–1227, Barcelona, Spain, November 2011.

[4] D. Min, J. Lu, and M. N. Do, "Depth video enhancement based on weighted mode filtering," *IEEE Transactions on Image Processing*, vol. 21, no. 3, pp. 1176–1190, 2012.

[5] J. Fu, S. Wang, Y. Lu, S. Li, and W. Zeng, "Kinect-like depth denoising," in *Proceedings of the IEEE International Symposium on Circuits and Systems (ISCAS '12)*, pp. 512–515, Seoul, South Korea, May 2012.

[6] M. Camplani and L. Salgado, "Efficient spatio-temporal hole filling strategy for kinect depth maps," in *Three-Dimensional Image Processing (3DIP) and Applications II*, vol. 8920 of *Proceedings of SPIE*, Burlingame, Calif, USA, 2012.

[7] O. Mac Aodha, N. D. Campbell, A. Nair, and G. J. Brostow, "Patch based synthesis for single depth image super resolution," in *Proceedings of the 12th European Conference on Computer Vision (ECCV '12)*, pp. 71–84, Florence, Italy, October 2012.

[8] J. Shen and S.-C. S. Cheung, "Layer depth denoising and completion for structured-light RGB-D cameras," in *Proceedings of the 26th IEEE Conference on Computer Vision and Pattern Recognition (CVPR '13)*, pp. 1187–1194, Portland, Ore, USA, June 2013.

[9] Y. Wang, F. Zhong, Q. Peng, and X. Qin, "Depth map enhancement based on color and depth consistency," *Visual Computer*, vol. 30, no. 10, pp. 1157–1168, 2014.

[10] J. Li, Z. Lu, G. Zeng, R. Gan, and H. Zha, "Similarity-aware patchwork assembly for depth image super-resolution," in *Proceedings of the 27th IEEE Conference on Computer Vision and Pattern Recognition (CVPR '14)*, pp. 3374–3381, Columbus, Ohio, USA, June 2014.

[11] T.-W. Hui and K. N. Ngan, "Motion-depth: RGB-D depth map enhancement with motion and depth in complement," in *Proceedings of the 27th IEEE Conference on Computer Vision and Pattern Recognition (CVPR '14)*, pp. 3962–3969, IEEE, Columbus, Ohio, USA, June 2014.

[12] J. Yang, X. Ye, K. Li, C. Hou, and Y. Wang, "Color-guided depth recovery from RGB-D data using an adaptive autoregressive model," *IEEE Transactions on Image Processing*, vol. 23, no. 8, pp. 3443–3458, 2014.

[13] S. Schuon, C. Theobalt, J. Davis, and S. Thrun, "Lidarboost: depth superresolution for TOF 3D shape scanning," in *Proceedings of the IEEE Computer Society Conference on Computer Vision and Pattern Recognition Workshops (CVPR Workshops '09)*, pp. 343–350, Miami, Fla, USA, June 2009.

[14] Y. Li, T. Xue, L. Sun, and J. Liu, "Joint example-based depth map super-resolution," in *Proceedings of the 13th IEEE International Conference on Multimedia and Expo (ICME '12)*, pp. 152–157, IEEE, Melbourne, Australia, July 2012.

[15] M. Kiechle, S. Hawe, and M. Kleinsteuber, "A joint intensity and depth co-sparse analysis model for depth map super-resolution," in *Proceedings of the 14th IEEE International Conference on Computer Vision (ICCV '13)*, pp. 1545–1552, Sydney, Australia, December 2013.

[16] H. Zheng, A. Bouzerdoum, and S. L. Phung, "Depth image super-resolution using multi-dictionary sparse representation," in *Proceedings of the IEEE International Conference on Image Processing*, pp. 957–961, IEEE, Melbourne, Australia, September 2013.

[17] J. Xie, C.-C. Chou, R. Feris, and M.-T. Sun, "Single depth image super resolution and denoising via coupled dictionary learning with local constraints and shock filtering," in *Proceedings of the IEEE International Conference on Multimedia and Expo (ICME '14)*, pp. 1–6, Chengdu, China, July 2014.

[18] S. Lu, X. Ren, and F. Liu, "Depth enhancement via low-rank matrix completion," in *Proceedings of the 27th IEEE Conference on Computer Vision and Pattern Recognition (CVPR '14)*, pp. 3390–3397, Columbus, Ohio, USA, June 2014.

[19] M. Elad and M. Aharon, "Image denoising via sparse and redundant representations over learned dictionaries," *IEEE Transactions on Image Processing*, vol. 15, no. 12, pp. 3736–3745, 2006.

[20] J. Yang, Z. Wang, Z. Lin, S. Cohen, and T. Huang, "Coupled dictionary training for image super-resolution," *IEEE Transactions on Image Processing*, vol. 21, no. 8, pp. 3467–3478, 2012.

10

Ferrography Wear Particles Image Recognition based on Extreme Learning Machine

Qiong Li, Tingting Zhao, Lingchao Zhang, Wenhui Sun, and Xi Zhao

College of Computer Science and Information Engineering, Tianjin University of Science and Technology, Tianjin 300457, China

Correspondence should be addressed to Xi Zhao; xi.zhao@tust.edu.cn

Academic Editor: Hui Cheng

The morphology of wear particles reflects the complex properties of wear processes involved in particle formation. Typically, the morphology of wear particles is evaluated qualitatively based on microscopy observations. This procedure relies upon the experts' knowledge and, thus, is not always objective and cheap. With the rapid development of computer image processing technology, neural network based on traditional gradient training algorithm can be used to recognize them. However, the feedforward neural network based on traditional gradient training algorithms for image segmentation creates many issues, such as needing multiple iterations to converge and easy fall into local minimum, which restrict its development heavily. Recently, extreme learning machine (ELM) for single-hidden-layer feedforward neural networks (SLFN) has been attracting attentions for its faster learning speed and better generalization performance than those of traditional gradient-based learning algorithms. In this paper, we propose to employ ELM for ferrography wear particles image recognition. We extract the shape features, color features, and texture features of five typical kinds of wear particles as the input of the ELM classifier and set five types of wear particles as the output of the ELM classifier. Therefore, the novel ferrography wear particle classifier is founded based on ELM.

1. Introduction

Wear particle analysis for machine condition monitoring and fault diagnosis is not a new topic in tribology. Roylance et al. [1] pioneered computer wear pattern recognition research direction by establishing the computer aided visual engineering (CAVE) to extract the wear particles' Fourier factor, roundness factor, and edge details. Then, they used these features to classify portion wear particles [2–7]. Hamblin and Stachowiak proposed the spike parameter to extract characteristics of abrasive particles via adequate description of wear debris boundary in multiscale [8]. Xu et al. [9] developed a set of wear particle image analysis system through the integration of neural network and expert system, which can realize the interactive wear particles image automatic recognition. Tian et al. [10] and Peng and Kirk [11] employed laser scanning confocal microscopy to acquire three-dimensional particle images. The scanning method provides surface information from the analysis of wear surface morphology, which facilitates better understanding of wear features and wear level. Stachowiak et al. [12] utilized an automated classification system to analyze and identify fatigue, abrasive, and adhesive wear particles. Such analysis and identification operations were fulfilled mainly based on the analyses of the area perimeter, elongation parameters, convexity, and surface texture. van Otterlo [13] distinguished different points on the curve of an object boundary by means of the proposed parameters, which might represent points with respect to a reference position in polar coordinate system. This approach provided a unified theoretical basis for analyzing shape similarity according to the proposed parametric contour representations. Laghari and Memon developed an automatic analysis system, KBWPAS, to reduce the dependence on domain experts [14]. Wang combined principal component analysis and gray relational analysis to optimize the parameters of features of wear particles [15]. Wang and Yin used wear particle shape, color, and surface texture parameters as input vectors and introduced radial basis function neural network to conduct automatic classification and identification on wear particle [16]. Gu applied support vector machine to ferrography wear particle pattern recognition and built wear particle classifier based on supper vector machine [17]. Yuan and Yan took surface texture and surface roughness as the important indicators of three-dimensional surface characteristic and adopted artificial intelligent neural network method to identify wear particle type, which efficiently improved accuracy of wear particle classification [18]. Luo et al. proposed to use principal component analysis to extract characteristic parameters of wear particles and then introduced BP neural network to conduct automatic classification on wear particles. This method optimized parameters firstly, which yielded better classification rate compared to traditional BP neural network [19]. Zhou et al. used the same method to optimize parameters of wear particle features and applied the improved genetic algorithms to the LS-SVM's parameter optimization, thus improving wear particle recognition rate [20].

At present, a lot of image classification algorithms have been proposed, among which the neural network based image classification algorithms are widely used. However, the traditional feedforward neural networks need many iterations to converge and are easy to fall into the local optimum in the procession of image classification, which seriously limit its development and application. On the other hand, the extreme learning machine as a new machine learning algorithm has become more and more popular due to its few adjustable parameters, fast learning speed, and good generalization performance. However, the effectiveness of the application of ELM to the field of image processing of iron spectrum abrasive particles is open to be investigated. In this paper, we present the method to apply the extreme learning machine to the recognition of wear particles.

The rest of the paper is organized as follows. Section 2 introduces principles and methods used in this paper. In Section 3, we describe the proposed method. In Section 4, we describe the experiments. Section 5 finalizes the paper with the conclusions.

2. Principle and Method

2.1. Characteristic Parameters of Wear Particles. Generally, characteristic parameters play important role in the final result of wear particles recognition. Although the obtained ferroscope images of wear particles are not three-dimensional, the size, texture, and color features are enough for typical wear particles recognition.

Among the shape and size factors of wear particles, area, perimeter, equivalent circle diameter, circularity, aspect ratio, concavity, and so on are commonly used features to be extracted from the color images, which are captured via the microscope. These shape and size parameters can be further approximated by converting the sample color image into binary image with the values of 0 and 1.

Besides the shape and size parameters, texture is also very important in image analysis. The texture is the description of the pixel distribution of a picture in gray space. Machines or equipment with a pair of relatively moving contact surfaces for long-time operation tends to produce wear particles. Usually, the surface texture provides statistical information of the wear state in the machinery operation process. Textural features are investigated by using the distribution function to analyze the gray level of the characteristic region of wear particles. Among the parameters of wear particles, two of the characteristic features are employed to distinguish the surface texture, namely, roughness and directivity. Gray level cooccurrence matrix (GLCM) [21] is commonly used to describe the surface textural factors of wear particles. The textural features, including energy, entropy, dependency, inertia moment, and stability, can be extracted by employing the GLCM. However, the connection between the obtained textural features and human vision perceptual system has not been established. Based on the psychological representation of human vision perception to the texture, Tamura et al. proposed the expression of the textural feature, Tamura textural feature [22], which can be used to extract six properties of the textural features, contrast ratio, roughness, direction level, regularity, linearity, and coarseness. Moreover, we have found that HoG gives a good simulation of the variation of the particle images [23]. The histogram of oriented gradient for the local region of the wear particle images can be calculated by using HoG.

Color feature is another kind of significant properties to analyze wear particle images. The color factor is important for the identification of the red oxides, black oxides, and fatigued copper wear particles. By analyzing the color of the wear particles, mechanical abrasion degree can be predicted. The image commonly consisted of three-primary colors of red (R), green (G), and blue (B). In the RGB space, the three-color factors of $R(i, j)$, $G(i, j)$, and $B(i, j)$ denote the color features of wear particles. After the images of wear particles are processed, the first-order and second-order statistical values (mean and variance) of $R(i, j)$, $G(i, j)$, and $B(i, j)$ and the third-order color matrix of the HSV color space are extracted as the color characteristic parameters.

2.2. Extreme Learning Machine. The extreme learning machine (ELM) algorithm was originally proposed by Huang et al. [24–27]. The method is proved to be a universal approximator given enough hidden neurons [28]. It works as follows.

Consider a set of N distinct samples (x_i, t_i) with $x_i \in R^d$ and $t_i \in R^c$. Then a single layer feedforward network with L hidden neurons is modeled as

$$\sum_{i=1}^{L} \beta_i \phi\left(\mathbf{w_i x_j} + b_i\right), \quad j \in [1, N] \qquad (1)$$

with ϕ being an activation function, $\mathbf{w_i}$ the input weights, b_i the biases, and β_i the output weights.

In the case where the single layer feedforward network would perfectly approximate the data, the errors between the estimated outputs y_i and the targets t_i are zero, and the relation between inputs, weights, and targets is

$$\sum_{i=1}^{L} \beta_i \phi\left(\mathbf{w_i x_j} + b_i\right) = t_i, \quad j \in [1, N] \qquad (2)$$

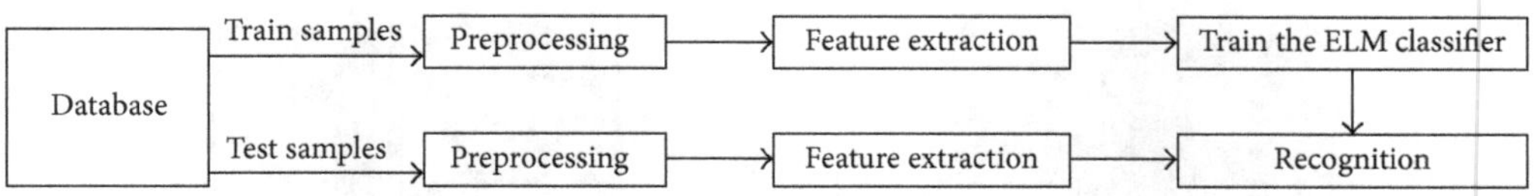

FIGURE 1: Wear particles recognition based on ELM

which can be written compactly as $\mathbf{H}\beta = \mathbf{T}$, with

$$\mathbf{H} = \begin{pmatrix} \phi(\mathbf{w}_1\mathbf{x}_1 + b_1) & \cdots & \phi(\mathbf{w}_L\mathbf{x}_1 + b_L) \\ \vdots & \ddots & \vdots \\ \phi(\mathbf{w}_1\mathbf{x}_N + b_1) & \cdots & \phi(\mathbf{w}_L\mathbf{x}_N + b_L) \end{pmatrix}$$

$$\beta = (\beta_1^T \cdots \beta_L^T),$$

$$T = (t_1^T \cdots t_N^T). \quad (3)$$

Finding the output weights β from the hidden layer outputs $\mathbf{H}$ and targets is a linear regression problem. In the general case of $N \neq d$, a minimum L^2-norm solution is given by the Moore-Penrose generalized inverse, or pseudoinverse, of the matrix $\mathbf{H}$ denoted as $\mathbf{H}\dagger$ [29]. The training of ELM requires no iterations, and the most computational efficient part is the calculation of the pseudoinverse of the matrix $\mathbf{H}_{(NsL)}$, which makes ELM an extremely fast artificial neural networks method.

3. Proposed Method

Extreme learning machine (ELM) is a new machine learning algorithm. It is increasingly favored by many researchers due to its simple structure, fast learning speed, and good generalization performance. This part will introduce extreme learning machine (ELM) method into the wear particles image recognition. The specific flow chart is shown in Figure 1

The exact procedures are as follows: The first four steps are image preprocessing as shown in Figure 2, which include the steps of 3D media filter, K-means clustering segmentation, region growing, morphology expansion, and erosion; the fifth step is feature extraction and the last step is recognition.

Step 1 (the three-dimensional median filter [30]). Apply the median filter to the three components of RGB color images; then use the relevant algorithms incorporating three components to get three-dimensional image of median filter. The change of the picture is shown in Figure 3.

Step 2 (K-means clustering segmentation [31]). K-means clustering is used to segment wear particles images in Lab color mode. The calculation of 3D color images needs more computer memory and is time consuming, but the characteristic information of 3D color images contributes to the wear particles recognition. Compared with three-dimensional color images, two-dimensional color images not only have quick computing speed, but also can be well segmented. We apply K-means clustering to the two-dimensional color images; then we can get the wear particle image that contains the color information. We choose the Lab color mode for its relatively small mutual association between three components. As (a, b) component represents

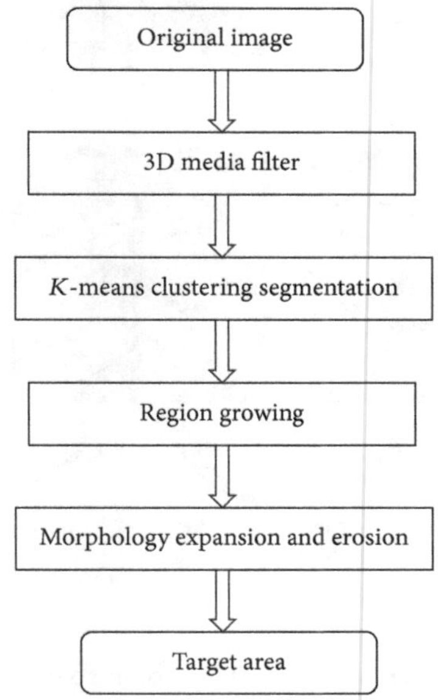

FIGURE 2: Image preprocessing.

the color change and L component represents the change of brightness in a Lab mode image, we use (a, b) component for K-means clustering. Firstly, the input of wear particle image is transformed from RGB space to Lab space. Randomly generate three initial clustering points for K-means clustering algorithm. At last, the cluster region with the most complete target region is selected as the target segmentation region.

Step 3 (region growing [32]). K-means clustering cannot extract the target area and just simply labeled it. After clustering, we use region growing method to extract the interested area and remove the grains around it.

Step 4 (morphology expansion and erosion). Corrosion and expansion are two important operations of mathematical morphology, which can make the segmentation target area closing, remove the holes in the target area, and make the segmentation region more even. The area that we are interested in extracted by region growing may have holds and rough edges. It requires using mathematical morphology method for further processing in order to get a complete wear particle image. The changes of the processed image are shown in Figure 4.

Step 5 (feature extraction). Due to the fact that different wear particles have different characteristics as shown in Figure 3, different characteristic parameters need to be extracted for recognition. The most obvious characteristic of red oxide is color, so we extract the first-order origin matrix, second-order center matrix, and third-order center matrix of H, S,

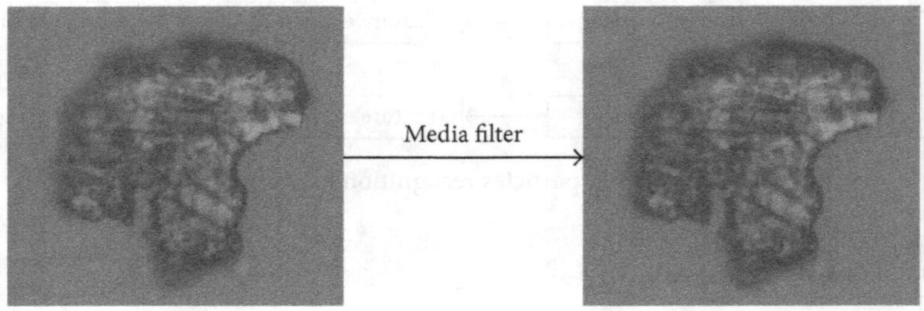

FIGURE 3: The three-dimensional median filter.

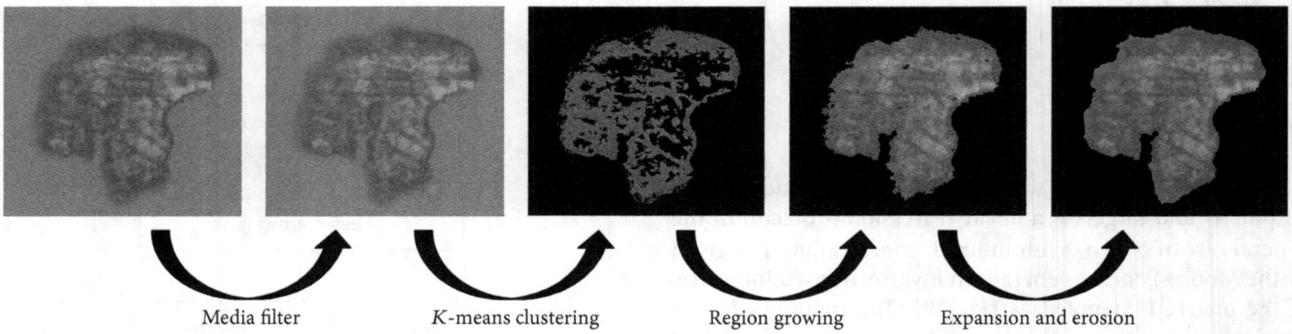

FIGURE 4: Changes of the image in the preprocessing.

and V components in the HSV color space as red oxide's feature. Fatigue wear particles showed very similar morphological characteristics. The majority of fatigue wear particles are laminar particles. They had smooth flat surfaces and irregular contours. Based on the characteristics of fatigue wear particle, its HoG feature is extracted for classification. For the cutting wear particle, its shape feature is obvious. We get its area, perimeter, equivalent diameter, roundness, aspect ratio, and concavity which is due to the reflection of light and another important basis to recognize the spherical particles. This method uses the following characteristics of spherical particles to identify them: area, perimeter, equivalent diameter, roundness and slenderness ratio, concavity, energy, entropy, moment of inertia, local stability, R mean value, G mean value, and B average value as well as R, G, and B standard deviation. The surface of fatigue wear is relatively rough, characterized by severe galling, parallel scratches, or partial oxidation and irregular edge contour. As scratch is an important feature to recognize fatigue wears, we extracted its Tamura features.

Step 6 (using extreme learning machine for recognition). We use the extracted features to train the ELM classifier and then use the trained network for wear particle recognition.

4. Experiments

4.1. Experimental Database. In this paper, the experimental database was collected from Guangdong provincial Key Laboratory of Petrochemical Equipment Fault Diagnosis. Lubricating oil samples of petrochemical plant machinery were adopted to make ferrogram through using analytical ferrograph. Through a double light microscope, the morphological features and distribution were observed and the database was set up by shooting featured particles' images trough CDD. The database consists of five typical ferrography abrasive particles, with totally 149 particles pictures. There are 33 red oxide images, 42 fatigue wear particles images, 29 cutting wear particles images, 18 spherical wear particles images, and 27 severe sliding wear particles images. The size of the images is 1024 × 768. Five kinds of wear particles are shown in Figure 5.

TABLE 1: Average recognition rate.

Training percentage (%)	Average recognition rate (%)		
	70	80	90
Red oxide	90.00	100	100
Fatigue	80.77	91.25	100
Cutting	77.78	88.33	100
Spherical	100	100	100
Severe sliding	85.00	100	100

4.2. Results of Experiments. We conduct five experiments; each experiment only recognizes one of the five wear particles. Take the red oxide as an example; we extract color characteristics of all particles and use the ELM to classify them into two categories, red oxide and non-red oxide; thus we recognize the red oxide. According to this method, we recognize fatigue, cutting, and spherical and severe sliding wear particles.

To test the performance of the proposed method, three sets of experiments had been carried out for one kind wear particle. 70%, 80%, and 90% of each type of wear particles images are used for training and the rest of them for test. The result of the experiments is shown in Table 1.

Table 2: Average recognition rate of three classifiers (%).

classifier and training percentage (%)	BP				SVM				ELM			
	70	80	90	Average training time (s)	70	80	90	Average training time (s)	70	80	90	Average training time (s)
Red oxide	99.44	100	100	0.26	84.62	87.50	90.00	0.19	90.00	100	100	**0.14**
Fatigue	82.00	88.78	90	1.43	76.00	76.00	80.00	0.22	80.77	91.25	100	**0.18**
Cutting	85.00	92.00	95.00	0.19	87.50	90.00	90.00	0.09	77.78	88.33	100	**0.07**
Spherical	87.40	95.00	97.00	0.28	80.00	82.00	85.00	0.15	100	100	100	**0.13**
Severe sliding	83.00	88.00	94.05	0.21	78.95	88.89	91.57	0.11	85.00	100	100	**0.07**

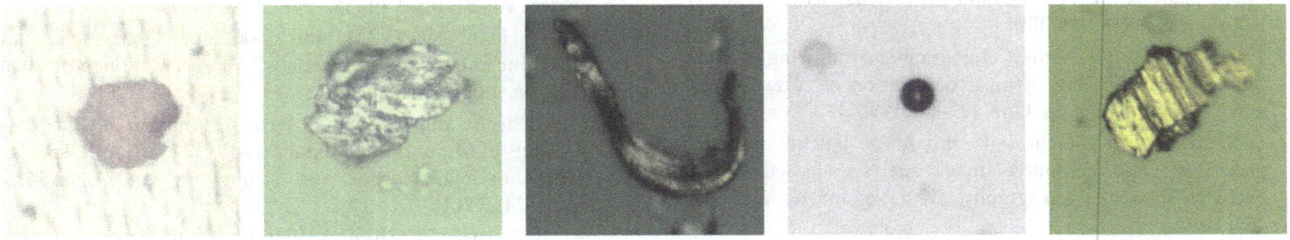

Red oxide Fatigue Cutting Spherical Severe sliding

Figure 5: Five kinds of wear particles.

We compared the proposed method with two other classification methods, which are BP [33] and SVM [34]. We use the same features which we used to train the ELM to train the BP and SVM and then test them with the same samples. Table 2 shows that the recognition rates for training images and the average training time.

4.3. Discussion. Table 1 shows the results of the experiments. The recognition rate of red oxide, spherical wear particles, and severe sliding wear particles is great and can reach 100%. The reason why the recognition rate of cutting wear particles is not as good as the above three classifications is the shape features which we have chosen. For some short, thick, and big cutting wear particles, its aspect ratio is not obvious. The computer may mix the cutting wear particles with parts of fatigue wear particles and severe sliding wear particles so that it leads to the low recognition rate. However, for the fatigue wear particles, because the scratched parts of wear particles may not be obvious or not too many and Tamura features cannot represent it well, it causes the relatively low recognition rate.

As we extract different features for different wear particles, the average training time of the five wear particles is different. we can see from Table 2 that ELM has a good time efficiency. For the same kind of wear particle, the BP's recognition rate is close to ELM, but its time efficiency is much lower than ELM. Among the three kinds of classifiers the SVM's performance is the lowest on both recognition rate and time efficiency.

5. Conclusion

In this paper, ELM is introduced into the wear particles image recognition. Different from traditional method, each time only one of the five wear particles is recognized. This avoids the simultaneous extraction of various wear particle characteristics, resulting in redundancy features, and achieves better classification results. However, as shown from the experimental results, the recognition rate of fatigue and cutting wear particles are to be improved. Future work will include how to get robust wear particle features and improve the recognition rate based on ELM.

Conflicts of Interest

The authors declare that they have no conflicts of interest.

Acknowledgments

This work was supported by the National Natural Science Foundation of China under Grant no. 61502338 and no. 61502339, the 2015 Key Projects of Tianjin Science and Technology Support Program no. 15ZCZDGX00200, the Open Fund of Guangdong Provincial Key Laboratory of Petrochemical Equipment Fault Diagnosis no. GDUPTKLAB201504, and the Fund of Tianjin Food Safety & Low Carbon Manufacturing Collaborative Innovation Center.

References

[1] B. J. Roylance, I. A. Albirdewi, and M. S. Laghari, "Computer-aided vision engineering (CAVE): quantification of wear particle morphology," *Lubrication Engineering*, vol. 50, no. 2, pp. 111–116, 1994.

[2] S. Raadnui and B. J. Roylance, "The classification of wear particle shape," *Lubrication Engineering*, vol. 51, no. 5, pp. 432–437, 1995.

[3] K. K. Yeung, A. J. McKenzie, D. Liew, and G. A. Luoma, "Development of computer-aided image analysis for filter debris analysis," *Lubrication Engineering*, vol. 50, no. 4, pp. 293–299, 1994.

[4] F. T. Barwell, "The contribution of particle analysis to the study of wear of metals," *Wear*, vol. 90, no. 1, pp. 167–181, 1983.

[5] A. D. H. Thomas, T. Davies, and A. R. Luxmoore, "Computer image analysis for identification of wear particles," *Wear*, vol. 142, no. 2, pp. 213–226, 1991.

[6] T. Pancewicz and I. Mruk, "Holographic contouring for determination of three-dimensional description of surface roughness," *Wear*, vol. 199, no. 1, pp. 127–131, 1996.

[7] T. B. Kirk, G. W. Stachowiak, and A. W. Batchelor, "Fractal parameters and computer image analysis applied to wear particles isolated by ferrography," *Wear*, vol. 145, no. 2, pp. 347–365, 1991.

[8] M. G. Hamblin and G. W. Stachowiak, "A multi-scale measure of particle abrasivity, and its relation to two-body abrasive wear," *Wear*, vol. 190, no. 2, pp. 190–196, 1995.

[9] K. Xu, A. R. Luxmoore, L. M. Jones, and F. Deravi, "Integration of neural networks and expert systems for microscopic wear particle analysis," *Knowledge-Based Systems*, vol. 11, no. 3-4, pp. 213–227, 1998.

[10] Y. Tian, J. Wang, Z. Peng, and X. Jiang, "A new approach to numerical characterisation of wear particle surfaces in three-dimensions for wear study," *Wear*, vol. 282-283, pp. 59–68, 2012.

[11] Z. Peng and T. B. Kirk, "Computer image analysis of wear particles in three-dimensions for machine condition monitoring," *Wear*, vol. 223, no. 1-2, pp. 157–166, 1998.

[12] G. P. Stachowiak, G. W. Stachowiak, and P. Podsiadlo, "Automated classification of wear particles based on their surface texture and shape features," *Tribology International*, vol. 41, no. 1, pp. 34–43, 2008.

[13] P. J. van Otterlo, *A Contour Oriented Approach to Digital Shape Analysis*, Kluver, Amsterdam, The Netherlands, 1988.

[14] M. S. Laghari and Q. A. Memon, "Knowledge based wear particle analysis," *International Journal of Information Technology*, vol. 1, no. 3, pp. 91–95, 2004.

[15] J. Q. Wang and X. L. Wang, "A wear particle identification method by combining principal component analysis and grey relational analysis," *Wear*, vol. 304, no. 1-2, pp. 96–102, 2013.

[16] W. H. Wang and Y. H. Yin, "Abrasive identification system based on radial basis function neural network," *Tribology*, vol. 123, no. 4, pp. 340–343, 2003.

[17] D. Q. Gu, "Ferrography wear pattern recognition based on support vector machine," *China Mechanical Engineering*, vol. 17, no. 13, pp. 1391–1394, 2006.

[18] C. Q. Yuan and X. P. Yan, "Recognition of the type of abrasive," *Lubrication*, vol. 32, no. 3, pp. 21–23, 2007.

[19] B. H. Luo, Y. W. Huang, and Y. F. Liu, "Wear particles identification based PCA-BP neural network," *Lubrication*, vol. 35, no. 6, pp. 117–120, 2010.

[20] W. Zhou, B. Jing, and S. Deng, "Aeroengine abrasive identification based on IGA and LS-SVM," *Lubrication*, vol. 38, no. 1, pp. 14–18, 2013.

[21] R. M. Haralick, I. Dinstein, and K. Shanmugam, "Textural features for image classification," *IEEE Transactions on Systems, Man and Cybernetics*, vol. 3, no. 6, pp. 610–621, 1973.

[22] B. H. Tamura, S. Mori, and T. Yamawaki, "Texture features corresponding to visual perception," *IEEE Transactions on Systems, Man, and Cybernetics*, vol. 8, no. 6, pp. 460–473, 1978.

[23] N. Dalal and B. Triggs, "Histograms of oriented gradients for human detection," in *Proceedings of the IEEE Computer Society Conference on Computer Vision and Pattern Recognition (CVPR '05)*, vol. 1, pp. 886–893, IEEE, San Diego, Calif, USA, June 2005.

[24] G.-B. Huang, Q.-Y. Zhu, K. Z. Mao, C.-K. Siew, P. Saratchandran, and N. Sundararajan, "Can threshold networks be trained directly?" *IEEE Transactions on Circuits and Systems II: Express Briefs*, vol. 53, no. 3, pp. 187–191, 2006.

[25] G.-B. Huang, Q.-Y. Zhu, and C.-K. Siew, "Extreme learning machine: theory and applications," *Neurocomputing*, vol. 70, no. 1–3, pp. 489–501, 2006.

[26] Y. Miche, M. van Heeswijk, P. Bas, O. Simula, and A. Lendasse, "TROP-ELM: a double-regularized ELM using LARS and Tikhonov regularization," *Neurocomputing*, vol. 74, no. 16, pp. 2413–2421, 2011.

[27] G.-B. Huang, H. Zhou, X. Ding, and R. Zhang, "Extreme learning machine for regression and multiclass classification," *IEEE Transactions on Systems, Man, and Cybernetics, Part B: Cybernetics*, vol. 42, no. 2, pp. 513–529, 2012.

[28] G.-B. Huang, L. Chen, and C.-K. Siew, "Universal approximation using incremental constructive feedforward networks with random hidden nodes," *IEEE Transactions on Neural Networks*, vol. 17, no. 4, pp. 879–892, 2006.

[29] C. R. Rao and S. K. Mitra, *Generalized Inverse of a Matrix and Its Applications*, John Wiley & Sons, Berkeley, Calif, USA, 1971.

[30] W. Yu, Y. Ma, L. Zheng et al., "Research of improved adaptive median filter algorithm," in *Proceedings of the 2015 International Conference on Electrical and Information Technologies for Rail Transportation (EITRT '15)*, Springer, Hunan Province, China, 2016.

[31] J. A. Hartigan and M. A. Wong, "A K-means clustering algorithm," *Applied Statistics*, vol. 28, no. 1, pp. 100–108, 2013.

[32] S.-Y. Wan and W. E. Higgins, "Symmetric region growing," *IEEE Transactions on Image Processing*, vol. 12, no. 9, pp. 1007–1015, 2003.

[33] D. E. Rumelhart, G. E. Hinton, and R. J. Williams, "Learning representations by back-propagating errors," in *Neurocomputing: Foundations of Research*, pp. 533–536, MIT Press, Cambridge, Mass, USA, 1986.

[34] C. Cortes and V. Vapnik, "Support-vector networks," *Machine Learning*, vol. 20, no. 3, pp. 273–297, 1995.

11

Design of IIR Digital Filters with Arbitrary Flatness using Iterative Quadratic Programming

Yasunori Sugita

Nagaoka University of Technology, 1603-1 Kamitomioka, Nagaoka City, Niigata 940-2188, Japan

Correspondence should be addressed to Yasunori Sugita; sugita@vos.nagaokaut.ac.jp

Academic Editor: Jar Ferr Yang

This paper presents a design method of Chebyshev-type and inverse-Chebyshev-type infinite impulse response (IIR) filters with an approximately linear phase response. In the design of Chebyshev-type filters, the flatness condition in the stopband is preincorporated into a transfer function, and an equiripple characteristic in the passband is achieved by iteratively solving the QP problem using the transfer function. In the design of inverse-Chebyshev-type filters, the flatness condition in the passband is added to the constraint of the QP problem as the linear matrix equality, and an equiripple characteristic in the stopband is realized by iteratively solving the QP problem. To guarantee the stability of the obtained filters, we apply the extended positive realness to the QP problem. As a result, the proposed method can design the filters with more high precision than the conventional methods. The effectiveness of the proposed design method is illustrated with some examples.

1. Introduction

Digital filters are needed in many applications in signal processing. Infinite impulse response (IIR) filters are useful for high-speed processing, and IIR filters with lower order can be realized that are comparable to the magnitude responses of finite impulse response (FIR) filters. However, its design is more difficult than FIR filters because IIR filters have a rational transfer function and are not always stable. For that reason, the study on IIR filter design is still a hot topic in the area of digital signal processing and many design methods on the IIR filters have been presented [1–8]. In this paper, we will treat two types of filters: Chebyshev type which is equiripple in the passband and flat in the stopband, Inverse Chebyshev type which is equiripple in the stopband and flat in the stopband. These filters are needed in many image processing applications to suppress ringing and chessboard distortion and are used in each situation appropriately. Moreover, in many signal processing applications, such as wavelet transform and channel equalization, the complex coefficient digital filters are required to meet some specifications that cannot be achieved by real coefficient filters, such as asymmetric spectral response [9–11].

In [12, 13], the design methods based on the Remez algorithm were proposed for the Chebyshev-type and inverse-Chebyshev-type IIR filters with an approximately linear phase response. These methods can design the filters with small computational complexity. However, the filters that can be designed using these methods are restricted greatly because of a condition imposed on setting the initial value. Moreover, these methods cannot guarantee the stability of the filter obtained. By using the linear semi-infinite programming and the extended positive realness, the design method of stable inverse Chebyshev-type IIR filters with an approximately linear phase response has been proposed [14]. This method needs to set an initial value appropriately and the performance of the filter obtained is dependent on the initial value.

In this paper, a design method based on quadratic programming (QP) is presented for stable IIR filters and for FIR filters with prescribed flatness in the passband or stopband and an approximately linear phase response in the passband. To guarantee the stability of the filter obtained, we apply the extended positive realness to the QP problem. As a result, the proposed method can design the filters that cannot be designed by the conventional methods. This paper is organized as follows: in Section 2, the flatness conditions in the passband and stopband are described. In Section 3, the design algorithm for inverse-Chebyshev-type filters is described. In Section 4, the design algorithm for Chebyshev-type filters is described. To verify the effectiveness of the proposed method, several design examples are given in Section 5. Section 6 is the conclusions of this work.

2. IIR Digital Filters and Flatness Conditions

The frequency response $H(e^{j\omega})$ of an IIR digital filter is defined as

$$H\left(e^{j\omega}\right) = \frac{A\left(e^{j\omega}\right)}{B\left(e^{j\omega}\right)} = \frac{\sum_{n=0}^{N} a_n e^{-jn\omega}}{\sum_{m=0}^{M} b_m e^{-jm\omega}}, \tag{1}$$

where N and M are the orders of the numerator and denominator, respectively. $a_n = a_{rn} + ja_{in}$ and $b_m = b_{rm} + jb_{im}$ are the filter coefficients and $b_0 = 1$ in general. The desired frequency response $H_d(e^{j\omega})$ can be expressed as

$$H_d\left(e^{j\omega}\right) = \begin{cases} e^{-j\tau_d\omega}, & (\text{passband}), \\ 0, & (\text{stopband}), \end{cases} \tag{2}$$

where τ_d is a desired group delay.

Then, the flatness conditions of the magnitude and group delay at $\omega = \omega_u$ in the passband are given as follows:

$$\left.\frac{\partial^i \left|H\left(e^{j\omega}\right)\right|}{\partial\omega^i}\right|_{\omega=\omega_u} = \begin{cases} 1, & (i = 0), \\ 0, & \left(i = 1, 2, \ldots, K_p - 1\right), \end{cases}$$
$$\left.\frac{\partial^i \tau(\omega)}{\partial\omega^i}\right|_{\omega=\omega_u} = \begin{cases} \tau_d, & (i = 0), \\ 0, & \left(i = 1, 2, \ldots, K_p - 2\right), \end{cases} \tag{3}$$

where K_p is a parameter expressing the flatness in the passband. The magnitude flatness condition at $\omega = \omega_t$ in the stopband is

$$\left.\frac{\partial^i \left|H\left(e^{j\omega}\right)\right|}{\partial\omega^i}\right|_{\omega=\omega_t} = 0, \quad (i = 0, 1, \ldots, K_s - 1), \tag{4}$$

where K_s is a parameter expressing the flatness in the stopband.

3. Design of Inverse-Chebyshev-Type Filters

3.1. Problem Formulation. Let $\widetilde{H}(e^{j\omega})$ be a noncausal shifted version of $H(e^{j\omega})$;

$$\begin{aligned}\widetilde{H}\left(e^{j\omega}\right) &= H\left(e^{j\omega}\right) e^{j\tau_d\omega} \\ &= \frac{\sum_{n=0}^{N} a_n e^{-j(n-\tau_d)\omega}}{\sum_{m=0}^{M} b_m e^{-jm\omega}} = \frac{\widetilde{A}\left(e^{j\omega}\right)}{B\left(e^{j\omega}\right)}.\end{aligned} \tag{5}$$

With (5), the flatness conditions in (3) become

$$\left.\frac{\partial^i \widetilde{H}\left(e^{j\omega}\right)}{\partial\omega^i}\right|_{\omega=\omega_u} = \begin{cases} 1, & (i = 0), \\ 0, & \left(i = 1, 2, \ldots, K_p - 1\right). \end{cases} \tag{6}$$

It is shown in Appendix that (6) is equivalent to (3). Equation (6) is the same as follows [15]:

$$\left.\frac{\partial^i \widetilde{A}\left(e^{j\omega}\right)}{\partial\omega^i}\right|_{\omega=\omega_u} = \left.\frac{\partial^i B\left(e^{j\omega}\right)}{\partial\omega^i}\right|_{\omega=\omega_u}, \quad \left(i = 0, 1, \ldots, K_p - 1\right). \tag{7}$$

Consequently, we can get the linear equation in matrix form:

$$\mathbf{U}\mathbf{h} = \mathbf{V}, \tag{8}$$

where $\mathbf{U} = [\mathbf{U}_1 \ \ \mathbf{U}_2]$, $\mathbf{V} = [1, 0, \ldots, 0]^T$, and $\mathbf{h} = [a_{r0}, \ldots, a_{rN}, a_{i0}, \ldots, a_{iN}, b_{r1}, \ldots, b_{rM}, b_{i1}, \ldots, b_{iM}]^T$ and $(\cdot)^T$ denotes the transpose of $(\cdot)$. $\mathbf{U}_1$ and $\mathbf{U}_2$ are in (9) and (10), respectively:

$$\mathbf{U}_1 = \begin{bmatrix} (0-\tau_d)^0 \cos((0-\tau_d)\omega_u) & \cdots & (N-\tau_d)^0 \cos((N-\tau_d)\omega_u) & (0-\tau_d)^0 \sin((0-\tau_d)\omega_u) & \cdots & (N-\tau_d)^0 \sin((N-\tau_d)\omega_u) \\ \vdots & \ddots & \vdots & \vdots & \ddots & \vdots \\ (0-\tau_d)^{K_p-1} \cos((0-\tau_d)\omega_u) & \cdots & (N-\tau_d)^{K_p-1} \cos((N-\tau_d)\omega_u) & (0-\tau_d)^{K_p-1} \sin((0-\tau_d)\omega_u) & \cdots & (N-\tau_d)^{K_p-1} \sin((N-\tau_d)\omega_u) \\ -(0-\tau_d)^0 \sin((0-\tau_d)\omega_u) & \cdots & -(N-\tau_d)^0 \sin((N-\tau_d)\omega_u) & (0-\tau_d)^0 \cos((0-\tau_d)\omega_u) & \cdots & (N-\tau_d)^0 \cos((N-\tau_d)\omega_u) \\ \vdots & \ddots & \vdots & \vdots & \ddots & \vdots \\ -(0-\tau_d)^{K_p-1} \sin((0-\tau_d)\omega_u) & \cdots & -(N-\tau_d)^{K_p-1} \sin((N-\tau_d)\omega_u) & (0-\tau_d)^{K_p-1} \cos((0-\tau_d)\omega_u) & \cdots & (N-\tau_d)^{K_p-1} \cos((N-\tau_d)\omega_u) \end{bmatrix}, \tag{9}$$

$$\mathbf{U}_2 = \begin{bmatrix} -1^0 \cos(1\omega_u) & \cdots & -M^0 \cos(M\omega_u) & -1^0 \sin(1\omega_u) & \cdots & -M^0 \sin(M\omega_u) \\ \vdots & \ddots & \vdots & \vdots & \ddots & \vdots \\ -1^{K_p-1} \cos(1\omega_u) & \cdots & -M^{K_p-1} \cos(M\omega_u) & -1^{K_p-1} \sin(1\omega_u) & \cdots & -M^{K_p-1} \sin(M\omega_u) \\ 1^0 \sin(1\omega_u) & \cdots & M^0 \sin(M\omega_u) & -1^0 \cos(1\omega_u) & \cdots & -M^0 \cos(M\omega_u) \\ \vdots & \ddots & \vdots & \vdots & \ddots & \vdots \\ 1^{K_p-1} \sin(1\omega_u) & \cdots & M^{K_p-1} \sin(M\omega_u) & -1^{K_p-1} \cos(1\omega_u) & \cdots & -M^{K_p-1} \cos(M\omega_u) \end{bmatrix}. \tag{10}$$

Let $\widetilde{H}_d(e^{j\omega})$ be the desired magnitude response; that is,

$$\widetilde{H}_d\left(e^{j\omega}\right) = \begin{cases} 1, & \text{(passband)}, \\ 0, & \text{(stopband)}. \end{cases} \quad (11)$$

Using (5) and (11), the weighted least squares design problem is

$$\min_{\mathbf{h}} J = \sum_{l=1}^{L} W(\omega_l) \left|\widetilde{H}\left(e^{j\omega_l}\right) - \widetilde{H}_d\left(e^{j\omega_l}\right)\right|^2, \quad (12)$$

where L is the total number of grid points in the passband and stopband, $W(\omega_l)$ is the weighting function, and ω_l ($l = 1, \ldots, L$) are the discrete frequency points used in the calculation. However, it is difficult to solve (12) directly because $\widetilde{H}(e^{j\omega_l})$ is a rational function. Thus, we use the following iterative design formula:

$$\min_{\mathbf{h}} J = \sum_{l=1}^{L} \frac{W(\omega_l) \left|\widetilde{A}\left(e^{j\omega_l}\right) - \widetilde{H}_d\left(e^{j\omega_l}\right) B\left(e^{j\omega_l}\right)\right|^2}{\left|B_{k-1}\left(e^{j\omega_l}\right)\right|^2}, \quad (13)$$

where k is the number of the iterations.

After some manipulation, (13) can be formulated as the following QP problem:

$$\min_{\mathbf{h}} \mathbf{h}^T \left(\mathbf{P}_R^T \mathbf{W} \mathbf{P}_R + \mathbf{P}_I^T \mathbf{W} \mathbf{P}_I\right) \mathbf{h} - 2\left(\mathbf{d}_R^T \mathbf{W} \mathbf{P}_R - \mathbf{d}_I^T \mathbf{W} \mathbf{P}_I\right) \mathbf{h}, \quad (14)$$

where $\mathbf{P}_R = [(\mathbf{e}_{NR} + \mathbf{e}_{NI}), -(\mathbf{D}_R\mathbf{e}_{MR} + \mathbf{D}_I\mathbf{e}_{MI})]$, $\mathbf{P}_I = [(\mathbf{e}_{NI} - \mathbf{e}_{NR}), (\mathbf{D}_I\mathbf{e}_{MR} - \mathbf{D}_R\mathbf{e}_{MI})]$, $\mathbf{e}_{NR} = [1, \ldots, \cos(\omega_l(N-\tau_d)), 0, \ldots, \sin(\omega_l(N-\tau_d))]$, $\mathbf{e}_{NI} = [0, \ldots, -\sin(\omega_l(N-\tau_d)), 1, \ldots, \cos(\omega_l(N-\tau_d))]$, $\mathbf{e}_{MR} = [\cos(\omega_l), \ldots, \cos(\omega_l M), \sin(\omega_l), \ldots, \sin(\omega_l M)]$, $\mathbf{e}_{MI} = [-\sin(\omega_l), \ldots, -\sin(\omega_l M), \cos(\omega_l), \ldots, \cos(\omega_l M)]$, $\mathbf{D} = \text{diag}(\mathbf{d}) = \mathbf{D}_R - j\mathbf{D}_I$, $\mathbf{d} = [\widetilde{H}_d(e^{j\omega_1}), \ldots, \widetilde{H}_d(e^{j\omega_L})] = \mathbf{d}_R - j\mathbf{d}_I$, and $\mathbf{W} = \text{diag}([W(\omega_1)/|B_{k-1}(\omega_0)|^2, \ldots, W(\omega_L)/|B_{k-1}(\omega_L)|^2])$.

3.2. Update of the Weighting Function $W(\omega)$. It has been well known that the filters obtained under weighted least square criterion have a large ripple near the band edges. So in order to realize the equiripple characteristics in the passband or stopband or both, the weighting function used at every iteration is adjusted using the modified Lawson's method [16] and the QP problem is solved to obtain the coefficients. In this paper, the weighting function $W(\omega)$ in kth iteration step is updated as follows:

$$W_{k+1}(\omega) = \frac{W_k(\omega)\beta_k(\omega)}{(1/L)\sum_{l=1}^{L} W_k(\omega_l)\beta_k(\omega_l)}, \quad (15)$$

where the envelope function $\beta_k(\omega)$ is given as the function of straight line formed by joining together all the extremal points of the same frequency band of interest on the error function which is expressed as

$$E_k(\omega) = \left|\widetilde{H}\left(e^{j\omega}\right) - \widetilde{H}_d\left(e^{j\omega}\right)\right|. \quad (16)$$

Using the extremal points $\widehat{\omega}_i$ of $E_k(\omega)$, $\beta_k(\omega)$ can be calculated by

$$\beta_k(\omega) = \frac{\omega - \widehat{\omega}_i}{\widehat{\omega}_{i+1} - \widehat{\omega}_i} E_k(\widehat{\omega}_{i+1}) + \frac{\widehat{\omega}_{i+1} - \omega}{\widehat{\omega}_{i+1} - \widehat{\omega}_i} E_k(\widehat{\omega}_i), \quad \text{for } \widehat{\omega}_i < \omega < \widehat{\omega}_{i+1}, \quad (17)$$

where $\widehat{\omega}_i$ denotes the ith extremal frequency of the error function $E_k(\omega)$.

3.3. Stability Constraint. To obtain the stable IIR filters, the stability condition based on a positive realness has been applied to many design methods. However, the use of this condition may exclude the candidate for the transfer function with excellent performance because this condition is a sufficient condition to assure the stability and is often too restrictive. In [17], an iterative method for the stability guarantee based on the positive realness was proposed in order to get a better transfer function. In this method, a stability condition is given by

$$\text{Re}\left\{B\left(e^{j\omega}\right)\right\} \geq \delta, \quad -\pi \leq \omega < \pi, \quad (18)$$

where $\delta < 1$. If the maximum pole radius $P_{\max}$ of the filter obtained using a given δ is greater than prescribed maximum allowable pole radius r_m, δ is updated and redesign is carried out using the updated δ. The update of δ and the redesign are repeated until satisfying $|P_{\max} - r_m| \leq \epsilon_R$, where ϵ_R is a positive small value. The update procedure of δ is described in Section 4.2.

Using the discrete angular frequency ω_i ($i = 1, \ldots, R$), (18) can be expressed as the linear matrix inequality

$$\mathbf{\Gamma h} \geq \boldsymbol{\lambda}, \quad (19)$$

where

$$\mathbf{\Gamma} = \begin{bmatrix} 0 & \cdots & 0 & \cos(\omega_1) & \cdots & \cos(M\omega_1) \\ \vdots & \ddots & \vdots & \vdots & \ddots & \vdots \\ 0 & \cdots & 0 & \cos(\omega_R) & \cdots & \cos(M\omega_R) \end{bmatrix}, \quad \boldsymbol{\lambda} = [\delta - 1, \ldots, \delta - 1]^T. \quad (20)$$

Thus, the design problem in which the stability constraint and the update of the weighting function were considered becomes a standard QP problem as below:

$$\begin{aligned} \min_{\mathbf{h}_k} \quad & \mathbf{h}_k^T \left(\mathbf{P}_R^T \mathbf{W}_k \mathbf{P}_R + \mathbf{P}_I^T \mathbf{W}_k \mathbf{P}_I\right) \mathbf{h}_k \\ & - 2\left(\mathbf{d}_R^T \mathbf{W}_k \mathbf{P}_R - \mathbf{d}_I^T \mathbf{W}_k \mathbf{P}_I\right) \mathbf{h}_k, \\ \text{sub. to} \quad & \mathbf{\Gamma h}_k \geq \boldsymbol{\lambda}, \\ & \mathbf{U h}_k = \mathbf{V}, \end{aligned} \quad (21)$$

where $\mathbf{W}_k = \text{diag}([W_k(\omega_1)/|B_{k-1}(\omega_0)|^2, \ldots, W_k(\omega_L)/|B_{k-1}(\omega_L)|^2])$.

This problem can be solved using a powerful QP tool, such as *quadprog* in MATLAB.

4. Design of Chebyshev-Type Filters

4.1. Problem Formulation. In this section, we consider the filters with a flatness in the stopband. With a noncausal shifted version $\widetilde{H}(e^{j\omega})$ in (5), the flatness condition in (4) becomes

$$\left.\frac{\partial^i \left|\widetilde{H}\left(e^{j\omega}\right)\right|}{\partial \omega^i}\right|_{\omega=\omega_t} = 0, \quad (i = 0, 1, \ldots, K_s - 1). \tag{22}$$

In order to meet the flatness condition of (22), it is necessary to place K_s multiple zeros at $\omega = \omega_t$. Hence, the frequency response $\widehat{H}(e^{j\omega})$ can be expressed as

$$\widehat{H}\left(e^{j\omega}\right) = \frac{\widehat{A}\left(e^{j\omega}\right)}{B\left(e^{j\omega}\right)} = \frac{\left(1+e^{-j\omega}\right)^{K_s} \sum_{n=0}^{N-K_s} c_n e^{-j(n-\tau_d)\omega}}{1+\sum_{m=1}^{M} b_m e^{-jm\omega}}, \tag{23}$$

where $c_n = c_{rn} + jc_{in}$. To obtain the equiripple responses in the passband, we consider the following iterative design formula:

$$\min_{\mathbf{c},\mathbf{b}} J = \sum_{l=1}^{L} \frac{W\left(\omega_l\right)}{\left|B_{k-1}\left(e^{j\omega_l}\right)\right|^2} \left|\widehat{A}\left(e^{j\omega_l}\right) - H_d\left(e^{j\omega_l}\right) B\left(e^{j\omega_l}\right)\right|^2. \tag{24}$$

Thus, the design problem in which the stability constraint and the update of the weighting function were considered becomes a standard QP problem as below:

$$\begin{aligned} \min_{\widehat{\mathbf{h}}_\mathbf{k}} \quad & \widehat{\mathbf{h}}_\mathbf{k}^T \left(\widehat{\mathbf{P}}_R^T \mathbf{W}_k \widehat{\mathbf{P}}_R + \widehat{\mathbf{P}}_I^T \mathbf{W}_k \widehat{\mathbf{P}}_I\right) \widehat{\mathbf{h}}_\mathbf{k} \\ & - 2\left(\mathbf{d}_R^T \mathbf{W}_k \widehat{\mathbf{P}}_R - \mathbf{d}_I^T \mathbf{W}_k \widehat{\mathbf{P}}_I\right) \widehat{\mathbf{h}}_\mathbf{k}, \\ \text{sub. to} \quad & \mathbf{\Gamma} \widehat{\mathbf{h}}_\mathbf{k} \geq \boldsymbol{\lambda}, \end{aligned} \tag{25}$$

where $\widehat{\mathbf{h}} = [c_{r0}, \ldots, c_{r(N-K_s)}, c_{i0}, \ldots, c_{i(N-K_s)}, b_{r1}, \ldots, b_{rM}, b_{i1}, \ldots, b_{iM}]^T$, $\widehat{\mathbf{P}}_R = [(\mathbf{G}_R \widehat{\mathbf{e}}_{NR} + \mathbf{G}_I \widehat{\mathbf{e}}_{NI}), -(\mathbf{D}_R \mathbf{e}_{MR} + \mathbf{D}_I \mathbf{e}_{MI})]$, $\widehat{\mathbf{P}}_I = [(\mathbf{G}_R \widehat{\mathbf{e}}_{NI} - \mathbf{G}_I \widehat{\mathbf{e}}_{NR}), (\mathbf{D}_I \mathbf{e}_{MR} - \mathbf{D}_R \mathbf{e}_{MI})]$, $\mathbf{G} = \operatorname{diag}((1-e^{-j(\omega_l-\omega_t)})^{K_s}) = \mathbf{G}_R - j\mathbf{G}_I$, $\widehat{\mathbf{e}}_{NR} = [1, \ldots, \cos(\omega_l(N-K_s)), 0, \ldots, \sin(\omega_l(N-K_s))]$, $\widehat{\mathbf{e}}_{NI} = [0, \ldots, -\sin(\omega_l(N-K_s)), 1, \ldots, \cos(\omega_l(N-K_s))]$, $\mathbf{e}_{MR} = [\cos(\omega_l), \ldots, \cos(\omega_l M), \sin(\omega_l), \ldots, \sin(\omega_l M)]$, $\mathbf{e}_{MI} = [-\sin(\omega_l), \ldots, -\sin(\omega_l M), \cos(\omega_l), \ldots, \cos(\omega_l M)]$, $\mathbf{D} = \operatorname{diag}(\mathbf{d}) = \mathbf{D}_R - j\mathbf{D}_I$, $\mathbf{d} = [H_d(e^{j\omega_1}), \ldots, H_d(e^{j\omega_L})] = \mathbf{d}_R - j\mathbf{d}_I$, and $\mathbf{W}_k = \operatorname{diag}([W_k(\omega_1)/|B_{k-1}(\omega_1)|^2, \ldots, W_k(\omega_L)/|B_{k-1}(\omega_L)|^2])$.

4.2. Design Procedure. The design procedure of the proposed method is summarized as follows.

Step 0. Set the design specifications.

Step 1. Solve the QP problem to obtain the filter coefficient $\mathbf{h}_k$.

Step 2. If $\operatorname{sum}(|\mathbf{h}_k - \mathbf{h}_{k-1}|)/\operatorname{sum}(|\mathbf{h}_k|) \leq \epsilon$ and $P_{\max} < r_m$, stop; if $\operatorname{sum}(|\mathbf{h}_k - \mathbf{h}_{k-1}|)/\operatorname{sum}(|\mathbf{h}_k|) \leq \epsilon$ and $P_{\max} > r_m$, go to Step 4; otherwise, go to Step 3.

Step 3. Update the weighting function $W(\omega)$ using (15)–(17) and go back to Step 1.

Step 4. Calculate $\delta' = \min_{\text{all } \omega} \operatorname{Re}\{B(\omega)\}$ from the obtained filter and then set to $\delta_l = \delta'$ and $\delta_u = 1$.

Step 5. Calculate $\delta = (\delta_l + \delta_u)/2$.

Step 6. Solve the QP problem using the updated δ.

Step 7. If $\operatorname{sum}(|\mathbf{h}_k - \mathbf{h}_{k-1}|)/\operatorname{sum}(|\mathbf{h}_k|) \leq \epsilon$ and $|P_{\max} - r_m| \leq \epsilon_R$, stop; if $\operatorname{sum}(|\mathbf{h}_k - \mathbf{h}_{k-1}|)/\operatorname{sum}(|\mathbf{h}_k|) \leq \epsilon$ and $|P_{\max} - r_m| > \epsilon_R$, go to Step 9; otherwise then go to Step 8.

Step 8. Update the weighting function $W(\omega)$ using (15)–(17) and go back to Step 6.

Step 9. If $P_{\max} < r_m$, set to $\delta_u = \delta$; if $P_{\max} > r_m$, set to $\delta_l = \delta$, and then go back to Step 5.

5. Design Examples

In this section, some design examples are given to illustrate the effectiveness of the proposed method. In all the following examples, $\epsilon = 10^{-7}$, $\epsilon_R = 10^{-5}$, and the initial value of δ is -10^2. The program of the proposed design algorithm was coded by MATLAB and was run on the PC which has Core i5-M520 processor and 2 GB memory.

5.1. Example 1. To compare with the conventional method [12], we design the Chebyshev-type IIR filters which have an equiripple characteristic in passband and a flat characteristic in stopband. The filter specifications are as follows:

$$\begin{gathered} N = 15, \qquad M = 6, \qquad K_s = 11, \\ H_d\left(e^{j\omega}\right) = \begin{cases} e^{-j\tau_d\omega}, & 0 \leq |\omega| \leq 0.3\pi, \\ 0, & \omega = \pi, \end{cases} \\ \tau_d = \{9, 10, 11, 12, 13\}. \end{gathered} \tag{26}$$

In the proposed method, we set the maximum allowable pole radius r_m to 0.98, and the number of grid points is $L = 500$ and $R = 1000$. Note that the conventional method [12] cannot guarantee the stability of the obtained filter.

The performance of the obtained filters is shown in Table 1 which are obtained by the conventional method [12] based on Remez algorithm. In Table 1, R_p is the maximum magnitude error in passband, G_{err} is the maximum group delay error in passband, and $P_{\max}$ is the maximum pole radius of the obtained filter. From Table 1, it is confirmed that the proposed method needs more computational cost (CPU time and iteration number) than the conventional method, but the characteristic of the obtained filters by the proposed method is about the same as or better than them by the conventional method. Particularly, the filter obtained by the conventional method is unstable when $\tau_d = 9$. The stability of the filters obtained by the method which cannot guarantee the stability depends on a given group delay specification. In fact, when the group delay τ_d is less than 9.5, all filters obtained by conventional method were unstable. In contrast, the proposed method can always obtain the stable filters

TABLE 1: Comparison with [12] in Example 1.

	τ_d	R_p	G_{err}	P_{max}	CPU time (s)	Iteration number
Proposed	9	5.622×10^{-5}	3.618×10^{-3}	0.980	1.334	16
	10	1.248×10^{-5}	7.741×10^{-4}	0.952	0.587	10
	11	4.416×10^{-5}	6.501×10^{-3}	0.890	0.537	10
	12	1.098×10^{-4}	3.224×10^{-2}	0.858	0.510	8
	13	2.060×10^{-4}	1.146×10^{-1}	0.847	0.525	9
Reference [12]	9	2.281×10^{-6}	2.657×10^{-5}	1.051	0.121	6
	10	1.269×10^{-5}	7.257×10^{-4}	0.953	0.134	6
	11	4.608×10^{-5}	6.499×10^{-3}	0.891	0.124	6
	12	1.168×10^{-4}	3.238×10^{-2}	0.858	0.124	6
	13	3.129×10^{-4}	1.122×10^{-1}	0.846	0.130	6

TABLE 2: Comparison with [13] in Example 2.

	τ_d	R_s (dB)	P_{max}	CPU time (s)	Iteration number
Proposed	7	27.85	0.980	1.812	45
	8	35.02	0.928	0.512	16
	9	41.38	0.850	0.372	14
	10	46.61	0.784	0.385	12
	11	50.89	0.746	0.351	11
Reference [13]	7	25.78	1.002	0.212	7
	8	34.07	0.924	0.205	7
	9	40.46	0.850	0.202	7
	10	45.76	0.784	0.207	7
	11	50.10	0.746	0.198	6

because the maximum pole radius of the filter obtained can be prespecified. Therefore, the proposed method can design the filters that cannot be designed by the conventional method.

5.2. Example 2. Next we design the inverse Chebyshev-type IIR filters with the real coefficients in order to compare with the conventional method [13]. The filter specifications are as follows:

$$N = 12, \qquad M = 5, \qquad K_p = 10,$$
$$H_d\left(e^{j\omega}\right) = \begin{cases} e^{-j\tau_d\omega}, & \omega = 0, \\ 0, & 0.5\pi \le |\omega| \le \pi, \end{cases} \tag{27}$$
$$\tau_d = \{7, 8, 9, 10, 11\}.$$

In the proposed method, we set the maximum allowable pole radius r_m to 0.98, and the number of grid points is $L = 500$ and $R = 1000$. Note that the conventional method [13] cannot guarantee the stability of the obtained filter.

The numerical performance of the obtained filters is shown in Table 2 which are obtained by the conventional method [13] based on Remez algorithm. In Table 2, R_s is the minimum stopband attenuation in dB. From Table 2, it is confirmed that the proposed method needs more computational cost than the conventional method but gives slightly better characteristics. Moreover, when the group delay τ_d is less than 7.2, the filters obtained by conventional method were unstable. In contrast, the proposed method can always obtain the stable filters. Therefore, the proposed method can design the filters that cannot be designed by the conventional method.

Next, we designed the filter with $\tau_d = 9$ for comparison with [14]. The minimum stopband attenuation of the filter obtained was 41.38 dB in the proposed method and 40.04 dB in [14]. It is thought that this difference is because the performance of the filter obtained by the conventional method is dependent on the given initial value.

5.3. Example 3. Next, as the example of the filter with the complex coefficients, we design the inverse Chebyshev-type IIR filter with the following specifications:

$$N = 12, \qquad M = 5,$$
$$H_d\left(e^{j\omega}\right) = \begin{cases} e^{-j9\omega}, & \omega = 0.1\pi, \\ 0, & -\pi \le \omega \le -0.6\pi,\ 0.7\pi \le \omega < \pi, \end{cases}$$
$$K_p = \{9, 10, 11\}. \tag{28}$$

In this example, we use $r_m = 1.0$, $L = 500$, and $R = 1000$.

The proposed method took an average of 0.78 seconds for convergence. The magnitude response and group delay response of the obtained filter are shown in Figures 1, 2, and 3. From these figures, it is confirmed that the magnitude and group delay responses have a flat characteristic at $\omega = 0.1\pi$ and the magnitude response in stopband is equiripple. Table 3 is the numerical performance of the obtained filter.

5.4. Example 4. Finally, we show the design examples of inverse-Chebyshev-type FIR filters. The filter specifications are as follows:

$$N = 1999, \qquad M = 0,$$
$$H_d\left(e^{j\omega}\right) = \begin{cases} e^{-j900\omega}, & \omega = 0, \\ 0, & 0.0080\pi \le |\omega| \le \pi, \end{cases} \tag{29}$$
$$K_p = \{10, 12, 14\}.$$

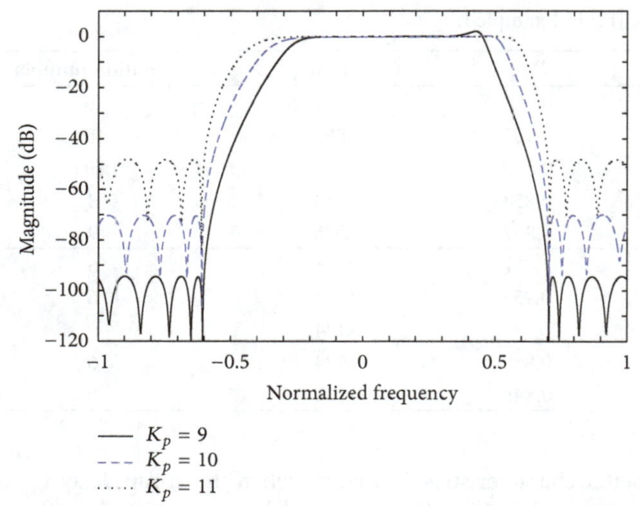

FIGURE 1: Overall magnitude response of the proposed IIR filters in Example 3.

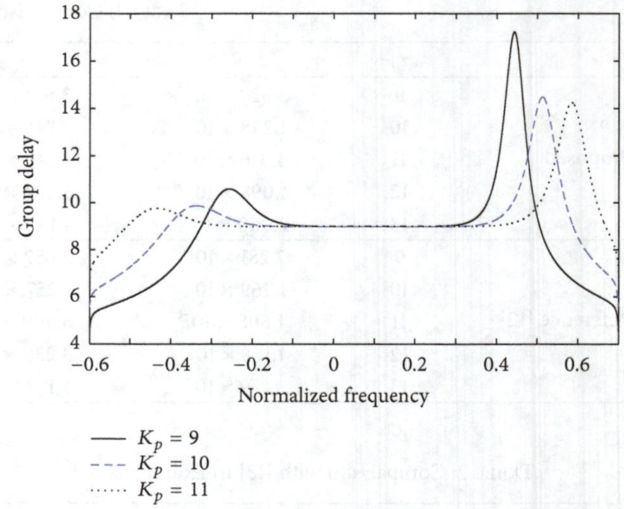

FIGURE 3: Group delay response of the proposed IIR filters in Example 3.

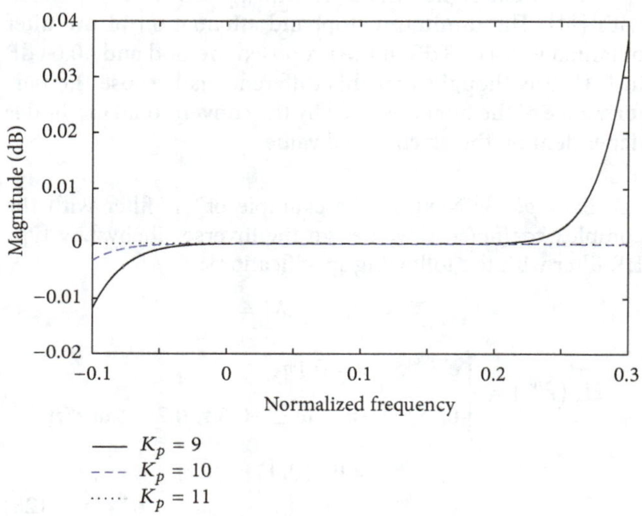

FIGURE 2: Passband magnitude response of the proposed IIR filters in Example 3.

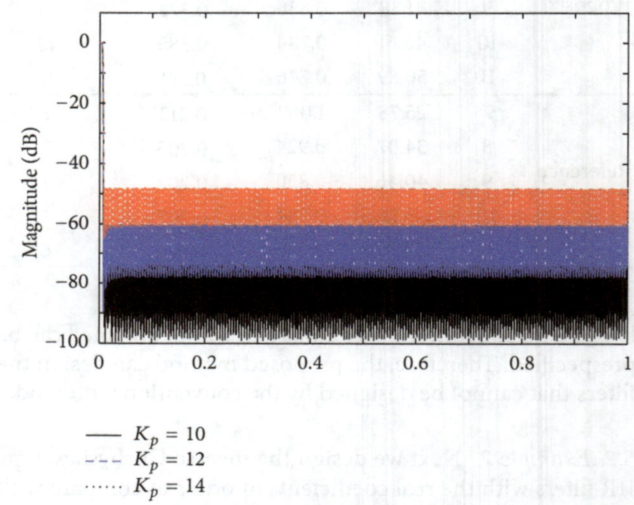

FIGURE 4: Overall magnitude response of the proposed FIR filters in Example 4.

In this example, we set the number of the grit points L to 10,000.

The proposed method took an average of 2,794 seconds for convergence. The magnitude response and group delay response of the obtained filter are shown in Figures 4, 5, and 6. From these figures, it is confirmed that the magnitude and group delay responses have a flat characteristic at $\omega = 0$ and the magnitude response in stopband is equiripple. The minimum stopband attenuation of the obtained filters was 72.98 (dB) for $K_p = 10$, 59.11 (dB) for $K_p = 12$, and 47.19 (dB) for $K_p = 14$. We confirmed that these filters cannot be designed by the conventional method [13]. Hence, the proposed method can also design the high-order filters that cannot be designed by the conventional method.

TABLE 3: The performance of the filters in Example 3.

K_p	R_s (dB)	$P_{\max}$
9	94.43	0.902
10	69.92	0.865
11	48.03	0.856

6. Conclusion

In this paper, a design method based on quadratic programming has been proposed for approximately linear phase IIR filters and FIR filters with prescribed flatness in passband or stopband. To guarantee the stability of the filter obtained, we applied the extended positive realness to the QP problem.

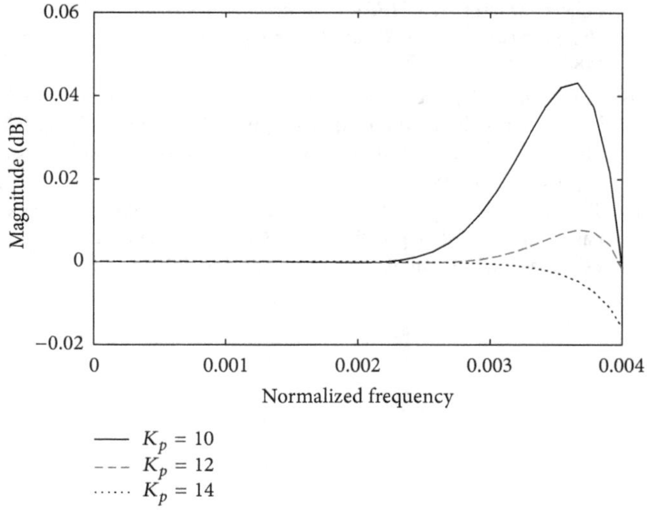

FIGURE 5: Passband magnitude response of the proposed FIR filters in Example 4.

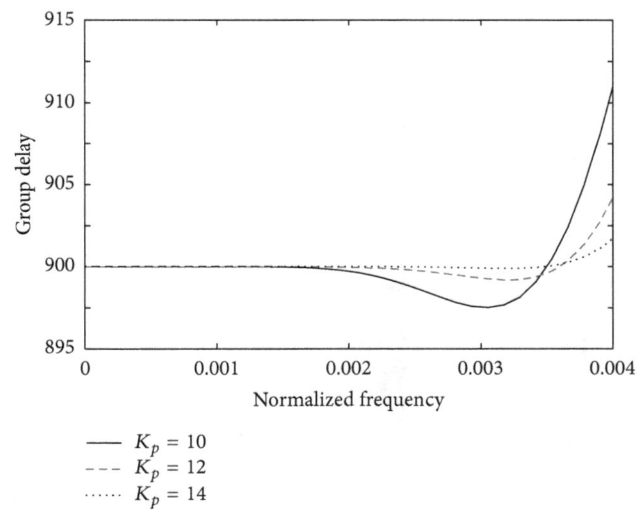

FIGURE 6: Group delay response of the proposed FIR filters in Example 4.

With this method, stable IIR filters and FIR filters that are Chebyshev type and inverse Chebyshev type can be easily designed. In the design examples, we showed that the performance of the filter obtained by the proposed method is about the same as or better than the conventional method, and the filters which cannot be designed by the conventional method can also be designed.

Appendix

In this appendix it is shown that (6) is equivalent to (3).

Equation (1) is expressed with the magnitude response $|H(e^{j\omega})|$ and group delay response $\tau(\omega)$ as follows:

$$H\left(e^{j\omega}\right) = \left|H\left(e^{j\omega}\right)\right| e^{-j\tau(\omega)\omega}. \qquad \text{(A.1)}$$

With (A.1), (5) is expressed as

$$\widetilde{H}\left(e^{j\omega}\right) = \left|H\left(e^{j\omega}\right)\right| e^{j\{\tau_d\omega-\tau(\omega)\omega\}} = \left|H\left(e^{j\omega}\right)\right| e^{j\widetilde{\theta}(\omega)}. \qquad \text{(A.2)}$$

Thus, $\widetilde{H}(e^{j\omega_u}) = 1$ is equivalent to $|H(e^{j\omega_u})| = 1$, $\tau(\omega_u) = \tau_d$ and $\widetilde{\theta}(\omega_u) = 0$.

The first-order derivative of (A.2) is expressed as follows:

$$\frac{\partial \widetilde{H}\left(e^{j\omega}\right)}{\partial\omega} = \frac{\partial\left|H\left(e^{j\omega}\right)\right|}{\partial\omega} e^{j\widetilde{\theta}(\omega)} + j\left|H\left(e^{j\omega}\right)\right| e^{j\widetilde{\theta}(\omega)} \frac{\partial\widetilde{\theta}(\omega)}{\partial\omega}. \qquad \text{(A.3)}$$

Because $|H(e^{j\omega_u})| = 1$ and $\widetilde{\theta}(\omega_u) = 0$, (A.3) is

$$\left.\frac{\partial \widetilde{H}\left(e^{j\omega}\right)}{\partial\omega}\right|_{\omega=\omega_u} = \left.\frac{\partial\left|H\left(e^{j\omega}\right)\right|}{\partial\omega}\right|_{\omega=\omega_u} + j\left.\frac{\partial\widetilde{\theta}(\omega)}{\partial\omega}\right|_{\omega=\omega_u}. \qquad \text{(A.4)}$$

Thus, $(\partial\widetilde{H}(e^{j\omega})/\partial\omega)|_{\omega=\omega_u} = 0$ is equivalent to $(\partial|H(e^{j\omega})|/\partial\omega)|_{\omega=\omega_u} = 0$ and $(\partial\widetilde{\theta}(\omega)/\partial\omega)|_{\omega=\omega_u} = 0$.

Next, we assume that (A.5) is equivalent to (A.6) and (A.7):

$$\left.\frac{\partial^i \widetilde{H}\left(e^{j\omega}\right)}{\partial\omega^i}\right|_{\omega=\omega_u} = \begin{cases} 1, & (i = 0), \\ 0, & \left(i = 1, 2, \ldots, K_p - 1\right), \end{cases} \qquad \text{(A.5)}$$

$$\left.\frac{\partial^i \left|H\left(e^{j\omega}\right)\right|}{\partial\omega^i}\right|_{\omega=\omega_u} = \begin{cases} 1, & (i = 0), \\ 0, & \left(i = 1, 2, \ldots, K_p - 1\right), \end{cases} \qquad \text{(A.6)}$$

$$\left.\frac{\partial^i \tau(\omega)}{\partial\omega^i}\right|_{\omega=\omega_u} = \begin{cases} \tau_d, & (i = 0), \\ 0, & \left(i = 1, 2, \ldots, K_p - 2\right). \end{cases} \qquad \text{(A.7)}$$

The kth-order derivative of $\widetilde{H}(e^{j\omega})$ is

$$\frac{\partial^k \widetilde{H}\left(e^{j\omega}\right)}{\partial\omega^k} = \sum_{i=0}^{k} {}_kC_i \frac{\partial^{k-i}\left|H\left(e^{j\omega}\right)\right|}{\partial\omega^{k-i}} \frac{\partial^i e^{j\widetilde{\theta}(\omega)}}{\partial\omega^i}. \qquad \text{(A.8)}$$

Using (A.6) and (A.7), (A.8) is

$$\begin{aligned} \left.\frac{\partial^k \widetilde{H}\left(e^{j\omega}\right)}{\partial\omega^k}\right|_{\omega=\omega_u} &= \left.\frac{\partial^k\left|H\left(e^{j\omega}\right)\right|}{\partial\omega^k}\right|_{\omega=\omega_u} + \left.\frac{\partial^k e^{j\widetilde{\theta}(\omega)}}{\partial\omega^k}\right|_{\omega=\omega_u} \\ &= \left.\frac{\partial^k\left|H\left(e^{j\omega}\right)\right|}{\partial\omega^k}\right|_{\omega=\omega_u} + j\left.\frac{\partial^k\widetilde{\theta}(\omega)}{\partial\omega^k}\right|_{\omega=\omega_u}. \end{aligned} \qquad \text{(A.9)}$$

Thus, $(\partial^k\widetilde{H}(e^{j\omega})/\partial\omega^k)|_{\omega=\omega_u} = 0$ is equivalent to $(\partial^k|H(e^{j\omega})|/\partial\omega^k)|_{\omega=\omega_u} = 0$ and $(\partial^k\widetilde{\theta}(\omega)/\partial\omega^k)|_{\omega=\omega_u} = 0$.

Because $\widetilde{\theta}(\omega) = \tau_d\omega - \tau(\omega)\omega$ and $\tau(\omega_u) = \tau_d$, $(\partial^k\widetilde{\theta}(\omega)/\partial\omega^k)|_{\omega=\omega_u} = 0$ is equivalent to

$$\left.\frac{\partial^k\{\tau(\omega)\}}{\partial\omega^k}\right|_{\omega=\omega_u} = \frac{\partial^k\{\tau_d\}}{\partial\omega^k}. \qquad \text{(A.10)}$$

Thus (6) is equivalent to (3).

Conflict of Interests

The author declares that there is no conflict of interests regarding to the publication of this paper.

References

[1] S. C. Chan, K. M. Tsui, and K. W. Tse, "Design of constrained causal stable IIR filters using a new second-order-cone-programming-based model-reduction technique," *IEEE Transactions on Circuits and Systems II*, vol. 54, no. 2, pp. 107–111, 2007.

[2] A. Jiang and H. K. Kwan, "Minimax design of IIR digital filters using SDP relaxation technique," *IEEE Transactions on Circuits and Systems. I. Regular Papers*, vol. 57, no. 2, pp. 378–390, 2010.

[3] K. Miyahara, Y. Sugita, and T. Yoshikawa, "Design of IIR digital filters with specified magnitude and group delay using differential evolution," *IEICE Transactions on Fundamentals of Electronics, Communications and Computer Sciences*, vol. J95-A, no. 2, pp. 189–197, 2012.

[4] S.-T. Pan, "Design of robust D-stable IIR filters using genetic algorithms with embedded stability criterion," *IEEE Transactions on Signal Processing*, vol. 57, no. 8, pp. 3008–3016, 2009.

[5] T. Miyata, N. Aikawa, Y. Sugita, and T. Yoshikawa, "A design method for 1-D IIR filters with a necessary and sufficient stability criterion," *International Journal of Innovative Computing, Information and Control*, vol. 7, no. 9, pp. 5605–5618, 2011.

[6] C. Dai, W. Chen, and Y. Zhu, "Seeker optimization algorithm for digital IIR filter design," *IEEE Transactions on Industrial Electronics*, vol. 57, no. 5, pp. 1710–1718, 2010.

[7] R. Kaur, M. S. Patterh, J. S. Dhillon, and D. Singh, "Heuristic search method for digital IIR filter design," *WSEAS Transactions on Signal Processing*, vol. 8, no. 3, pp. 121–134, 2012.

[8] B. Singh, J. S. Dhillon, and Y. S. Brar, "Predator prey optimization method for the design of IIR filter," *WSEAS Transactions on Signal Processing*, vol. 9, no. 2, pp. 51–62, 2013.

[9] X. P. Zhang, M. D. Desai, and Y. N. Peng, "Orthogonal complex filter banks and wavelets: some properties and design," *IEEE Transactions on Signal Processing*, vol. 47, no. 4, pp. 1039–1048, 1999.

[10] R. Perry, D. R. Bull, and A. Nix, "Efficient adaptive complex filtering algorithm with application to channel equalisation," *IEE Proceedings—Vision, Image and Signal Processing*, vol. 146, no. 2, pp. 57–64, 1999.

[11] Z. Lin and Y. Liu, "Design of complex FIR filters with reduced group delay error using semidefinite programming," *IEEE Signal Processing Letters*, vol. 13, no. 9, pp. 529–532, 2006.

[12] R. Takeuchi, X. Zhang, T. Yoshikawa, and Y. Takei, "Design of Chebyshev-type IIR filters with approximately linear phase characteristics," *Electronics and Communications in Japan, Part III: Fundamental Electronic Science*, vol. 87, no. 2, pp. 1–9, 2004.

[13] X. Zhang, M. Aizawa, and T. Yoshikawa, "Design of flat delay IIR filters with flat passband and equiripple stopband magnitude responses," *IEICE Transactions on Fundamentals*, vol. 90, no. 7, pp. 578–587, 2007.

[14] T. Matsumoto and K. Suyama, "Design of flat delay IIR filters with flat passband and equiripple stopband magnitude responses," *IEICE Transactions on Fundamentals*, vol. 94, no. 3, pp. 159–167, 2011.

[15] R. Hegde and B. A. Shenoi, "Magnitude approximation of digital filters with specified degrees of flatness and constant group delay characteristics," *IEEE Transactions on Circuits and Systems II: Analog and Digital Signal Processing*, vol. 45, no. 11, pp. 1476–1486, 1998.

[16] Y.-C. Lim, J.-H. Lee, C. K. Chen, and R.-H. Yang, "A weighted least squares algorithm for quasi-equiripple FIR and IIR digital filter design," *IEEE Transactions on Signal Processing*, vol. 4, no. 3, pp. 551–558, 1992.

[17] T. Yamazaki and K. Suyama, "Design of stable IIR filters based on extended positive realness," *IEICE Transactions on Fundamentals A*, vol. 93, no. 8, pp. 517–526, 2010.

12

Visual Sensor based Image Segmentation by Fuzzy Classification and Subregion Merge

Huidong He,[1] Xiaoqian Mao,[1] Wei Li,[2] Linwei Niu,[3] and Genshe Chen[4]

[1]*School of Electrical and Information Engineering, Tianjin University, Tianjin 300072, China*
[2]*Department of Computer & Electrical Engineering and Computer Science, California State University, Bakersfield, CA 93311, USA*
[3]*Department of Math and Computer Science, West Virginia State University, Institute, WV 25112, USA*
[4]*Intelligent Fusion Technology, Inc., Germantown, MD 20876, USA*

Correspondence should be addressed to Wei Li; wli@csub.edu

Academic Editor: Ping Feng Pai

The extraction and tracking of targets in an image shot by visual sensors have been studied extensively. The technology of image segmentation plays an important role in such tracking systems. This paper presents a new approach to color image segmentation based on fuzzy color extractor (FCE). Different from many existing methods, the proposed approach provides a new classification of pixels in a source color image which usually classifies an individual pixel into several subimages by fuzzy sets. This approach shows two unique features: the spatial proximity and color similarity, and it mainly consists of two algorithms: CreateSubImage and MergeSubImage. We apply the FCE to segment colors of the test images from the database at UC Berkeley in the RGB, HSV, and YUV, the three different color spaces. The comparative studies show that the FCE applied in the RGB space is superior to the HSV and YUV spaces. Finally, we compare the segmentation effect with Canny edge detection and Log edge detection algorithms. The results show that the FCE-based approach performs best in the color image segmentation.

1. Introduction

Image segmentation is a fundamental issue in image processing and computer vision as it is a preliminary process in many applications that need to segment an image into meaningful regions. Applications of segmenting gray-level images are very limited so studies on color image segmentation become interesting because color images provide much more information than gray-level ones. A number of image segmentation techniques have been proposed in past decades [1–3]. A physical model in [4] is proposed for color image understanding. A widely used technique for color image segmentation is based on the idea of region growing. A color segmentation algorithm combining region growing and region merging processes is discussed in [5]. The fuzzy growing and expanding approach presented in [6] uses histogram tables for fine segmentation. A method for segmentation of color texture region using the defined class maps is presented in [7]. An approach to contrast analysis using fuzzy region growing is documented in [8]. Color image segmentation using seeded region growing is presented in [9–11]. Most pixel clustering algorithms work in the RGB color space in which the pixels are represented by red, green, and blue components. Clustering methods [12, 13], for example, *c*-means fuzzy classification [14], are often considered as an unsupervised classification of pixels as a prior knowledge of the image. Both supervised and unsupervised segmentation of textured color images are presented in [15, 16]. The level setting approaches to process a color image for segmentation are discussed in [17, 18]. All of the proposed techniques above provide a crisp segmentation of images, where each pixel is classified into a unique subset. This classification may not reflect an understanding of images by humans very well, as the work [1] stated "the image segmentation problem is basically one of psychophysical perception, and therefore not susceptible to a purely analytical solution." Probably, one of

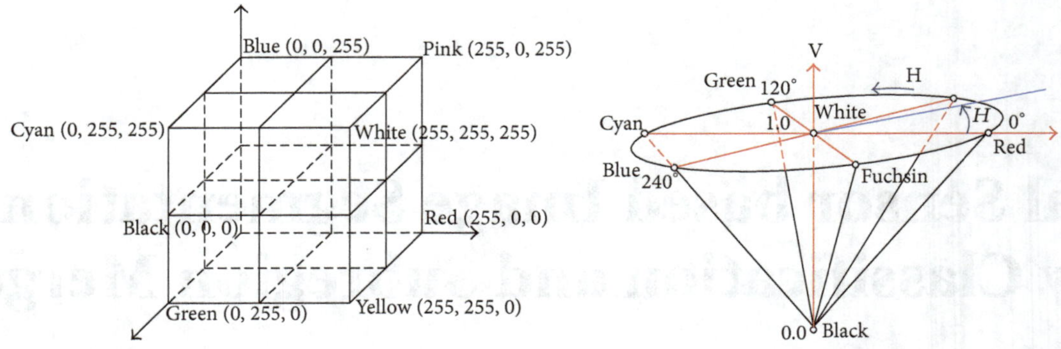

(a) RGB color space

(b) HSV color space

FIGURE 1: Color distribution in the RGB and HSV color spaces.

the difficulties in color image segmentation is how to define color conception, since an understanding of colors in an image varies by different observers.

Considering uncertainty and ambiguity in many real applications, this paper presents an approach to color image segmentation based on a fuzzy color extractor (FCE). Its preliminary version appears in [19]. The FCE is directly extended from the fuzzy gray-level extractor, which was applied to recognize landmarks on roads for autonomous vehicle navigation in [20–22]; that is, the fuzzy rules for extracting a road from a gray-level image are as follows:

> *If* the gray level of a pixel is close to the gray-level of the road,
>
> *Then* one can make it black, or
>
> *Else* one can make it to white.

If these are modified to the rules for extracting an object from a color image, then

> *if* the color components of a pixel closely matches the color components of an object,
>
> *then* one can extract this pixel as a *Matched* pixel of the object, or
>
> *else* one can extract this pixel as an *Unmatched* pixel.

The FCE viewed as a mapping operator of measuring colors is able to extract colors that are similar to a color pattern (seed). If a pixel is extracted it is defined as a *Matched* pixel; otherwise it is defined as an *Unmatched* pixel. In doing this, we develop procedures IniSeedList and UpdateSeedList for automatic selection of seeds. Our approach mainly consists of two algorithms: CreateSubImage and MergeSubImage. The CreateSubImage algorithm always extracts the seed as the first *Matched* pixel into a subimage and continually uses the FCE to check neighbors of each *Matched* pixel to grow the subimage. The subimage is created when neighbors' colors of all *Matched* pixels are "not close to" the seed's color. The CreateSubImage algorithm iteratively invokes UpdateSeedList to select new seeds to create next subimages until all the pixels in the image are processed. Different from all of the color segmentation techniques discussed above, the CreateSubImage algorithm most likely classifies an individual pixel into several subimages by fuzzy sets. As a result, two created subimages may share a common area, which can be considered as a color similarity metric for merging the subimages. The MergeSubImage algorithm first removes insignificant subimages from the set of subimages and merges significant subimages in order to extract meaningful segmented images. The insignificant subimages usually have a "too small" or a "too large" ratio of their area sizes over the source image. A "too small" insignificant subimage is mainly caused by a tiny fragment or a noise pixel in the source image, while a "too large" insignificant image is created by a "bad" seed. Then, the algorithm merges two subimages together according to their color similarity described by the size of their common pixels. We tested the proposed approach using exemplar color images from the database at UC Berkeley [23] in three different color spaces and discussed its effectiveness and robustness by comparing segmentation effect with two edge detection algorithms, namely, Canny edge detection and Log edge detection.

2. Three Color Spaces

In many color-based computer vision applications, it is very important to segment colors in an image, but it is very difficult due to the fact that colors are ambiguous in uncertain environments. In this study, we apply the FCE to segment colors in the image represented in the RGB, HSV, and YUV spaces.

2.1. RGB Color Space. RGB is the most traditional color space and is widely used in image segmentation because of its simplicity and fast processing [24, 25]. Colors consist of red (R), green (G), and blue (B) components, as shown in Figure 1(a).

2.2. HSV Color Space. HSV is another color space widely used for image processing, where a color is decomposed into hue (H), saturation (S), and value (V). The distribution of color in the HSV color space is shown in Figure 1(b). The color components in the RGB space can be transformed into the HSV space by

$$\begin{aligned}
\max &= \max(\mathrm{R},\mathrm{G},\mathrm{B}),\\
\min &= \min(\mathrm{R},\mathrm{G},\mathrm{B}),\\
\mathrm{V} &= \max\\
\mathrm{S} &= \frac{(\max-\min)}{\max}\times 255,\\
\mathrm{H} &= \begin{cases} \dfrac{(\mathrm{G}-\mathrm{B})}{(\max-\min)}\times 60 & \mathrm{R}=\max\\ 120+\dfrac{(\mathrm{B}-\mathrm{R})}{(\max-\min)}\times 60 & \mathrm{G}=\max\\ 240+\dfrac{(\mathrm{R}-\mathrm{G})}{(\max-\min)}\times 60 & \mathrm{B}=\max \end{cases}\\
\mathrm{H} &= \mathrm{H}+360,\quad \text{if } \mathrm{H}<0,\\
\mathrm{H} &= \mathrm{H}\times 255\div 360.
\end{aligned} \tag{1}$$

Usually, the RGB values are between 0 and 255, so the HSV values are also between 0 and 255 through the transformation by (1).

2.3. YUV Color Space. YUV is the color space which is used in an analog color TV system, which translates color image components obtained in the RGB space into the luminance Y and the colors B -Y (U) and R - Y (V) by

$$\begin{aligned}
\mathrm{Y} &= 0.299\mathrm{R}+0.587\mathrm{G}+0.114\mathrm{B},\\
\mathrm{U} &= (-0.1687\mathrm{R}-0.3313\mathrm{G}+0.500\mathrm{B})+128,\\
\mathrm{V} &= (0.500\mathrm{R}-0.4187\mathrm{G}-0.0813\mathrm{B})+128.
\end{aligned} \tag{2}$$

The YUV values are between 0 and 255 through the transformation by (2). The advantage of using the YUV color space for color segmentation is that the luminance Y is separated from the colors U and V.

3. Fuzzy Color Extractor

The fuzzy color extractor proposed here enables us to deal with such uncertainty. We will only discuss how to use the FCE-based algorithm defined in the RGB color space to segment color images, because the algorithm can directly process the color images in the HSV and YUV spaces, which are converted from the RGB space. Figure 1(a) shows the RGB space, where colors are represented by their red, green, and blue components in an orthogonal Cartesian space. The upper-left corner in the image is the origin. The color of each pixel $p(n,m)$ denoted by $p(n,m)_{\mathrm{RGB}}$ is processed to separate its red, green, and blue components ($p(n,m)_{\mathrm{R}}$, $p(n,m)_{\mathrm{G}}$, $p(n,m)_{\mathrm{B}}$). The FCE extracts a cluster of colors based on a defined color pattern (seed) denoted by $\mathrm{seed}_{\mathrm{RGB}}$, $\mathrm{seed}_{\mathrm{RGB}}$ and is either directly defined by its RGB components ($\mathrm{seed}_{\mathrm{R}}$, $\mathrm{seed}_{\mathrm{G}}$, $\mathrm{seed}_{\mathrm{B}}$) or determined by a pixel in the image. The color components' differences between a pixel $p(n,m)_{\mathrm{RGB}}$ and $\mathrm{seed}_{\mathrm{RGB}}$ are calculated as follows:

$$\begin{aligned}
\mathrm{dif}(n,m)_{\mathrm{R}} &= p(n,m)_{\mathrm{R}}-\mathrm{seed}_{\mathrm{R}},\\
\mathrm{dif}(n,m)_{\mathrm{G}} &= p(n,m)_{\mathrm{G}}-\mathrm{seed}_{\mathrm{G}},\\
\mathrm{dif}(n,m)_{\mathrm{B}} &= p(n,m)_{\mathrm{B}}-\mathrm{seed}_{\mathrm{B}},\\
&\quad 0\le m<M,\ 0\le n<N.
\end{aligned} \tag{3}$$

M and N represent the size of an array which holds an image. The corresponding fuzzy rules can be obtained as follows:

If $\mathrm{dif}(n,m)_{\mathrm{R}}$ and $\mathrm{dif}(n,m)_{\mathrm{G}}$ and $\mathrm{dif}(n,m)_{\mathrm{B}}$ are *Zero*, *Then* $p(n,m)$ is *Matched*.

If $\mathrm{dif}(n,m)_{\mathrm{R}}$ or $\mathrm{dif}(n,m)_{\mathrm{G}}$ or $\mathrm{dif}(n,m)_{\mathrm{B}}$ is *Negative* or *Positive*,

Then $p(n,m)$ is *Unmatched*.

Both the rules indicate that $p(n,m)$ belongs to the object to be extracted, if the Euclidean distances between $p(n,m)_{\mathrm{RGB}}$ and $\mathrm{seed}_{\mathrm{RGB}}$ along the three axes in RGB coordinate system are small enough; otherwise, $p(n,m)$ does not belong to the object.

Figure 2(a) shows the membership functions ($\mu_{\mathrm{N}}(x)$, $\mu_{\mathrm{Z}}(x)$, $\mu_{\mathrm{P}}(x)$) for the input fuzzy variables *(Negative, Zero, Positive)* defined by

$$\begin{aligned}
\mu_{\mathrm{N}}(x) &= \begin{cases} 1 & -255\le x<-\alpha_2\\ \dfrac{x+\alpha_1}{\alpha_1-\alpha_2} & -\alpha_2\le x<-\alpha_1\\ 0 & -\alpha_1\le x\le 255, \end{cases}\\
\mu_{\mathrm{Z}}(x) &= \begin{cases} 0 & -255\le x<-\alpha_2\\ \dfrac{x+\alpha_2}{\alpha_2-\alpha_1} & -\alpha_2\le x<-\alpha_1\\ 1 & -\alpha_1\le x\le\alpha_1\\ \dfrac{x-\alpha_1}{\alpha_2-\alpha_1} & \alpha_1\le x<\alpha_2\\ 0 & \alpha_2\le x\le 255, \end{cases}\\
\mu_{\mathrm{P}}(x) &= \begin{cases} 0 & -255\le x<-\alpha_1\\ \dfrac{x+\alpha_2}{\alpha_2-\alpha_1} & \alpha_1\le x<\alpha_2\\ 1 & \alpha_2\le x\le 255, \end{cases}
\end{aligned} \tag{4}$$

and Figure 2(b) shows the membership functions ($\mu_{\mathrm{M}}(x)$, $\mu_{\mathrm{U}}(x)$) for the output fuzzy variables *(Matched, Unmatched)* defined by

$$\begin{aligned}
\mu_{\mathrm{M}}(x) &= \begin{cases} \dfrac{\rho_{\mathrm{M}}-x}{\rho_{\mathrm{M}}} & 0\le x<\rho_{\mathrm{M}}\\ 0 & \rho_{\mathrm{M}}\le x\le 255, \end{cases}\\
\mu_{\mathrm{U}}(x) &= \begin{cases} 0 & 0\le x<\rho_{\mathrm{U}}\\ \dfrac{x-\rho_{\mathrm{U}}}{255-\rho_{\mathrm{U}}} & \rho_{\mathrm{U}}\le x\le 255, \end{cases}
\end{aligned} \tag{5}$$

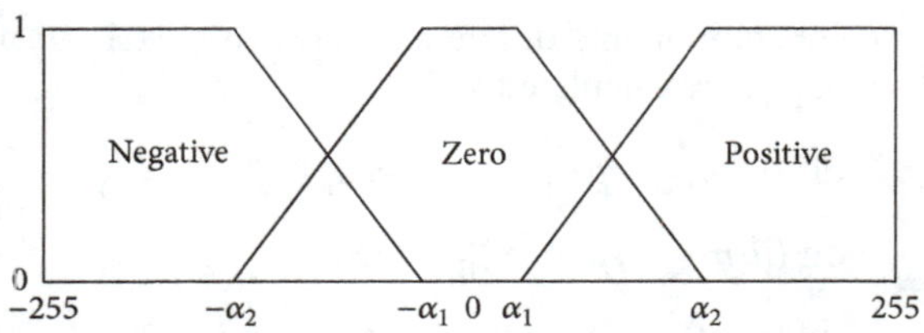

(a) Membership functions for color differences

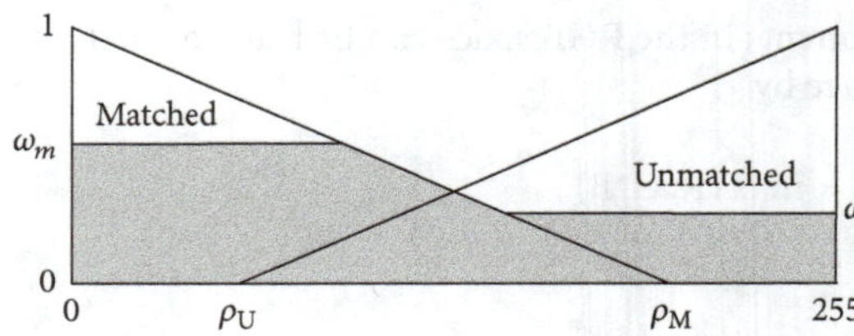

(b) Membership functions for defuzzification

FIGURE 2: Definition of membership functions for the FCE.

(a) I_{FLOWER}

(b) $\{I_{\mathrm{RED}}, I_{\mathrm{GREEN}}, I_{\mathrm{BLUE}}\}$

(c) $I_{\mathrm{BACKGROUND}}$

(d) I_{PETAL}

FIGURE 3: Extraction of colors in I_{FLOWER} image by FCE.

where $\rho_M + \rho_U = 255$. Based on $\mathrm{dif}(n,m)_R$, $\mathrm{dif}(n,m)_G$, and $\mathrm{dif}(n,m)_B$, the fuzzy rules produce the weight ω_m for *Matched* and the weight ω_u for *Unmatched* by

$$\omega_m = \min\{\mu_Z(R), \mu_Z(G), \mu_Z(B)\},$$

$$\omega_u = \max\{\mu_N(R), \mu_N(G), \mu_N(B), \mu_P(R), \mu_P(G), \mu_P(B)\}. \tag{6}$$

Figure 2(b) shows the produced areas in the output domain, while ω_m and ω_u cut $\mu_M(x)$ and $\mu_U(x)$. A crisp output value, $\Delta\rho_F$, is calculated by the centroid defuzzification method:

$$\Delta\rho_F = \frac{\int \mu_{\mathrm{out}}(x)\, x dx}{\int \mu_{\mathrm{out}}(x)\, dx}, \tag{7}$$

where $\mu_{\mathrm{out}}(x)$ represents the envelope function of the areas cut by ω_m and ω_u in fuzzy output domain. If $\Delta\rho_F < \sigma$, $p(n,m)$ is extracted, $p(n,m)$ is not extracted, where σ is a threshold. The FCE can be understood as a mapping operator between Euclidean distances $\{\mathrm{dif}(n,m)_R, \mathrm{dif}(n,m)_G, \mathrm{dif}(n,m)_B\}$ in the RGB space and a difference $\Delta\rho_F$ in the intensity space under a fuzzy metric.

The example in Figure 3 shows how to create some subimages with interesting colors from the source image I_{FLOWER} by the FCE. First, we define three seeds: $\mathrm{seed}_{\mathrm{RED}} = (255, 0, 0)$, $\mathrm{seed}_{\mathrm{GREEN}} = (0, 255, 0)$, and $\mathrm{seed}_{\mathrm{BLUE}} = (0, 0, 255)$ to create three subimages $\{I_{\mathrm{RED}}, I_{\mathrm{GREEN}}, I_{\mathrm{BLUE}}\}$ that hold the words "red," "green," and "blue" from the source image I_{FLOWER}, as shown in Figure 3(b). Also, we select a seed $p(0,0)_{\mathrm{RGB}} = (6, 6, 8)$ at the origin to remove the background from the source image I_{FLOWER} as shown in Figure 3(c). Figure 3(d) shows $I_{\mathrm{PETAL}} = I_{\mathrm{FLOWER}} - (I_{\mathrm{RED}} \cup I_{\mathrm{GREEN}} \cup I_{\mathrm{BLUE}} \cup I_{\mathrm{BACKGROUND}})$ that holds the color components of the flower petal.

TABLE 1: Definitions of the eight subspaces shown in Figure 1(a).

Vertex i	Subspace	Red-axis	Green-axis	Blue-axis
v^1_{RGB} (0, 0, 0)	ν_black	$[0, 127]$	$[0, 127]$	$[0, 127]$
v^2_{RGB} (0, 0, 255)	ν_blue	$[0, 127]$	$[0, 127]$	$[128, 255]$
v^3_{RGB} (0, 255, 0)	ν_green	$[0, 127]$	$[128, 255]$	$[0, 127]$
v^4_{RGB} (0, 255, 255)	ν_cyan	$[0, 127]$	$[128, 255]$	$[128, 255]$
v^5_{RGB} (255, 0, 0)	ν_red	$[128, 255]$	$[0, 127]$	$[0, 127]$
v^6_{RGB} (255, 0, 255)	ν_pink	$[128, 255]$	$[0, 127]$	$[128, 255]$
v^7_{RGB} (255, 255, 0)	ν_yellow	$[128, 255]$	$[128, 255]$	$[0, 127]$
v^8_{RGB} (255, 255, 255)	ν_white	$[128, 255]$	$[128, 255]$	$[128, 255]$

```
IniSeedList (I_IMG)
  for i ← 1 to 8 do
      pixel_num_i ← 0;
  for n ← 1 to N do
      for m ← 1 to M do
            Calculate d_i(n, m) in (8);
            d_k(n, m) ← min{d_i(n, m)};
            pixel_num_k ← pixel_num_k + 1;
   Find seed_cand^(i) in each subspace;
      // with the shortest distance to its vertex i.
   Push seed_cand^(i) into SeedList;
   Sort seed_cand^(i);
      // according to pixel number classified in the subspaces.
return SeedList;
```

PSEUDOCODE 1

4. Creating Subimages Using FCE

This section discusses the CreateSubImage algorithm for creating a set of subimages from a color image using the FCE in the RGB space. A critical step in this algorithm is to select seeds for the FCE. For this purpose, we evenly divide the RGB space into the eight subspaces {ν_black, ν_red, ν_green, ν_blue, ν_yellow, ν_pink, ν_cyan, ν_white}, which are defined by the well-defined colors: {black, red, green, blue, yellow, pink, cyan, white}, at their corresponding vertices of the RGB space, as shown in Figure 1(a). Table 1 lists their definitions; for example, ν_black is defined by a cubic $[0, 127] \times [0, 127] \times [0, 127]$ on the red-axis, green-axis, and blue-axis.

We calculate the distances of each pixel's RGB components to the eight well-defined colors by

$$d_i(n, m) = \sum \left[p(n, m)_{RGB} - v^i_{RGB}\right]^2 \quad i = 1, 2, \ldots, 8, \quad (8)$$

where n and m are the index of a pixel in the image, $p(n, m)_{RGB}$ is the color components of the pixel located at (n, m), v^i_{RGB} is the color components of vertex i in the RGB space, and $d_i(n, m)$ is the distance between the pixel's color and vertex i.

We developed two procedures, IniSeedList and UpdateSeedList, to initialize a seed list and to update it. For an input image, IniSeedList classifies each pixel into one of the eight subspaces according to its minimum distance to vertex i that is a well-defined color in the RGB space. In each subspace, IniSeedList selects such a pixel as a seed candidate, seed_cand$^{(i)}$, that has the shortest distance to its vertex i. Then, IniSeedList inserts seed_cand$^{(i)}$ into the seed list in order of the number of pixels classified in the subspaces and returns it. UpdateSeedList checks if seed_cand$^{(i)}$ is popped out or is extracted by the FCE. If so, UpdateSeedList updates seed_cand$^{(i)}$ in its corresponding subspace. The pseudocode of the IniSeedList procedures is shown in Pseudocode 1 and the pseudocode of the UpdateSeedList is shown in Pseudocode 2.

Table 2 shows an example of a seed list generated by IniSeedList from the color image, I_{BIRD}. Columns 1 and 2 in Table 2 indicate the defined subspaces and the number of the classified pixels in the subspaces, column 3 shows the order of seed_cand$^{(i)}$ in the list, and columns 4 and 5 list the RGB color components and locations of seed_cand$^{(i)}$ in I_{IMG}, respectively. The top seed in Table 2 is seed_cand$^{(1)}$ = (244, 251, 243) with location (269, 271) from the subspace ν_white holding 100,369 pixels. After the algorithm pops out this seed for creating $I_M{}^{(i)}$, UpdateSeedList produces a new seed candidate in the subspace ν_white from $I_U{}^{(1)}$ and inserts it on the top of the list. The new seed_cand$^{(1)}$ = (240, 251, 247) is located at (283, 267) in $I_U{}^{(1)}$ and is also from the subspace ν_white with 105,045 pixels. Please note that the

TABLE 2: The seed list generated from the color image, I_{BIRD}.

Subspace	Number of pixels	seed_cand$^{(i)}$ order	seed_cand$^{(i)}$ colors	seed_cand$^{(i)}$ locations (n, m)
v_white	100369	1	(244, 251, 243)	(269, 271)
v_black	19393	2	(0, 6, 0)	(193, 233)
v_cyan	70	3	(151, 100, 57)	(216, 190)
v_red	26	4	(206, 166, 107)	(225, 204)
v_blue	0	N/A	N/A	N/A
v_yellow	0	N/A	N/A	N/A
v_green	0	N/A	N/A	N/A
v_pink	0	N/A	N/A	N/A

```
UpdateSeedList (I_IMG, SeedList)
    subspace_index (i) ← 0;
    for i ← 1 to 8 do
        if (seed_cand^(i) ∈ I_IMG)
            subspace_index (i) ← 1;
        else
            Remove seed_cand^(i) from SeedList;
    for i ← 1 to 8 do
        if subspace_index (i) = 0
            Find seed_cand^(i) in subspace i;
                // with the shortest distance to its vertex i.
            Push seed_cand^(i) into SeedList;
    Sort seed_cand^(i);
        // according to pixel number classified in the subspaces.
return SeedList;
```

PSEUDOCODE 2

input image to IniSeedList is $I_{IMG} = I_{BIRD} - I_{CENTER}$, instead of I_{BIRD} itself. I_{CENTER} is an image extracted by $\text{seed}_{CENTER} = (127, 127, 127)$ that is the center of the RGB space. It is difficult to determine in which subspace the center color is classified, so any seed close to the center may not be able to differentiate colors very well, as shown in Figure 4(a). In order to guarantee to get a "good" seed, our approach does not select any seed falling in I_{CENTER}. It is also possible that there is no pixel in a subspace; that is, the seed candidates are less than eight. For example, the seed candidates for I_{BIRD} are not available in the subspaces v_blue, v_yellow, v_green, and v_pink.

The pseudocode of the CreateSubImage procedures is listed as shown in Pseudocode 3.

CreateSubImage creates a set of subimages from a color source image. CreateSubImage invokes mainly three procedures: IniSeedList, UpdateSeedList, and FuzExtractor. The pseudocode of the FuzExtractor is shown in Pseudocode 4.

CreateSubImage has two arguments: I_{SRC} and center, where I_{SRC} is a source image and center is a Boolean variable. If center = true, I_{CENTER} is created by seed_{CENTER}, otherwise I_{CENTER} is assigned by the empty image ø. FuzExtractor is mainly implemented based on (3) and extracts a subimage I_{TEMP}. The FuzExtractor procedure starts with seed_{RGB} to generate I_{MATCH} with region growing. Because seed_{RGB} belongs to its subimage I_{MATCH}, FuzExtractor pushes its four neighbor pixels into a queue and checks the first-in element, pixel_{FIRST}, in the queue to see if it belongs to the subimage I_{MATCH}. If pixel_{FIRST} belongs to I_{MATCH}, its new four neighbors are pushed into the queue. The FuzExtractor procedure repeats this process until the queue is empty, which indicates that there are no more pixels in the source image I_{SRC} matching the seed_{RGB} color. After FuzExtractor returns I_{MATCH}, CreateSubImage assigns I_{MATCH} to I_{TEMP} as described.

CreateSubImage uses two conditions to check if I_{TEMP} is significant or not. The first condition is to check I_{TEMP}/I_{SRC} that it is a ratio of I_{TEMP} area over I_{SRC} one. If this ratio is too small, CreateSubImage ignores I_{TEMP} as I_{TEMP} holds too few pixels extracted from the source image I_{SRC}. I_{TEMP} is most likely affected by noise and is not significant for any object in the source image. CreateSubImage also uses the condition, (I_{TEMP}/I_{SRC}) & var (I_{TEMP}), to check I_{TEMP}, where var (I_{TEMP}) is a variance defined by

$$\varpi = E\left[(I_{TEMP})^2\right],$$
$$\text{var}(I_{TEMP}) = E\left[(I_{TEMP} - \varpi)^2\right]. \tag{9}$$

A big I_{TEMP}/I_{SRC} indicates a large area size of I_{TEMP} and a big var (I_{TEMP}) implies that I_{TEMP} contains the color components

(a) I_{CENTER}

(b) I_{TEMP}

(c) $I_{\text{UNCLASSIED}}$

FIGURE 4: (a) The image I_{CENTER} extracted by the seed, $\text{seed}_{\text{CENTER}} = (127, 127, 127)$. (b) An insignificant image, I_{TEMP}, extracted by a "bad" seed. (c) The image $I_{\text{UNCLASSIED}}$.

```
CreateSubImage (I_SRC, center)
    i ← 0; I_M^(i+1) ← ø; I_U^(i) ← I_SRC;
    If (center = true)
        I_CENTER ← FuzExtractor (I_SRC, seed_CENTER);
    else I_CENTER ← ø;
    I_IMG ← I_U^(i) − I_CENTER;
    SeedList ← IniSeedList (I_IMG);
    while (SeedList ≠ ø) do
        seed_RGB ← The first element in SeedList;
        I_TEMP ← FuzExtractor (I_SRC, seed_RGB);
        if I_TEMP/I_SRC > ε
            if (I_TEMP/I_SRC < γ1 & var (I_TEMP) < γ2)
                I_M^(i+1) ← I_TEMP;
        I_U^(i+1) ← I_SRC − (I_M^(i+1) ∪ I_M^(i−1) ... I_M^(1));
        I_IMG ← I_U^(i+1) − I_CENTER;
        SeedList ← IniSeedList (I_IMG);
        i ← i + 1;
    end
return {I_M^(i+1)};
```

PSEUDOCODE 3

with a wide range distribution, as shown in Figure 4(b). Such a subimage is also considered as insignificant. If I_{TEMP} is significant, CreateSubImage assigns I_{TEMP} to a *Matched* subimage $I_{\text{M}}^{(i+1)}$ and creates an *Unmatched* $I_{\text{U}}^{(i+1)}$, where $I_{\text{M}}^{(i+1)}$ holds the color components matching the color of seed_{RGB} and $I_{\text{U}}^{(i+1)}$ holds unextracted pixels in the source image. CreateSubImage iteratively processes $I_{\text{U}}^{(i)}$ until $I_{\text{U}}^{(i)} = \text{ø}$; that is, each pixel in the image is processed.

There may exist tiny holes in the subimage $I_{\text{M}}^{(i)}$. Because these tiny fragments may be insignificant for object pattern

```
FuzExtractor (I_SRC, seed_RGB)
    Move seed_RGB into I_MATCH;
    Push four neighbor pixels of seed_RGB into queue;
    while (queue ≠ ø) do
            pixel_FIRST ← the top element in queue;
            Calculate Δρ_F using (3);
            if (Δρ_F < σ)
                Move pixel_FIRST into I_MATCH;
                Push unduplicated neighbors of pixel_FIRST into queue;
        end
    return I_MATCH;
```

PSEUDOCODE 4

```
MergeSubImage ({I_M^(i)})
    K ← I;
    for i ← 1 to I do
        for l ← i − 1 to I do
            if (I_M^(i) ∩ I_M^(l))/I_M^(i) > τ;
                I_M^(l) ← I_M^(i) ∪ I_M^(l);
                Remove I_M^(i) from the list;
                K ← K − 1;
                break;
    for k ← 1 to K do
        I_OBJ^(k) ← I_M^(k);
        I_SEG ← I_OBJ^(k) ∪ I_OBJ^(k−1) ··· I_OBJ^(1);
return I_SEG;
```

PSEUDOCODE 5

TABLE 3: Subimages created from I_{BIRD}.

ID	Locations	RGB colors	Pixels	Variance
(1)	(269, 271)	(244, 251, 243)	14308	148
(2)	(283, 267)	(240, 251, 247)	9674	114
(3)	(267, 243)	(230, 252, 239)	64924	150
(4)	(193, 233)	(0, 6, 0)	15643	917
(5)	(358, 174)	(228, 246, 232)	10492	157
(6)	(276, 280)	(229, 232, 221)	15307	240
(7)	(292, 302)	(209, 222, 213)	10894	372
(8)	(58, 141)	(197, 216, 212)	13815	590
(9)	(292, 296)	(1, 12, 4)	3003	1419
(10)	(290, 302)	(11, 14, 5)	3529	1602
(11)	(464, 285)	(7, 22, 25)	3941	1591
(12)	(275, 252)	(17, 30, 13)	17645	1311
(13)	(320, 261)	(20, 38, 26)	4831	1961
(14)	(68, 138)	(54, 39, 16)	26009	2416
(15)	(216, 190)	(151, 100, 57)	2916	2394
(16)	(357, 158)	(50, 53, 46)	26769	2703
(17)	(293, 294)	(185, 198, 189)	11018	534
(18)	(352, 171)	(206, 191, 162)	67427	375
(19)	(49, 238)	(89, 72, 42)	28510	3311
(20)	(1, 320)	(58, 86, 108)	26845	4777

features, the algorithm removes them by filling up such tiny holes with their original pixels. Usually, these tiny fragments are caused by a small RGB distance of two pixels that have the same RGB values and are located in the same row or in the same column, so the algorithm checks such a distance if it is less than a threshold, $\delta = 7$. If so, the algorithm fills the tiny holes when creating the subimage, $I_M^{(i)}$. The algorithm generates a seed from the *Unmatched* subimage $I_U^{(i)}$ but always creates a subimage $I_M^{(i)}$ from the source image I_{SRC} so it is very probable that the algorithm extracts an individual pixel into multiple subimages. We use a list to maintain $\{I_M^{(i)}\}$ according to a sending order of their sizes. Table 3 shows an example of the proposed algorithm that creates 20 subimages from the source image I_{BIRD} in Figure 5(a). Column 1 in Table 3 contains $I_M^{(i)}$ ID, columns 2 and 3 list the locations and the color components of the seeds for extracting the subimages $I_M^{(i)}$ using the FCE, column 4 shows the number of pixels extracted into the subimages $I_M^{(i)}$, and column 5 records the variance of $I_M^{(i)}$.

5. Merging Subimages

The proposed algorithm merges significant subimages into a set of meaningful segmented objects by the procedure MergeSubImage. The pseudocode of the MergeSubImage procedure is shown in Pseudocode 5.

The inputs to the procedure MergeSubImage are the subimages $\{I_M^{(i)}\}$, and the output from MergeSubImage is the segmented image I_{SEG}. Figures 5(a)-5(b) show a color image, I_{BIRD}, and the segmented image in the RGB space, I_{SEG}, which is a union of 6 objects, $I_{OBJ}^{(i)}$, $i = 1, \ldots, 6$, shown in Figures 5(f)–5(k). Figure 5(e) shows the manually made benchmark, $I_{BECHMARK}$.

MergeSubImage merges the subimages according to the overlap sizes of their color components to construct some objects in the source image. In the process of the Merge-SubImage, the FCE usually classifies an individual pixel into multiple subimages. It causes these subimages to share a common color area in the RGB space, and even probably a

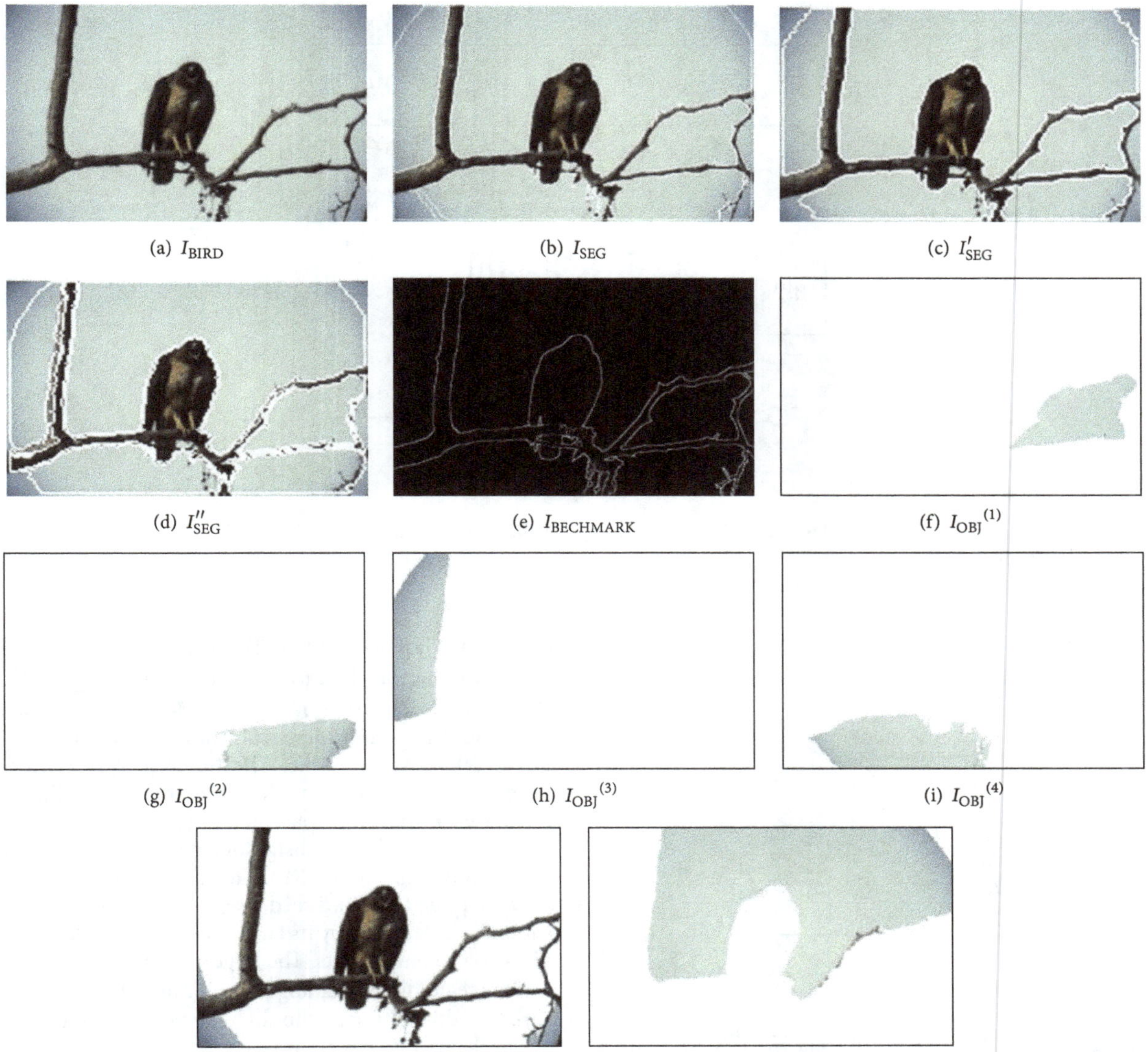

(a) I_{BIRD} (b) I_{SEG} (c) I'_{SEG}

(d) I''_{SEG} (e) I_{BECHMARK} (f) $I_{\mathrm{OBJ}}{}^{(1)}$

(g) $I_{\mathrm{OBJ}}{}^{(2)}$ (h) $I_{\mathrm{OBJ}}{}^{(3)}$ (i) $I_{\mathrm{OBJ}}{}^{(4)}$

(j) $I_{\mathrm{OBJ}}{}^{(5)}$ (k) $I_{\mathrm{OBJ}}{}^{(6)}$

FIGURE 5: Source image, I_{BIRD}; segmented images in the RGB, HSV, and YUV space, I_{SEG}, I'_{SEG}, and I''_{SEG}; manually made benchmark, I_{BECHMARK}; and partitioned objects in the RGB space, $I_{\mathrm{OBJ}}{}^{(i)}$.

subimage may fully cover another subimage. The common area of two subimages implies two features: spatial proximity and color similarity, because the two subimages connect together and have a partition with the same color if they share the common area. Therefore, our algorithm uses this unique characteristic to merge subimages. MergeSubImage uses the union $(I_{\mathrm{M}}{}^{(i)} \cap I_{\mathrm{M}}{}^{(l)})/I_{\mathrm{M}}{}^{(i)}$ to check the common area size of two subimages $I_{\mathrm{M}}{}^{(i)}$ and $I_{\mathrm{M}}{}^{(l)}$ in the RGB space. A bigger value of $(I_{\mathrm{M}}{}^{(i)} \cap I_{\mathrm{M}}{}^{(l)})/I_{\mathrm{M}}{}^{(i)}$ means that $I_{\mathrm{M}}{}^{(i)}$ and $I_{\mathrm{M}}{}^{(l)}$ share a larger common area; that is, their color components are more similar. A data structure for maintaining the subimages $\{I_{\mathrm{M}}{}^{(i)}\}$ is a list that sorts $\{I_{\mathrm{M}}{}^{(i)}\}$ in a sending order of their sizes. Technical implementation of merging $\{I_{\mathrm{M}}{}^{(i)}\}$ is shown in Figure 6. The merging process starts with the last element $I_{\mathrm{M}}{}^{(i)}$ with $i = I$, located at the list tail, which is the subimage with the smallest area size, and checks $I_{\mathrm{M}}{}^{(l)}$ with each subimage $\{I_{\mathrm{M}}{}^{(l)}, l = i-1, \ldots, 1\}$ via $(I_{\mathrm{M}}{}^{(i)} \cap I_{\mathrm{M}}{}^{(l)})/I_{\mathrm{M}}{}^{(i)}$. If $(I_{\mathrm{M}}{}^{(i)} \cap I_{\mathrm{M}}{}^{(l)})/I_{\mathrm{M}}{}^{(i)} > \tau$, the algorithm merges $I_{\mathrm{M}}{}^{(i)}$ and $I_{\mathrm{M}}{}^{(l)}$ together, removes $I_{\mathrm{M}}{}^{(i)}$ from the list, and replaces $I_{\mathrm{M}}{}^{(i)}$ with $I_{\mathrm{M}}{}^{(i)} \cup I_{\mathrm{M}}{}^{(l)}$, where $I_{\mathrm{M}}{}^{(l)}$ size is bigger than $I_{\mathrm{M}}{}^{(i)}$ one. After processing all the subimages in the list, the list holds the partitioned objects $\{I_{\mathrm{OBJ}}{}^{(i)}\}$ from the input source image I_{SRC}. MergeSubImage returns the partitioned objects $\{I_{\mathrm{OBJ}}{}^{(k)}\}$ as well as its union, $I_{\mathrm{OBJ}}{}^{(k)} \cup I_{\mathrm{OBJ}}{}^{(k-1)} \cdots I_{\mathrm{OBJ}}{}^{(1)}$, as the segmented image I_{SEG}.

Table 4 lists the extracted subimages from the source image I_{BIRD} in Figure 5(a). MergeSub- Image merges these

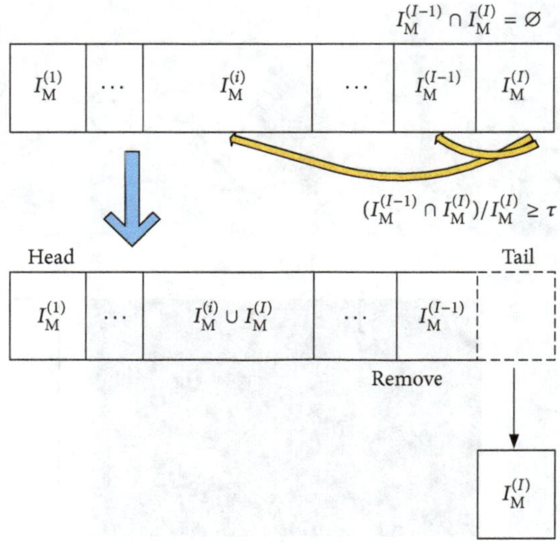

FIGURE 6: Process for merging subimages ${I_M}^{(i)}$ and ${I_M}^{(l)}$.

TABLE 4: ${I_{OBJ}}^{(i)}$ merged by subimages created from I_{BIRD}.

Segmented objects	Sub-images
${I_{OBJ}}^{(1)}$	${I_M}^{(11)}$
${I_{OBJ}}^{(2)}$	${I_M}^{(2)}, {I_M}^{(19)}, {I_M}^{(22)}$
${I_{OBJ}}^{(3)}$	${I_M}^{(22)}$
${I_{OBJ}}^{(4)}$	${I_M}^{(22)}, {I_M}^{(22)}$
${I_{OBJ}}^{(5)}$	${I_M}^{(22)}, {I_M}^{(22)}, {I_M}^{(22)}, {I_M}^{(22)}, {I_M}^{(22)}, {I_M}^{(22)}, {I_M}^{(22)}, {I_M}^{(22)}, {I_M}^{(22)}, {I_M}^{(22)}, {I_M}^{(22)}$
${I_{OBJ}}^{(6)}$	${I_M}^{(22)}, {I_M}^{(22)}$

20 subimages into 6 partitioned objects $\{{I_{OBJ}}^{(i)}\}$ in Figures 5(f)–5(k). Table 4 also shows information on which and how many subimages are merged into which object. For example, the partitioned object ${I_{OBJ}}^{(1)}$ shown in Figure 5(f) comes from a single subimage ${I_M}^{(11)}$, while the partitioned object ${I_{OBJ}}^{(5)}$ in Figure 5(j) is produced by merging 11 subimages $\{{I_M}^{(9)}, {I_M}^{(22)}, {I_M}^{(23)}, {I_M}^{(27)}, {I_M}^{(29)}, {I_M}^{(32)}, {I_M}^{(34)}, {I_M}^{(36)}, {I_M}^{(38)}, {I_M}^{(41)}, {I_M}^{(43)}\}$. Figure 5(b) shows the segmented image I_{SEG} by a union operation of ${I_{OBJ}}^{(6)} \cup {I_{OBJ}}^{(5)} \cdots {I_{OBJ}}^{(1)}$.

6. Discussions and Conclusions

In this section, we present our selection of the membership functions' parameters of the FCE defined in (4)–(7) and settings of the CreateSubImage and MergeSubImage procedures used for our tests. We empirically selected the parameters of the membership functions as $\alpha_1 = 60$, $\alpha_2 = 200$, $\rho_M = 20$, $\rho_M + \rho_U = 255$, and $\sigma = 90$. A smaller α_1 may more precisely segment color components in an input image but produces more extracted subimages. A slight change of α_2 does not affect the segmentation result, so it is not sensitive. ρ_M should be at least less than 127.5. The threshold, σ, is important for classifying pixels into a *Matched* subimage ${I_M}^{(i)}$ or an *Unmatched* subimage ${I_U}^{(i)}$. We select the parameters in CreateSubImage and MergeSubImage as $\varepsilon = 0.01$, $\gamma_1 = 0.25$, $\gamma_2 = 500$, and $\tau = 0.6$. The function of ε is to remove insignificant subimages with a "too small" area that may be caused by noise. Both parameters γ_1 and γ_2 are defined to remove another type of insignificant subimages, for example, I_{TEMP} shown in Figure 4(b). This subimage is not extracted very well due to a "bad seed," $\text{seed}_{RGB} = (64, 28, 28)$. Such a subimage is featured by its large area size and a large variance of its color components. The large variance indicates that the subimage holds a wide range of color distribution in the RGB space, so this subimage should be removed. τ is the threshold to check the overlap area over the subimage with a smaller area. In this paper, $\tau = 0.6$ herein means that the overlap area is 60% of the subimage with the smaller area. The bigger τ is, the more similar the color components of the two subimages are. Changing τ affects the number of partitioned objects generated by MergeSubImage. As an extreme example, $\tau = 1.0$ means that only if the larger subimage fully overlaps the smaller subimage, the two subimages will be merged together.

In our implementation, the algorithm invokes the CreateSubImage procedure twice. For the first invocation, an input image, I_{IMG}, to IniSeedList and UpdateSeedList, is $I_{SRC} - I_{CENTER}$, where I_{CENTER} is extracted by $\text{seed}_{CENTER} = (127, 127, 127)$. Because the first call is to process a source image I_{SRC}, the algorithm does not select a seed close to the RGB space center. As discussed in Section 4, the seed seed_{CENTER} cannot differentiate color components which are far from the well-defined eight colors. For example, the extracted image I_{CENTER} shown in Figure 4(a) holds many color components of six corn cobs. Because the algorithm does not select any seed from I_{CENTER}, it is very possible that some of the pixels in the source image remain unclassified. In

TABLE 5: Statistics of subimages and partitioned objects of five color images.

ID	I_{BIRD}	I_{CORN}	I_{LAKE}	I_{MAN}	I_{BABOON}
$I_M^{(i)}$	20	41	16	22	85
$I_{OBJ}^{(i)}$	6	13	6	54	6

order to process the unclassified pixels, we create an image $I_{UNCLASSIED} = I_{SRC} - \cup I_{OBJ}^{(i)}$, as shown in Figure 4(c). The algorithm calls the CreateSubImage a second time to process $I_{UNCLASSIED}$, but the input image to IniSeedList and UpdateSeedList is $I_{UNCLASSIED}$ itself. The variance of $I_{UNCLASSIED}$ is smaller than that of its original source image I_{SRC} as most pixels with a large range of color distribution are classified after first calling CreateSubImage. For this reason, we halve the parameter $\alpha_1 = 30$ of the membership functions to process $I_{UNCLASSIED}$. The proposed approach segments the color image I_{CORN} in Figure 7(a) into 12 objects. Figure 7(b) shows the segmented image. Figures 7(f)–7(q) display the first 12 partitioned objects. $I_{OBJ}^{(3)}$ and $I_{OBJ}^{(6)}$ in Figures 7(h) and 7(k) are extracted from $I_{UNCLASSIED}$ in Figure 4(c).

In order to test the robustness of the proposed algorithm, we applied the algorithm with the same settings defined in the first paragraph of this section to process the five color images $\{I_{BIRD}, I_{CORN}, I_{LAKE}, I_{MAN}, I_{BABOON}\}$, in the RGB color space as shown in Figures 5(a), 7(a), 8(a), 9(a), and 10(a), respectively. Their corresponding segmented images are displayed in Figures 5(b), 7(b), 8(b), 9(b), and 10(b), respectively. The first four color images are from the database at UC Berkeley [23]. These selected color images represent different features and diverse scenes that might be taken under different light conditions. The source image I_{BIRD} in Figure 5(a) holds a bird on a tree trunk under the sky. The source image I_{CORN} in Figure 7(a) holds six corns with many small multicolor kernels and its segmented image holding the six colorful corns. The source image I_{LAKE} in Figure 8(a) shows a lake and a mountain under the sky at evening. Figures 9(a) and 10(a) show the color images I_{MAN} and I_{BABOON}. Table 5 lists the subimages, and the partitioned objects from the 5 color images. 85 subimages are extracted from I_{BABOON}, but the partitioned objects are only 6.

There is a lack of a strong mathematical model for precisely expressing the human's understanding of colors due to psychophysical perception. For a given color, different observers may perceive its color definition differently. That is why the FCE-based approach classifying an individual pixel into several subimages is reasonable. The proposed approach to color image segmentation is easy to use, and it does not apply any procedure where one needs to preprocess source images. Removing subimages with "too small" areas is filtering short edges and noise in a color image. The approach automatically partitions a color image into an appropriate number of meaningful objects or regions. The manually made benchmarks are downloaded from the database at UC Berkeley [23], as shown in Figures 5(e), 7(e), 8(e), 9(e), 11(e), and 12(e), respectively. These benchmarks display desired segmentation results represented by the images' boundaries. Although the proposed approach does not reach the benchmarks' results, it is able to partition the color images into meaningful objects, for example, the I_{CORN} segmentation results as shown in Figure 7.

In this study, we also compared the performance of the FCE-based algorithm by segmenting the same test images $\{I_{BIRD}, I_{CORN}, I_{LAKE}, I_{MAN}, I_{HORSE}, I_{VEGGI}\}$ in the RGB, HSV, and YUV spaces, respectively. In this case, we only transformed the color image representation from the RGB space into the HSV and YUV spaces according to (1) and (2). The test images are selected randomly from the database at UC Berkeley [23], as shown in Figures 5(a), 7(a), 8(a), 9(a), 11(a), and 12(a). Their corresponding segmented images in the RGB, HSV, and YUV spaces are displayed in Figures 5(b)–5(d), 7(b)–7(d), 8(b)–8(d), 9(b)–9(d), 11(b)–11(d), and 12(b)–12(d). We find that the segmented images in the HSV space I'_{SEG} are the most different from $I_{BENCHMARK}$ by comparison. So the FCE-based algorithm defined in the HSV space is not adopted. In some cases, the segmentation results in the YUV space are superior to the RGB spaces, such as the "face" of I_{MAN} in Figure 9 and the "boxes" in I_{VEGGI} in Figure 12. But in most cases, the segmented images in the RGB space I_{SEG} are the most similar to $I_{BENCHMARK}$. It is proven that the FCE-based algorithm defined in the RGB space is the most robust. The possible reason for the results is the transformation of the color image representation from the RGB space into the HSV and YUV spaces may cause color errors so that the segmentation is influenced.

In order to show the advantages of the new approach to color image segmentation, we applied Canny edge detection and Log edge detection algorithms to segment the test images $\{I_{LAKE}, I_{MAN}, I_{HORSE}, I_{VEGGI}\}$, respectively, and the segmented images were shown in Figures 8(f)-8(g), 9(f)-9(g), 11(f)-11(g), and 12(f)-12(g). Then we compared I_{Canny} and I_{Log} with the segmented images through the FCE-based approach, namely, I_{SEG}, I'_{SEG}, and I''_{SEG}, and meanwhile $I_{BECHMARK}$ can be used as a metric to evaluate the segmentations. As shown in Figures 8 and 9, the edges in I_{Canny} and I_{Log} are "too many" or "too few" compared with $I_{BECHMARK}$, which proves that the segmentation effect with Canny edge detection and Log edge detection is inferior to the FCE-based approach. Although the segmentation effect with Log edge detection is superior to the FCE-based approach applied in the HSV and YUV spaces, it performs not well as the proposed approach applied in the RGB color space, as shown in Figures 11 and 12. Overall, compared with Canny edge detection and Log edge detection, the new approach based on FCE defined in the RGB color space presents the best performance in the color image segmentation.

The approach is effective in tracking or following a certain color because the FCE-based approach does not need a precise color pattern to extract a set of colors. We apply this algorithm to extract chemical plumes in images taken undersea at very poor illumination conditions [26]. In this paper, we start selecting seeds near the eight well-defined colors, that is, at the vertices of the RGB space, and move their selection from the vertices to the RGB center. Our further research will investigate more effective methods for selecting seeds because seed selection plays an important

(a) I_{CORN} (b) I_{SEG} (c) I'_{SEG}

(d) I''_{SEG} (e) $I_{BENCHMARK}$ (f) $I_{OBJ}^{(1)}$

(g) $I_{OBJ}^{(2)}$ (h) $I_{OBJ}^{(3)}$ (i) $I_{OBJ}^{(4)}$

(j) $I_{OBJ}^{(5)}$ (k) $I_{OBJ}^{(6)}$ (l) $I_{OBJ}^{(7)}$

(m) $I_{OBJ}^{(8)}$ (n) $I_{OBJ}^{(9)}$ (o) $I_{OBJ}^{(10)}$

(p) $I_{OBJ}^{(11)}$ (q) $I_{OBJ}^{(12)}$

FIGURE 7: Source image, I_{CORN}; segmented images in the RGB, HSV, and YUV space, I_{SEG}, I'_{SEG}; and I''_{SEG}; manually made benchmark, $I_{BECHMARK}$; and partitioned objects in the RGB space, $I_{OBJ}^{(i)}$.

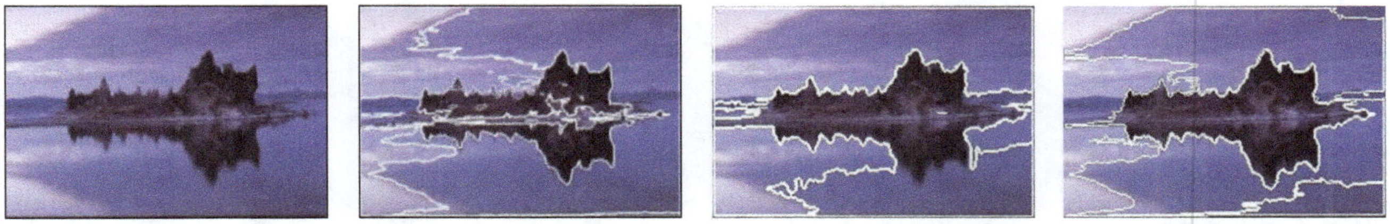

(a) I_{LAKE} (b) I_{SEG} (c) I'_{SEG} (d) I''_{SEG}

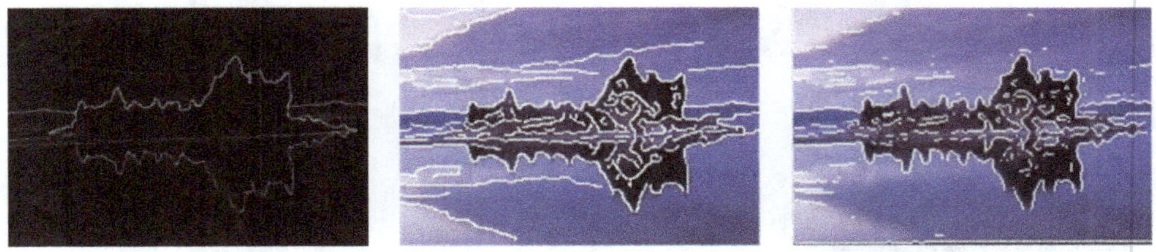

(e) $I_{\mathrm{BENCHMARK}}$ (f) I_{Canny} (g) I_{Log}

FIGURE 8: Source image, I_{LAKE}; segmented images in the RGB, HSV, and YUV space, I_{SEG}, I'_{SEG}, and I''_{SEG}; manually made benchmark, I_{BECHMARK}; segmented images with Canny edge detection and Log edge detection, I_{Canny}, I_{Log}.

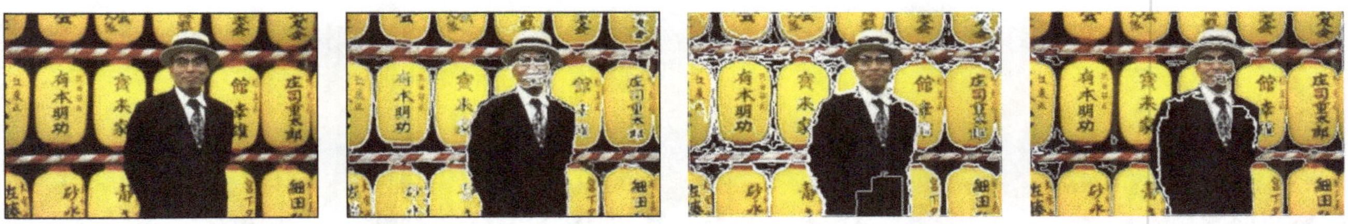

(a) I_{MAN} (b) I_{SEG} (c) I'_{SEG} (d) I''_{SEG}

(e) $I_{\mathrm{BENCHMARK}}$ (f) I_{Canny} (g) I_{Log}

FIGURE 9: Source image, I_{MAN}; segmented images in the RGB, HSV, and YUV space, I_{SEG}, I'_{SEG}, and I''_{SEG}; manually made benchmark, I_{BECHMARK}; segmented images with Canny edge detection and Log edge detection, I_{Canny}, I_{Log}.

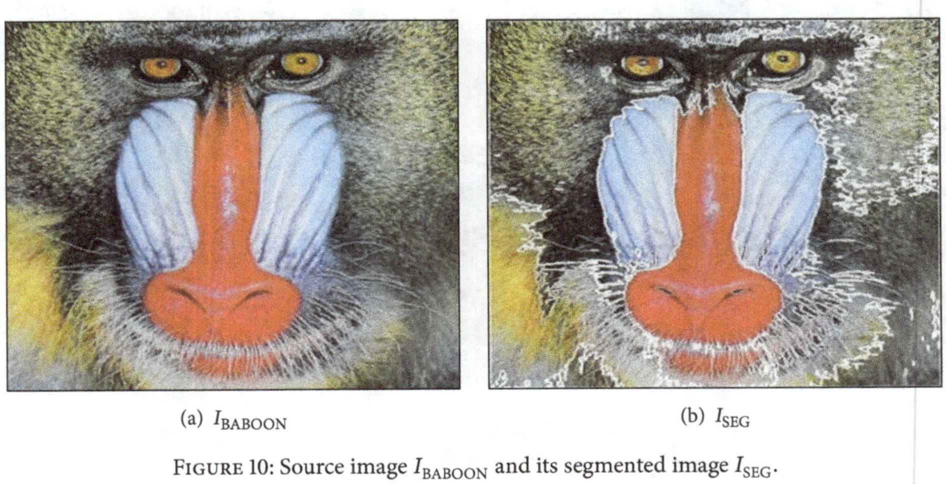

(a) I_{BABOON} (b) I_{SEG}

FIGURE 10: Source image I_{BABOON} and its segmented image I_{SEG}.

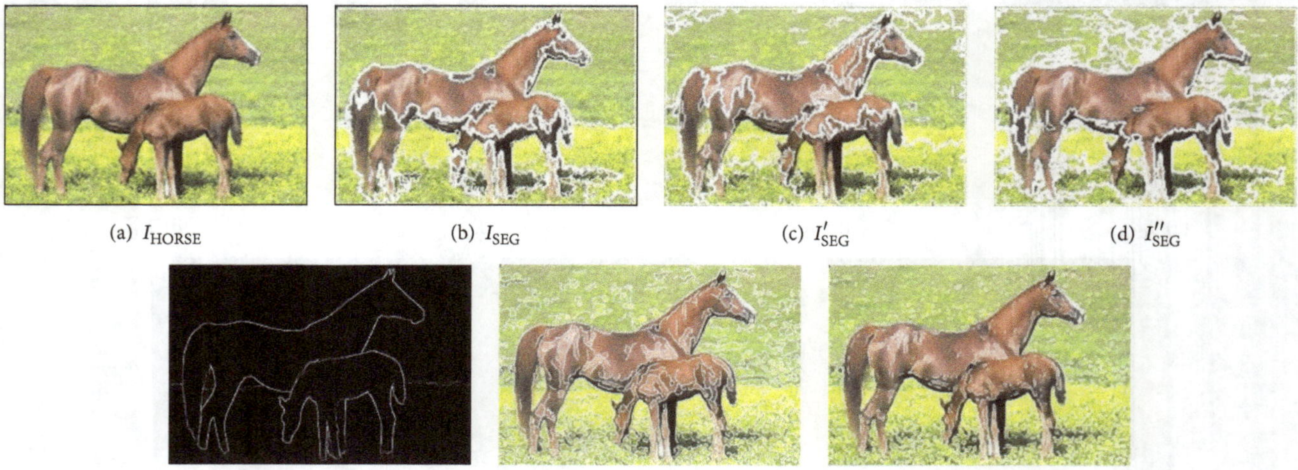

(a) I_{HORSE} (b) I_{SEG} (c) I'_{SEG} (d) I''_{SEG}

(e) $I_{\mathrm{BENCHMARK}}$ (f) I_{Canny} (g) I_{Log}

FIGURE 11: Source image, I_{HORSE}; segmented images in the RGB, HSV, and YUV space, I_{SEG}, I'_{SEG}, and I''_{SEG}; manually made benchmark, I_{BECHMARK}; segmented images with Canny edge detection and Log edge detection, I_{Canny}, I_{Log}.

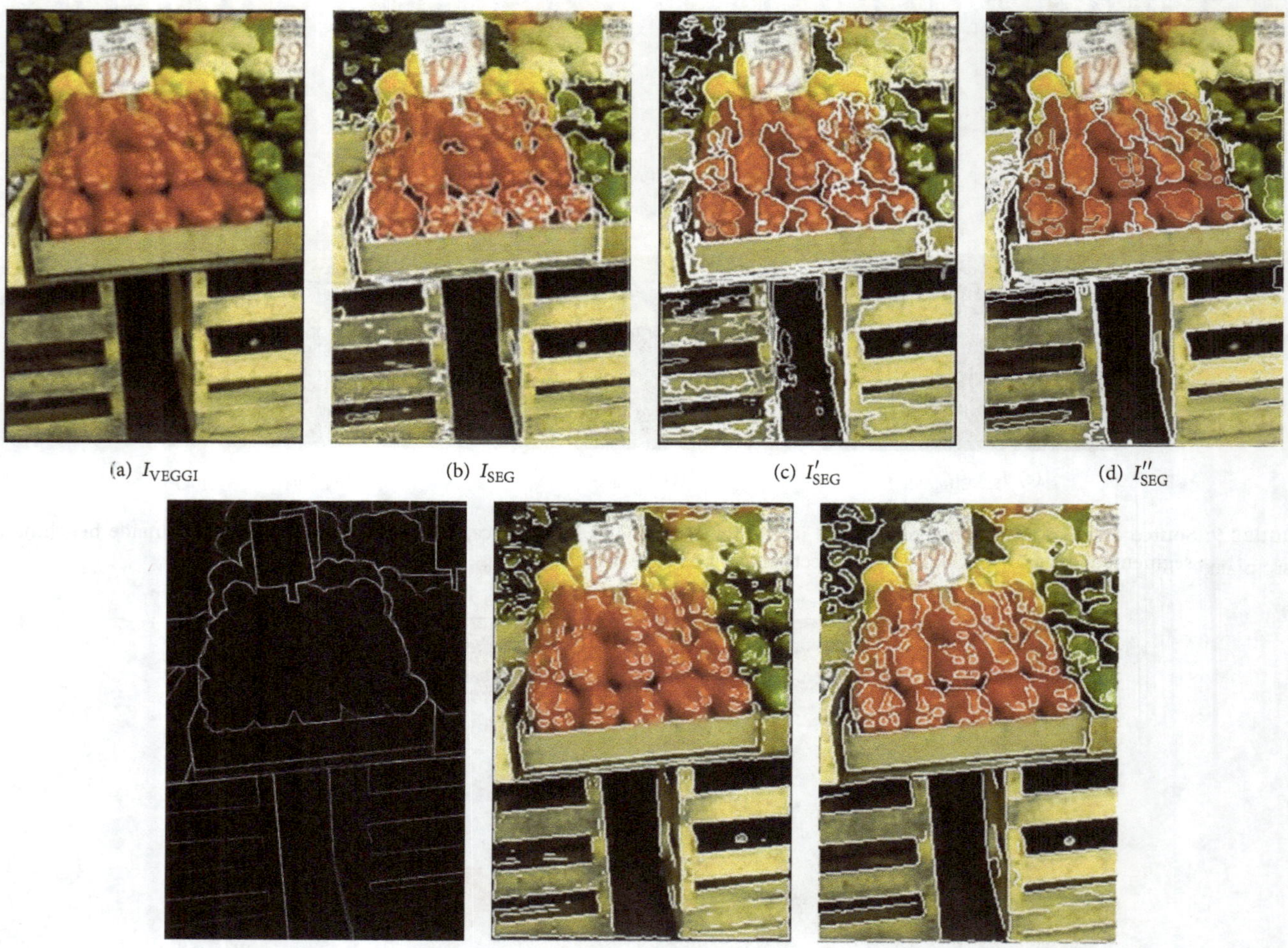

(a) I_{VEGGI} (b) I_{SEG} (c) I'_{SEG} (d) I''_{SEG}

(e) $I_{\mathrm{BENCHMARK}}$ (f) I_{Canny} (g) I_{Log}

FIGURE 12: Source image, I_{VEGGI}; segmented images in the RGB, HSV, and YUV space I_{SEG}, I'_{SEG}, and I''_{SEG}; manually made benchmark, I_{BECHMARK}; segmented images with Canny edge detection and Log edge detection, I_{Canny}, I_{Log}.

role in achieving a good segmentation performance. Another issue we will address in our study is how to automatically determine the parameters of fuzzy sets based on an input color image.

Conflicts of Interest

The authors declare no conflicts of interest.

Authors' Contributions

All the authors contributed equally to this work.

Acknowledgments

This research work was funded by the National Natural Science Foundation of China under Grant no. 61473207.

References

[1] K. S. Fu and J. K. Mui, "A survey on image segmentation," *The Journal of the Pattern Recognition Society*, vol. 13, no. 1, pp. 3–16, 1981.

[2] N. R. Pal and S. K. Pal, "A review on image segmentation techniques," *Pattern Recognition*, vol. 26, no. 9, pp. 1277–1294, 1993.

[3] H. D. Cheng et al., "Color image segmentation: Advances & Prospects," *Pattern Recognition*, vol. 34, no. 12, pp. 2257–2281, 2001.

[4] G. J. Klinker, S. A. Shafer, and T. Kanade, "A physical approach to color image understanding," *International Journal of Computer Vision*, vol. 4, no. 1, pp. 7–38, 1990.

[5] A. Tremeau and N. Borel, "A region growing and merging algorithm to color segmentation," *Pattern Recognition*, vol. 30, no. 7, pp. 1191–1203, 1997.

[6] A. Moghaddamzadeh and N. Bourbakis, "A fuzzy region growing approach for segmentation of color images," *Pattern Recognition*, vol. 30, no. 6, pp. 867–881, 1997.

[7] Y. Deng and B. S. Manjunath, "Unsupervised segmentation of color-texture regions in images and video," *IEEE Transactions on Pattern Analysis and Machine Intelligence*, vol. 23, no. 8, pp. 800–810, 2001.

[8] Y. F. Ma and H. J. Zhang, "Contrast-based image attention analysis by using fuzzy growing," in *Proc. Eleventh ACM International Conference on Multimedia.(ACM '03)*, pp. 374–381, New York, NY, USA, 2003.

[9] R. Adams and L. Bischof, "Seeded region growing," *IEEE Transactions on Pattern Analysis and Machine Intelligence*, vol. 16, no. 6, pp. 641–647, 1994.

[10] J. Fan, D. K. Y. Yau, A. K. Elmagarmid, and W. G. Aref, "Automatic image segmentation by integrating color-edge extraction and seeded region growing," *IEEE Transactions on Image Processing*, vol. 10, no. 10, pp. 1454–1466, 2001.

[11] F. Y. Shih and S. Cheng, "Automatic seeded region growing for color image segmentation," *Image and Vision Computing*, vol. 23, no. 10, pp. 877–886, 2005.

[12] Y. W. Lim and S. U. Lee, "On the color image segmentation algorithm based on the thresholding and the fuzzy c-means techniques," *Pattern Recognition*, vol. 23, no. 9, pp. 935–952, 1990.

[13] D. Mújica-Vargas, F. J. Gallegos-Funes, and A. J. Rosales-Silva, "A fuzzy clustering algorithm with spatial robust estimation constraint for noisy color image segmentation," *Pattern Recognition Letters*, vol. 34, no. 4, pp. 400–413, 2013.

[14] J. C. Bezdek, *Pattern Recognition with Fuzzy Objective Function Algorithms*, Advanced Applications in Pattern Recognition, New York, NY, USA, 1981.

[15] N. Paragios and R. Deriche, "Geodesic active regions and level set methods for supervised texture segmentation," *International Journal of Computer Vision*, vol. 46, no. 3, pp. 223–247, 2002.

[16] D. K. Panjwani and G. Healey, "Markov Random Field Models for Unsupervised Segmentation of Textured Color Images," *IEEE Transactions on Pattern Analysis and Machine Intelligence*, vol. 17, no. 10, pp. 939–954, 1995.

[17] L. A. Vese and T. F. Chan, "A multiphase level set framework for image segmentation using the Mumford and Shah model," *International Journal of Computer Vision*, vol. 50, no. 3, pp. 271–293, 2002.

[18] J. Liu and Y. H. Yang, "Multiresolution color image segmentation," *IEEE Trans. Pattern Analysis and Machine Intelligence*, vol. 16, no. 7, pp. 689–700, Jul 1994.

[19] G. Zhao, Y. Li, G. Chen et al., "A fuzzy-logic based approach to color segmentation," in *Proceedings of the SPIE 8739, Sensors and Systems for Space Applications VI, 873910*, Baltimore, Md, USA, 2013.

[20] W. Li, G. Lu, and Y. Wang, "Recognizing white line markings for vision-guided vehicle navigation by fuzzy reasoning," *Pattern Recognition Letters*, vol. 18, no. 8, pp. 771–780, 1997.

[21] W. Li, X. Jiang, and Y. Wang, "Road recognition for vision navigation of an autonomous vehicle by fuzzy reasoning," *Fuzzy Sets and Systems*, vol. 93, no. 3, pp. 275–280, 1998.

[22] W. Li, F. M. Wahl, J. Z. Zhou, H. Wang, and K. Z. He, "Vision-Based Behavior Control of Autonomous Systems by Fuzzy Reasoning," *Sensor Based Intelligent Robots*, vol. 1724, pp. 311–325, 1999.

[23] http://www.eecs.berkeley.edu/Research/Projects/CS/vision/grouping/segbench/.

[24] X. C. Li, H. K. Liu, F. Wang, and C. Y. Bai, "The survey of fuzzy clustering method for image segmentation," *Journal of Image and Graphics*, vol. 17, no. 4, pp. 447–458, 2012.

[25] J. Zhang, M. Li, and Z. L. Yang, "Color image segmentation based on fuzzy color extraction," *Journal of Tianjin University of Technology*, no. 6, pp. 33–38, 2014.

[26] Y. Tian, X. Kang, Y. Li et al., "Identifying rhodamine dye plume sources in near-shore oceanic environments by integration of chemical and visual sensors," *Sensors*, vol. 13, no. 3, pp. 3776–3798, 2013.

13

An Image Encryption Scheme of Logistic Modulation using Computer-Generated Hologram and Chaotic Map

Hui Ren, Jun Wang, and Qiong-Hua Wang

School of Electronics and Information Engineering, Sichuan University, Chengdu 610065, China

Correspondence should be addressed to Jun Wang; jwang@scu.edu.cn

Academic Editor: Jit S. Mandeep

We introduce an image encryption method based on computer-generated hologram (CGH) and two-dimensional Sine Logistic modulation map (2D-SLMM). We combine CGH and 2D-SLMM to improve encryption security. During the encryption process, the hologram needs to be logistically modulated by 2D-SLMM. This logistic modulation technique can avoid complex algorithms. Simulation results and security analysis demonstrate that the proposed approach has a high security level, good invisibility of image information in ciphertext, large key space, and strong robustness.

1. Introduction

Since Refregier and Javidi proposed a double random-phase encryption technique in 1995 [1], various improved optical image encryption methods have been presented [2–6]. However, data information of optical encryption is difficult to be transmitted through the network. And without the digital processing or conversion, the information must be reconstructed with an optical method. The digitization of encryption information favors the preservation and transmission of information. One of the most effective methods of the digitization of encrypted information is CGH [7–10]. Besides the fact that the virtual object, which does not yet exist in nature, can be recorded, CGH also permits any wavelength to be selected and system parameters to be adjusted arbitrarily, which offers more flexibility. Since this encryption method has only two secret keys of wavelength and diffraction distance, it is necessary to expand the key space of cipher system.

In the field of image encryption, many encryption methods involved chaos theory [11–16]. Chaotic maps have the properties of unpredictability and sensitivity to their parameters and initial values. They can generate different random sequences with different settings of parameters or initial values. Existing chaotic maps can be classified into two categories: one-dimensional (1D) chaotic maps and high-dimensional (HD) chaotic maps. 1D chaotic maps usually contain one variable and a few parameters. However, the structures and chaotic orbits of 1D chaotic maps are rather simple and may be estimated. And parameters or/and initial values of them may be predicted, too. Compared with 1D chaotic maps, HD chaotic maps usually have more complex structures and better chaotic performance. These make their chaotic orbits much difficult to predict [17, 18]. 2D Sine Logistic modulation map (2D-SLMM) which belongs to a kind of HD chaotic map has been proposed [17]. Performance analysis is provided to show that 2D-SLMM has wider chaotic range, better ergodicity, and hyperchaotic properties than the existing chaotic maps. If only 2D-SLMM is used in logistic modulation, the encrypted information cannot be completely hidden. Therefore, it is desired to increase the invisibility of image information in ciphertext.

In order to expand secret key space further and make the encrypted information totally invisible, we proposed an encryption method combining CGH and 2D-SLMM in this paper. Firstly, the mathematical model of CGH is built to get the hologram of original image. Then, the various parameters and initial values of the chaotic system are determined to acquire 2D-SLMM by them. Lastly, 2D-SLMM is modulated by the obtained hologram with a very simple way to get

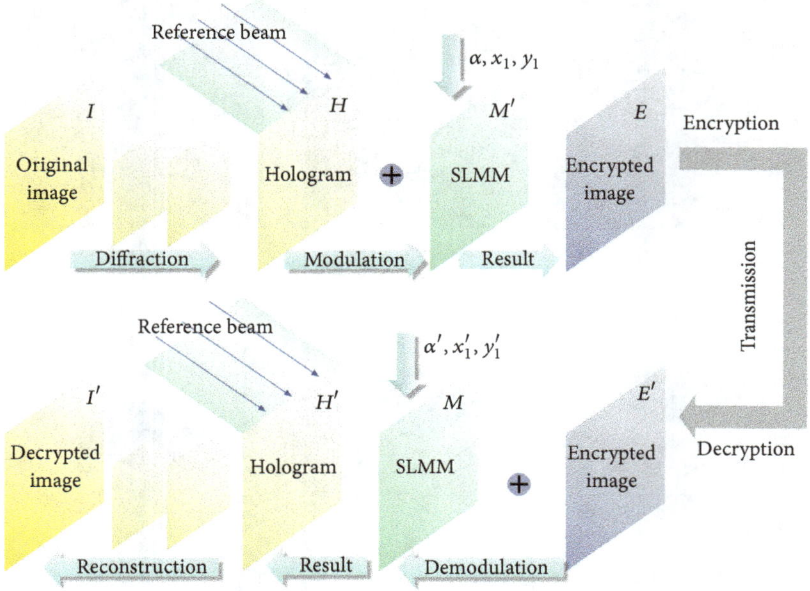

FIGURE 1: Encryption and decryption process.

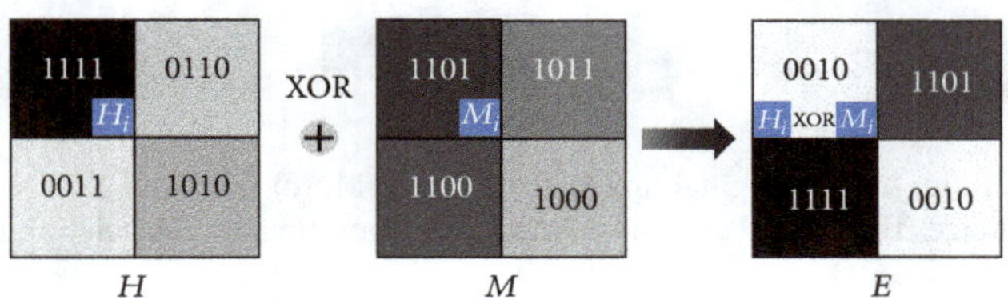

FIGURE 2: Process of logistic modulation.

the encrypted image. This modulation technique is logistic modulation, which can be implemented in not only software but hardware. Therefore, it is beneficial to greatly expand the application scope of this method. Simulation and security analysis verify the feasibility and security of this method.

2. Proposed Encryption Method

The proposed encryption process is shown in Figure 1. At the beginning, we get the hologram marked with H. Then, acquire 2D-SLMM marked with M by setting parameters and initial values. In the end, the obtained hologram is modulated logistically by 2D-SLMM to get the encrypted image marked with E. It should be emphasized that because the decryption is the inverse progress of encryption, the detailed description of decryption is omitted.

First, we need to obtain the hologram of the original image. The schematic diagram of the generation of a hologram can be referenced in [19], where wave of object and reference beam interference with each other. And the interference pattern as a hologram is to be recorded on the observation plane.

Next, 2D-SLMM is acquired with different settings of parameters and initial values. It is defined by

$$\begin{aligned} x_{i+1} &= \alpha\left(\sin\left(\pi y_i\right) + \beta\right) x_i \left(1 - x_i\right) \\ y_{i+1} &= \alpha\left(\sin\left(\pi x_{i+1}\right) + \beta\right) y_i \left(1 - y_i\right), \end{aligned} \tag{1}$$

where α and β are control parameters. $\alpha \in [0, 1]$ and $\beta \in [0, 3]$.

2D-SLMM is derived from two 1D chaotic maps, the Logistic and Sine maps. A combination of the Sine map and parameter β is used to modulate the output of the Logistic map to enhance its nonlinearity and randomness. The result is then extended from one-dimension to two-dimension to obtain 2D-SLMM. Two output values of 2D-SLMM x_{i+1} and y_{i+1} intertwine each other. Thus, its orbits and iteration values are difficult to predict. When parameter β is close to 3, 2D-SLMM shows good chaotic performance. For simplicity, we set $\beta = 3$ in the rest of this paper [17]. In addition to control parameters, the initial values x_1 and y_1 can also be used as secret keys, which can expand the key space significantly.

Finally, the obtained hologram H is logistically modulated by 2D-SLMM M to get encrypted image E. The process is shown in Figure 2, where H_i and M_i represent number i pixel of H and M, respectively. The numbers in the figure are the binary representation of pixel values. It should be emphasized that, to achieve better effect, we can also make

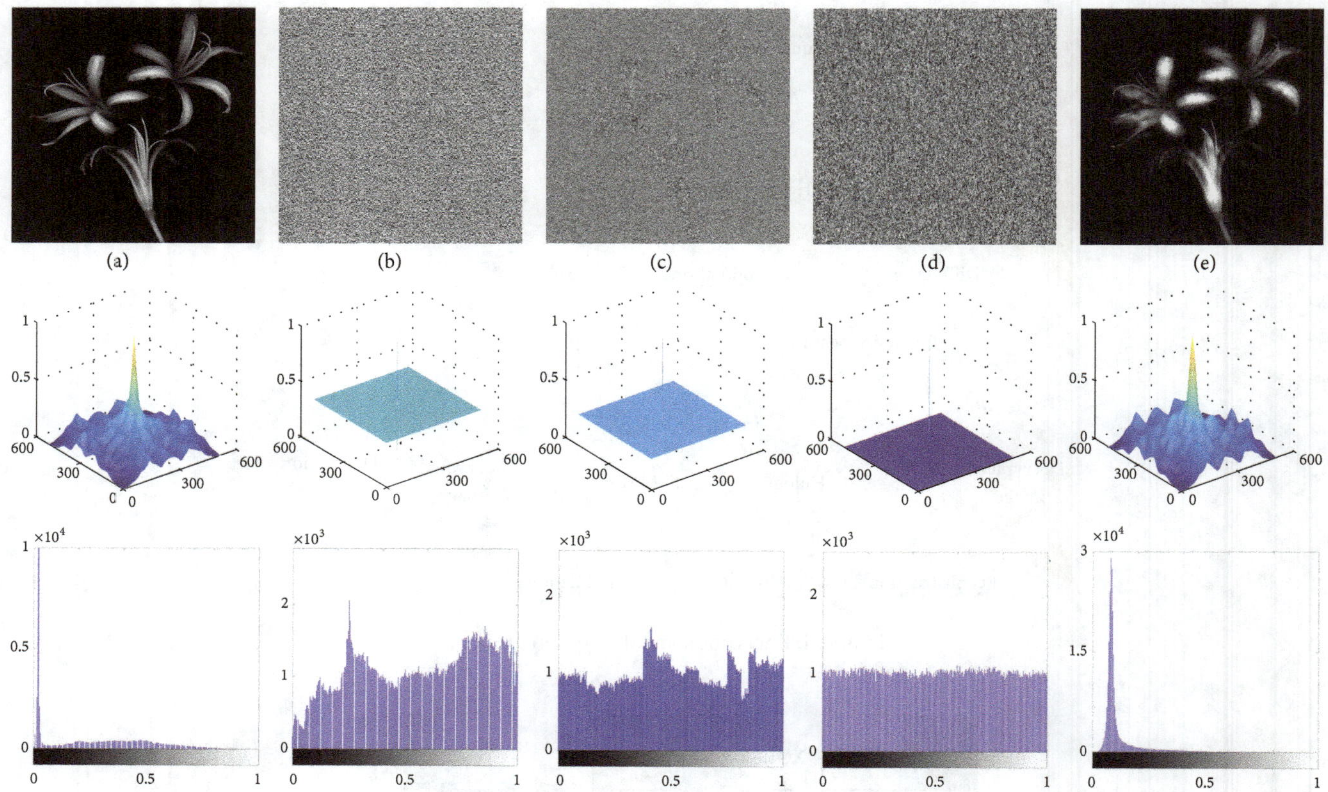

FIGURE 3: Simulation of proposed method: (a) the original image, (b) 2D-SLMM, (c) 1st round XOR encrypted image, (d) 2nd round XOR encrypted image, and (e) decrypted image. The second row shows their corresponding autocorrelation images and the third row shows their corresponding histograms.

twice logistic modulation and the detailed analysis is depicted in Section 3.

3. Simulated Results and Security Analysis

The object to be encrypted is a gray image of "Flowers" with 512 × 512 pixels as shown in Figure 3(a) and 2D-SLMM is shown in Figure 3(b). After the first round of logistic modulation, as can be seen from Figure 3(c), there is still a little bit of effective information. Therefore, we make the second round of modulation to complete the encryption. Then the encrypted image is shown in Figure 3(d). Obviously, all the information is hidden in the noise-like image. Finally, the decrypted image is in Figure 3(e). The rest of images of Figure 3 show their corresponding autocorrelation images and histograms.

The proposed method can apply to many other types of images. There are two sets of comparison experiment, "Camera man" and "Chess," respectively, in Figures 4 and 5. In addition to proposed method as shown in Figure 4(a), there are three other encryption methods as shown in Figures 4(c), 4(d), and 4(e). They are SBE-LBP [20], BPE-XOR [21], and the method of using 2D-SLMM to modulate original image instead of hologram. The rest images of Figure 4 show their corresponding histograms and autocorrelation images. By comparison, we can find that the encryption effect of the proposed method is the best. It needs to be emphasized that we just take only 1st round XOR operation to ensure fairness. The analysis of Figure 5 is the same as that of Figure 4.

3.1. Robustness Test. Figure 6 shows experimental results of noise and data loss attacks to the cipher-images. Some attacked on holograms with noise are shown in Figures 6(a)–6(c) and their corresponding decrypted images are shown in Figures 6(g)–6(i). The image quality assessment, structural similarity (SSIM) index, can assess the similarity between two images [22]. So we calculate the SSIM indexes between decrypted images Figures 6(g)–6(i) and original image Figure 3(a). When SIMM index is 1, it means that the two images are exactly the same. Less than 1, it means that the similarity between the two images is weak. By calculation, the SIMM index between Figures 6(g) and 3(a) is 0.44 rather than 1. It is because that there still exists the error of algorithm even without noise interference. Through these SSIM indexes, we can draw a conclusion that this method can resist noise attack when the noise level is not too high.

Then, the situation of data loss in public media transmission is simulated. Some encrypted images of data loss are shown in Figures 6(d)–6(f) and their corresponding decrypted images are shown in Figures 6(j)–6(l). At the same time, the SSIM indexes between decrypted images and original image Figure 3(a) are on the top of the decrypted

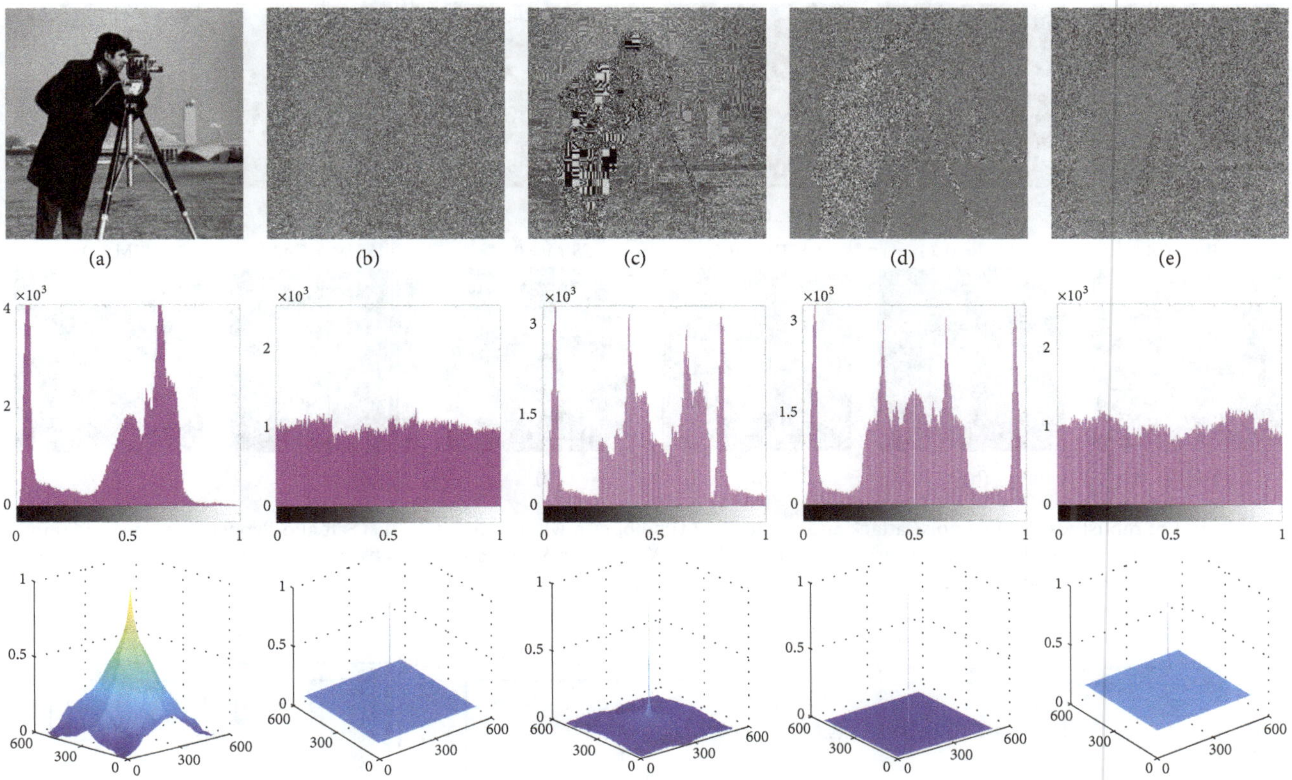

FIGURE 4: Comparative experiment of "Camera man": (a) original image, (b) 2D-SLMM, (c) SBE-LBP, (d) BPE-XOR, and (e) method of using 2D-SLMM to modulate original image. The second row shows their corresponding histograms and the third row shows their corresponding autocorrelation images.

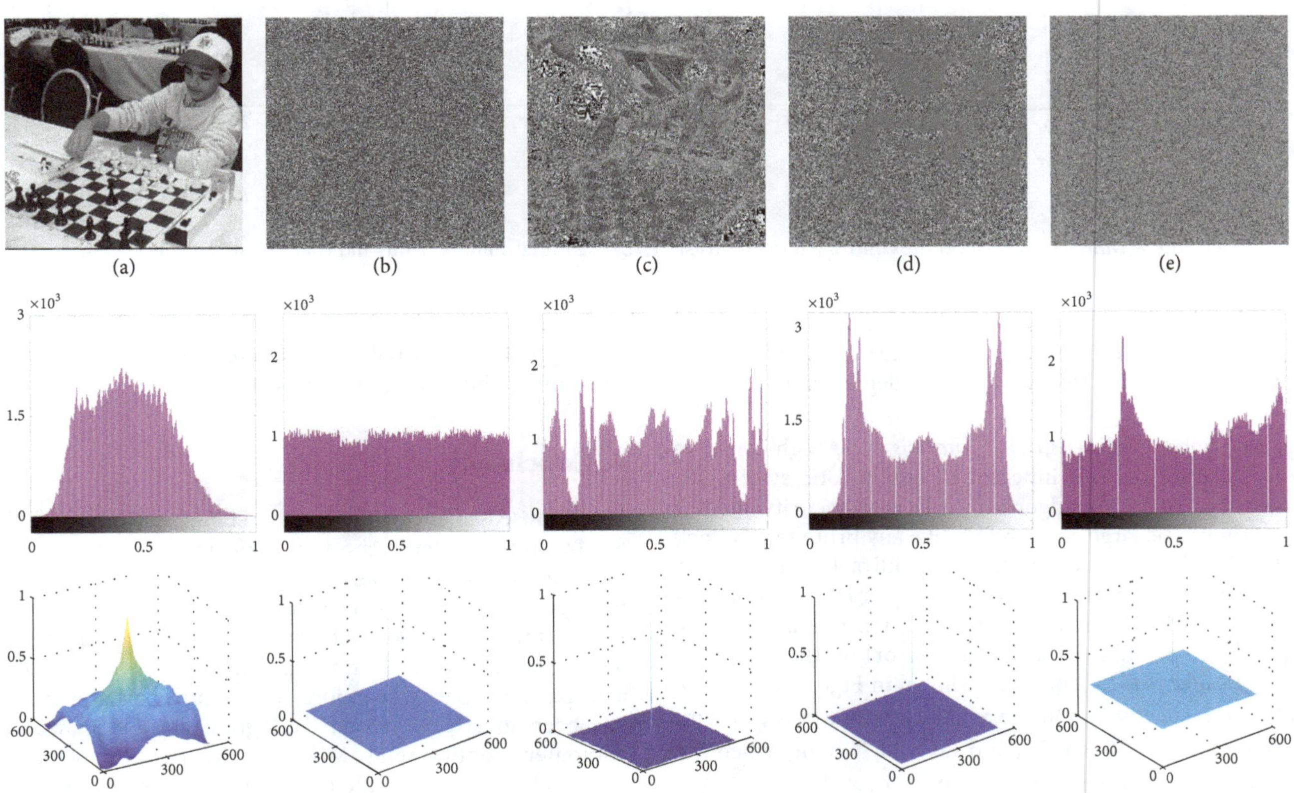

FIGURE 5: Comparative experiment of "Chess": (a) original image, (b) 2D-SLMM, (c) SBE-LBP, (d) BPE-XOR, and (e) method of using 2D-SLMM to modulate original image. The second row shows their corresponding histograms and the third row shows their corresponding autocorrelation images.

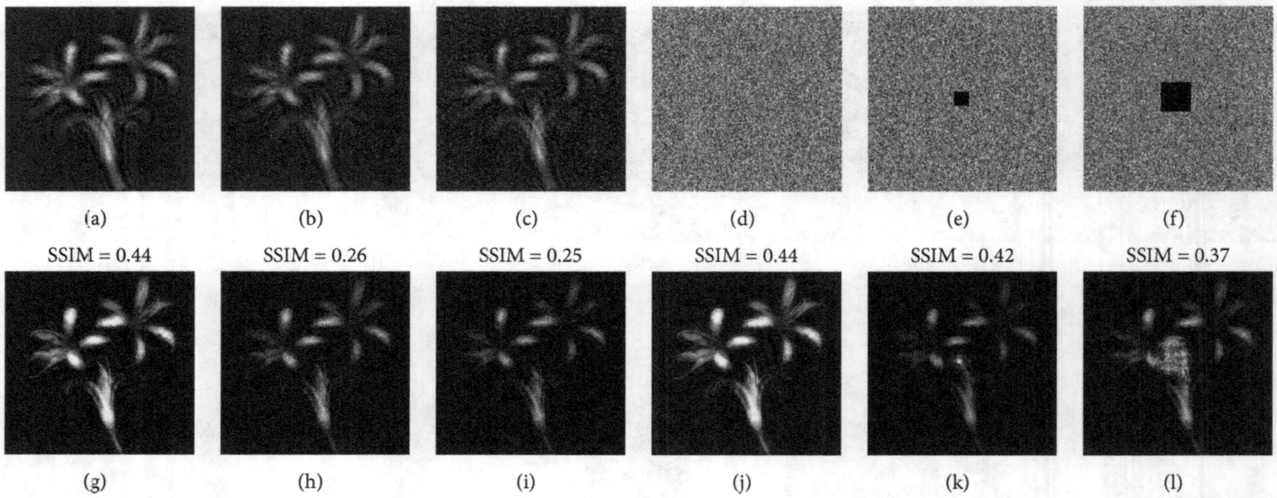

FIGURE 6: Test of robustness against noise attack and occlusion: (a) hologram without noise, (b) 1% Salt and Pepper noise on hologram, (c) 1.5% Salt and Pepper noise on hologram, (d) encrypted image, (e) 0.6% data loss on (d), and (f) 2.44% data loss on (d). The second row shows their corresponding decrypted images.

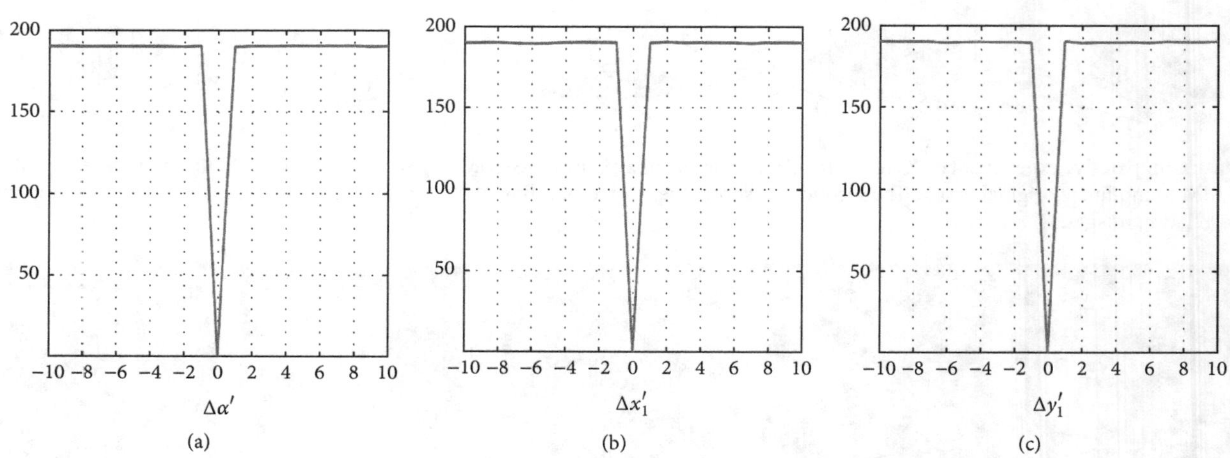

FIGURE 7: MSE varies with (a) the control parameter α' shift, (b) the initial value x_1' shift, and (c) the initial value y_1' shift.

images. It can be seen from SSIM indexes of Figures 6(j)–6(l) that this method can tolerant a certain degree of data loss.

3.2. Key Space and Sensitivity Analysis. The high sensitive to initial conditions is inherent in any chaotic system. To provide an encryption algorithm with high security, the key space should be large enough to make any brute force attack ineffective. When wavelength λ and diffraction distance z are given certain values, control parameter α and and initial values x_1 and y_1 can also be used as encryption keys. The dependence of MSE (Mean Square Error) on the change of decryption keys α', x_1', and y_1' is shown in Figures 7(a), 7(b), and 7(c), respectively. On the x-axis, the unit of α', x_1', and y_1' is 9.0×10^{-17}, 1.0×10^{-16}, and 1.0×10^{-15}. Only when the values of α', x_1', and y_1' are the same as α, x_1, and y_1 do the MSE reach a very small value. In the case of small derivations, the MSE increases quickly and there is a failure to recognize the original image visually. The high sensitivity results in great difficulty in duplicating the decryption system.

4. Conclusion

We proposed an encryption method based on CGH and 2D-SLMM. It has better encryption performance than the compared methods. The experiments also verify the resistance to multiple attacks: noise, occlusion, and the tiny change of keys. Furthermore, beside keys of λ and z, the other parameters of α, x_1, and y_1 are used as additional secret keys to provide a larger key space. Therefore, the security of the system has been greatly improved. Hologram can be realized in computer or optical setup. Moreover, the algorithm of logistic modulation is very simple to implement in software, hardware, and all-optical XOR logic gate. Therefore, the proposed encryption method is not only effective but also flexible. In the future, it could be applied in the field of national defense science and technology or other places requiring a higher level of image security and system flexibility.

Conflicts of Interest

The authors declare that they have no conflicts of interest.

References

[1] P. Refregier and B. Javidi, "Optical image encryption based on input plane and Fourier plane random encoding," *Optics Expresss*, vol. 20, no. 7, pp. 767–769, 1995.

[2] A. Alfalou and A. Mansour, "Double random phase encryption scheme to multiplex and simultaneous encode multiple images," *Applied Optics*, vol. 48, no. 31, pp. 5933–5947, 2009.

[3] J. Yi and G. Tan, "Optical compression and encryption system combining multiple measurement matrices with fractional Fourier transform," *Applied Optics*, vol. 54, no. 36, pp. 10650–10658, 2015.

[4] X.-W. Li and I.-K. Lee, "Modified computational integral imaging-based double image encryption using fractional Fourier transform," *Optics and Lasers in Engineering*, vol. 66, pp. 112–121, 2015.

[5] X. Li, D. Xiao, and Q.-H. Wang, "Error-free holographic frames encryption with CA pixel-permutation encoding algorithm," *Optics and Lasers in Engineering*, vol. 100, pp. 200–207, 2018.

[6] C. Li, D. Lin, and J. Lu, "Cryptanalyzing an Image-Scrambling Encryption Algorithm of Pixel Bits," *IEEE MultiMedia*, vol. 24, no. 3, pp. 64–71, 2017.

[7] J. Liu, H. Jin, L. Ma, Y. Li, and W. Jin, "Optical color image encryption based on computer generated hologram and chaotic theory," *Optics Communications*, vol. 307, pp. 76–79, 2013.

[8] S.-X. Xi, X. Wang, X. Sun, S. Chang, and L. Lin, "Three random phase encryption technology in the Fresnel diffraction system based on computer-generated hologram," *Optical Engineering*, vol. 53, no. 1, Article ID 011004, 2014.

[9] S. Xi, X. Wang, L. Song et al., "Experimental study on optical image encryption with asymmetric double random phase and computer-generated hologram," *Optics Express*, vol. 25, no. 7, pp. 8212–8222, 2017.

[10] H. Yoshikawa and T. Yamaguchi, "Review of Holographic Printers for Computer-Generated Holograms," *IEEE Transactions on Industrial Informatics*, vol. 12, no. 4, pp. 1584–1589, 2016.

[11] G. Ye and K.-W. Wong, "An efficient chaotic image encryption algorithm based on a generalized Arnold map," *Nonlinear Dynamics*, vol. 69, no. 4, pp. 2079–2087, 2012.

[12] Y. Zhou, L. Bao, and C. L. P. Chen, "Image encryption using a new parametric switching chaotic system," *Signal Processing*, vol. 93, no. 11, pp. 3039–3052, 2013.

[13] A. Kanso, H. Yahyaoui, and M. Almulla, "Keyed hash function based on a chaotic map," *Information Sciences*, vol. 186, pp. 249–264, 2012.

[14] X. Wang, W. Zhang, W. Guo, and J. Zhang, "Secure chaotic system with application to chaotic ciphers," *Information Sciences*, vol. 221, pp. 555–570, 2013.

[15] Y.-Q. Zhang and X.-Y. Wang, "A symmetric image encryption algorithm based on mixed linear-nonlinear coupled map lattice," *Information Sciences*, vol. 273, pp. 329–351, 2014.

[16] Y. Zhou, L. Bao, and C. L. P. Chen, "A new 1D chaotic system for image encryption," *Signal Processing*, vol. 97, pp. 172–182, 2014.

[17] Z. Hua, Y. Zhou, C.-M. Pun, and C. L. P. Chen, "2D Sine Logistic modulation map for image encryption," *Information Sciences*, vol. 297, pp. 80–94, 2015.

[18] Z. Hua and Y. Zhou, "Image encryption using 2D Logistic-adjusted-Sine map," *Information Sciences*, vol. 339, pp. 237–253, 2016.

[19] X. Su, *Optical Information*, Science Press, Beijing, China, 2nd edition, 2010.

[20] Y. Zhou, K. Panetta, S. Agaian, and C. L. P. Chen, "Image encryption using P-Fibonacci transform and decomposition," *Optics Communications*, vol. 285, no. 5, pp. 594–608, 2012.

[21] J. A. C. Gallas, "Structure of the parameter space of the Hénon map," *Physical Review Letters*, vol. 70, no. 18, pp. 2714–2717, 1993.

[22] Z. Wang, A. C. Bovik, H. R. Sheikh, and E. P. Simoncelli, "Image quality assessment: from error visibility to structural similarity," *IEEE Transactions on Image Processing*, vol. 13, no. 4, pp. 600–612, 2004.

14

Log-PF: Particle Filtering in Logarithm Domain

Christian Gentner, Siwei Zhang, and Thomas Jost

German Aerospace Center (DLR), Institute of Communications and Navigation, Oberpfaffenhofen, 82234 Wessling, Germany

Correspondence should be addressed to Christian Gentner; christian.gentner@dlr.de

Academic Editor: Víctor Elvira

This paper presents a particle filter, called Log-PF, based on particle weights represented on a logarithmic scale. In practical systems, particle weights may approach numbers close to zero which can cause numerical problems. Therefore, calculations using particle weights and probability densities in the logarithmic domain provide more accurate results. Additionally, calculations in logarithmic domain improve the computational efficiency for distributions containing exponentials or products of functions. To provide efficient calculations, the Log-PF exploits the Jacobian logarithm that is used to compute sums of exponentials. We introduce the weight calculation, weight normalization, resampling, and point estimations in logarithmic domain. For point estimations, we derive the calculation of the minimum mean square error (MMSE) and maximum a posteriori (MAP) estimate. In particular, in situations where sensors are very accurate the Log-PF achieves a substantial performance gain. We show the performance of the derived Log-PF by three simulations, where the Log-PF is more robust than its standard particle filter counterpart. Particularly, we show the benefits of computing all steps in logarithmic domain by an example based on Rao-Blackwellization.

1. Introduction

Many scientific problems involve dynamic systems, for example, in navigation applications. Dynamic systems can be described by state-space models where the state is only observable by noisy measurements. Recursive Bayesian filters are algorithms to estimate an unknown probability density function (PDF) of the state recursively by measurements over time. Such a filter consists of two steps: prediction and update. In the prediction step, the PDF of the state is calculated based on the system model. During the update step, the current measurement is used to correct the prediction based on the measurement model. In this way, the posterior PDF of the state is estimated recursively over time. Particle filters (PFs) are implementations of recursive Bayesian filters which approximate the posterior PDF by a set of random samples, called particles, with associated weights. Several types of PFs have been developed over the last few years [1–8]. They differ in their choice of the importance sampling density and the resampling step.

A common way is to choose the importance sampling density to be equal to the prior, for example, the bootstrap filtering algorithm [1]. However, if the width of the likelihood distribution is too small in comparison to the width of the prior distribution or if measurements are located in the tail of the prior distribution, this choice may fail; see [4]. These situations may arise when sensors are very accurate or measurements rapidly change over time such that the particle states after the prediction step might be located in the tail of the likelihood. Additionally, numerical representation of numbers may limit the computational accuracy by floating point errors. In these situations, a common way is to use the likelihood particle filter (LPF) [3, 5]. The LPF uses the likelihood distribution for the importance sampling density and the prior for the weight update. The LPF is recommended when the width of the likelihood distribution is much smaller compared to the one of the prior and accordingly, the posterior density function is more similar to the likelihood than to the prior. However, in many situations, it is impossible to draw samples from the likelihood distribution. Furthermore, the LPF is not suitable for an underdetermined system where the number of measurements is lower than the number of states per time instant. Additionally, using the likelihood as proposal distribution might increase the variance of the simulated samples according to [5].

In this paper, we derive a PF that operates in logarithmic domain (log-domain), called Log-PF. The Log-PF represents the weights in log-domain which enables a more accurate representation of low weights with a limited number of bits. Particularly, when the involved distributions contain exponentials or products of functions, the log-domain representation is computationally more efficient [9]. The derived Log-PF uses the Jacobian logarithm [10–12] to describe all steps of the PF, including weight update, weight normalization, resampling, and point estimations in log-domain. In this paper, we derive the minimum mean square error (MMSE) and the maximum a posteriori (MAP) point estimators.

The paper is structured as follows: First, we describe in Section 2 standard PFs; thereafter we derive the proposed Log-PF in Section 2. Afterwards, we derive in Section 4 two point estimators in log-domain: Section 4.1 describes the MMSE estimator and Section 4.2 the MAP estimator. We evaluate the Log-PF by simulations and compare the results to standard PF implementations and Kalman filters (KFs) in Section 5. Particularly, we show by an example based on Rao-Blackwellization the benefits by computing all steps in log-domain. For distributed particle filters like [13] similar results are expected. Finally, Section 6 concludes the paper. Throughout the paper, we will use the following notations:

(i) All vectors are interpreted as column vectors.

(ii) $\mathbf{I}$ denotes an identity matrix.

(iii) Matrices are denoted by bold capital letters and vectors by bold small letters.

(iv) $[\mathbf{x}]_l$ denotes the lth element of vector $\mathbf{x}$.

(v) $a \sim \mathcal{N}(\mu_a, \sigma_a^2)$ denotes a Gaussian distributed or multivariate random variable a with mean μ_a and variance σ_a^2.

(vi) $\mathbf{E}[x]$ stands for expectation or sample mean of x.

(vii) $1 : k$ stands for all integer numbers starting from 1 to k, thus $1, 2, \ldots, k$.

(viii) $\hat{x}$ denotes the estimate of x.

(ix) $\{x^{(i)}\}_{i=1}^{N}$ defines the set for x_i with $i = 1, \ldots, N$.

2. Particle Filtering

PFs represent the probability density of the state vector $\mathbf{x}_k$ at time step k by N_p particles. According to [1–3, 5] and assuming a first-order hidden Markov model, the posterior filtered density $\mathsf{p}(\mathbf{x}_k \mid \mathbf{z}_{0:k})$ is approximated as

$$\mathsf{p}(\mathbf{x}_k \mid \mathbf{z}_{0:k}) \approx \sum_{j=1}^{N_p} w_k^{(j)} \delta\left(\mathbf{x}_k - \mathbf{x}_k^{(j)}\right), \tag{1}$$

where $\mathbf{z}_{0:k}$ defines the measurement vector for the time steps $0, \ldots, k$, $\delta(\cdot)$ stands for the Dirac distribution, $\mathbf{x}_k^{(j)}$ denotes particle state, and $w_k^{(j)}$ denotes the normalized weight with

$$w_k^{(j)} = \frac{w_k^{*(j)}}{\sum_{i=1}^{N_p} w_k^{*(i)}}. \tag{2}$$

The unnormalized weight is denoted by $w_k^{*(j)}$, while the weight update is calculated as

$$w_k^{*(j)} = w_{k-1}^{(j)} \frac{\mathsf{p}\left(\mathbf{z}_k \mid \mathbf{x}_k^{(j)}\right) \mathsf{p}\left(\mathbf{x}_k^{(j)} \mid \mathbf{x}_{k-1}^{(j)}\right)}{\mathsf{q}\left(\mathbf{x}_k^{(j)} \mid \mathbf{x}_{k-1}^{(j)}, \mathbf{z}_k\right)}, \tag{3}$$

with the importance density $\mathsf{q}(\mathbf{x}_k \mid \mathbf{x}_{k-1}^{(j)}, \mathbf{z}_k)$, the likelihood distribution $\mathsf{p}(\mathbf{z}_k \mid \mathbf{x}_k^{(j)})$, and the transition prior distribution $\mathsf{p}(\mathbf{x}_k^{(j)} \mid \mathbf{x}_{k-1}^{(j)})$; see [1–3, 5]. For $N_p \to \infty$, the approximation used in (33) approaches $\mathsf{p}(\mathbf{x}_k \mid \mathbf{z}_{0:k})$. By (33), (2), and (3), the sequential importance sampling (SIS) PF can be described which is the basis of most PFs [2].

For numerical stability reasons, weights are often computed and stored in the log-domain, which is also computationally efficient when the distributions involved contain exponentials or products. Thus, we obtain from (3) the update equation in log-domain, with

$$\begin{aligned}\widehat{w}_k^{*(j)} = {} & \widehat{w}_{k-1}^{(j)} + \ln\left(\mathsf{p}\left(\mathbf{z}_k \mid \mathbf{x}_k^{(j)}\right)\right) + \ln\left(\mathsf{p}\left(\mathbf{x}_k^{(j)} \mid \mathbf{x}_{k-1}^{(j)}\right)\right) \\ & - \ln\left(\mathsf{q}\left(\mathbf{x}_k \mid \mathbf{x}_{k-1}^{(j)}, \mathbf{z}_k\right)\right),\end{aligned} \tag{4}$$

where we define with

$$\widehat{w}_k^{(j)} = \ln\left(w_k^{(j)}\right) \tag{5}$$

the log-domain weight (log-weight) $\widehat{w}_k^{(j)}$ for particle j. After calculating the weights $\widehat{w}_k^{*(j)}$ in log-domain, the weights are transferred for further processing to the linear domain (lin-domain) with $w_k^{*(j)} = e^{\widehat{w}_k^{*(j)}}$ for $j = 1, \ldots, N_p$, where the numerical accuracy is lost due to floating point representation. In order to obtain a more stable PF implementation, the weights $\widehat{w}_k^{*(j)}$ can be transferred to the lin-domain by

$$w_k^{+(j)} = e^{\widehat{w}_k^{*(j)} - \max_l(\widehat{w}_k^{*(l)})}, \tag{6}$$

such that $w_k^{(j)} = w_k^{+(j)} / \sum_{i=1}^{N_p} w_k^{+(i)}$; see, for example, [9].

In the following we investigate a different approach, where the transformation from the log-domain to the lin-domain is not necessary. Hence, we show that all steps of the PF can be computed in log-domain.

3. Algorithm Derivation

To compute all steps of the PF in log-domain, we obtain for the approximation of the posterior filtered density from (33)

$$\mathsf{p}(\mathbf{x}_k \mid \mathbf{z}_{0:k}) \approx \sum_{j=1}^{N_p} e^{\widehat{w}_k^{(j)}} \delta\left(\mathbf{x}_k - \mathbf{x}_k^{(j)}\right) \tag{7}$$

using (5). The normalization of the log-weight can be calculated directly in log-domain as a simple subtraction, with

$$\widehat{w}_k^{(j)} = \widehat{w}_k^{*(j)} - \widehat{W}_k, \tag{8}$$

$\ln(\sum_{l=1}^{n} e^{\delta_l}) = \text{Jacob}(\{\delta_l\}_{l=1}^{n})$
(1) Init: $\Delta_1 = \delta_1$;
(2) **for** $l = 2 : n$ **do**
(3) $\quad \Delta_l = \max(\delta_l, \Delta_{l-1}) + \ln(1 + e^{-|\delta_l - \Delta_{l-1}|})$;
(4) $\ln(\sum_{l=1}^{n} e^{\delta_l}) = \Delta_n$;

ALGORITHM 1: Iterative Jacobian algorithm.

where $\widehat{W}_k$ denotes the normalization factor with

$$\widehat{W}_k = \ln\left(\sum_{i=1}^{N_p} e^{\widehat{\omega}_k^{*(i)}}\right). \tag{9}$$

To compute the normalization factor $\widehat{W}_k$ of (9) without transferring the log-weights to the lin-domain, the Jacobian logarithm [10, 11] can be used. The Jacobian logarithm computes the logarithm of a sum of two exponentials $\ln(e^{\delta_1} + e^{\delta_2})$ using the max $(\cdot)$ operator and adding a correction term; that is,

$$\ln\left(e^{\delta_1} + e^{\delta_2}\right) = \max\left(\delta_1, \delta_2\right) + \ln\left(1 + e^{-|\delta_2 - \delta_1|}\right). \tag{10}$$

With (10) and as derived in [12], the expression $\ln(\sum_{l=1}^{n} e^{\delta_l})$ can be calculated iteratively as

$$\begin{aligned} \ln\left(e^{\delta_1} + \cdots + e^{\delta_n}\right) &= \ln\left(\Delta + e^{\delta_n}\right) \\ &= \max\left(\ln\left(\Delta\right), \delta_n\right) \\ &\quad + \ln\left(1 + e^{-|\ln(\Delta) - \delta_n|}\right), \end{aligned} \tag{11}$$

where $\delta = \ln(e^{\delta_1} + \cdots + e^{\delta_{n-1}})$ and $\Delta = e^{\delta_1} + \cdots + e^{\delta_{n-1}}$. Hence, using the Jacobian logarithm allows computing operations such as summations like in (9) efficiently in the log-domain. For later conveniences, we express (11) by an iterative algorithm shown in Algorithm 1 by a pseudocode; that is,

$$\ln\left(\sum_{l=1}^{n} e^{\delta_l}\right) = \text{Jacob}\left(\{\delta_l\}_{l=1}^{n}\right), \tag{12}$$

where $\{\delta_l\}_{l=1}^{n}$ defines the set for δ_l with $l = 1, \ldots, n$. Thus, the normalization factor $\widehat{W}_k$ of (9) can be calculated iteratively by

$$\widehat{W}_k = \text{Jacob}\left(\left\{\widehat{\omega}_k^{*(i)}\right\}_{i=1}^{N_p}\right). \tag{13}$$

Hence, we obtain for the log-weight normalization of (8),

$$\widehat{\omega}_k^{(j)} = \widehat{\omega}_k^{*(j)} - \text{Jacob}\left(\left\{\widehat{\omega}_k^{*(i)}\right\}_{i=1}^{N_p}\right). \tag{14}$$

Please note that a complexity reduction can be obtained if the term $\ln(1 + e^{-|\delta_l - \Delta_{l-1}|})$ of Algorithm 1 is read from a hard-coded lookup table as a function of $|\delta_l - \Delta_{l-1}|$.

By using (14), the SIS PF can be described in log-domain as shown in Algorithm 2 by a pseudocode. Algorithm 2 is

$\{\mathbf{x}_k^{(j)}, \widehat{\omega}_k^{(j)}\}_{j=1}^{N_p} = \text{Log-SIS}(\{\mathbf{x}_{k-1}^{(j)}, \widehat{\omega}_{k-1}^{(j)}\}_{j=1}^{N_p})$
(1) **for** $j = 1 : N_p$ **do**
(2) $\quad$ Draw: $\mathbf{x}_k^{(j)} \sim q(\mathbf{x}_k^{(j)} \mid \mathbf{x}_{k-1}^{(j)}, \mathbf{z}_k)$;
(3) $\quad$ Calculate: $\widehat{\omega}_k^{*(j)}$ according to (4);
(4) Calculate: $\widehat{W}(k) = \text{Jacob}(\{\widehat{\omega}_k^{*(j)}\}_{j=1}^{N_p})$;
(5) **for** $j = 1 : N_p$ **do**
(6) $\quad$ Normalize: $\widehat{\omega}_k^{(j)} = \widehat{\omega}_k^{*(j)} - \widehat{W}(k)$;

ALGORITHM 2: SIS-PF in log-domain (SIS Log-PF).

evaluated at each time step k, where $\{\mathbf{x}_k^{(j)}, \widehat{\omega}_k^{(j)}\}_{j=1}^{N_p}$ denotes the set for the particle states $\mathbf{x}_k^{(j)}$ and log-weights $\widehat{\omega}_k^{(j)}$ with $j = 1, \ldots, N_p$ for time step k.

One of the crucial problems of the SIS PF is degeneracy (another problem which is not discussed in this paper is the selection of the importance density $q(\mathbf{x}_k \mid \mathbf{x}_{k-1}^{(j)}, \mathbf{z}_k)$ of (4); see e.g., [2]).

After a few time steps all particles except for one have low weights and do not contribute anymore to the computation of the posterior PDF; that is, the distribution estimation degenerates. A suitable measure of degeneracy is the effective sample size N_{eff} [1–3, 5]. A widely used approximation for the effective sample size is

$$N_{\text{eff}} \approx P_{N_p}^{(2)}(\mathbf{w}_k) = \frac{1}{\sum_{j=1}^{N_p} \left(w_k^{(j)}\right)^2} = \frac{\left(\sum_{j=1}^{N_p} \left(w_k^{*(j)}\right)\right)^2}{\sum_{j=1}^{N_p} \left(w_k^{*(j)}\right)^2} \tag{15}$$

with $\mathbf{w}_k = (w_k^{(1)}, \ldots, w_k^{(N_p)})^T$ and $1 \leq P_{N_p}^{(2)}(\mathbf{w}_k) \leq N_p$. A small value of $P_{N_p}^{(2)}(\mathbf{w}_k)$ indicates a severe degeneracy. By using the Jacobian logarithm of (12), we obtain from (15) the effective sample size in log-domain, with

$$\begin{aligned} \ln\left(N_{\text{eff}}\right) \approx \ln\left(P_{N_p}^{(2)}\left(\mathbf{w}_k\right)\right) &= -\ln\left(\sum_{j=1}^{N_p} e^{2\cdot\widehat{\omega}_k^{(j)}}\right) \\ &= -\text{Jacob}\left(\left\{2 \cdot \widehat{\omega}_k^{(j)}\right\}_{j=1}^{N_p}\right) \\ &= 2 \cdot \text{Jacob}\left(\left\{\widehat{\omega}_k^{*(j)}\right\}_{j=1}^{N_p}\right) \\ &\quad - \text{Jacob}\left(\left\{2 \cdot \widehat{\omega}_k^{*(j)}\right\}_{j=1}^{N_p}\right). \end{aligned} \tag{16}$$

Alternative effective sample size approximations as introduced in [14] can also be represented in log-domain. Table 1 summarizes four generalized alternative effective sample size approximations in lin-domain and log-domain which depend on the parameter r. Please note $P_{N_p}^{(2)}(\mathbf{w}_k)$ as in (15) is obtained from $P_{N_p}^{(r)}(\mathbf{w}_k)$ with $r = 2$.

The Generic PF extends the SIS PF by a resampling step to prevent degeneration as shown in Algorithm 3 by a pseudocode. The basic idea of resampling is to eliminate

TABLE 1: Generalized effective sample size functions $P^{(r)}_{N_p}(\mathbf{w}_k)$, $D^{(r)}_{N_p}(\mathbf{w}_k)$, $V^{(r)}_{N_p}(\mathbf{w}_k)$, $S^{(r)}_{N_p}(\mathbf{w}_k)$ to approximate N_{eff} in lin-domain as defined in [14] and log-domain with their coefficients $a^{(r)}_{\{\cdot\}}$ and $b^{(r)}_{\{\cdot\}}$. Please note $P^{(2)}_{N_p}(\mathbf{w}_k)$ as in (15) is obtained from $P^{(r)}_{N_p}(\mathbf{w}_k)$ with $r=2$.

Function	Coefficient
$P^{(r)}_{N_p}(\mathbf{w}_k) = \dfrac{1}{a_P^{(r)}\sum_{j=1}^{N_p}\left(w_k^{(j)}\right)^r + b_P^{(r)}}$	$a_P^{(r)} = \dfrac{1-N_p}{N_p^{2-r}-N_p}$
$\ln\left(P^{(r)}_{N_p}(\mathbf{w}_k)\right) = -\text{Jacob}\left(\left\{\left\{\ln\left(a_P^{(r)}\right) + r\cdot\varpi_k^{(j)}\right\}_{j=1}^{N_p}, \ln\left(b_P^{(r)}\right)\right\}\right)$	$b_P^{(r)} = \dfrac{N_p^{2-r}-1}{N_p^{2-r}-N_p}$
$D^{(r)}_{N_p}(\mathbf{w}_k) = \dfrac{1}{a_D^{(r)}\left[\sum_{j=1}^{N_p}\left(w_k^{(j)}\right)^r\right]^{1/r} + b_D^{(r)}}$	$a_D^{(r)} = \dfrac{N_p-1}{N_p-N_p^{1/r}}$
$\ln\left(D^{(r)}_{N_p}(\mathbf{w}_k)\right) = -\text{Jacob}\left(\left\{\ln\left(a_D^{(r)}\right) + \dfrac{1}{r}\text{Jacob}\left(\left\{r\cdot\varpi_k^{(j)}\right\}_{j=1}^{N_p}\right), \ln\left(b_D^{(r)}\right)\right\}\right)$	$b_D^{(r)} = \dfrac{1-N_p^{1/r}}{N_p-N_p^{1/r}}$
$V^{(r)}_{N_p}(\mathbf{w}_k) = a_V^{(r)}\sum_{j=1}^{N_p}\left(w_k^{(j)}\right)^r + b_V^{(r)}$	$a_V^{(r)} = \dfrac{N_p^{r-1}(N-1)}{1-N_p^{r-1}}$
$\ln\left(V^{(r)}_{N_p}(\mathbf{w}_k)\right) = \text{Jacob}\left(\left\{\left\{\ln\left(a_V^{(r)}\right) + r\cdot\varpi_k^{(j)}\right\}_{j=1}^{N_p}, \ln\left(b_V^{(r)}\right)\right\}\right)$	$b_V^{(r)} = \dfrac{N_p^{r}-1}{N_p^{r-1}-1}$
$S^{(r)}_{N_p}(\mathbf{w}_k) = a_S^{(r)}\left[\sum_{j=1}^{N_p}\left(w_k^{(j)}\right)^r\right]^{1/r} + b_r b_S^{(r)}$	$a_S^{(r)} = \dfrac{N_p-1}{N_p^{(1-r)/r}-1}$
$\ln\left(S^{(r)}_{N_p}(\mathbf{w}_k)\right) = \text{Jacob}\left(\left\{\ln\left(a_S^{(r)}\right) + \dfrac{1}{r}\text{Jacob}\left(\left\{r\cdot\varpi_k^{(j)}\right\}_{j=1}^{N_p}\right), \ln\left(b_S^{(r)}\right)\right\}\right)$	$b_S^{(r)} = \dfrac{N_p^{(1-r)/r}-N_p}{(1-r)/r-1}$

$\{\mathbf{x}_k^{(j)}, \widehat{\varpi}_k^{(j)}\}_{j=1}^{N_p}$ = Generic-Log($\{\mathbf{x}_{k-1}^{(j)}, \widehat{\varpi}_{k-1}^{(j)}\}_{j=1}^{N_p}$)
(1) **for** $j = 1 : N_p$ **do**
(2) Draw: $\mathbf{x}_k^{(j)} \sim q(\mathbf{x}_k^{(j)} \mid \mathbf{x}_{k-1}^{(j)}, \mathbf{z}_k)$;
(3) Calculate: $\varpi_k^{*(j)}$ according to (4);
(4) Calculate: $\widehat{W}_k = \text{Jacob}(\{\varpi_k^{*(j)}\}_{j=1}^{N_p})$;
(5) **for** $j = 1 : N_p$ **do**
(6) Normalize: $\widehat{\varpi}_k^{(j)} = \varpi_k^{*(j)} - \widehat{W}_k$;
(7) Calculate $\ln(N_{\text{eff}})$ according to Table 1;
(8) **if** $\ln(N_{\text{eff}}) < \ln(N_{\text{thr}})$ **then**
(9) Resample with Algorithm 4: Obtaining $\{\mathbf{x}_k^{(j)}, \widehat{\varpi}_k^{(j)}\}_{j=1}^{N_s}$;

ALGORITHM 3: Generic-PF in log-domain (Generic Log-PF).

$\{\overline{\mathbf{x}}_k^{(j)}, \overline{\varpi}_k^{(j)}\}_{j=1}^{N_p}$ = Log-Resampling($\{\mathbf{x}_k^{(j)}, \widehat{\varpi}_k^{(j)}\}_{j=1}^{N_p}$)
(1) Initialize the log-CDF: $[\mathbf{c}]_1 = \widehat{\varpi}_k^{(j)}$;
(2) **for** $j = 2 : N_p$ **do**
(3) Construct log-CDF using the Jacobian logarithm: $[\mathbf{c}]_j = \max(\widehat{\varpi}_k^{(j)}, [\mathbf{c}]_{j-1}) + \ln(1 + e^{-|[\mathbf{c}]_{j-1} - \widehat{\varpi}_k^{(j)}|})$;
(4) $i = 1$;
(5) Draw starting point: $[\mathbf{u}]_1 \sim \mathcal{U}[0, N_p^{-1}]$;
(6) **for** $j = 1 : N_p$ **do**
(7) $[\mathbf{u}]_j = \ln([\mathbf{u}]_1 + N_p^{-1}(j-1))$;
(8) **while** $[\mathbf{u}]_j > [\mathbf{c}]_i$ **do**
(9) $i = i + 1$;
(10) Assign: $\{\overline{\mathbf{x}}_k^{(j)}, \overline{\varpi}_k^{(j)}\} = \{\mathbf{x}_k^{(i)}, -\ln(N_p)\}$;

ALGORITHM 4: Resampling in log-domain.

particles with low weights and reproduce particles with high weights. Whenever a significant degeneracy is observed in the Generic PF, that is, $\ln(N_{\text{eff}})$ is less than a threshold $\ln(N_{\text{thr}})$, the particles are resampled. Algorithm 4 shows a pseudocode of the systematic resampling algorithm [15] transferred into log-domain. In Algorithm 4, $\mathcal{U}[0, N_p^{-1}]$ denotes the uniform distribution on the interval $[0, N_p^{-1}]$ (cf. Algorithm 4 Line 5). Similarly to the descriptions before, the Jacobian logarithm is used to construct the estimated sampled cumulative distribution function in log-domain (log-CDF); see Algorithm 4 Line 3. The estimated sampled log-CDF is presented by a vector $\mathbf{c}$ with length N_p and element $[\mathbf{c}]_j$ with $j = 1, \ldots, N_p$. By $[\mathbf{x}]_j$, we denote the jth element of the vector $\mathbf{x}$. According to the estimated sampled log-CDF, particles with high weights are reproduced and particles with low weights are eliminated.

In Section 5, we use the sequential importance resampling (SIR) PF; see [1], as an example for comparing the performance of the linear domain PF (Lin-PF) and Log-PF. Therefore, Algorithm 5 shows a pseudocode of the SIR PF in log-domain, which is derived from the Generic PF by setting the importance density to be equal to the transitional prior

$\{\mathbf{x}_k^{(j)}, \widehat{\omega}_k^{(j)}\}_{j=1}^{N_p} = \text{Log-SIR}(\{\mathbf{x}_{k-1}^{(j)}, \widehat{\omega}_{k-1}^{(j)}\}_{j=1}^{N_p})$
(1) **for** $j = 1 : N_p$ **do**
(2) Draw: $\mathbf{x}_k^{(j)} \sim \mathrm{p}(\mathbf{x}_k^{(j)} \mid \mathbf{x}_{k-1}^{(j)})$;
(3) Calculate: $\widehat{\omega}_k^{*(j)} = \ln(\mathrm{p}(\mathbf{z}_k \mid \mathbf{x}_k^{(j)}))$;
(4) Calculate: $\widehat{W}_k = \text{Jacob}(\{\widehat{\omega}_k^{*(j)}\}_{j=1}^{N_p})$;
(5) **for** $j = 1 : N_p$ **do**
(6) Normalize: $\widehat{\omega}_k^{(j)} = \widehat{\omega}_k^{*(j)} - \widehat{W}_k$;
(7) Resample with Algorithm 4: Obtaining $\{\mathbf{x}_k^{(j)}, \widehat{\omega}_k^{(j)}\}_{j=1}^{N_s}$;

ALGORITHM 5: SIR-PF in log-domain (SIR Log-PF).

$\{\mathbf{x}_k^{(j)}, w_k^{(j)}\}_{j=1}^{N_p} = \text{Generic-Lin-Log}(\{\mathbf{x}_{k-1}^{(j)}, w_{k-1}^{(j)}\}_{j=1}^{N_p})$
(1) **for** $j = 1 : N_p$ **do**
(2) Draw: $\mathbf{x}_k^{(j)} \sim \mathrm{q}(\mathbf{x}_k^{(j)} \mid \mathbf{x}_{k-1}^{(j)}, \mathbf{z}_k)$;
(3) Calculate: $\widehat{\omega}_k^{*(j)}$ according to (4);
(4) **for** $j = 1 : N_p$ **do**
(5) Transfer and Normalize: $w_k^{+(j)} = e^{\widehat{\omega}_k^{*(j)} - \max_l(\widehat{\omega}_k^{*(l)})}$;
(6) Calculate: $W_k = \sum_{j=1}^{N_p} w_k^{+(j)}$;
(7) **for** $j = 1 : N_p$ **do**
(8) Normalize: $w_k^{(j)} = w_k^{+(j)}/W_k$;
(9) Calculate N_{eff} according to Table 1;
(10) **if** $N_{\text{eff}} < N_{\text{thr}}$ **then**
(11) Resample with Algorithm 4: Obtaining $\{\mathbf{x}_k^{(j)}, w_k^{(j)}\}_{j=1}^{N_s}$;

ALGORITHM 6: Generic-PF in lin-domain with weight calculation in log-domain (Generic Lin-Log-PF).

distribution, with $\mathrm{q}(\mathbf{x}_k \mid \mathbf{x}_{k-1}^{(j)}, \mathbf{z}_k) = \mathrm{p}(\mathbf{x}_k^{(j)} \mid \mathbf{x}_{k-1}^{(j)})$ and using $N_{\text{eff}} = 1$, that is, performing resampling at each time step [2].

Additionally, we compare in Section 5 the proposed Log-PF to the PF implementation which computes the weights in log-domain and uses (6) to obtain the weights in lin-domain, called Lin-Log-PF in the following. A pseudocode of the Generic Lin-Log-PF is shown in Algorithm 6: the weights are calculated in log-domain according to (4) and normalized and transferred to the lin-domain according to $w_k^{+(j)} = e^{\widehat{\omega}_k^{*(j)} - \max_l(\widehat{\omega}_k^{*(l)})}$ as mentioned in Section 3 and, for example, [9]. Please note further improvements can be obtained if the weights $\widehat{\omega}_k^{*(j)}$ are directly propagated in log-domain if resampling is not necessary.

4. Log-PF Point Estimators

In many applications, we are interested in a point estimate of the state instead of its a posteriori PDF. In this section we derive the MMSE and MAP point estimators based on the a posteriori density estimated by the Log-PF.

4.1. Minimum Mean Square Error Estimate. The MMSE point estimate using the approximated a posteriori density, see, for example, [16], is defined by

$$\widehat{\mathbf{x}}_k^{\text{MMSE}} = \sum_{j=1}^{N_p} e^{\widehat{\omega}_k^{(j)}} \mathbf{x}_k^{j}, \tag{17}$$

where the lth element of the vector $\widehat{\mathbf{x}}_k^{\text{MMSE}}$ can expressed as

$$\left[\widehat{\mathbf{x}}_k^{\text{MMSE}}\right]_l = \sum_{j=1}^{N_p} e^{\widehat{\omega}_k^{(j)}} \left[\mathbf{x}_k^{(j)}\right]_l. \tag{18}$$

In order to use the Jacobian logarithm to compute (18) in log-domain, we separate the positive and negative values of $[\mathbf{x}_k^{(j)}]_l$ with

$$\begin{aligned} \mathcal{A}_{+,l} &= \left\{j \mid j \in \left\{1, \ldots, N_p\right\} \wedge \left[\mathbf{x}_k^{(j)}\right]_l > 0\right\}, \\ \mathcal{A}_{-,l} &= \left\{j \mid j \in \left\{1, \ldots, N_p\right\} \wedge \left[\mathbf{x}_k^{(j)}\right]_l < 0\right\} \end{aligned} \tag{19}$$

and the corresponding log-weight $\widehat{\omega}_k^{(j)}$ accordingly. Please note $[\mathbf{x}_k^{(j)}]_l = 0$ is not considered in (19) because $(e^{\widehat{\omega}_k^{(j)}} \cdot [\mathbf{x}_k^{(j)}]_l) = 0$. Thus, we obtain from (18) for the MMSE estimate,

$$\begin{aligned} \left[\widehat{\mathbf{x}}_k^{\text{MMSE}}\right]_l &= e^{\ln(\sum_{j\in\mathcal{A}_{+,l}} e^{\widehat{\omega}_k^{(j)} + \ln(|[\mathbf{x}_k^{(j)}]_l|)})} \\ &\quad - e^{\ln(\sum_{j\in\mathcal{A}_{-,l}} e^{\widehat{\omega}_k^{(j)} + \ln(|[\mathbf{x}_k^{(j)}]_l|)})} \\ &= e^{\ln(\sum_{j\in\mathcal{A}_{+,l}} e^{m^{j,l}})} - e^{\ln(\sum_{j\in\mathcal{A}_{-,l}} e^{m^{j,l}})} \\ &= e^{\text{Jacob}(\{m^{j,l}\}_{j\in\mathcal{A}_{+,l}})} - e^{\text{Jacob}(\{m^{j,l}\}_{j\in\mathcal{A}_{-,l}})}, \end{aligned} \tag{20}$$

where we introduced $m^{j,l}$ with

$$m^{j,l} = \widehat{\omega}_k^{(j)} + \ln\left(\left|\left[\mathbf{x}_k^{(j)}\right]_l\right|\right). \tag{21}$$

4.2. Maximum A Posteriori Estimate. The MAP point estimate is defined as

$$\widehat{\mathbf{x}}_k^{\text{MAP}} = \arg\max_{\mathbf{x}_k} \mathrm{p}\left(\mathbf{x}_k \mid \mathbf{z}_{1:k}\right), \tag{22}$$

which can be approximated, see [17], by

$$\widehat{\mathbf{x}}_k^{\text{MAP}} \approx \arg\max_{\mathbf{x}_k^{(j)}} \mathrm{p}\left(\mathbf{z}_k \mid \mathbf{x}_k^{(j)}\right) \sum_{i=1}^{N_p} \mathrm{p}\left(\mathbf{x}_k^{(j)} \mid \mathbf{x}_{k-1}^{(i)}\right) w_{k-1}^{(i)}, \tag{23}$$

for $j = 1, \ldots, N_p$. The corresponding MAP state estimator using weights in log-domain can be calculated using the Jacobian logarithm of (12) with

$$\begin{aligned} \widehat{\mathbf{x}}_k^{\text{MAP}} &\approx \arg\max_{\mathbf{x}_k^{(j)}} \mathrm{p}\left(\mathbf{z}_k \mid \mathbf{x}_k^{(j)}\right) \sum_{i=1}^{N_p} e^{\ln(\mathrm{p}(\mathbf{x}_k^{(j)}|\mathbf{x}_{k-1}^{(i)})) + \widehat{\omega}_{k-1}^{(i)}} \\ &= \arg\max_{\mathbf{x}_k^{(j)}} \mathrm{p}\left(\mathbf{z}_k \mid \mathbf{x}_k^{(j)}\right) \\ &\quad \times \text{Jacob}\left(\left\{\ln\left(\mathrm{p}\left(\mathbf{x}_k^{(j)} \mid \mathbf{x}_{k-1}^{(i)}\right)\right) + \widehat{\omega}_{k-1}^{(i)}\right\}_{i=1}^{N_p}\right). \end{aligned} \tag{24}$$

5. Simulations

In this section, we demonstrate the performance of the Log-PF using floating point 64-bit number accuracy according to IEEE Standard 754 for double precision with three simulations.

5.1. Linear Processes. First, we simulate a linear Gaussian model. The KF introduced in [18] is an optimal recursive Bayesian filter which can be used if the considered system is linear and the probabilistic model is Gaussian. Hence, we compare the Log-PF and the Lin-PF to the KF as benchmark. The simulation considers the linear transition model

$$\mathbf{x}_{k+1} = \mathbf{F}\mathbf{x}_k + \mathbf{v}_k = \begin{pmatrix} 1 & 0.5 \\ 0 & 1 \end{pmatrix} \mathbf{x}_k + \mathbf{v}_k, \tag{25}$$

with the transition matrix $\mathbf{F}$, the state vector $\mathbf{x}_k = (a_k, b_k)^T$, and the zero-mean multivariate Gaussian distributed process noise $\mathbf{v}_k \sim \mathcal{N}(\mathbf{0}, \sigma_v \mathbf{I})$ with standard deviation σ_v and the identity matrix $\mathbf{I}$. The measurement model is defined by

$$z_k = \mathbf{H}\mathbf{x}_k + n_k = \begin{pmatrix} 1 & 1 \end{pmatrix} \mathbf{x}_k + n_k, \tag{26}$$

with the measurement matrix $\mathbf{H}$ and the zero-mean Gaussian distributed measurement noise $n_k \sim \mathcal{N}(0, \sigma_n)$ with standard deviation σ_n. Based on the measurements z_k, the state sequence $\mathbf{x}_k$ for $k = 1, \ldots, 60$ is estimated using a KF, the Lin-PF and the Log-PF with $N_p = 200$ particles and known initial state $\mathbf{x}_1 = (2, 2)^T$. For the Lin-PF, we use the standard PF implementation as well as the Lin-Log-PF.

First, we compare the KF to the SIR Lin-PF, SIR Lin-Log-PF, and the SIR Log-PF. In order to see the robustness of the SIR Log-PF, we variate the measurement noise standard deviation σ_n from 10^8 down to 10^{-150}. We simulate 1000 different realizations with known initial state for each run. Figure 1 shows the root mean square error (RMSE) averaged over all time steps and simulations versus the decreasing measurement noise standard deviation σ_n. The abbreviation SIR Log-PF MAP stands for the MAP point estimate and SIR Log-PF MMSE for the MMSE point estimate of the SIR Log-PF. Respectively, the abbreviations SIR Lin-PF MAP, SIR Lin-Log-PF MAP, SIR Lin-PF MMSE, and SIR Lin-Log-PF MMSE stand for the SIR Lin-PF and SIR Lin-Log-PF point estimates. We see that the KF obtains the best estimation results followed by the SIR Log-PF and SIR Lin-Log-PF. Figure 1 shows additionally an enlarged subfigure of the region for $10^8 < \sigma_n < 10^{-8}$. For $\sigma_n > 10^{-2}$, all SIR PFs obtain equivalent performance. As soon as σ_n decreases, the RMSE decreases for the SIR PFs until $\sigma_n = 10^{-2}$. For lower measurement noise standard deviations, the RMSE of the SIR Lin-PF and increases up to a limit of 0.7 whereas the accuracy of the SIR Log-PF and SIR Lin-Log-PF are limited by the number of particles. This effect is caused by the number representation of the particle weights of the SIR Lin-PF. For $\sigma_n < 10^0$, the particle weights of the SIR Lin-PF are small and at similar order as numerical errors due to number representation. Therefore, numerical errors dominate the resampling step such that the resampling algorithm draws

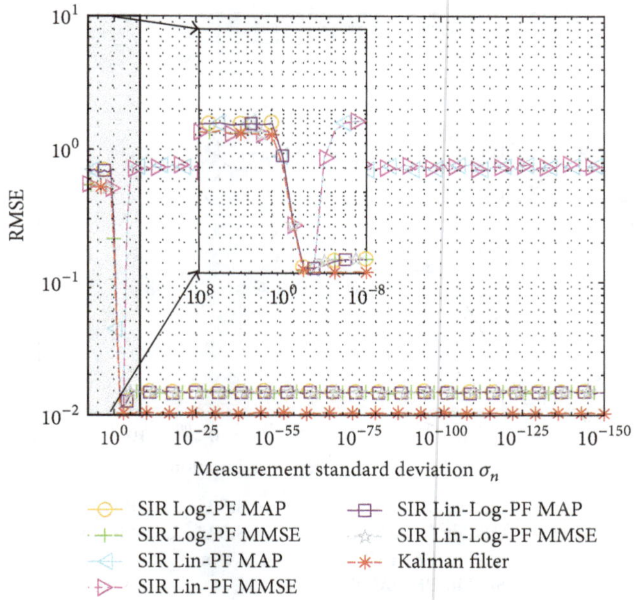

FIGURE 1: Performance evaluation of the SIR Log-PF, SIR Lin-PF, SIR Lin-Log-PF, and KF by the RMSE versus a decreasing measurement standard deviation σ_n.

particles based on numerical inaccuracies. Thus, the update step of the SIR Lin-PF loses its effect and we obtain 0.7 as the expected error of the process model in (25). For lower standard deviations, the accuracy of the SIR Log-PF and Lin-Log-PF is limited by the number of particles. The RMSE of the SIR Log-PF and Lin-Log-PF stay constant around 0.05 for the standard deviations between $10^{-1} > \sigma_n > 10^{-150}$. The KF is the optimal filter for this simulation; hence, the RMSE of the KF is limited by the process noise; hence, the RMSE is constant around 0.01.

To summarize, we obtain for all considered standard deviations equal or better estimation results using the SIR Log-PF compared to the SIR Lin-PF. However, the Lin-Log-PF which computes the weights in log-domain using (6) obtains similar simulation results compared to the SIR Log-PF. Hence, in the following we show the benefits of the Log-PF by using SIS PFs, where no resampling is preformed. The resampling step is a key point for the success of a PF which is applied to avoid the degeneracy. However, resampling yields to a loss of diversity in the propagation of particles and entails an additional computational cost [14]. Similar to the SIR PF, we set the importance density to be equal to the transitional prior distribution, with $q(\mathbf{x}_k \mid \mathbf{x}_{k-1}^{(j)}, \mathbf{z}_k) = p(\mathbf{x}_k^{(j)} \mid \mathbf{x}_{k-1}^{(j)})$.

Figure 2 shows the RMSE averaged over all time steps and simulations versus the decreasing measurement noise standard deviation σ_n. The curve looks similar to Figure 1. For $\sigma_n > 10^{-2}$, the SIS PFs obtain equivalent performance. As soon as σ_n decreases, the RMSE decreases for the SIS PFs until $\sigma_n = 10^{-2}$. For lower measurement noise standard deviations, the RMSE of the SIS Lin-PF and Lin-Log-PF increase up to the limit of 0.7. The RMSE of the SIS Log-PF stays constant

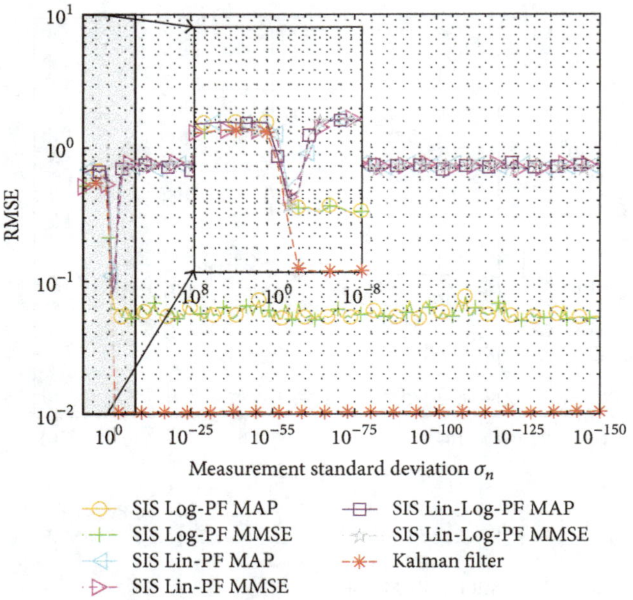

FIGURE 2: Performance evaluation of the SIS Log-PF, SIS Lin-PF, SIS Lin-Log-PF, and KF by the RMSE versus a decreasing measurement standard deviation σ_n.

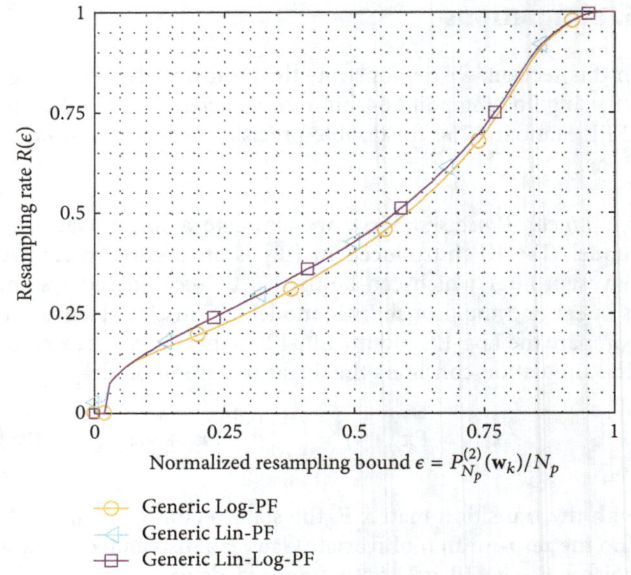

FIGURE 3: Performance evaluation of the Generic Log-PF, Generic Lin-PF, and Generic Lin-Log-PF by the resampling rate $R(\epsilon)$ versus the normalized resampling bound $\epsilon = P^{(2)}_{N_p}(\mathbf{w}_k)/N_p$.

around 0.03 for the standard deviations between $10^{-2} > \sigma_n > 10^{-150}$.

However, since the SIS PF uses no resampling step, the transformation of the weights from the log-domain to the lin-domain is not essential. Hence, if the weights of the SIS Lin-Log-PF are not transferred to the lin-domain and the weights are normalized with $\widehat{\varpi}_k^{(j)} = \widehat{\varpi}_k^{*(j)} - \max_l(\widehat{\varpi}_k^{*(l)})$ and propagated to the next time instant, the Lin-Log-PF obtains equivalent estimation results than the SIS Log-PF. Therefore, the transformation to the lin-domain in Line 6 in Algorithm 6 may introduce numerical inaccuracies. As long as no normalization is needed, the Lin-Log-PF can be computed completely in the log-domain.

5.2. Nonlinear Processes: Generic Particle Filter. In this section, we use the Generic PF which decides adaptively when the resampling step is performed. Similar to the SIR PF, we set the importance density to be equal to the transitional prior distribution, with $\mathrm{q}(\mathbf{x}_k \mid \mathbf{x}_{k-1}^{(j)}, \mathbf{z}_k) = \mathrm{p}(\mathbf{x}_k^{(j)} \mid \mathbf{x}_{k-1}^{(j)})$. We compare the Generic Log-PF to the Generic Lin-PF and Generic Lin-Log-PF. As mentioned before, a pseudocode of the Generic Lin-Log-PF is shown in Algorithm 6. In all algorithms, resampling is performed whenever the effective sample size N_{eff} falls below a threshold N_{thr}. We use the approximation $P^{(2)}_{N_p}(\mathbf{w}_k)$ for the effective sample size in the lin-domain and respectively $\ln(P^{(2)}_{N_p}(\mathbf{w}_k))$ in the log-domain. Similar to [14], we consider a stochastic volatility model

$$x_k = \alpha x_{k-1} + v_k,$$
$$z_k = e^{x_k/2} n_k \tag{27}$$

with $\alpha = 0.99$ and the zero-mean Gaussian distributed process and measurement noise $v_k \sim \mathcal{N}(0, \sigma_v)$ and $n_k \sim \mathcal{N}(1, \sigma_n)$, where n_k is a multiplicative noise. Based on the measurements z_k, the state sequence x_k for $k = 1, \ldots, 3000$ is estimated using the Generic PFs with $N_p = 100$ particles. We set the measurement noise standard deviation to $\sigma_n = 0.01$ and the process noise standard deviation to $\sigma_v = 0.01$. For performance evaluation, we try to recreate situations with rapidly changing measurements, that is, a certain model mismatch between the true likelihood of the process and the likelihood representation inside the PF. Inside the Generic PFs, we use the measurement noise standard deviation $\sigma_{n,P} = 10^{-4}$. Hence, taking the model mismatch into account it is more likely that particle states are located in the tail of the PF's likelihood after the prediction step. For the simulations, we variate the normalized resampling bound $\epsilon = P^{(2)}_{N_p}(\mathbf{w}_k)/N_p$ from 0 to 1 using a grid of 0.01 resolution (equivalently for the log-domain). We simulate 1000 different realizations with known initial state and count the number of performed resampling for each run. The resampling rate is afterwards calculated as $R(\epsilon) = \mathbf{E}[\text{Number Resampling using } \epsilon/3000]$, where $\mathbf{E}[x]$ stands for the sample mean of x.

Figure 3 shows the resampling rate $R(\epsilon)$ averaged over all time steps and simulations versus the normalized resampling bound ϵ. For $0.2 < \epsilon < 0.8$, the Generic Log-PF has a lower resampling rate than the Generic Lin-PF and the Generic Lin-Log-PF. Figure 4 shows the RMSE versus the resampling rate $R(\epsilon)$, where the RMSE decreases with increasing resampling rate for all Generic PFs. However, we can observe that, for a resampling rate of $0.15 < R(\epsilon) < 0.5$, we obtain a slightly lower RMSE with the Generic Log-PF than using the Generic Lin-PF or Generic Lin-Log-PF.

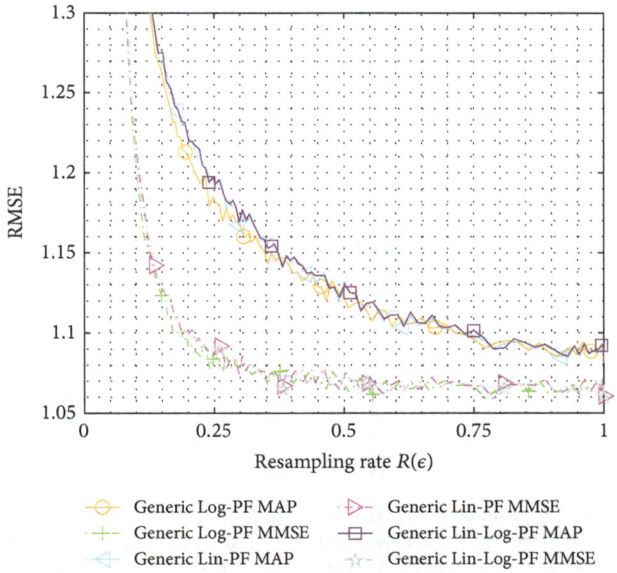

FIGURE 4: Performance evaluation of the Generic Log-PF, Generic Lin-PF, and Generic Lin-Log-PF by the RMSE versus the resampling rate $R(\epsilon)$.

5.3. Rao-Blackwellization. In this section we consider a simultaneous localization and mapping (SLAM) example with radio signals indicated in Figure 5 according to the system model in [19] (similar results are expected in, e.g., belief propagation [20] or distributed PFs [13]). A receiver which is moving along an arbitrary trajectory is measuring the distances

$$\mathbf{d}_k = \left[d_1(k), \dots, d_{N_k}(k)\right]^T, \tag{28}$$

with

$$d_i(k) = \left\|\mathbf{r}_{u,k} - \mathbf{r}_{T,i,k}\right\| + c_{i,k} + n_{i,k}, \tag{29}$$

between the receiver at location $\mathbf{r}_{u,k}$ and $i = 1, \dots, N_k$ transmitters at location $\mathbf{r}_{T,i,k}$ with the distance offset $c_{i,k}$ and zero-mean Gaussian distributed measurement noise $n_{i,k} \sim \mathcal{N}(0, \sigma_i)$ with standard deviation σ_i. The receiver uses the control input $\mathbf{u}_k$ to move from state $\mathbf{x}_{u,k-1}$ to state $\mathbf{x}_{u,k}$. In order to use the distance measurements $\mathbf{d}_k$ for positioning, the positioning algorithm estimates the receiver and transmitter states simultaneously. The state vector $\mathbf{x}_k$ describing the complete system at time instant k is

$$\mathbf{x}_k = \left(\mathbf{x}_{u,k}^T, \mathbf{x}_{T,k}^T\right)^T, \tag{30}$$

with the receiver $\mathbf{x}_{u,k}$ and the transmitter states $\mathbf{x}_{T,k}$ which are also unknown. The receiver state $\mathbf{x}_{u,k} = (\mathbf{r}_{u,k}^T, \mathbf{v}_{u,k}^T)^T$ includes the receiver position $\mathbf{r}_{u,k}$ and the receiver velocity $\mathbf{v}_{u,k}$, while the transmitter states are defined by

$$\mathbf{x}_{T,k} = \left(\mathbf{x}_{T,1,k}^T, \dots, \mathbf{x}_{T,N_k,k}^T\right)^T \tag{31}$$

with

$$\mathbf{x}_{T,i,k} = \left(\mathbf{r}_{T,i,k}^T, c_{i,k}\right)^T, \tag{32}$$

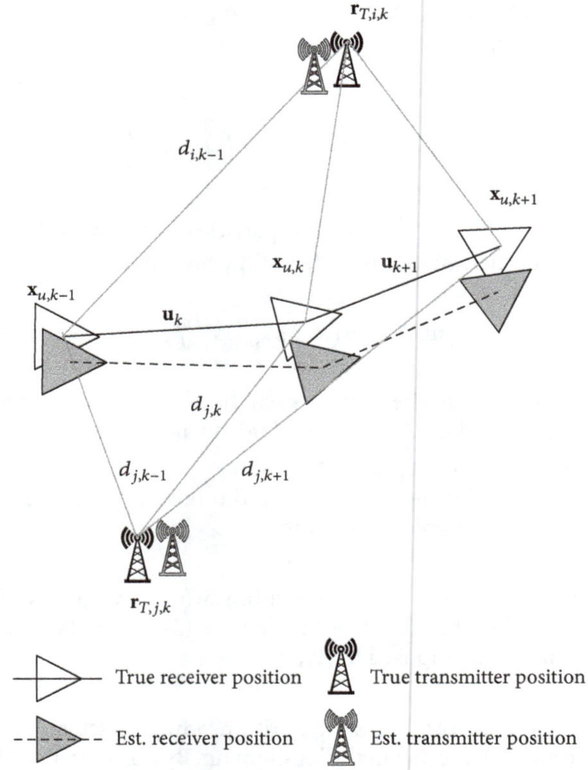

FIGURE 5: Overview of the SLAM example: the moving receiver simultaneously estimates its position and the location of the transmitters.

for $i = 1, \dots, N_k$ transmitters. We consider a static environment with a fixed number of transmitters and a receiver moving along an arbitrary trajectory. However, for notational convenience, a time dependence on k is introduced here for the transmitter positions $\mathbf{r}_{T,i,k}$. Additionally, we assume based on [19] that the distance offset $c_{i,k}$ is constant.

The state space is estimated by an algorithm according to [19, 21] based on Rao-Blackwellization [22], where the states space of $\mathbf{x}_k$ is partitioned into subspaces. Hence, we use PFs to estimate the subspaces representing the transmitters inside a PF. The reason to use a PF instead of a low complexity extended Kalman filter (EKF) is the nonlinearity of the measurements in (29). The algorithm of [19, 21] is based on a superordinate particle filter (superPF) and subordinate particle filters (subPFs). Each particle $j = 1, \dots, N_p$ of the superPF with the state vector $\mathbf{x}_{u,k}^{(j)} = (\mathbf{r}_{u,k}^{(j)\,T}, \mathbf{v}_{u,k}^{(j)\,T})^T$ holds N_k subPFs. Each subPF is represented by the particles $\mathbf{x}_{T,i,k}^{(j,a)}$ with $a = 1, \dots, N_{P,j,i}$ where $N_{P,j,i}$ stands for the number of particles in the ith subPF with $i = 1, \dots, N_k$, estimating $\mathbf{x}_{T,i,k}^{(i)}$.

Similar to [19, 21], the posterior distribution $\mathrm{p}(\mathbf{x}_{u,k}, \mathbf{x}_{T,k} \mid \mathbf{d}_{1:k}, \mathbf{u}_{1:k}, \mathbf{x}_{u,0})$ can be approximated by importance samples, as

$$\mathrm{p}\left(\mathbf{x}_{u,k}, \mathbf{x}_{T,k} \mid \mathbf{d}_{1:k}, \mathbf{u}_{1:k}, \mathbf{x}_{u,0}\right) \approx \sum_{j=1}^{N_p} w_k^{(j)} \delta\left(\mathbf{x}_{u,k} - \mathbf{x}_{u,k}^{(j)}\right), \tag{33}$$

where $w_k^{(j)}$ defines the weight for the jth particle at time instant k with

$$w_k^{(j)} \propto \mathsf{p}\left(\mathbf{d}_k \mid \mathbf{x}_{u,k}^{(j)}, \mathbf{d}_{k-1}\right) \propto \prod_{i=1}^{N_k} \sum_{a=1}^{N_{P,i,j}} w_{i,k}^{(j,a)} \tag{34}$$

and the weight $w_{i,k}^{(j,a)}$ of the ath particle of the ith subPF for the jth particle of the superPF at time instant k with

$$w_{i,k}^{(j,a)} \triangleq \mathsf{p}\left(d_{i,k} \mid \mathbf{x}_{u,k}^{(j)}, \mathbf{x}_{T,i,k}^{(j,a)}\right). \tag{35}$$

Resampling is performed at each time instant to prevent degeneration; hence, (34) and (35) do not depend on the weights $w_{k-1}^{(j)}$ and $w_{i,k-1}^{(j,a)}$, respectively.

In the following we compare the position accuracy of three different implementations:

(i) Lin-PF-SLAM. SLAM algorithm which calculates the posterior distribution according to (33), (34), and (35) in the lin-domain as proposed in [19, 21].

(ii) Log-PF-SLAM. SLAM algorithm which calculates the posterior distribution in the log-domain. By transferring (33), (34), and (35) to the log-domain, we obtain

$$\widehat{w}_k^{(j)} \propto \ln\left(\mathsf{p}\left(\mathbf{d}_k \mid \mathbf{x}_{u,k}^{(j)}, \mathbf{d}_{k-1}\right)\right)$$
$$\propto \sum_{i=1}^{N_k} \ln\left(\sum_{a=1}^{N_{P,i,j}} e^{\widehat{w}_{i,k}^{(j,a)}}\right) = \sum_{i=1}^{N_k} \text{Jacob}\left(\left\{\widehat{w}_{i,k}^{(j,a)}\right\}_{a=1}^{N_{P,i,j}}\right). \tag{36}$$

(iii) Lin-Log-PF-SLAM. SLAM algorithm which calculates the weights in the log-domain and transfers the weights afterwards to the lin-domain using (6). Hence, using (6) for (34), we obtain

$$w_k^{+(j)} = e^{\widehat{w}_k^{(j)} - \max_l\left(\widehat{w}_k^{(l)}\right)}, \tag{37}$$

where

$$\widehat{w}_k^{(j)} \propto \sum_{i=1}^{N_k} \max_m\left(\widehat{w}_{i,k}^{(j,m)}\right) + \ln\left(\sum_{a=1}^{N_{P,i,j}} e^{\widehat{w}_{i,k}^{(j,a)} - \max_m\left(\widehat{w}_{i,k}^{(j,m)}\right)}\right). \tag{38}$$

We evaluate the performance of the different algorithms using a two dimensional scenario with three static transmitters and a moving receiver indicated in Figure 6. The transmitters are located at constant locations summarized in Table 2. The receiver is moving on a random pathway for 60 s with a system sampling interval of 1 s.

The transition model of the receiver states uses a standard discrete white noise acceleration model [23], with

$$\mathbf{x}_{u,k} = \mathbf{A}\left(\dot{\Psi}_k, \Delta_k\right)\mathbf{x}_{u,k-1} + \mathbf{b}\mathbf{n}_{t,k}^T, \tag{39}$$

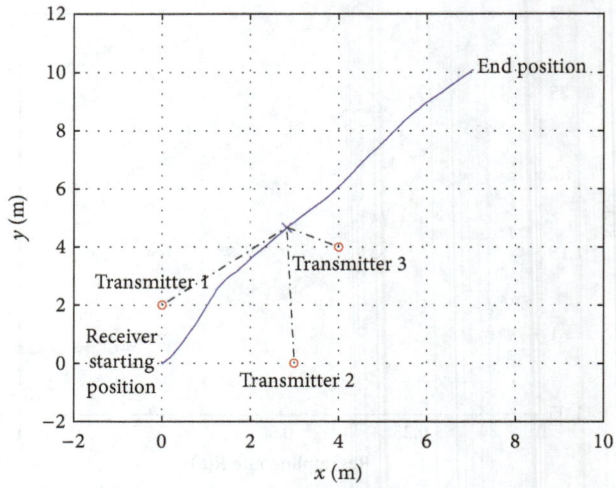

FIGURE 6: Simulated scenario with three transmitters indicated by the red circles and a receiver moving on the blue track. The figure indicates the propagation paths from the transmitters to the receiver for $k = 30$ s.

TABLE 2: Transmitter properties.

Transmitter i	x-pos [m]	y-pos [m]	$c_{i,k}$ [m]	σ_i [m]
Transmitter 1	0	2	1	0.001
Transmitter 2	3	0	1	0.01
Transmitter 3	4	0	0	0.01

with

$$\mathbf{A}\left(\dot{\Psi}_k, \Delta_k\right) = \begin{pmatrix} 1 & 0 & \Delta_k & 0 \\ 0 & 1 & 0 & \Delta_k \\ 0 & 0 & \cos\left(\dot{\Psi}_k\right) & -\sin\left(\dot{\Psi}_k\right) \\ 0 & 0 & \sin\left(\dot{\Psi}_k\right) & \cos\left(\dot{\Psi}_k\right) \end{pmatrix}, \tag{40}$$

$$\mathbf{b} = \begin{pmatrix} 0 \\ 0 \\ 1 \\ 1 \end{pmatrix}, \tag{41}$$

in a two-dimensional Cartesian coordinate system with the zero-mean multivariate Gaussian distributed process noise $\mathbf{n}_{t,k} \sim \mathcal{N}(\mathbf{0}, \sigma_v \mathbf{I})$ with standard deviation σ_v. The receiver state vector $\mathbf{x}_{u,k} = (\mathbf{r}_{u,k}, \mathbf{v}_{u,k})^T$ consists of the x-y positions $\mathbf{r}_{u,k} = (r_{u,x,k}, r_{u,y,k})^T$ and the velocities $\mathbf{v}_{u,k} = (v_{u,x,k}, v_{u,y,k})^T$ where $v_{u,x,k}, v_{u,y,k}$ are the corresponding velocities in x-y direction. The transition matrix in (40) includes a rotation matrix with the heading changes $\dot{\Psi}_k$, which are used as control input $\mathbf{u}_k$. The transmitter states $\mathbf{x}_{T,i,k}$ are time-invariant; hence, we obtain for the transition prior $\mathsf{p}(\mathbf{x}_{T,i,k} \mid \mathbf{x}_{T,i,k-1})$ of the ith transmitter $\mathsf{p}(\mathbf{x}_{T,i,k} \mid \mathbf{x}_{T,i,k-1}) = \delta(\mathbf{x}_{T,i,k} - \mathbf{x}_{T,i,k-1})$.

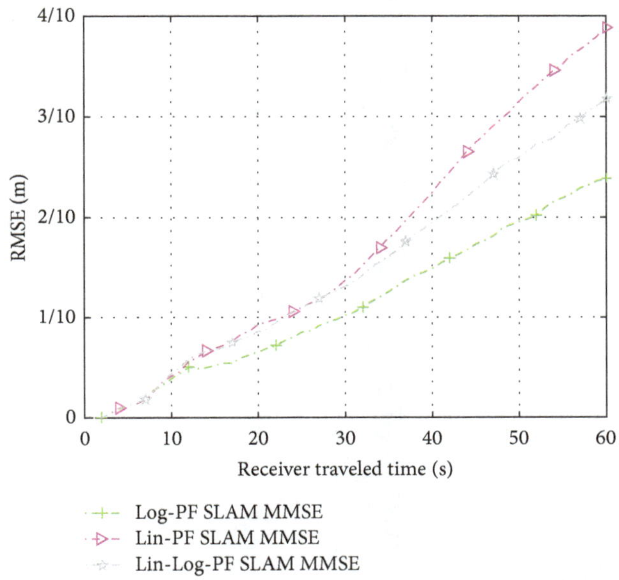

FIGURE 7: RMSEs of the estimated receiver positions versus receiver traveled time for different algorithms.

The simulations are performed using $N_p = 3000$ and $N_{P,j,i} = 1000$ particles for all transmitters $i = 1, \ldots, N_k$. For the initialization, we use prior information $\mathbf{x}_{u,0}$ which is the knowledge on the starting position and velocity. Please note that an unknown starting position and direction or larger initial uncertainties may result in a biased and rotated coordinate system for the estimation. For simplicity, we use also prior information on the transmitter states $\mathbf{x}_{T,i,0}$ (please see [19, 21], e.g., on unknown transmitter states). The prior information includes a Gaussian distribution with standard deviation of 0.5 m centered around the true transmitter states, that is, position and distance offset. For computing the position estimate $\hat{\mathbf{r}}_{u,k}$, we use the MMSE estimate as introduced in [19, 21].

Figure 7 shows the RMSE versus the receiver traveled time for the different implementations with $\mathrm{RMSE}_{u,k} = \sqrt{\mathbf{E}\{\|\mathbf{r}_{u,k} - \hat{\mathbf{r}}_{u,k}\|^2\}}$ and 200 independent evaluations. At the starting time, the RMSE for all algorithms are similar because of the identical initialization. Afterwards, the RMSE increases caused by the dilution of precision (DOP). However, we can clearly see that we obtain a higher accuracy using the Log-PF compared to the Lin-PF and the Lin-Log-PF.

6. Conclusion

In this paper we derived a particle filter representation in logarithmic domain, called Log-PF. The derivations show that the weight calculation, weight normalization, resampling, and point estimations can be expressed in logarithmic domain using the Jacobian logarithm. Representing the weight of each particle in logarithmic domain allows reducing the effect of numerical issues. Furthermore, the algorithm derived in this paper can be generalized to multidimensional nonparametric marginalization.

Conflicts of Interest

The authors declare that there are no conflicts of interest regarding the publication of this paper.

Acknowledgments

The authors would like to thank Patrick Robertson and Luigi Bruno for many interesting discussions on this problem. This work has been performed in the framework of the DLR project *Navigation 4.0* and the European Union Horizon 2020 research and innovation programme under Grant Agreement no. 636537 *HIGHTS (High precision positioning for Cooperative-ITS applications)*.

References

[1] N. J. Gordon, D. J. Salmond, and A. F. M. Smith, "Novel approach to nonlinear/non-gaussian Bayesian state estimation," *IEE proceedings Radar Signal Processing*, vol. 140, no. 2, pp. 107–113, 1993.

[2] M. S. Arulampalam, S. Maskell, N. Gordon, and T. Clapp, "A tutorial on particle filters for online nonlinear/non-Gaussian Bayesian tracking," *IEEE Transactions on Signal Processing*, vol. 50, no. 2, pp. 174–188, 2002.

[3] B. Ristic, S. Arulampalam, and N. Gordon, *Beyond the kalman filter: particle filters for tracking applications*, Artech House, 2004.

[4] R. van der, N. Merwe, A. Doucet, and E. Wan, "The Unscented Particle Filter," Tech. Rep., Cambridge University Engineering Department, August 2000.

[5] Z. Chen, "Bayesian Filtering: From Kalman Filters to Particle Filters, and Beyond," Tech. Rep., McMaster University, 2003.

[6] F. Gustafsson, "Particle filter theory and practice with positioning applications," *IEEE Aerospace and Electronic Systems Magazine*, vol. 25, no. 7, pp. 53–81, 2010.

[7] L. Martino, J. Read, V. Elvira, and F. Louzada, "Cooperative parallel particle filters for online model selection and applications to urban mobility," *Digital Signal Processing*, vol. 60, pp. 172–185, 2017.

[8] C. C. Drovandi, J. M. McGree, and A. N. Pettitt, "A sequential Monte Carlo algorithm to incorporate model uncertainty in Bayesian sequential design," *Journal of Computational and Graphical Statistics*, vol. 23, no. 1, pp. 3–24, 2014.

[9] J. Lien, U. J. Ferner, W. Srichavengsup, H. Wymeersch, and M. Z. Win, "A comparison of parametric and sample-based message representation in cooperative localization," *International Journal of Navigation and Observation*, Article ID 281592, 2012.

[10] J. Erfanian, S. Pasupathy, and G. Gulak, "Reduced Complexity symbol detectors with parallel structures for isi channels," *IEEE Transactions on Communications*, vol. 42, no. 234, pp. 1661–1671, 1994.

[11] W. Koch and A. Baier, "Optimum and sub-optimum detection of coded data disturbed by time-varying intersymbol interference," in *Proceedings of the IEEE Global Telecommunications Conf.*, vol. 3, pp. 1679–1684, December 1990.

[12] P. Robertson, E. Villebrun, and P. Hoeher, "A comparison of optimal and sub-optimal MAP decoding algorithms operating in the log domain," in *Proceedings of the IEEE Int. Conf. on Communications*, vol. 2, pp. 1009–1013, June 1995.

[13] S. Zhang, E. Staudinger, W. Wang, C. Gentner, A. Dammann, and E. Sandgren, "DiPLoc: Direct signal domain particle filtering for network localization," in *Proceedings of the 28th International Technical Meeting of the Satellite Division of the Institute of Navigation, ION GNSS 2015*, pp. 2267–2274, usa, September 2015.

[14] L. Martino, V. Elvira, and F. Louzada, "Effective sample size for importance sampling based on discrepancy measures," *Signal Processing*, vol. 131, pp. 386–401, 2017.

[15] G. Kitagawa, "Monte Carlo filter and smoother for non-Gaussian nonlinear state space models," *Journal of Computational and Graphical Statistics*, vol. 5, no. 1, pp. 1–25, 1996.

[16] A. Doucet, N. J. Gordon, and V. Krishnamurthy, "Particle filters for state estimation of jump Markov linear systems," *IEEE Transactions on Signal Processing*, vol. 49, no. 3, pp. 613–624, 2001.

[17] S. Saha, Y. Boers, H. Driessen, P. K. Mandal, and A. Bagchi, "Particle based map state estimation: a comparison," in *Proceedings of the In Proc. of the 12th Int. Conf*, pp. 278–283, July 2009.

[18] R. E. Kalman, "A new approach to linear filtering and prediction problems," *Journal of Basic Engineering*, vol. 82, no. 1, pp. 35–45, 1960.

[19] C. Gentner, T. Jost, W. Wang, S. Zhang, A. Dammann, and U.-C. Fiebig, "Multipath assisted positioning with simultaneous localization and mapping," *IEEE Transactions on Wireless Communications*, vol. 15, no. 9, pp. 6104–6117, 2016.

[20] A. T. Ihler, J. W. Fisher, R. L. Moses, and A. S. Willsky, "Non-parametric belief propagation for self-localization of sensor networks," *IEEE Journal on Selected Areas in Communications*, vol. 23, no. 4, pp. 809–819, 2005.

[21] C. Gentner, R. Pöhlmann, M. Ulmschneider, T. Jost, and S. Zhang, "Positioning using terrestrial multipath signals and inertial sensors," *Mobile Information Systems*, vol. 2017, pp. 1–18, 2017.

[22] A. Doucet, N. de Freitas, P. Kevin, and J. Stuart, "Rao-Blackwellised Particle Filtering for Dynamic Bayesian Networks," in *Proc. Conf. Uncertainty in Artificial Intelligence*, pp. 176–183, UAI, 2000.

[23] Y. Bar-Shalom, Li. Rong, and T. Kirubarajan, *Estimation with Applications to Tracking And Navigation: Theory Algorithms and Software*, Wiley Sons, 2004.

15

A Novel Framework for Interactive Visualization and Analysis of Hyperspectral Image Data

Johannes Jordan, Elli Angelopoulou, and Andreas Maier

Pattern Recognition Lab, University of Erlangen-Nuremberg, Erlangen, Germany

Correspondence should be addressed to Johannes Jordan; johannes.jordan@cs.fau.de

Academic Editor: Sos Agaian

Multispectral and hyperspectral images are well established in various fields of application like remote sensing, astronomy, and microscopic spectroscopy. In recent years, the availability of new sensor designs, more powerful processors, and high-capacity storage further opened this imaging modality to a wider array of applications like medical diagnosis, agriculture, and cultural heritage. This necessitates new tools that allow general analysis of the image data and are intuitive to users who are new to hyperspectral imaging. We introduce a novel framework that bundles new interactive visualization techniques with powerful algorithms and is accessible through an efficient and intuitive graphical user interface. We visualize the spectral distribution of an image via parallel coordinates with a strong link to traditional visualization techniques, enabling new paradigms in hyperspectral image analysis that focus on interactive raw data exploration. We combine novel methods for supervised segmentation, global clustering, and nonlinear false-color coding to assist in the visual inspection. Our framework coined Gerbil is open source and highly modular, building on established methods and being easily extensible for application-specific needs. It satisfies the need for a general, consistent software framework that tightly integrates analysis algorithms with an intuitive, modern interface to the raw image data and algorithmic results. Gerbil finds its worldwide use in academia and industry alike with several thousand downloads originating from 45 countries.

1. Introduction

Multispectral imaging allows capturing of rich reflectance information that is not available with traditional RGB cameras. It has played a key role in the field of remote sensing for decades, ever since aircrafts and satellites became equipped with hyperspectral sensors systems. One such notable example is the Airborne Visible/Infrared Imaging Spectrometer (AVIRIS) [1]. Hyperspectral sensors are less widely used "on the ground," though they are becoming more popular in areas like astronomy, cultural heritage, agriculture, and medical imaging.

In a multispectral image, each pixel is a vector of intensity values, where each value corresponds to the scene radiance over a small range of wavelengths. The resulting vectors typically are of lengths ranging between 31 and 200. The high-dimensional nature of the data and its strong interband correlations pose challenges to computer vision algorithms. Adaptations are needed to expose and exploit the information contained in the data. A similar problem arises when one tries to manually process such high-dimensional data. There is a lack of intuition which demands a presentation that guides the user during data exploration. Effectively, one needs two components: simultaneous visualization of different aspects of the information and strong user interaction.

As the popularity of multispectral sensors has been increasing, so has the need for computer-aided, interactive analysis. For example, one of the first popular analysis frameworks for providing a graphical interface for inspection was the Spectral Image Processing System which was introduced in 1992 [2]. However, almost all available multispectral analysis software is tailored to specific applications. As expected, most of the available software focuses on remote sensing and is designed to foster a workflow specific to this domain.

In recent years, the availability of cheaper, easier-to-use multispectral cameras further opened this modality to

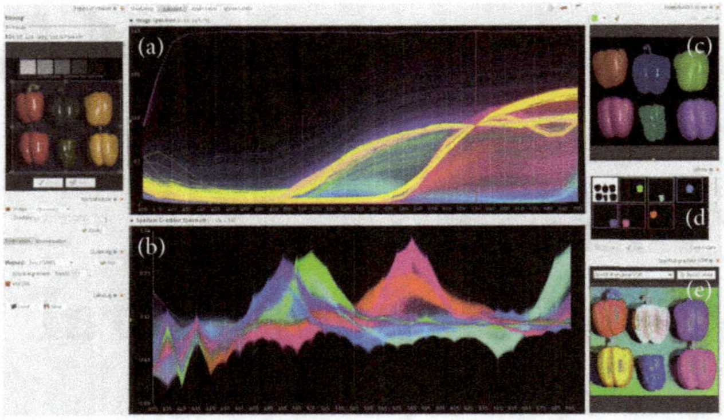

FIGURE 1: Gerbil user interface. (a) Spectral distribution view, (b) spectral gradient distribution view, (c) spatial view of a single image band with segment overlays, (d) label manager, and (e) false-color display.

new applications. Various devices based on electronically tunable filters [3] or imaging spectrometers [4] are commercially available from several vendors. An example of sensor targeting emerging applications is MuSIS [5], which is currently in use in several museums and libraries. However, the users of these sensors lack a software solution that enables them to explore the data in an interactive and intuitive fashion.

To address this limitation we developed the Gerbil software platform which incorporates established interactive visualization concepts to provide a new presentation of spectral images as well as a new workflow for inspecting them. Figure 1 depicts a screenshot of Gerbil in use for inspecting a multispectral image. Gerbil is developed by an open-source software project in the tradition of free software frameworks in the signal processing research community such as OpenCV [8] or Weka [9]. In this paper we provide a scientific reference for Gerbil and describe the core methodologies it encompasses. We also provide a detailed presentation of the feasibility and performance of the algorithms in Gerbil and show how our novel approach is a significant improvement compared to established practice.

The paper bundles and expands on the following contributions:

(i) Interactive visualization of spectral distributions based on efficient parallel coordinates [6]

(ii) Supervised segmentation of hyperspectral data [10]

(iii) Fast global clustering with superpixel support [11]

(iv) Fast nonlinear false-color visualization [12]

These methods were derived by adapting established algorithms for hyperspectral data and interactive time constraints. They allow us to introduce new paradigms in hyperspectral image analysis that focus on interactive raw data exploration, built on the tight incorporation of the aforementioned techniques into a comprehensive open-source software framework. The software is publicly available at http://gerbilvis.org/ under a free software license.

2. Previous Work

One of the first widely recognized software packages for multispectral analysis, LARSYS, became available in the 1960s. It was operated via a text terminal. Several frameworks for graphical, interactive multispectral, or hyperspectral data analysis that are still of broad influence today date back to the early 1990s. Earlier frameworks focus on a specific application, most prominently in the field of remote sensing. Boardman et al. [13] provided an overview of the history of established software systems for remote sensing hyperspectral data. Biehl [14] created an updated list on the web in 2007. In this section we give first a brief overview of domain-specific software packages before reviewing more widely applicable software. We focus on the interactive inspection of images.

2.1. Remote Sensing. The Spectral Image Processing System (SIPS) was presented by Kruse et al. [2] in 1993. It introduced the Spectral Angle Mapper (SAM), a spectrum-comparison tool still popular in the field of spectral matching [15]. Data was presented either as single bands or in a false-colored composite of three user-selectable bands. The user could also view a selected pixel's spectrum. Color-coded stacked spectra were provided for a selected slice (a vertical or horizontal line scan or an arbitrary path in the image). This set of visualization forms are still typical of popular software for hyperspectral capture or analysis.

The MultiSpec freeware package by Biehl and Landgrebe [16] can analyze multispectral images from various sources, using the powerful GDAL library [17] for data I/O. The focus of this software is classification. The target audience is the general Earth science community. Besides providing all common visualizations already mentioned, it can generate biplots that relate selected regions in a pair of bands and a statistics image display, which depicts the correlation between these bands.

A widely used commercial software is ENVI [18], initially developed by Boardman et al. [13]. Several innovations in hyperspectral analysis were introduced in ENVI, including, for example, the Pixel Purity Index (PPI) [19]. PPI finds the most spectrally pure pixels in an image. These pixels typically correspond to endmembers (constituent spectra of an image). A notable innovation in visualization is the *n-Dimensional Visualizer*, which uses PPI as input. It is an interactive data visualization method that allows real-time rotation of scatterplots in n-dimensions. The presentation of n-dimensional scatterplots in 2D can be somewhat unintuitive. Yet the tool is valuable to experts for identification of endmembers based on the depicted point clouds.

Two other notable pieces of software for hyperspectral remote sensing data analysis are HyperCube [23] and Opticks [24]. HyperCube is released by the US Army Corps of Engineers and contains functions to filter, warp (register two images), calibrate and undistort, photometrically project, and arithmetically manipulate the data. Opticks also includes many common tools like GIS annotations, false coloring, and histograms. Additionally, third parties can provide functionality through an external module, adding further capabilities to the software. For example, the Spectral Processing Extension [25] includes typical tools for hyperspectral data analysis.

Within the remote sensing context, many works exist on visualizing a multispectral image by employing dimensionality reduction. The goal is to present the image in false-color RGB by reducing dimensionality from spectral vectors of length D to length 3, which are then used as r, g, and b values. Based on the resulting pixel colors, the user should be able to discriminate regions of the image according to his specific interests. Methods to achieve this include variants of the principal component analysis (PCA), independent component analysis (ICA), and nonlinear methods. Some variants also incorporate classification results. A recent approach by Cui et al. [26] focuses on the interactive adaptation of such visualization. However, it still uses the common pixel map representation of the image.

2.2. Astronomy. A field with a long history of hyperspectral imaging is astronomy. The most notable software framework in this area is ds9 [27], available from the Smithsonian Astrophysical Observatory. This software is very powerful in the spatial representation of astronomical imagery. However, it provides a limited 3D visualization of the data cube that is rarely viable for nonastronomical data.

Recently, Li et al. [28] explicitly tackled the question of how to present multiband data. They draw image bands in 3D either as an image-stack or as a volume-rendered model, for example, horseshoe. Their volume rendering handles the obvious problem of clutter by applying transparency to individual data points based on their intensity or on a user-adjusted mask. However, one cannot generally assume that large image regions can be faded out to still unobstructedly display more relevant data.

2.3. Other Application Domains. Multispectral imaging has become increasingly popular in the preservation and analysis of artwork as well as historical documents. Colantoni et al. [29] analyzed multispectral images of paintings from the perspective of human color perception. From the image data, a representation in the CIE XYZ color space [30] is computed under controlled virtual illumination. Several tools can then be applied for visualization of trichromatic data. The original spectra are not considered in the analysis.

In 2010, Kim et al. [31] presented a solution for interactive visualization of historical documents. They provided a nicely designed self-contained analysis tool and incorporated innovations in how the data is presented. Some are specific to document analysis, for example, dealing with aging-related artifacts, while others are more general. The *spectral lens* feature, for example, presents data from two spectral bands in a single display. Similarity maps can be computed based on the L_2 norm between the mean spectral value of a selected region and all image pixels. However, similarity measures like SAM [15] would be better suited for spectra comparison. A 3D histogram plot is used to compare two spectra.

The National ICT Australia (NICTA) offers the Scyven software free of charge which features the reflectance recovery and material classification pipeline developed at NICTA [7, 32]. Its visualization includes false coloring and an adaptation of the parallel coordinates visualization introduced in Gerbil (see Section 3.2).

Labitzke et al. [33] introduced an interactive framework for linear spectral unmixing of multispectral datasets. Spectral unmixing is especially popular in remote sensing, where the spatial resolution is often low. Many algorithms exist for finding endmembers and performing spectral unmixing [34]. Labitzke et al. introduced an incremental method that can semiautomatically find endmembers. Then, visual feedback is provided by their complementary visualization that reflects the quality of the characterizing set. This set can be interactively modified in order to improve the unmixing. The authors explicitly differentiated their approach from Gerbil [33]. Spectral unmixing is integral to their interactive visualization approach, while in our methods we visualize the raw data. The algorithm and workflow proposed by Labitzke et al. are not in opposition to our approach, but instead they nicely complement the visualization capabilities of Gerbil.

3. Visual Inspection

In general, existing approaches share the same basic set of data representations and user interactions, while individual extensions typically follow a specific application scenario. However, the wider range of applications of multispectral imaging necessitates an interactive visualization framework that is both sufficiently general for a broader range of applications and more versatile than existing basic representations. In our framework, Gerbil, we follow a novel concept that enables an entirely new workflow in exploring a multispectral image. It revolves around presentation and exploration that make the image data more apparent to the user, effectively allowing a more direct interpretation of the data. The user does not rely on, but is merely guided by, automatic processing.

3.1. Main Concept. Figure 2 gives an overview of the key features that constitute Gerbil. In order to have a framework that is suitable for use in a large variety of applications, we incorporate several well-understood hyperspectral analysis tools into Gerbil, while adding concepts from the data visualization and computer vision communities. The value of our approach lies in employing these features in a fashion that enriches the user experience while investigating a multispectral image. One such example is the incorporation of parallel coordinates, a technique for display of high-dimensional data. Parallel coordinates are used in visualizing spectral distributions. The visualization is highly dynamic and interacts with the topological representations of the image (e.g., a grayscale depiction of a single image band). Another contributing factor is the emphasis on the *spectral gradient* descriptor [35]. Its distribution is displayed next to the spectral distribution. The spectral gradient is a multispectral descriptor that focuses on data aspects, which are directly related to material and reflectance properties.

Figure 1 depicts the three main components of the graphical user interface: the visualization of original spectra (Figure 1(a)) and the spectral gradient spectra (Figure 1(b)) both via parallel coordinates and the spatial image view

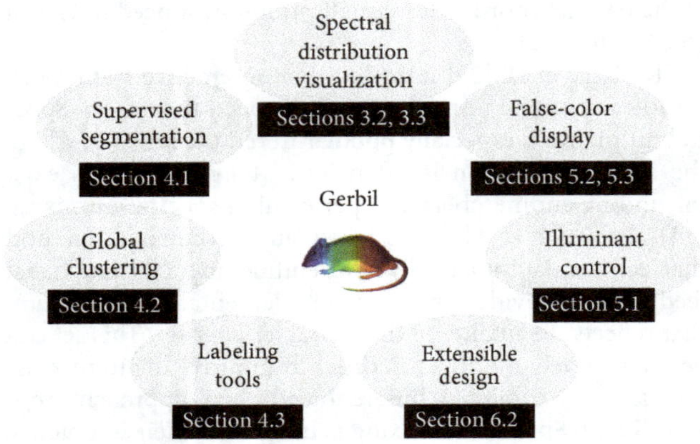

FIGURE 2: Overview of the Gerbil feature set and corresponding article sections.

(Figure 1(c)). The interaction between these elements, as is explained in detail in Section 3.3, is augmented by tools that give further guidance to the user. One such example is dimensionality reduction via PCA or nonlinear methods, as discussed in Section 5. Another tool is feature-space clustering of data, described in Section 4.2. Especially important in our workflow is supervised segmentation, where the user provides input depending on the goal of their investigation or on prior knowledge. In Section 4.1, we describe how we employ a graph-based segmentation algorithm for this task.

3.2. Spectral Distribution View. Existing approaches on depicting a multispectral image in its entirety are limited by the spatial layout of the image. The image data is viewed as a cube, with the z-axis corresponding to spectral bands, which are stacked on top of each other. Use of modern volume rendering techniques can make this representation useful in some scenarios [28]. However in most cases, where there is no sparsity in pixels of interest, a very cluttered view is obtained that reveals the shortcomings of a simple 3D arrangement. Other methods, such as scatter plots or false coloring, rely on dimensionality reduction. While the visualization of reduced data is helpful in many applications, it is hard to preserve subtle details. In contrast, a good visualization of the original data helps the observer evaluate how well a dimensionality reduction method fits a specific application.

One way of addressing this issue is to defer from the topological relations in the image, for the time-being, and concentrate instead on a graspable representation of the spectral distribution. To do this efficiently, we employ the parallel coordinates [36] method as explained below. This is a well established technique for visualizing high-dimensional geometry and analyzing multivariate data. It has been widely used, for example, in financial applications and geographic information systems. One can see the traditional spectral visualization as a specific instantiation of parallel coordinates visualization. By building on the more general concept, we can incorporate tools from the visualization community for high-dimensional data presentation.

According to the parallel coordinates concept, a D-dimensional feature space (resulting from D spectral bands) is projected onto a two-dimensional view as follows. D parallel vertical lines denote the D axes, that is, the D spectral bands. The y-coordinate on the ith axis corresponds to a spectrum's value at band i. To display the spectral vector of a pixel, a polyline is drawn with its vertices lying on the corresponding vertical axes. The resulting display follows the layout of a plot where the x-axis would denote wavelength and the y-axis denotes intensity.

3.2.1. Optimized Parallel Coordinates. Drawing a polyline for each single multispectral vector has several drawbacks: it is time consuming; and the display may easily get cluttered in which case single polylines may not separate well from the rest of the data. A solution to both clutter and speed concerns is to draw fewer polylines, where a polyline can represent a set of pixels. With this distinction, polylines that represent more pixels appear stronger.

We realize this representation by introducing a histogram in the feature space with B evenly distributed bins in each dimension. B is user adjustable between 2 and the dynamic range of the captured data (typically 2^8 to 2^{16}). For example, the histogram for a 31-band multispectral image with $B = 256$ would hold $B^D = (2^8)^{31} = 2^{248}$ bins. Building a dense histogram of this size is not feasible; however a sparse histogram can be created by using an ordinary hashing algorithm. The key idea here is that the amount of occupied bins is bound by the spatial resolution of the input image. For example, a spatial resolution of 512×512 leaves only 2^{18} possible distinct values, effectively giving an upper bound to the amount of bins filled in a sparse histogram, or the amount of hash keys needed.

For each populated bin, a polyline is drawn in the parallel coordinates visualization. The strength of the polyline is manipulated by assigning it an opacity α. α is determined by the relationship between the number of pixels represented by the bin, n_{bin}, and the total number of processed pixels, n_{total}.

$$\alpha = c_\alpha \frac{f(n_{\text{bin}} + 1)}{f(n_{\text{total}})}, \tag{1}$$

where c_α is a user-adjustable factor and $f(\cdot)$ is a logarithmic function. The logarithm emphasises bins with fewer pixels. The idea is that even a single pixel should be perceptible. The logarithm also ensures that the resulting dynamic range of alpha values can be represented reasonably well with an 8-bit alpha channel. Nowadays, Gerbil overcomes this limitation by drawing onto a floating-point framebuffer and the user can choose $f(\cdot)$ to be linear.

3.2.2. Drawing Refinements. The newer design of Gerbil includes several measures to further enhance the visual quality and accuracy of the spectral distribution display. By dividing the feature space into a fixed number of equally spaced bins, the histogram applies a nonadaptive quantization of a spectral vector $\mathbf{x}$. A possible strategy to reduce the introduced quantization error is to employ a binning that is adapted to the observed data. A straight-forward method is to

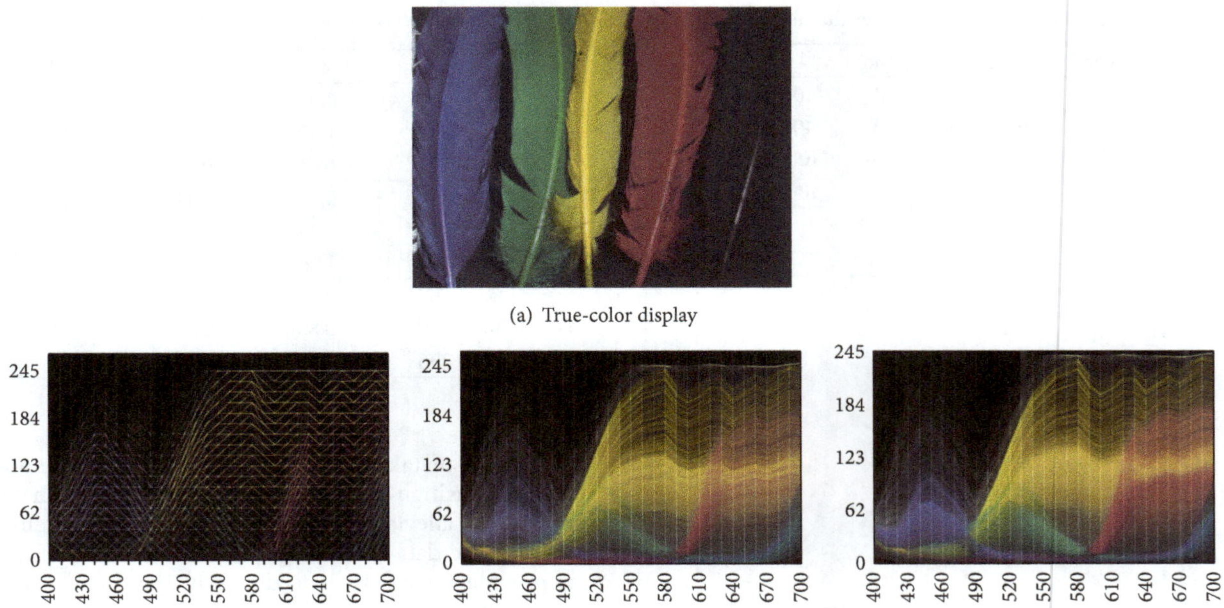

(a) True-color display

(b) Previous work [6], $B = 20$ (c) Proposed method, $B = 20$ (d) Proposed method, $B = 256$

FIGURE 3: Spectral distribution view of cropped *feathers* image with different bin parameters. Spectra in true color.

perform a separate histogram equalization in each dimension i, which enforces a uniform PDF of the mapped vector values x_i [37]. While it works well for big clusters of similar pixels, spectra that are sparsely represented in the image will suffer from such an operation. It may increase the average accuracy at a great expense in the precision of single pixel representations, which is not desirable.

We employ an alternative strategy by adjusting how a bin is drawn. Our method achieves an improved general accuracy without a significant expense in the accuracy for any single pixel. When drawing a bin, we draw the mean vector of its contents instead of its midrange values. This can be computed on the fly while adding new vectors, with a final normalization based on the already stored number of entries. Figure 3 shows an example of the visual quality gain. A broad description of the spectral distribution is already possible with a low B.

Another hindrance in visual quality is the mutual obstruction of pixel representations. In many use-cases, pixels are color coded (see Section 3.3). This involves effectively drawing several distributions on top of each other. In highly populated intensity ranges this can lead to extensive occlusions. There exists work on clutter reduction in parallel coordinate plots that tackles this problem [38]. The key idea is to only render a randomly sampled subset of the data points. Instead, we significantly reduce clutter by drawing data in a randomly shuffled order, without dismissing any information.

3.2.3. Evaluation. The error introduced by the histogram-based quantization can be measured by the average root mean squared error (RMSE) as well as the maximum absolute deviation (MAD), between original vectors $\mathbf{x}_i$ and their quantized counterparts $\hat{\mathbf{x}}_i$. This gives us the measure

$$\text{RMSE} = \sum_{i=1}^{N} \frac{\sqrt{\sum_{j=1}^{D} \left(\hat{x}_{ij} - x_{ij}\right)^2 / D}}{N}, \quad (2)$$

for the average RMSE, and

$$\text{MAD} = \max \left(\max\left(\left|\hat{x}_{ij} - x_{ij}\right|,\ j \leq D\right),\ i \leq N\right), \quad (3)$$

where $\max(\cdot)$ is the sample maximum.

The number of bins per dimension B is a crucial parameter. It lets the user choose between drawing speed, viewing quality, and accuracy. Even a considerably low B should provide acceptable accuracy, and the speed-up by lowering B should be effective. We evaluate RMSE and MAD for both the previously published [6] and the newly proposed drawing methods on several datasets from different application domains. Table 1 provides statistics on the datasets. The CAVE dataset consists of objects in a laboratory settings, while Foster captured natural (outdoor) scenes. Indian Pines and D.C. Mall are widely used remote sensing images.

We use a desktop machine equipped with a quad-core Intel Core i7 CPU running at 2.80 GHz and a GeForce GTX 550 Ti consumer graphics card for testing the computational performance. We measure the time needed for drawing operations via `GL_TIMESTAMP` and draw in WUXGA resolution.

In Figure 4 we plot running times against accuracy for varying B. *Bin Center* describes the originally published drawing method, while *Mean* denotes our refined drawing method. We can observe that the average RMSE becomes

Table 1: Datasets used for evaluation.

Name	Size	Resolution	Bands	λ (nm)	Source
CAVE	32	512×512	31	400–700	[20]
Foster	8	1340×1020	33	400–720	[21]
Indian Pines	1	145×145	220	400–2500	[22]
D.C. Mall	1	307×1280	191	400–2475	[22]

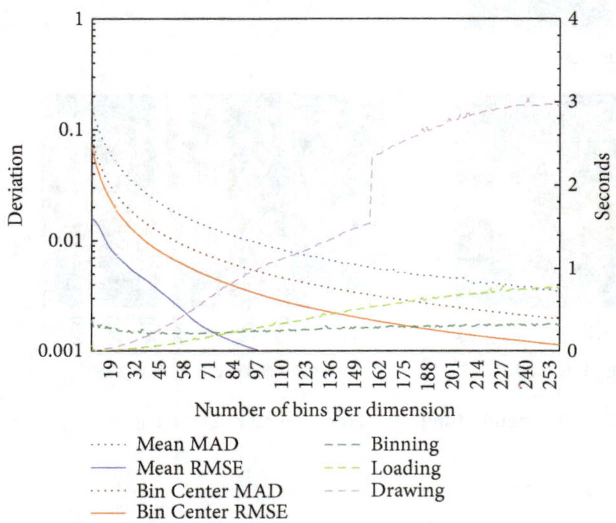

Figure 4: Experimental results on a scene from the Foster dataset. Quantization errors are plotted for both the old (Bin Center) and the refined (Mean) drawing method.

negligible with higher B for both methods. However, our method achieves low RMSE values for considerably lower settings of B. Due to outliers present in some of the histogram bins, the MAD for the refined method is somewhat higher than the original MAD.

The time needed to build the histogram is denoted as *binning* and is not determined by B. The time needed for preparing the geometry and loading it on the GPU (*Loading*) slowly grows with B. In contrast, B plays an important role for the time needed for the drawing operation (*Drawing*). For higher B values, the time needed for drawing grows to multiples of the time needed for preparation. Hence the histogramming plays an important role in achieving interactivity.

In Table 2 the running times and accuracy measures for the refined drawing method are listed on all datasets except Indian Pines, where, due to the very low resolution of the image, running times are negligible. The times shown for preparation are the combined histogram building and geometry loading times. As in Figure 4, it is observed that a low B can already achieve a small quantization error. Setting $B = 64$ is a reasonable compromise between speed and accuracy on the tested datasets. It provides an effective speed-up in comparison to a high histogram resolution without a perceivable loss in drawing accuracy.

It can be seen in Table 2 that on our test machine, even with a moderate setting, drawing the spectra from a large

Table 2: Average drawing times and accuracy.

Dataset	# bins	Prep. (s)	Draw. (s)	RMSE	MAD
CAVE	32^{31}	0.089	0.161	0.002	0.027
	64^{31}	0.115	0.240	0.001	0.014
	256^{31}	0.177	0.390	0.000	0.003
Foster	32^{33}	0.274	0.146	0.005	0.031
	64^{33}	0.358	0.554	0.002	0.014
	256^{33}	1.085	2.955	0.000	0.004
D.C. Mall	32^{191}	0.585	1.789	0.001	0.026
	64^{191}	0.881	3.171	0.000	0.013
	256^{191}	1.020	3.554	0.000	0.003

image may still take several seconds. Typically it is impractical to work on a high-resolution image without a region of interest. Yet, we alleviate longer drawing times by incrementally drawing the data (disabled in the benchmarking) in order to provide direct visual feedback in the form of a rough approximation of the full data (as pixels are drawn in a random order).

3.3. Interactive Inspection. An important aspect of today's visualization approaches for multivariate data is interactive manipulation of the presentation. A single view most often cannot provide the full understanding that may be gained by a series of user-controlled depictions. User input is vital to parallel coordinates in particular. We provide several mechanisms for both transient (cursor highlights) and persistent (color labels) interactive viewing.

In the parallel coordinate plots, the user can dynamically highlight specific data points, that is, spectral vectors that represent pixels. These are displayed in yellow and with full opacity as an overlay over nonhighlighted spectra. We realize this in OpenGL with layered frame buffers. Updates to the highlight only need to redraw highlighted spectra. While the highlight constantly follows the keyboard and/or mouse input; the corresponding pixels are instantly highlighted in the spatial view. While scrolling through the spectra, the topological view always reveals which pixels contribute to the highlighted spectra.

Two modes of operation exist for highlighting in the spectral distribution: single-band limited and multiband limited. The single-band limited highlight is formed by all spectral vectors falling into bins that share a specific intensity range in one band (see Figure 5(a)). The coarseness of this selection is therefore directly related to the binning parameter. The user selects the band and intensity range via mouse click or cursor keys. The multiband limited highlight provides a higher level of control in exchange for more detailed user input. Here, in each band, separate lower and upper intensity bounds can be set (see Figure 5(b)).

Another method of highlighting exists in the topological view. Here, the user can direct a cursor over individual pixels in the spatial view. In the corresponding viewports, the spectrum, and spectral gradient data of the pixel under the cursor are presented as a yellow overlay.

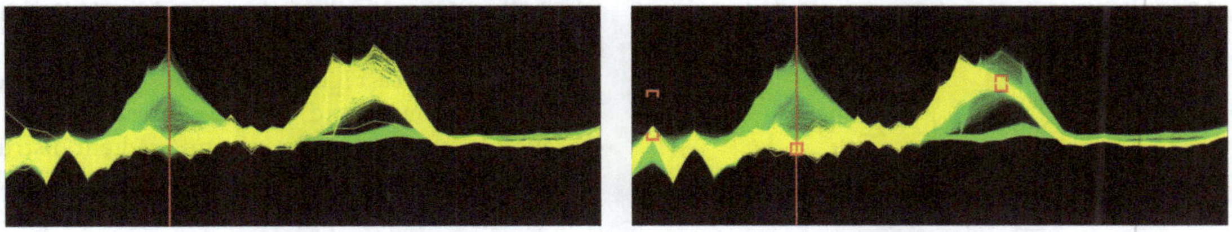

(a) Single-band limited (b) Multiband limited

Figure 5: Highlighting modes used in the spectral gradient distribution view of a scene with two objects. In (a), one objects is highlighted. In (b), further discrimination is performed by adding intensity bounds in two further bands.

While highlights give instant feedback, they constantly change as the user investigates the data. It is often desired to keep a selection of pixels distinguished from the rest of the data, for example, for comparison purposes. We call this pixel set a label; each pixel can be part of at most one label. A label can be seen as a permanent highlight. For each label, a distinct sparse histogram is created as described in Section 3.2.1. It is then drawn in the label color. When a histogram bin is part of the current transient highlight, the color is significantly shifted towards yellow.

When the user has selected pixels within the transient highlight, they may add this set of pixels to a label or delete their label association. By doing so, they can iteratively refine the labeling of the data to concentrate on specific details. Another way to alter the labeling is to use a "label" brush in the topological view or to use automated segmentation methods as discussed in Section 4. Labelings can also be stored for later use.

Labels are important because they serve as a memory in the connection between different representations of the multispectral image such as between a single band and the spectral distribution. A selection, or temporary highlight in one representation, is instantly propagated to the others. By labeling this highlight, it becomes permanent. The user can then continue his investigation within another representation, for example, the spectral gradient view, that may reveal new insights within this labeling. For example, a user may start by hand-labeling parts of a scene in the topological view. Then, they may restrict the spectral distribution view to this label. The parallel coordinates visualization could reveal a certain variance within the labeled pixels. By using selection tools inside this viewport, the user could separate parts of the pixels and assign them to a second label. The topological view instantly reveals which parts of the object contribute to which portion of the distinguished spectra. Next, the user may initialize a multiband limited highlight based on this second label. This reveals other regions in the scene that share spectral characteristics with the labeled pixels.

As a result, we facilitate a workflow of inspecting an image that is not possible with a traditional hyperspectral analysis framework. It is based on concurrent, concerted work with both topological and spectral views and allows a smooth and instantaneous switch in attention between them. Such a step-by-step exploration enables the user to quickly discover and grasp underlying information. In the visualization domain this procedure is considered a valuable tool for understanding complex data [39]. A narrated video demonstration of this workflow can be accessed at http://video.gerbilvis.org/.

4. Segmentation and Labeling

An interactive interface to the multispectral data is no replacement for automatic processing. In fact, the two approaches together form a powerful combination. Within our framework, it is easy to interpret and assess the results of algorithms used in automated analysis. These results can be a good starting point for further interactive analysis. Gerbil is equipped with two powerful methods that segment the data either according to spectral characteristics on a global level or based on topological relation and local similarity. In the latter case we bring supervised segmentation to the multispectral domain especially for the purpose of interactive inspection.

4.1. User-Guided Segmentation. Supervised segmentation incorporates user-provided prior knowledge. A user specifies a set of background and foreground pixels. All other pixels are then determined as belonging to either the background or the foreground. We make this concept a powerful tool within our interactive workflow.

The early version of Gerbil [6] included a rudimentary version of supervised segmentation. We have now adapted an existing algorithm family specifically to the multispectral domain [10].

4.1.1. Graph-Based Supervised Segmentation. In recent years, graph-based algorithms have dominated supervised segmentation on both grayscale and RGB images. In 2011, Couprie et al. introduced a framework for supervised segmentation that incorporates several key methods based on graph theory [40]. Their *power watersheds* integrate graph cuts, random walker, and watersheds algorithms in a single mathematical framework.

For this algorithm family, the input consists of two sets of pixels, the foreground seeds $\mathscr{F}$ and background seeds $\mathscr{B}$, as well as the pixel values $\mathbf{x}_i$, $1 \leq i \leq N$, which are strictly used in a differential manner. The topological relation of pixels is reflected in a graph structure. Each pixel i corresponds to a vertex v_i. A neighboring pixel j with corresponding v_j is connected to v_i via an edge e_{ij}. The edge weight w_{ij} of an edge e_{ij} is a function of the similarity between $\mathbf{x}_i$ and $\mathbf{x}_j$. We compute an N-element vector $\mathbf{y}$, where $y_i \in [0,1]$ indicates that a pixel i belongs to foreground or background via

$$\mathbf{y}^* = \underset{\mathbf{y}}{\operatorname{argmin}} \sum_{e_{ij}\in E} w_{ij}^p \left|y_i - y_j\right|^q + \sum_{v_i} w_{Fi}^p \left|y_i\right|^q + \sum_{v_i} w_{Bi}^p \left|y_i - 1\right|^q, \tag{4}$$

$$\text{s.t.} \quad \forall i \in \mathcal{F} : y_i = 1, \ \forall i \in \mathcal{B} : y_i = 0,$$

where w_{Fi} and w_{Bi} denote unary weights penalizing foreground and background affinity. Simple thresholding leads to a binary segmentation s with $s_i = 1$ (foreground pixel) if $y_i \geq 1/2$, and 0 (background pixel) otherwise. Based on the selection of parameters p and q, this minimization matches the graph cuts, random walker, or shortest paths algorithms. With $p = \infty$, $q \geq 1$, the power watershed algorithm is obtained. See Couprie et al. [40] for details.

Couprie et al. define the edge weights for grayscale and RGB images;

$$w_{ij} = \exp\left(-\beta\left(d\left(\mathbf{x}_i, \mathbf{x}_j\right)\right)^2\right), \tag{5}$$

where β is a constant and $d(\cdot)$ is the L_2 norm or the L_∞ norm [40]. For hyperspectral data, these choices are reasonable if a single band or the PCA of the image is used as input. However when operating on the full spectra of the image, a more appropriate distance function should be employed.

4.1.2. Adaptation to Hyperspectral Data. In our previous work, we evaluated a range of distance functions on multispectral data, including established similarity measures for spectral mapping [41, 42] and a data-driven measure [10]. In spectral mapping, captured spectra are compared to a database of known spectra to determine a material. Hence, spectral mapping similarity measures are well suited for hyperspectral distance functions.

The best performing measurement in our experiments is the spectral angle from the Spectral Angle Mapper [15]. The spectral angle has the property that it disregards pure intensity changes. It is defined for two spectra $\mathbf{u}$ and $\mathbf{v}$ as

$$\text{SA}(\mathbf{u}, \mathbf{v}) = \cos^{-1}\left(\frac{\langle \mathbf{u}, \mathbf{v}\rangle}{\|\mathbf{u}\|_2 \cdot \|\mathbf{v}\|_2}\right), \tag{6}$$

where $\langle\cdot,\cdot\rangle$ denotes the inner product and $\|\cdot\|_2$ the L_2 norm.

4.1.3. Example Use-Case. We demonstrate the use of supervised segmentation on the *superballs* image. It shows plastic balls of various color that highly reflect on each other. In Figure 6, the user investigates a red ball specifically. With an RGB camera we could only spot the reflection from the yellow ball (see Figure 6(a)). Calculating the PCA on the spectral gradient (as discussed in Section 5.2) also reveals the reflection from the green ball (see Figure 6(b)). However, the blue balls also reflect on the red ball, which is not revealed by these depictions. The first labeling, Figure 6(c), is obtained by setting one foreground seed in the middle of the ball and drawing a circle around the ball as background seeds. When looking at the spectral gradient distribution of that segment (see Figure 6(d)), we see the deviations from the distribution caused by these interreflections. Browsing through these parts of the spectral gradient distribution plot with the interactive highlight reveals their spatial locations. By setting corresponding seed points again, we can find three additional segments within the ball (see Figure 6(e)). The spectral gradient distribution plot (see Figure 6(f)) reveals how the pixels from these three segments contribute to the deviations in the distribution. To further understand them, the user might continue by also segmenting the surrounding balls, effectively adding them to the plot.

Our user-guided segmentation method is especially useful in interactive analysis. Segmentation results are helpful, quick to obtain, and easy to refine within the provided user interface.

4.2. Unsupervised Segmentation. While graph-based segmentation is valuable within an interactive session, it is object-driven in a sense that spatially connected pixels are grouped. However, often it is desired to explore the image as a whole, which means connecting pixels that share no spatial relation in the image, for example, to form a segment consisting of all scene objects that share the same albedo. Note that while the multiband thresholding detailed in Section 3.3 and illustrated in Figure 5(b) often allows finding such spatially disconnected material clusters in the spectral gradient manually, it can be a tedious task, especially when several material clusters in the scene are to be revealed. A global clustering method method however can reveal all material clusters in the scene and be used as a starting point for further analysis.

For this task we incorporate the *mean shift* algorithm into Gerbil. Mean shift has been used for global clustering of multispectral data for several tasks [43, 44]. Mean shift is a good choice due to its general purpose design. For example, while the well-studied and fast *k-means* algorithm works best on clusters with circular shape and the number of clusters needs to be defined a priori, mean shift is a density gradient estimator and it neither requires prior knowledge of the number of clusters nor constrains the shape of the clusters [45], which is desired property for analysis of arbitrary image data. A drawback of the original mean shift is computational complexity. The Fast Adaptive Mean Shift (FAMS) algorithm by Georgescu et al. [46] provides an important improvement to computation time by introducing an approximate nearest-neighbor search. However, the method still takes minutes to hours to segment a hyperspectral image. To facilitate the use of mean shift in interactive inspection, we introduced a new variant that combines mean shift with a superpixel presegmentation [11].

4.2.1. Superpixel Mean Shift Clustering. The mean shift algorithm works in feature space alone to obtain a global clustering rather than spatially constrained segments. It, thus, disregards any topological context. We reintroduce the use of spatial information by incorporating superpixel detection.

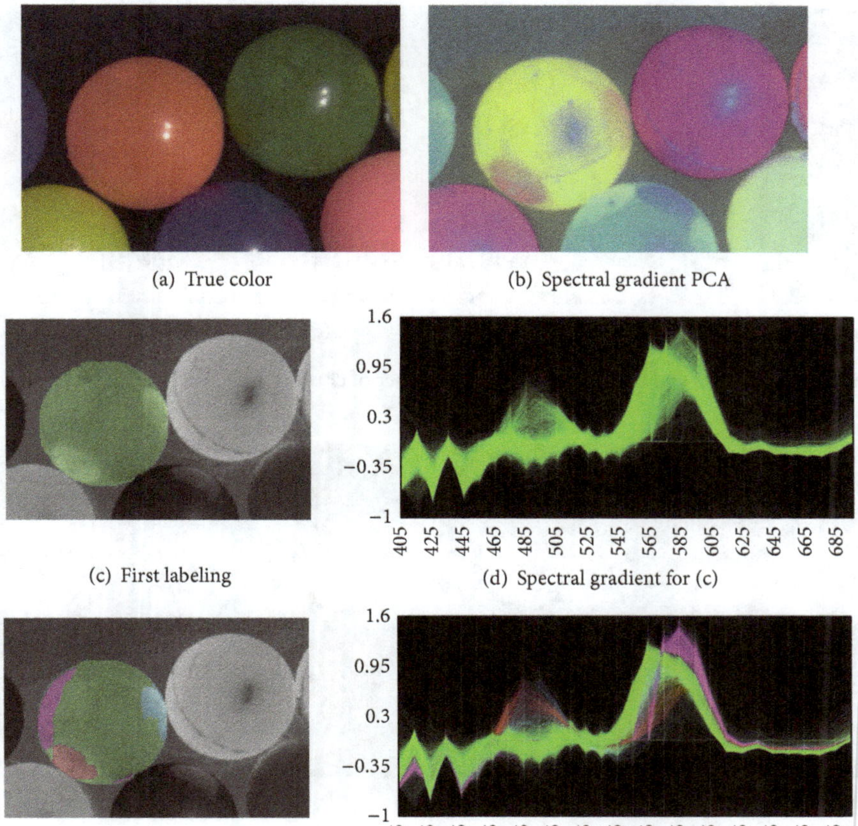

(a) True color (b) Spectral gradient PCA

(c) First labeling (d) Spectral gradient for (c)

(e) Second labeling (f) Spectral gradient for (e)

FIGURE 6: Inspection of cropped *superballs* image. Labelings are shown as overlay on the spectral gradient at 490–500 nm.

Both spatial context and superpixels gained attention recently as means to improve the performance of hyperspectral image processing tasks [47, 48].

In our algorithm, we first obtain a superpixel segmentation by adapting the method of Felzenszwalb and Huttenlocher [49] to hyperspectral data. We obtain a set of small, homogeneous regions. Then, FAMS is performed using these regions as data points. This significantly reduces the complexity of the mean shift operation. Overall computation times are in the range of five to ten seconds for images from the CAVE database [20]. The reliance on superpixels impacts the spatial resolution of the segmentation, in a way that often improves the spatial coherence of segments. A detailed presentation of this work can be found in [11].

4.2.2. Clustering Results and Refinement. In Figure 7, we show an example result on an image from the CAVE database. Clusters are assigned random colors with an equal spacing in the hue range. Our superpixel mean shift algorithm, denoted as MSSP, uses an established spectral matching measure [41] to form superpixels in the original feature space and then performs mean shift for each superpixel in the spectral gradient feature space. It is compared to FAMS operating in both the original feature space as well as the spectral gradient feature space (denoted as SG-FAMS). The results reveal that the spectral gradient distribution is more helpful for material clustering as compared to the original spectral distribution affected by geometry and that we can obtain a very good approximation of the SG-FAMS result with MSSP.

In Figure 8, we show another example result on an image from the Foster database, which other than the previous image captures a natural scene. Due to its high spatial resolution, clustering is a time-consuming task on this image. We compare our approach to the method by Huynh and Robles-Kelly [7], denoted HRK. Due to memory constraints, HRK was applied on a down-sampled (by a factor of two) version of the image and took 6 minutes. Note that, for HRK, the number of desired clusters needs to be set in advance. Our method took 154 seconds on the original image and 15 seconds on the down-sampled version.

Within Gerbil it is easy to further refine clustering results. The spectral distribution visualization reveals the compactness of a cluster. It helps the user to spot undersegmentation, where a single cluster contains several spectral profiles (as seen in Figure 6(d)). Oversegmentation, as is the case in Figure 8(c), can also be easily spotted by comparing the spectral distribution of two clusters. Gerbil provides an intuitive interface to merge such segments. To obtain

(a) True color (b) FAMS (41) (c) SG-FAMS (24) (d) MSSP (30)

FIGURE 7: Clustering results on *fake and real food* image. The number of clusters is shown in parentheses in the respective caption.

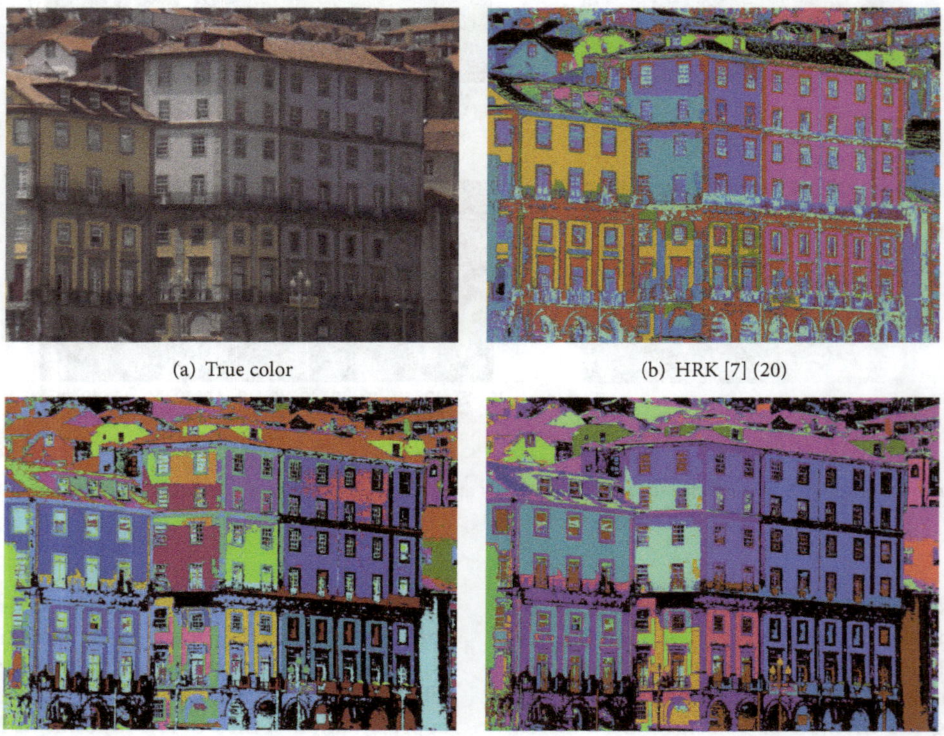

(a) True color (b) HRK [7] (20) (c) MSSP (27) (d) MSSP refined (20)

FIGURE 8: Clustering results on *Ribeira* image. The number of clusters is shown in parentheses in the respective caption.

the result in Figure 8(d), the user performed four simple merging operations. The user may also run a supervised segmentation to refine the clustering, especially in the case of an undersegmented area.

4.3. Label Inspection and Manipulation. The segmentation methods in Gerbil provide beneficial input in a typical usage scenario. However, a good presentation of the results is equally important. Depending on the underlying data, mean shift clustering may produce a high number of meaningful segments. In the supervised scenario, often follow-up segmentations are used to locally refine obtained segments. Several measures aid in discerning and manipulating segmentations.

The first measure is the automatic determination of practical label colors. A preset of label colors consists of the primary and secondary palette, excluding yellow, which is reserved for temporary highlights. For more than five labels, colors are determined by an equal spacing in the hue range excluding yellow. Figures 1, 6, 8, and 10 show examples of automatically labeled segmentations. In cases where a temporary highlight falls onto labeled pixels, the corresponding pixels are drawn in their label colors significantly shifted towards yellow.

The second measure is a specific mode of operation where existing single labels can be selected through mouse hover in the topological view. In this mode, the user can better examine which pixels and spectra are part of a specific label.

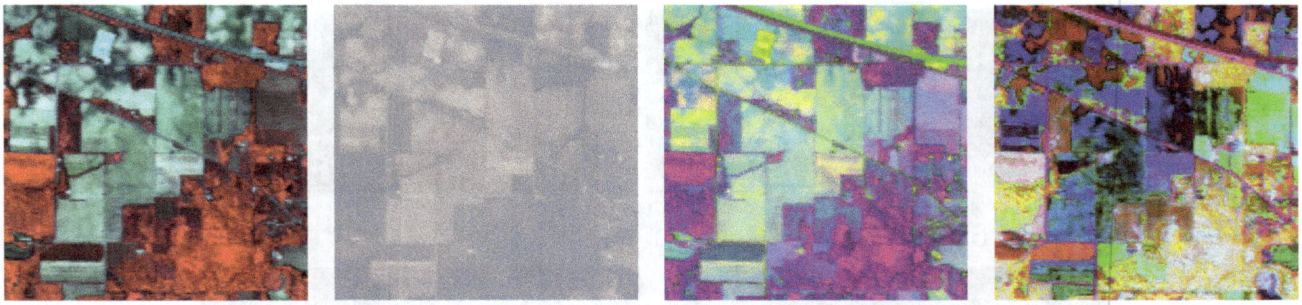

(a) Three-band composite (b) True color (c) PCA false color (d) SOM false color

FIGURE 9: Color visualizations of *Indian Pines* image. SOM visualization is more feature-rich when compared to PCA.

Several labels can be selected at once and then be either deleted or merged using the label manager (Figure 1(e)).

Finally, a label can be used to initialize a multiband limited highlight. The limits on each band are set to include all pixels contained in the label. This helps both in finding similar spectra that are not yet included in the label and in separating several clusters within the label by adjusting the limits. The remaining selection can then be added to another label or form a new label.

Gerbil is a very versatile tool for the creation of labelings for further use. A common use-case is the creation of ground truth data for evaluation of algorithms. For this tedious task, the user can start from a high-quality clustering or segmentation and quickly incorporate the information contained in all bands.

5. Color Coding and Color Display

A popular and useful method of visualizing multispectral and hyperspectral data is the generation of a color image that preserves the spatial relations in the image. While a grayscale image can present a single piece of information per pixel, a color image displays three data values within the r, g, b color triplet. There are four common categories of color displays for hyperspectral image data. (a) *Band composite*: three image bands are selected and then mapped to r, g, b color channels. (b) *Data fusion*: a subset of bands are fused according to a specific criterion. Most often, this is the human color perception, modeled by the CIE *XYZ* color-matching functions [30]. (c) *Dimensionality reduction*: the entire spectral information is reduced to three dimensions, which are mapped to r, g, b. (d) *Feature extraction*: a combination of application-specific indicators (e.g., contributions of three user-selected endmembers to each spectrum) forms a false-color image. For example, in [50], false-color images can depict the occurrence of forest types to assess forest structure.

In line with other analysis frameworks, Gerbil provides color displays based on human color perception (true color) and on linear dimensionality reduction via principal component analysis (PCA). Additionally, a fast nonlinear dimensionality reduction method based on a self-organizing map (SOM) was developed. Figure 9 depicts the results of the three methods on an example image.

5.1. Human Color Perception. For images whose spectrum lies within the range of visible light, an intuitive natural representation can be formed by mimicking human color perception. This representation is often referred to as *true color*. The CIE 1931 colorimetric system defines three color-matching functions $\overline{x}(\lambda), \overline{y}(\lambda), \overline{z}(\lambda)$ [30]. These functions describe the spectral sensitivity of a human observer, called the CIE 1931 standard colorimetric observer. The tristimulus values X, Y, and Z are then obtained by integrating the product of observed intensities and the respective color-matching function over the wavelength λ. In practice, the integrations are replaced by summations over bands of equal width $\Delta\lambda$ centered at wavelength λ [30]. Popular filter-based multispectral sensors capture the incoming light with $\Delta\lambda = 10$. The X, Y, and Z values are directly used in the CIE *XYZ* color space. They can be transformed to RGB given a reference white and a gamma correction value. We display true-color images in sRGB space as it comes with default values for both and is specifically designed for computer displays [51].

Scene illumination plays an important role in true-color visualization. As hyperspectral images provide a complete spectral response of each pixel and homogeneous lighting conditions are present in most data capture scenarios, illumination is easy to manipulate. Gerbil offers removal and exchange of illuminants [6] and includes a set of reference illuminants modeled as black-body radiators [30]. Apart from a true-color image with applied reillumination, the effect of the illuminant is also illustrated within the spectral distribution plot.

5.2. Principal Component Analysis. Objects of varied reflectance properties can yield the same color sensation, rendering true-color representations and their variants unsuitable for several applications. As an alternative, a *false-color* image can be created that puts emphasis on different characteristics of the data. A popular approach uses principal component analysis (PCA) [52], where the three components with maximum variance become the r, g, b values of a color image. Due to the nature of PCA, the per-component variance differs significantly. Therefore, an automatic white balancing is performed for display purposes [52].

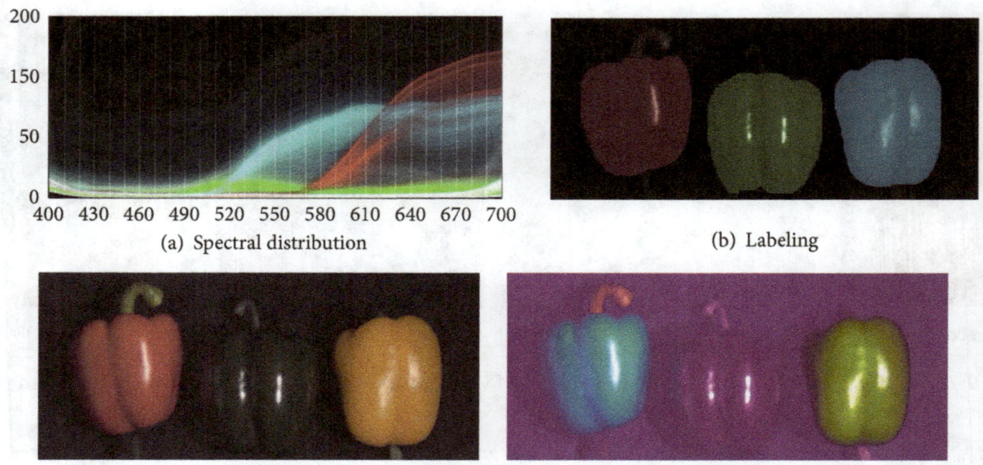

(a) Spectral distribution (b) Labeling

(c) True color (d) PCA false color

FIGURE 10: Displays of cropped *fake and real peppers* image. The spectral distribution (a) is label colored. (b) shows the respective labeling as an overlay on the image band at 530 nm.

Using three bases that explain most of the variance in the data is not always the most helpful representation for discerning relevant spectra. Figure 10 shows a counter example. The contrast of the green peppers against the background in Figure 10(d) is very low. It is a direct effect of the low object reflectance when compared to the red and yellow peppers in the image, as can be seen in Figure 10(a). This is unintuitive to human observers (Figure 10(c)) who have a high sensitivity for green. We would expect the false-color display to better distinguish the green pepper from the background.

The simultaneous availability of the parallel coordinates representation and the PCA-based false-color visualization allows one to easily judge the quality of the dimensionality reduction in comparison to the original image data.

5.3. Self-Organizing Map. A self-organizing map (SOM) converts the nonlinear statistical relationship between high-dimensional data into simpler geometric relationships [53]. In other words, a SOM provides a topological representation of the spectral vector distribution of a multispectral image. A typical SOM consists of a 1D array or 2D grid of *model vectors* (also referred to as neurons) that represent vectors in the original data space. The 3D SOM variant extends a 4-connected 2D lattice to a 6-connected 3D lattice [54]. The third dimension often enables the SOM to learn a more accurate topology on hyperspectral data. We exploit this topology for false coloring by mapping the 3D coordinates to the RGB cube [12, 55, 56].

5.3.1. Training and Traditional Approach. A SOM consists of n model vectors $\mathbf{m}_j \in \mathbb{R}^D$, with D corresponding to the number of bands in the input image. The 3D SOM is a cube of side length $n' = \sqrt[3]{n}$. For any input vector $\mathbf{x}$ (a pixel's captured spectrum), we can find the *best-matching unit* (BMU) $\mathbf{m}_c$ with index

$$c^{(\mathbf{x})} = \underset{j}{\operatorname{argmin}} \quad d\left(\mathbf{x}, \mathbf{m}_j\right), \tag{7}$$

where $d(\cdot)$ is the Euclidean distance. In our experiments we did not observe any advantage by changing the distance function, for example, to the spectral angle. The *location* $\mathbf{r}^{(j)} \in \mathbb{Z}^3$ of a neuron $\mathbf{m}_j$ describes its position in the 3D lattice topology and is a bijective mapping of j.

Our SOM training is unsupervised and uses s unlabeled pixels from the input image. Model vectors are initialized with random values. In each iteration $1 \le t \le s$, we randomly draw a spectrum $\mathbf{x}$. We first determine the BMU $\mathbf{m}_c$ of input $\mathbf{x}$. Then, all model vectors are updated as $\mathbf{m}_j^* = h_{c,j}(t) \cdot (\mathbf{x} - \mathbf{m}_j)$, where $h_{c,j}(t)$ defines the influence of $\mathbf{x}$ on the model vectors $\mathbf{m}_j$ based on the SOM topology. As suggested by Kohonen [53],

$$h_{c,j}(t) = \alpha(t) \cdot \exp\left(-\frac{\left\|\mathbf{r}^{(c)} - \mathbf{r}^{(j)}\right\|^2}{2\sigma^2(t)}\right). \tag{8}$$

The learning-rate factor $\alpha(t)$ is a user-adjustable parameter that is monotonically decreasing. The kernel width $\sigma(t)$ describes the sphere of influence of a sample vector in the SOM topology and is also monotonically decreasing. During the early training phase a SOM seeks a rough global ordering. In later iterations, local regions are smoothed out. We train the SOM on the input image itself on the fly and do not depend on any prior knowledge.

The trained SOM is then used as input for false coloring. In previous work [55], a false-color image has been created by finding BMU $c^{(\mathbf{x})}$ for each image pixel $\mathbf{x}$ (7). Then, an r, g, b triplet is created by scaling $\mathbf{r}^{(c)}$ by $255/n'$. However, this method leads to suboptimal visual quality and achieves even lower entropy values than PCA false coloring. The reason is the limited amount of model vectors, resulting in a strongly quantized output. In the configuration of [55], only 64 different color values could be produced. While we use SOMs of size $n = 10^3$, we are still far from the capabilities of an 8-bit color display. Training a SOM of larger sizes, however, would become infeasible both due to longer learning times and due to the limited amount of available training samples.

5.3.2. Improved Look-Up for False Coloring. To avoid the aforementioned quantized output, we introduce a novel look-up method [12]. Firstly, instead of using a single best-matching unit, we develop a set of best-matching units (BMUs). Furthermore, we order the BMUs according to the L_2 distance and assign a set of predetermined weights to the ordered set. These rank-based weights are crucial. While a simple unweighed combination would only smooth the result, using the L_2 distances directly as weights is not reliable [57]. In our high-dimensional space, distances would appear very close to each other and the weights would not discern well. We define rank weights that ensure both a majority contribution by the first BMU and significant contributions by the additional BMUs.

Consider, as in [58], a vector of BMU indices

$$\mathbf{c}^{(\mathbf{x})} = \underset{\mathbf{j}}{\operatorname{argmin}} \sum_{k=1}^{C} d\left(\mathbf{x}, \mathbf{m}_{j_k}\right), \tag{9}$$

where C is the number of desired BMUs. For each pixel value $\mathbf{x}$, we calculate a *representative location* $\mathbf{r}'$ as

$$\mathbf{r}' = \sum_{k=1}^{C} w_k \cdot \mathbf{r}^{(c_k^{(\mathbf{x})})}, \tag{10}$$

given weights w_k with the properties

$$\begin{gathered} w_k = 2w_{k+1}, \\ \sum_{k=1}^{C} w_k = 1, \\ \forall w_k,\ k < C, \\ d\left(\mathbf{x}, \mathbf{m}_{c_k}\right) < d\left(\mathbf{x}, \mathbf{m}_{c_{k+1}}\right), \\ \forall \mathbf{m}_{c_k},\ k < C, \end{gathered} \tag{11}$$

which expresses that the BMUs are sorted according to distance to the query vector and weighed by their rank, where the weight for rank k is always twice as high as the weight for subsequent rank $k + 1$. The representative location $\mathbf{r}' = \left(r_1' \ r_2' \ r_3'\right)$ describes the position in the learned topology that best represents $\mathbf{x}$. We finally obtain

$$\begin{aligned} \mathrm{r} &= \frac{r_1'}{n'}, \\ \mathrm{g} &= \frac{r_2'}{n'}, \\ \mathrm{b} &= \frac{r_3'}{n'}. \end{aligned} \tag{12}$$

Our method gives a high-quality, nonlinear dimensionality reduction. The advantage over previous false-color work based on SOMs is that we do not suffer from quantization effects. The advantage to other nonlinear methods, for example, ISOMAP [59], is speed. While ISOMAP takes minutes to hours to compute, we train a SOM with $n = 10^3$ in less than 20 seconds on a typical hyperspectral remote sensing image. As most computation is spent in training, execution time is not significantly affected by image size. Rather, it mostly depends on the SOM size and the dimensionality of the data.

The SOM itself can be visualized within Gerbil during, as well as after, its training. The topological views correspond to the traditional SOM-based data visualization and reveal the SOM topology. The parallel coordinate plots depict the distribution of the model vectors, and make it visually comparable to the distribution of the training data. These visualizations help in both assessing the training and in fine-tuning its parameters. Figure 11 depicts a remote sensing image and its spectral distribution view as compared to the spectral distribution of a SOM trained on the image, as well as the ten principal components computed by PCA.

The true-color and false-color display capabilities of Gerbil provide and extend the state of the art in hyperspectral visualization. They are especially helpful for selecting regions of interest (ROIs) that should be examined. In large images, due to clutter and computational expense, it is beneficial to select a ROI before performing further visualization tasks. Additionally, the false-color displays can give a good first impression of the data, for example, to spot the same reflectance in several regions of the image or to find inhomogeneities within a depicted object. The user is directed to specific regions for starting an investigation. An illustrating example is the *fake and real peppers* image, as depicted in Figure 12. In the SOM false-color display pixels with the same material or reflectance properties are consistently colored. One such example are the specular highlights. Another instance is the stems of the plastic and real peppers as annotated in Figure 12.

6. Software Framework

Gerbil is built around the idea of being able to form a base for other research projects. We ensure this by working with code libraries that are already popular in image processing and by a modular and easily expandable design.

6.1. Software Foundation. Gerbil is written in C++. It relies on OpenCV [8] for all image processing tasks and for the

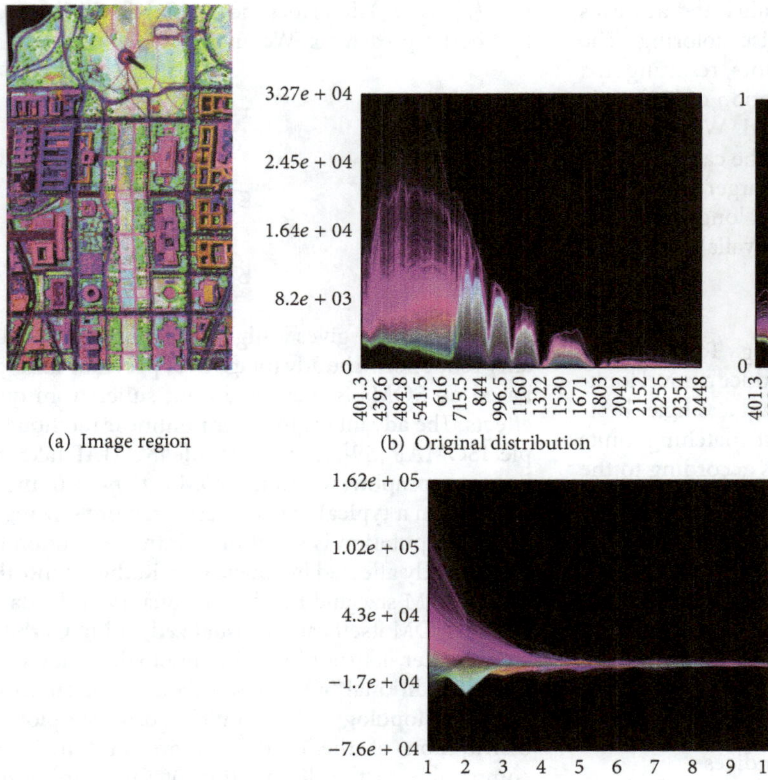

(a) Image region (b) Original distribution (c) SOM distribution

(d) Image PCA distribution

FIGURE 11: Center part of the *D.C. Mall* image and parallel coordinates visualizations computed from the data. Pixels and spectral vectors are colored using the SOM visualized in (c).

FIGURE 12: Figure 1(e), spectral gradient SOM display. The occurrences of two materials in their respective color coding (blue and light gray) are annotated with circles.

internal representation of multispectral data. Due to the tight integration of Gerbil and OpenCV it is very easy to apply OpenCV's extensive functionality on multispectral data within Gerbil. Furthermore, when available, the software utilizes CUDA through OpenCV for some of its common calculations. For image I/O we incorporate the Geospatial Data Abstraction Library (GDAL) [17]. It reads a large variety of image formats from various sources, including FITS, LAN, GIS, and the ENVI.hdr header formats.

For all operations related to the graphical user interface, the Qt framework [60] is used. Together with a CMake-based build system we maintain a truly cross-platform software that runs on a variety of operating systems, including all recent versions of Windows, MacOS X, and GNU/Linux.

6.2. Internal Design. Our software framework consists of both a powerful core and a modular system of extensions. In its core, Gerbil provides a flexible, easy to use programming interface to the image data. It is built on an internal representation of multispectral images that combines both per-band and per-pixel data structures [6].

6.2.1. Modularity. Gerbil is organized in modules that work independently of each other and of the user interface. A module can (optionally) depend on external libraries and other modules. Both the graphical user interface of Gerbil and a sophisticated command-line interface are stand-alone units that expose the functionality of the other modules to the user. The command-line interface enables batch processing of image analysis tasks.

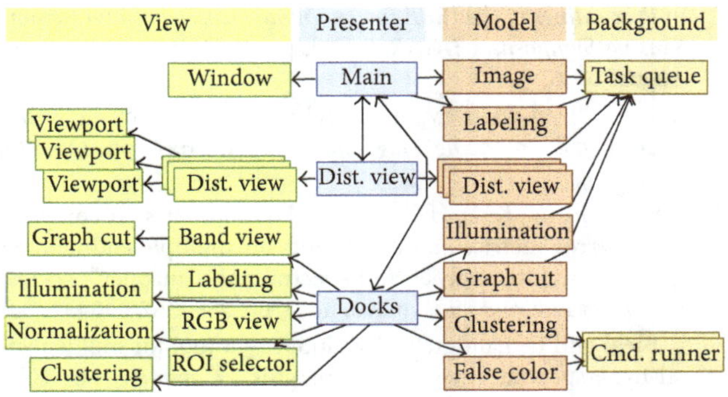

FIGURE 13: GUI software design. Presenter components control both view and model components.

6.2.2. Parallelization and Backgrounding. Performance and responsiveness are both crucial for interactive analysis. OpenCV utilizes vector instruction sets in CPU computation. In order to employ all CPU cores, we parallelize computationally demanding tasks via Intel Thread Building Blocks [61]. We also heavily recycle previously computed data. For example, image representations and distribution histograms are incrementally updated when the current region of interest is changed.

Computations need to run in the background to keep the interface responsive to the user and enable the user to cancel a running task. We introduce a custom shared datatype with locking facilities and an atomic swap operation. Background tasks calculate new data on their own shadow copy and lock the data for a single swap when finished. By using a sequential background queue and defined data swaps at the end of each task, we guarantee that at most two copies of the data are held in memory at once. An exception to this are tasks that can run for a longer time on their own and are not part of the regular user interaction. Examples are global clustering or computation of false-color representations. These tasks run independently on their own data copy.

6.2.3. GUI Software Design. A significant level of sophistication is required in creating Gerbil's GUI in order to expose the framework's functionality to the user while efficiently reacting to user input. We model the GUI module after the model–view–presenter [62] design pattern. Figure 13 gives an overview of the GUI components. *View* components display data and receive user input. Image data, as well as other representations, for example, distribution histograms, is handled by the respective *model* components. These components have access to both the background queue to enqueue tasks and may spawn command runners which can run in the background. The *presenter* components handle user requests and trigger calculations. They are in control of both view and model components, which do not see each other. For data synchronization, Qt's signal/slot messaging [60] is employed.

With this design we ensure a strict separation between models and views and reduce unnecessary overhead in data synchronization code where signals can be passed through.

6.2.4. Extensibility. With our software design it is easy to extend the functionality of Gerbil. New methods can be integrated as modules in the framework. They benefit from a defined interface for parameter setting and command execution through the command-line interface. They may also (optionally) depend on other modules which provide a standardized method call. To add a new method in the GUI, a new dock widget can be created as a view component, while interfacing data calculation with the corresponding module in a model component. The high level of abstraction on several levels keeps interfaces simple and prevents side effects of changes in other components.

7. Conclusions

Gerbil contains several key components that are novel and essential to interactive visual analysis of multispectral and hyperspectral image data. We introduce interactive visualization via parallel coordinates, supervised segmentation via a graph-cut algorithm family, fast global clustering using a combination of superpixel segmentation and mean shift density estimation, and fast nonlinear false coloring based on a 3D self-organizing map. In Gerbil, we tightly integrate these algorithms into a consistent framework that allows the user to explore the data in a novel way. Existing research and commercial software are often comprised of a rich toolbox of useful algorithms for specific applications. However, the goal of a general software framework that also provides intuitive access to the raw image data is unmet. Gerbil fills this gap within a modern architecture and GUI.

In recent years, multispectral capture has gained wider adoption across a range of applications, in particular in cultural heritage. Domain experts who did not work with multispectral data before need assistance in understanding and utilizing the modality. The new intuition in working with multispectral images that Gerbil strives to provide may therefore broaden the scope of multispectral analysis and foster adoption.

Gerbil is an active open-source project and accepts contributions on Github [63]. As of May 2016, it includes 25 902 physical source lines of code (SLOC) in 326 source code files with a total of 37 273 lines. In the past 12 months, 3622 downloads of the Gerbil software package were counted, originating from 45 countries.

For future work, more methods that are established in remote sensing need to be incorporated and adapted to interactive analysis, as done in [33]. Drawing techniques for the spectral distribution views should be further investigated, for example, for clutter reduction. Gerbil is an evolving project that constantly seeks more input from diverse application domains.

Competing Interests

The authors declare that they have no competing interests.

Acknowledgments

The authors gratefully acknowledge funding of this work by the German Research Foundation (GRK 1773 Heterogeneous Image Systems and Open Access Publishing Programme) and funding of the Gerbil open-source project by the European Space Agency (ESA SOCIS 2012–2016).

References

[1] R. O. Green, M. L. Eastwood, C. M. Sarture et al., "Imaging spectroscopy and the Airborne Visible/Infrared Imaging Spectrometer (AVIRIS)," *Remote Sensing of Environment*, vol. 65, no. 3, pp. 227–248, 1998.

[2] F. A. Kruse, A. B. Lefkoff, J. W. Boardman et al., "The Spectral Image Processing System (SIPS)—interactive visualization and analysis of imaging spectrometer data," *Remote Sensing of Environment*, vol. 44, no. 2-3, pp. 145–163, 1993.

[3] N. Gat, "Imaging spectroscopy using tunable filters: a review," in *Proceedings of the Society of Photo-Optical Instrumentation Engineers (SPIE) Conference Series*, vol. 4056 of *SPIE Society of Photo-Optical Instrumentation Engineers*, pp. 50–64, Bellingham, Wash, USA, 2000.

[4] J. Fisher, M. M. Baumback, J. H. Bowles, J. M. Grossmann, and J. A. Antoniades, "Comparison of low-cost hyperspectral sensors," in *Proceedings of the SPIE's International Symposium on Optical Science, Engineering, and Instrumentation*, vol. 3438, pp. 23–30, San Diego, Calif, USA, July 1998.

[5] Forth Photonics, "MuSIS," July 2012, http://archive.is/musis.forth-photonics.com.

[6] J. Jordan and E. Angelopoulou, "Gerbil—a novel software framework for visualization and analysis in the multispectral domain," in *Vision, Modeling and Visualization*, pp. 259–266, 2010.

[7] C. P. Huynh and A. Robles-Kelly, "A probabilistic approach to spectral unmixing," in *Structural, Syntactic, and Statistical Pattern Recognition*, pp. 344–353, Springer, Berlin, Germany, 2010.

[8] G. Bradski, "Open Source Computer Vision Library," Dr. Dobb's Journal of Software Tools, 2016, http://opencv.org/.

[9] M. Hall, E. Frank, G. Holmes, B. Pfahringer, P. Reutemann, and I. H. Witten, "The WEKA data mining software: an update," *ACM SIGKDD Explorations Newsletter*, vol. 11, no. 1, pp. 10–18, 2009.

[10] J. Jordan and E. Angelopoulou, "Supervised multispectral image segmentation with power watersheds," in *Proceedings of the 19th IEEE International Conference on Image Processing (ICIP '12)*, pp. 1585–1588, IEEE, Orlando, Fla, USA, September-October 2012.

[11] J. Jordan and E. Angelopoulou, "Mean-shift clustering for interactive multispectral image analysis," in *Proceedings of the 20th IEEE International Conference on Image Processing (ICIP '13)*, pp. 3790–3794, IEEE, Melbourne, Australia, September 2013.

[12] J. Jordan and E. Angelopoulou, "Hyperspectral image visualization with a 3-D self-organizing map," in *Proceedings of the Workshop on Hyperspectral Image and Signal Processing: Evolution in Remote Sensing (WHISPERS '13)*, pp. 1–4, IEEE, 2013.

[13] J. W. Boardman, L. L. Biehl, R. N. Clark et al., "Development and implementation of software systems for imaging spectroscopy," in *Proceedings of the IEEE International Geoscience and Remote Sensing Symposium (IGARSS '06)*, pp. 1969–1973, IEEE, July-August 2006.

[14] L. Biehl, "Eval-ware: hyperspectral imaging [best of the web]," *IEEE Signal Processing Magazine*, vol. 24, no. 4, pp. 125–126, 2007.

[15] P. E. Dennison, K. Q. Halligan, and D. A. Roberts, "A comparison of error metrics and constraints for multiple endmember spectral mixture analysis and spectral angle mapper," *Remote Sensing of Environment*, vol. 93, no. 3, pp. 359–367, 2004.

[16] L. Biehl and D. Landgrebe, "MultiSpec—a tool for multispectral-hyperspectral image data analysis," *Computers & Geosciences*, vol. 28, no. 10, pp. 1153–1159, 2002.

[17] F. Warmerdam, "The geospatial data abstraction library," in *Open Source Approaches in Spatial Data Handling*, G. B. Hall, M. G. Leahy, S. Balram, and S. Dragicevic, Eds., vol. 2, pp. 87–104, Springer, Berlin, Germany, 2008.

[18] Harris Geospatial Solutions, "ENVI," 2016, http://www.harrisgeospatial.com/ProductsandSolutions/GeospatialProducts/ENVI.aspx.

[19] J. Boardman, F. Kruse, and R. Green, "Mapping target signatures via partial unmixing of AVIRIS data," in *Proceedings of the Summaries, 5th JPL Airborne Earth Science Workshop*, vol. 1, pp. 23–26, 1995.

[20] F. Yasuma, T. Mitsunaga, D. Iso, and S. Nayar, "Generalized assorted pixel camera: post-capture control of resolution, dynamic range and spectrum," Tech. Rep., Columbia University, 2008.

[21] D. H. Foster, K. Amano, S. M. C. Nascimento, and M. J. Foster, "Frequency of metamerism in natural scenes," *Journal of the Optical Society of America A*, vol. 23, no. 10, pp. 2359–2372, 2006.

[22] Purdue Research Foundation, "Hyperspectral images," May 2016, https://engineering.purdue.edu/_biehl/MultiSpec/hyperspectral.html.

[23] United States Army Corpse of Engineers, "Hypercube," May 2016, http://www.erdc.usace.army.mil/Media/FactSheets/FactSheetArticleView/tabid/9254/Article/610433/hypercube.aspx.

[24] Ball Aerospace & Technologies Corp, "Opticks," May 2016, https://opticks.org/.

[25] "Opticks-Spectral Processing Extension," July 2012, https://opticks.org/display/opticksExt.

[26] M. Cui, A. Razdan, J. Hu, and P. Wonka, "Interactive hyperspectral image visualization using convex optimization," *IEEE Transactions on Geoscience and Remote Sensing*, vol. 47, no. 6, pp. 1673–1684, 2009.

[27] W. A. Joye and E. Mandel, "New features of SAOImage DS9," in *Astronomical Data Analysis Software and Systems XII*, vol. 295, pp. 489–492, 2003.

[28] H. Li, C.-W. Fu, and A. J. Hanson, "Visualizing multiwavelength astrophysical data," *IEEE Transactions on Visualization and Computer Graphics*, vol. 14, no. 6, pp. 1555–1562, 2008.

[29] P. Colantoni, R. Pillay, C. Lahanier, and D. Pitzalis, "Analysis of multispectral images of paintings," in *Proceedings of the European Signal Processing Conference*, Florence, Italy, 2006.

[30] G. Wyszecki and W. S. Stiles, *Color Science: Concepts and Methods, Quantitative Data and Formulae*, Wiley-Interscience, New York, NY, USA, 2nd edition, 2000.

[31] S. J. Kim, S. Zhuo, F. Deng, C.-W. Fu, and M. Brown, "Interactive visualization of hyperspectral images of historical documents," *IEEE Transactions on Visualization and Computer Graphics*, vol. 16, no. 6, pp. 1441–1448, 2010.

[32] National ICT Australia Limited, "Scyven," May 2016, http://www.scyven.com.

[33] B. Labitzke, S. Bayraktar, and A. Kolb, "Generic visual analysis for multi- and hyperspectral image data," *Data Mining and Knowledge Discovery*, vol. 27, no. 1, pp. 117–145, 2013.

[34] N. Keshava and J. F. Mustard, "Spectral unmixing," *IEEE Signal Processing Magazine*, vol. 19, no. 1, pp. 44–57, 2002.

[35] E. Angelopoulou, S. W. Lee, and R. Bajcsy, "Spectral gradient: a material descriptor invariant to geometry and incident illumination," in *Proceedings of the 7th IEEE International Conference on Computer Vision (ICCV '99)*, vol. 2, pp. 861–867, Kerkyra, Greece, September 1999.

[36] A. Inselberg and B. Dimsdale, "Parallel coordinates: a tool for visualizing multi-dimensional geometry," in *in IEEE Conference on Visualization*, pp. 361–378, 1990.

[37] R. Gonzalez, R. Woods, and S. Eddins, *Digital Image Processing*, Prentice Hall Press, New York, NY, USA, 3rd edition, 2008.

[38] G. Ellis and A. Dix, "Enabling automatic clutter reduction in parallel coordinate plots," *IEEE Transactions on Visualization and Computer Graphics*, vol. 12, no. 5, pp. 717–724, 2006.

[39] R. Fuchs and H. Hauser, "Visualization of multi-variate scientific data," *Computer Graphics Forum*, vol. 28, no. 6, pp. 1670–1690, 2009.

[40] C. Couprie, L. Grady, L. Najman, and H. Talbot, "Power watershed: a unifying graph-based optimization framework," *IEEE Transactions on Pattern Analysis and Machine Intelligence*, vol. 33, no. 7, pp. 1384–1399, 2011.

[41] Y. Du, C.-I. Chang, H. Ren, C.-C. Chang, J. O. Jensen, and F. M. D'Amico, "New hyperspectral discrimination measure for spectral characterization," *Optical Engineering*, vol. 43, no. 8, pp. 1777–1786, 2004.

[42] A. E. Gutiérrez-Rodríguez, M. A. Medina-Pérez, J. F. Martínez-Trinidad, J. A. Carrasco-Ochoa, and M. García-Borroto, "dissimilarity measures for ultraviolet spectra identification," in *Advances in Pattern Recognition: Second Mexican Conference on Pattern Recognition, MCPR 2010, Puebla, Mexico, September 27–29, 2010. Proceedings*, vol. 6256 of *Lecture Notes in Computer Science*, pp. 220–229, Springer, New York, NY, USA, 2010.

[43] S. Bo, L. Ding, H. Li, F. Di, and C. Zhu, "Mean shift-based clustering analysis of multispectral remote sensing imagery," *International Journal of Remote Sensing*, vol. 30, no. 4, pp. 817–827, 2009.

[44] X. Huang and L. Zhang, "An adaptive mean-shift analysis approach for object extraction and classification from urban hyperspectral imagery," *IEEE Transactions on Geoscience and Remote Sensing*, vol. 46, no. 12, pp. 4173–4185, 2008.

[45] D. Comaniciu and P. Meer, "Mean shift: a robust approach toward feature space analysis," *IEEE Transactions on Pattern Analysis and Machine Intelligence*, vol. 24, no. 5, pp. 603–619, 2002.

[46] B. Georgescu, I. Shimshoni, and P. Meer, "Mean shift based clustering in high dimensions: a texture classification example," in *Proceedings of the 9th IEEE International Conference on Computer Vision*, pp. 456–463, October 2003.

[47] M. Fauvel, J. A. Benediktsson, J. Chanussot, and J. R. Sveinsson, "Spectral and spatial classification of hyperspectral data using SVMs and morphological profiles," *IEEE Transactions on Geoscience and Remote Sensing*, vol. 46, no. 11, pp. 3804–3814, 2008.

[48] D. R. Thompson, L. Mandrake, M. S. Gilmore, and R. Castaño, "Superpixel endmember detection," *IEEE Transactions on Geoscience and Remote Sensing*, vol. 48, no. 11, pp. 4023–4033, 2010.

[49] P. F. Felzenszwalb and D. P. Huttenlocher, "Efficient graph-based image segmentation," *International Journal of Computer Vision*, vol. 59, no. 2, pp. 167–181, 2004.

[50] S. L. Ustin and A. Trabucco, "Using hyperspectral data to assess forest structure," *Journal of Forestry*, vol. 98, no. 6, pp. 47–49, 2000.

[51] M. Stokes, M. Anderson, S. Chandrasekar, and R. Motta, "A standard default color space for the internet-sRGB," Tech. Rep., Microsoft and Hewlett-Packard, 1996, http://www.color.org/sRGB.xalter.

[52] N. P. Jacobson and M. R. Gupta, "Design goals and solutions for display of hyperspectral images," *IEEE Transactions on Geoscience and Remote Sensing*, vol. 43, no. 11, pp. 2684–2692, 2005.

[53] T. Kohonen, *Self-Organizing Maps*, vol. 30 of *Springer Series in Information Sciences*, Springer, Berline, Germany, 3rd edition, 2001.

[54] J. Jordan and E. Angelopoulou, "Edge detection in multispectral images using the n-dimensional self-organizing map," in *Proceedings of the 18th IEEE International Conference on Image Processing (ICIP '11)*, pp. 3181–3184, IEEE, Brussels, Belgium, September 2011.

[55] J. Gorricha and V. Lobo, "Improvements on the visualization of clusters in geo-referenced data using self-organizing maps," *Computers & Geosciences*, vol. 43, pp. 177–186, 2012.

[56] J. M. Fonville, C. L. Carter, L. Pizarro et al., "Hyperspectral visualization of mass spectrometry imaging data," *Analytical Chemistry*, vol. 85, no. 3, pp. 1415–1423, 2013.

[57] K. Beyer, J. Goldstein, R. Ramakrishnan, and U. Shaft, "When is 'nearest neighbor' meaningful?" in *Database Theory—ICDT'99*, C. Beeri and P. Buneman, Eds., vol. 1540 of *Lecture Notes in Computer Science*, pp. 217–235, 1999.

[58] M. Sjöberg and J. Laaksonen, "Optimal combination of SOM search in best-matching units and map neighborhood," in *Advances in Self-Organizing Maps*, vol. 5629, pp. 281–289, Springer, Berlin, Germany, 2009.

[59] C. M. Bachmann, T. L. Ainsworth, and R. A. Fusina, "Improved manifold coordinate representations of large-scale hyperspectral scenes," *IEEE Transactions on Geoscience and Remote Sensing*, vol. 44, no. 10, pp. 2786–2803, 2006.

[60] The Qt Company, "Qt-cross-platform application and UI framework," 2016, http://qt-project.org/.

[61] J. Reinders, *Intel Threading Building Blocks: Outfitting C++ for Multi-Core Processor Parallelism*, O'Reilly Media, 2010.

[62] M. Potel, *MVP: Model-View-Presenter the Taligent Programming Model for c++ and Java*, Taligent, 1996.

[63] J. Jordan, *Gerbil Hyperspectral Visualization and Analysis Framework*, 2016, http://github.com/gerbilvis/gerbil.

Object Tracking via 2DPCA and ℓ_2-Regularization

Haijun Wang,[1,2] Hongjuan Ge,[1] and Shengyan Zhang[2]

[1]*College of Civil Aviation, Nanjing University of Aeronautics and Astronautics, Nanjing 211106, China*
[2]*Aviation Information Technology R & D Center, Binzhou University, Binzhou 256603, China*

Correspondence should be addressed to Haijun Wang; whjlym@163.com

Academic Editor: Jiri Jan

We present a fast and robust object tracking algorithm by using 2DPCA and ℓ_2-regularization in a Bayesian inference framework. Firstly, we model the challenging appearance of the tracked object using 2DPCA bases, which exploit the strength of subspace representation. Secondly, we adopt the ℓ_2-regularization to solve the proposed presentation model and remove the trivial templates from the sparse tracking method which can provide a more fast tracking performance. Finally, we present a novel likelihood function that considers the reconstruction error, which is concluded from the orthogonal left-projection matrix and the orthogonal right-projection matrix. Experimental results on several challenging image sequences demonstrate that the proposed method can achieve more favorable performance against state-of-the-art tracking algorithms.

1. Introduction

Visual tracking is one of the fundamental topics in computer vision and plays an important role in numerous researches and practical applications such as surveillance, human computer interaction, robotics, and traffic control. Existing object tracking algorithms can be divided into two categories, that is, discriminative or generative. Discriminative methods treat tracking as a binary classification problem with local search which estimates the decision boundary between an object image patch and the background. Babenko et al. [1] proposed an online multiple instance learning (MIL), which treats ambiguous positive and negative samples as bags to learn a discriminative classifier. Zhang et al. [2] propose a fasting compressive tracking algorithm which employs nonadaptive random projections that preserve the structure of the image feature.

Generative methods typically learn a model to represent the target object and incrementally update the appearance model to search for the image region with minimal reconstruction error. Inspired by the success of sparse representation in face recognition [3], supersolution [4], and inpainting [5], recently, sparse representation based visual tracking [6–9] has also attracted increasing interests. Mei and Ling [10] first extend sparse representation to object tracking and cast the tracking problem as determining the likeliest patch with a sparse representation of templates. The method can handle partial occlusion by treating the error term as sparse noise. However, it requires solving a series of complicated ℓ_1 norm related minimization problems many times and the time complexity is quite significant. Although some modified ℓ_1 norm methods have been proposed to speed up tracker, they are still far away from being real time.

Recently, many object tracking algorithms have been proposed to exploit the power of subspace representation from different points. Ross et al. [11] present a tracking method that incrementally learns a PCA low-dimensional subspace representation, efficiently adapting online to changes in the appearance of the target. However, this method is sensitive to partial occlusion. Zhong et al. [8] proposed a robust object tracking algorithm via sparse collaborative appearance model that exploits both holistic templates and local representations to account for appearance changes. Zhuang et al. [12] cast the tracking problem as finding the candidate that scores the highest in the evaluation model based upon a matrix called discriminative sparse similarity map. Qian et al. [13] exploit an appearance model based on extended incremental nonnegative matrix factorization for visual tracking. Wang and Lu [14] present a novel online object tracking algorithm by using 2DPCA and ℓ_1-regularization. This method can achieve good performance among many scenes. However, the coefficients and the sparse error matrix used in this method need an iterative algorithm to compute and the space and time complexity are too large to meet the real-time tracking.

Motivated by the aforementioned work, this paper presents a robust and fast ℓ_2 norm tracking algorithm with adaptive appearance model. The contributions of this work are threefold: (1) we exploit the strength of 2DPCA subspace representation using ℓ_2-regularization; (2) we remove the trivial templates from the sparse tracking method; (3) we present a novel likelihood function that considers the reconstruction error, which is concluded from the orthogonal left-projection matrix and the orthogonal right-projection matrix. Both qualitative and quantitative evaluation on video sequences demonstrate that the proposed method can handle occlusion, illumination changes, scale changes, and no-rigid appearance changes effectively in a lower computation complexity and can run in real time.

2. Object Representation via 2DPCA and ℓ_2-Regularization

Principal component analysis (PCA) is a classical feature extraction and data representation technique widely used in the areas of pattern recognition and computer vision. Compared with PCA, two-dimensional principal component analysis (2DPCA) [15] is based on 2D matrices rather than 1D vectors. So the image matrix does not need to be previously transformed into vector. That is, the extraction of image features is computationally more efficient using 2DPCA than PCA. In this paper, we represent the object by using 2D basis matrices. Given a series of image matrices $[\mathbf{B}_1, \mathbf{B}_2, \ldots, \mathbf{B}_K]$, the projection coefficients matrices $[\mathbf{A}_1, \mathbf{A}_2, \ldots, \mathbf{A}_K]$ can be got by solving the following function:

$$\min_{\mathbf{U},\mathbf{V},\mathbf{A}} \frac{1}{K}\sum_{i=1}^{K} \left\|\mathbf{B} - \mathbf{U}\mathbf{A}\mathbf{V}^{\mathrm{T}}\right\|_F^2, \tag{1}$$

where $\|\cdot\|_F$ denotes the Frobenius norm; $\mathbf{U}$ represents the orthogonal left-projection matrix; $\mathbf{V}$ represents the orthogonal right-projection matrix.

The cost function is set as an ℓ_2-regularization quadratic function:

$$\begin{aligned} J(\mathbf{A}) &= \left\|\mathbf{B} - \mathbf{U}\mathbf{A}\mathbf{V}^{\mathrm{T}}\right\|_F^2 + \lambda\|\mathbf{A}\|_F^2 \\ &= \frac{1}{2}\operatorname{tr}\left\{\left(\mathbf{B} - \mathbf{U}\mathbf{A}\mathbf{V}^{\mathrm{T}}\right)^{\mathrm{T}}\left(\mathbf{B} - \mathbf{U}\mathbf{A}\mathbf{V}^{\mathrm{T}}\right)\right\} + \lambda\|\mathbf{A}\|_F^2 \\ &= \frac{1}{2}\operatorname{tr}\left\{\left(\mathbf{B}^{\mathrm{T}} - \mathbf{V}\mathbf{A}^{\mathrm{T}}\mathbf{U}^{\mathrm{T}}\right)\left(\mathbf{B} - \mathbf{U}\mathbf{A}\mathbf{V}^{\mathrm{T}}\right)\right\} + \lambda\|\mathbf{A}\|_F^2 \\ &= \frac{1}{2}\operatorname{tr}\left\{\mathbf{B}^{\mathrm{T}}\mathbf{B} - \mathbf{B}^{\mathrm{T}}\mathbf{U}\mathbf{A}\mathbf{V}^{\mathrm{T}} - \mathbf{V}\mathbf{A}^{\mathrm{T}}\mathbf{U}^{\mathrm{T}}\mathbf{B} + \mathbf{V}\mathbf{A}^{\mathrm{T}}\mathbf{U}^{\mathrm{T}}\mathbf{U}\mathbf{A}\mathbf{V}^{\mathrm{T}}\right\} \\ &\quad + \lambda\|\mathbf{A}\|_F^2 \\ &= \frac{1}{2}\operatorname{tr}\left\{\mathbf{B}^{\mathrm{T}}\mathbf{B}\right\} - \operatorname{tr}\left\{\mathbf{B}^{\mathrm{T}}\mathbf{U}\mathbf{A}\mathbf{V}^{\mathrm{T}}\right\} + \frac{1}{2}\left\{\mathbf{V}\mathbf{A}^{\mathrm{T}}\mathbf{U}^{\mathrm{T}}\mathbf{U}\mathbf{A}\mathbf{V}^{\mathrm{T}}\right\} \\ &\quad + \lambda\|\mathbf{A}\|_F^2. \end{aligned} \tag{2}$$

Here, λ is a constant. The solution of (2) is easily derived as follows:

$$\begin{gathered} \frac{\partial J(\mathbf{A})}{\partial \mathbf{A}} = -\mathbf{U}^{\mathrm{T}}\mathbf{B}\mathbf{V} + \mathbf{U}^{\mathrm{T}}\mathbf{U}\mathbf{A}\mathbf{V}^{\mathrm{T}}\mathbf{V} + 2\lambda\mathbf{A} = 0 \\ \Downarrow \\ \mathbf{U}^{\mathrm{T}}\mathbf{U}\mathbf{A}\mathbf{V}^{\mathrm{T}}\mathbf{V} + 2\lambda\mathbf{A} = \mathbf{U}^{\mathrm{T}}\mathbf{B}\mathbf{V} \\ \mathbf{U}^{\mathrm{T}}\mathbf{U}\mathbf{A}\mathbf{V}^{\mathrm{T}}\mathbf{V} + 2\lambda\mathbf{I}_1\mathbf{A}\mathbf{I}_2 = \mathbf{U}^{\mathrm{T}}\mathbf{B}\mathbf{V} \\ \left\{\mathbf{V}^{\mathrm{T}}\mathbf{V}\otimes\mathbf{U}^{\mathrm{T}}\mathbf{U} + \mathbf{I}_2^{\mathrm{T}}\otimes 2\lambda\mathbf{I}_1\right\}^{-1}\operatorname{vec}(\mathbf{A}) = \operatorname{vec}\left(\mathbf{U}^{\mathrm{T}}\mathbf{B}\mathbf{V}\right) \\ \operatorname{vec}(\mathbf{A}) = \left\{\mathbf{V}^{\mathrm{T}}\mathbf{V}\otimes\mathbf{U}^{\mathrm{T}}\mathbf{U} + \mathbf{I}_2^{\mathrm{T}}\otimes 2\lambda\mathbf{I}_1\right\}\operatorname{vec}\left(\mathbf{U}^{\mathrm{T}}\mathbf{B}\mathbf{V}\right). \end{gathered} \tag{3}$$

Here, $\mathbf{I}_1$ and $\mathbf{I}_2$ mean the identity matrix. $\otimes$ stands for Kronecker product. vec($\mathbf{A}$) means the vector-version of the matrix $\mathbf{A}$. Therefore, we can get the projection coefficients matrix $\mathbf{A}$. Let $\mathbf{P} = \mathbf{V}^{\mathrm{T}}\mathbf{V}\otimes\mathbf{U}^{\mathrm{T}}\mathbf{U} + \mathbf{I}_2^{\mathrm{T}}\otimes 2\lambda\mathbf{I}_2$. Obviously, the projection matrix $\mathbf{P}$ is independent from $\mathbf{B}$ and we can precalculate it in each frame before circulation for all candidates. When a new candidate comes, we can simply calculate $\mathbf{P}\operatorname{vec}(\mathbf{U}^{\mathrm{T}}\mathbf{B}\mathbf{V})$ to obtain vec($\mathbf{A}$), which makes the proposed method very fast.

Here, we abandon the trivial templates completely, which makes the target able to be represented by the 2DPCA subspace fully. The error matrix can be obtained by the following equation after we get the projection coefficients matrix $\mathbf{A}$ from (3):

$$\mathbf{E} = \mathbf{B} - \mathbf{U}\mathbf{A}\mathbf{V}^{\mathrm{T}}. \tag{4}$$

So, the error matrix can be calculated once.

3. Tracking Framework Based on 2DPCA and ℓ_2-Regularization

Visual tracking is treated as a Bayesian inference task in a Markov model with hidden state variables. Given a series of image matrices $B = [\mathbf{B}_1, \mathbf{B}_2, \ldots, \mathbf{B}_K]$, we aim to estimate the hidden state variable $\mathbf{x}_t$ recursively:

$$\begin{aligned} &p(\mathbf{x}_t \mid B) \\ &\quad\propto p(\mathbf{B}_t \mid \mathbf{x}_t)\int p(\mathbf{x}_t \mid \mathbf{x}_{t-1})\, p(\mathbf{x}_{t-1} \mid B_{t-1})\, d\mathbf{x}_{t-1}, \end{aligned} \tag{5}$$

where $p(\mathbf{x}_t \mid \mathbf{x}_{t-1})$ is the motion model that represents the state transition between two consecutive states. $p(\mathbf{B}_t \mid \mathbf{x}_t)$ is the observation model which indicates the likelihood function.

Motion Model. We apply an affine image warp to model the target motion between consecutive states. Six parameters of the affine transform are used to model $p(\mathbf{x}_t \mid \mathbf{x}_{t-1})$ of a tracked target. Let $\mathbf{x}_t = \{x_t, y_t, \theta_t, s_t, \alpha, \phi_t\}$, where x_t, y_t, θ_t, s_t, α, and ϕ_t denote x and y translations, rotation angle, scale, and aspect ration and skew, respectively. The state transition is formulated by random walk; that is,

$p(\mathbf{x}_t \mid \mathbf{x}_{t-1}) = N(\mathbf{x}_t; \mathbf{x}_{t-1}, \Sigma)$, where Σ is a diagonal covariance matrix which indicates the variances of affine parameters.

Observation Model. If no occlusion occurs, an image observation $\mathbf{B}_t$ can be generated by a 2DPCA subspace (spanned by $\mathbf{U}$ and $\mathbf{V}$ and centered at μ). Here, we consider the partial occlusion in the appearance model for robust tracking. Thus, we assume that the centered image matrices $\overline{\mathbf{B}}_t$ ($\overline{\mathbf{B}}_t = \mathbf{B}_t - \mu$) can be represented by the linear combination of the projection matrices $\mathbf{U}$ and $\mathbf{V}$. Then, we draw N candidates in the state $\mathbf{x}_t$. For each of the observed image matrices, we solve a ℓ_2-regularization problem:

$$\min_{\mathbf{A}^i} \left\|\overline{\mathbf{B}}^i - \mathbf{U}\mathbf{A}^i\mathbf{V}^{\mathrm{T}}\right\|_F^2 + \lambda \left\|\mathbf{A}^i\right\|_2^2, \tag{6}$$

where i denotes the ith sample of the state $\mathbf{x}$. Thus, we obtain $\mathbf{A}^i$ and the likelihood can be measured by the reconstruction error:

$$p\left(\overline{\mathbf{B}}^i \mid \mathbf{B}^i\right) = \exp\left(-\left\|\overline{\mathbf{B}}^i - \mathbf{U}\mathbf{A}^i\mathbf{V}^{\mathrm{T}}\right\|_F^2\right). \tag{7}$$

However, it is noted that, just by penalizing error level, the precise location of the tracked target can be benefited. Therefore, we present a novel likelihood function, which considers both the reconstruction error and the level of error matrix:

$$p\left(\overline{\mathbf{B}}^i \mid \mathbf{B}^i\right) = \exp\left(-\left\|\overline{\mathbf{B}}^i - \mathbf{U}\mathbf{A}^i\mathbf{V}^{\mathrm{T}} - \mathbf{E}^i\right\|_F^2 - \lambda \left\|\mathbf{E}^i\right\|_1\right), \tag{8}$$

where $\mathbf{E}^i$ can be calculated by (9):

$$\mathbf{E}^i = \overline{\mathbf{B}}^i - \mathbf{U}\mathbf{A}^i\mathbf{V}^{\mathrm{T}}. \tag{9}$$

Here, $\mathbf{A}^i$ is calculated by (3).

Online Update. In order to handle the appearance change of tracked target, it is necessary to update the observation model. If some imprecise samples are used to update, the tracked model may degrade. Therefore, we present an occlusion-radio-based update mechanism. After obtaining the best candidate state of each frame, we compute the corresponding error matrix and the occlusion ratio γ. Two thresholds $\mathrm{thr}_1 = 0.1$ and $\mathrm{thr}_2 = 0.6$ are introduced to define the degree of occlusion. If $\gamma < \mathrm{thr}_1$, the tracked target is not occluded or a small part of it is occluded by the noise. Therefore, the model with sample is updated directly. If $\mathrm{thr}_1 < \gamma < \mathrm{thr}_2$, the tracked target is partially occluded. The occluded part is replaced by the average observation and the recovered candidate is used for update. If $\gamma > \mathrm{thr}_2$, most part of the tracked target is occluded. Therefore, the sample is discarded without update. After we cumulate enough samples, we use an incremental 2DPCA algorithm to update the tracker (left- and right-projection matrices).

4. Experiments

The proposed tracking algorithm is implemented in MATLAB which runs on a computer with Intel i5-3210 CPU (2.5 GHz) with 4 GB memory. The regularization λ is set to 0.05. The image observation is resized to pixels for the proposed 2DPCA representation. For each sequence, the location of the tracked target object is manually labeled in the first frame. 600 particles are adopted for the proposed algorithm accounting for the trade-off between effectiveness and speed. Our tracker is incrementally updated every 5 frames.

To demonstrate the effectiveness of the proposed tracking algorithm, we select six state-of-the-art trackers: the ℓ_1 tracker [10], the PN tracker [16], the VTD tracker [17], the MIL tracker [1], the Frag tracker [18], and the 2DPCAℓ_1 tracker [14] for comparison on several challenging image sequences including *Occlusion 1, David Outdoor, Caviar 2, Girl, Car 4, Car 11, Singer 1, Deer, Jumping,* and *Lemming*. The challenging factors include severe occlusion, pose change, motion blur, illumination variation, and background clutter.

4.1. Qualitative Evaluation

Severe Occlusion. We test four sequences (*Occlusion 1, David-Outdoor, Caviar 2,* and *Girl*) with long time partial or heavy occlusion and scale change. Figure 1(a) demonstrates that ℓ_1 algorithm, Frag algorithm, 2DPCAℓ_1, and our algorithms perform better, since these methods take partial occlusion into account. ℓ_1 algorithm, 2DPCAℓ_1, and our algorithms can handle occlusion by avoiding updating occluded pixels into the PCA basis and 2DPCA basis separately. Frag algorithm can work well on some simple occlusion cases (e.g., Figure 1(a), *Occlusion 1*) via the part-based representation. However, this method performs poorly on some more challenging videos (e.g., Figure 1(b), *DavidOutdoor*). MIL tracker is not able to track the occluded target in *DavidOutdoor* and *Caviar 2,* since the Harr-like features the MIL method adopted are less effective in distinguishing the similar objects. For the *Girl* video, the in- and out-of-plane rotation, partial occlusion, and the scale change make it difficult to track. It can be seen that the Frag and the proposed tracker work better than the other methods.

Illumination Change. Figures 1(e) and 1(f) present tracking results using the *Car 4, Car 11,* and *Singer 1* sequences with significant change of illumination and scale as well as background clutter. The ℓ_1 tracker, 2DPCAℓ_1 tracker, and the proposed tracker perform well in the *Car 4* sequences whereas the other trackers drift away when the target vehicle goes underneath the overpass or the trees. For *Car 11* sequences, 2DPCAℓ_1 and the proposed tracker can achieve robust tracking results whereas the other trackers drift away when drastic illumination change occurs or when similar object appears. In the *Singer 1* sequence, the drastic illumination and scale change make it difficult to track. It can be seen that the proposed tracker performs better than the other methods.

Motion Blur. It is difficult for tracking algorithms to predict the location of the tracked objects when the target moves abruptly. Figures 1(h) and 1(i) demonstrate the tracking results in the *Deer* and *Jumping* sequences. In *Deer* sequences,

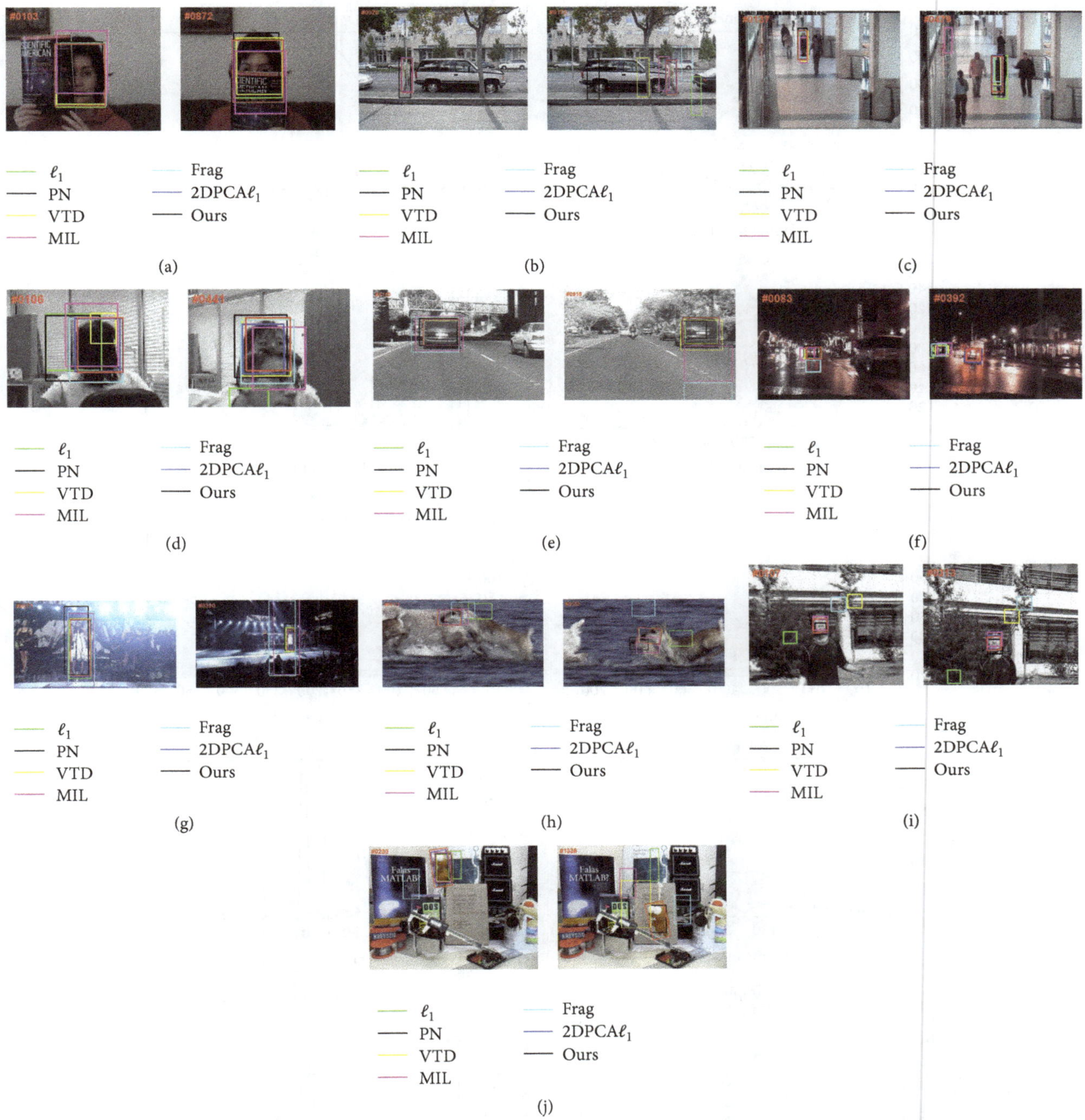

FIGURE 1: Sampled tracking results of partial evaluated algorithms on ten challenging sequences.

the animal appearance is almost indistinguishable due to the fast motion and most methods lost the target right at the beginning of the video. At frame 53, PN tracker locates the similar deer instead of the right object. From the results, we can see that the VTD tracker and our tracker perform better than the other algorithms. 2DPCAℓ_1 tracker may be able to track the target again by chance after failure. The appearance changes of the *Jumping* sequences are drastic such that the ℓ_1, Frag, and VTD tracker drift away from the object. Our tracker successfully keeps track of the object with small errors whereas the MIL, PN, and 2DPCAℓ_1 can track the target in some frames.

Background Clutter. Figure 1(j) illustrates the tracking results in the *Lemming* sequences with scale and pose change, as well as severe occlusion in cluttered background. Frag tracker lost the target object at the beginning of the sequence and when the target object moves quickly or rotates, the VTD tracker fails too. In contrast, the proposed method can adapt to the heavy occlusion, in-plane rotation, and scale change.

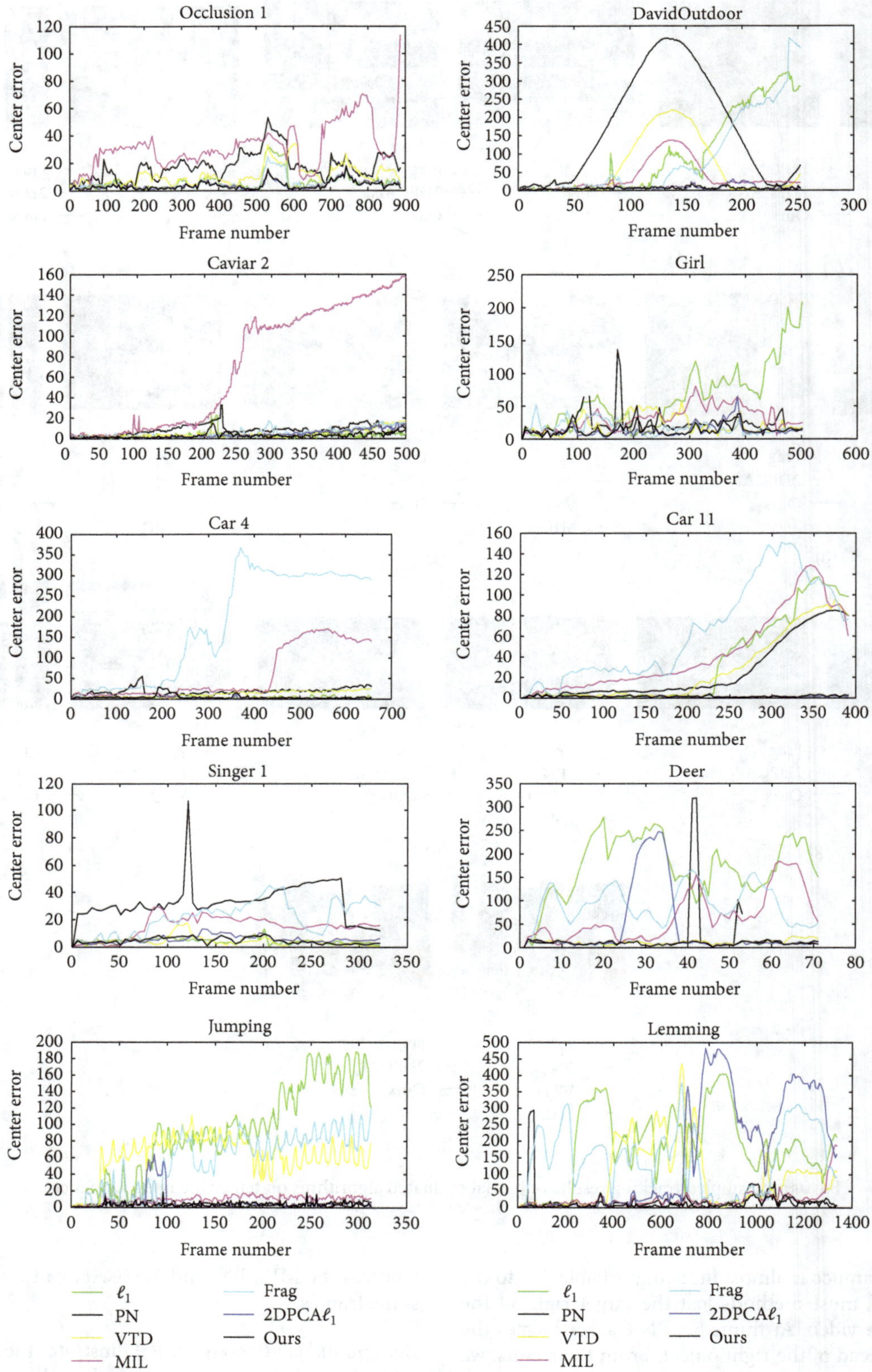

FIGURE 2: Center location error (in pixels) evaluation.

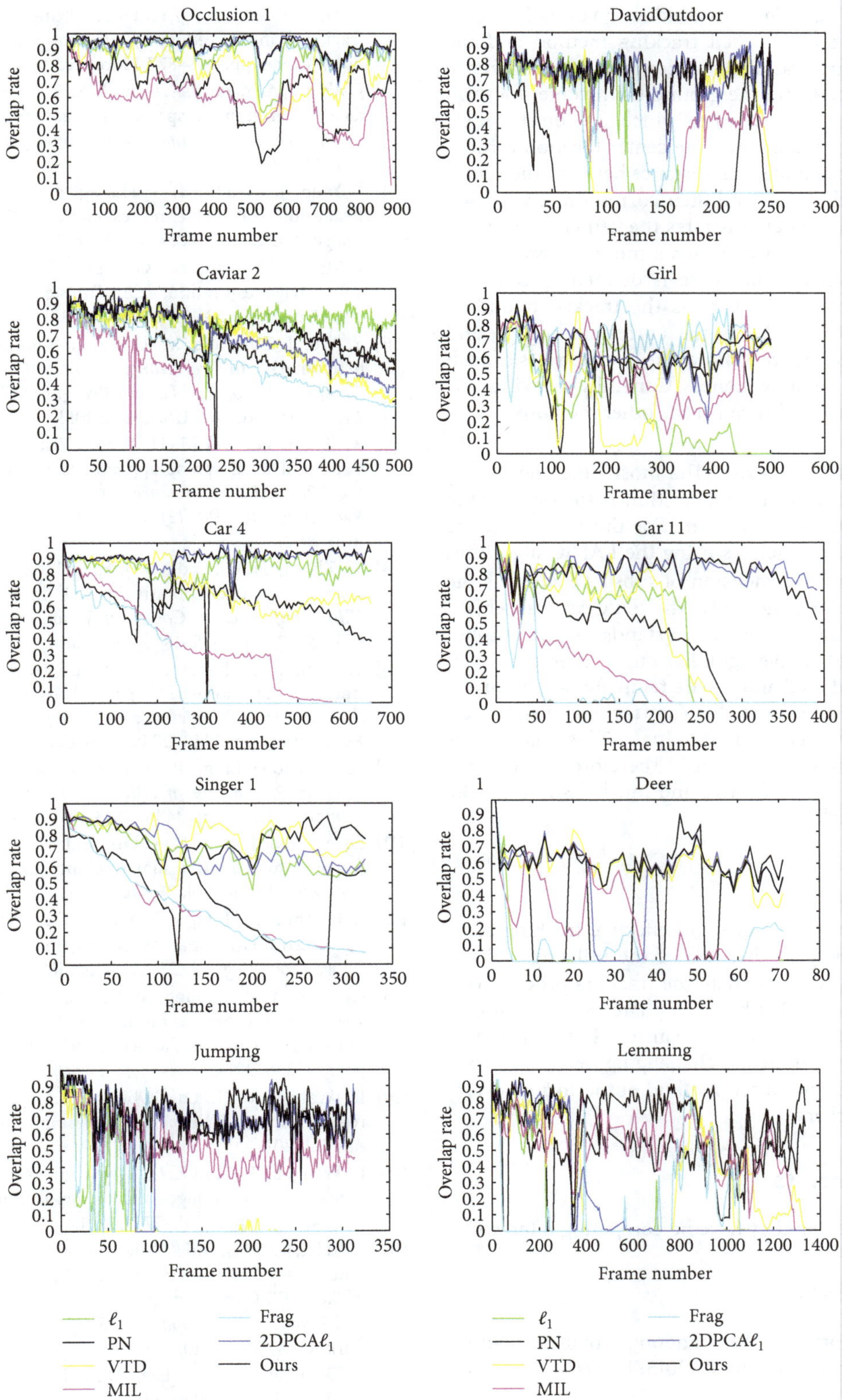

FIGURE 3: Overlap rate evaluation.

4.2. Quantitative Evaluation. To conduct quantitative comparisons between the proposed tracking method and the other sate-of-the-art trackers, we compute the difference between the predicted and the ground truth center location error in pixels and overlap rates which are most widely used in quantitative evaluation. The center location error is usually defined as the Euclidean distance between the center locations of tracked objects and their corresponding labeled ground truth. Figure 2 demonstrates the center error plots, where a smaller center error means a more accurate result in each frame. Overlap rate score is defined as score $=$ area$(R_t \cap R_g)$/area$(R_t \cup R_g)$. R_t is the tracked bounding box of each frame and R_g is the corresponding ground truth bounding box. Figure 3 shows the overlap rates of each tracking algorithm for all sequences. Generally speaking, our tracker performs favorably against the other methods.

4.3. Computational Complexity. The most time consuming part of the generative tracking algorithm is to compute the coefficients using the basis vectors. For the ℓ_1 tracker, the computation of the coefficients using the LASSO algorithm is $O(d^2 + dk)$. d is the dimension of subspace and k is the number of basis vectors. The load of the 2DPCAℓ_1 tracker [10] with ℓ_1 regularization is $O(mdk)$. m stands for the number of iterations (e.g., 10 on average). For our tracker, the trivial templates are abandoned and square templates are not used. So the load of our tracker is dk. The tracking speed of ℓ_1, 2DPCAℓ_1, and our method is 0.25 fps, 2.2 fps, and 5.2 fps separately (fps, frame per second). Therefore, we can get that our tracker is more effective and much faster than the aforementioned trackers.

5. Conclusion

In this paper, we present a fast and effective tracking algorithm. We first clarify the benefits of the utilizing 2DPCA basis vectors. Then, we formulate the tracking process with the ℓ_2-regularization. Finally, we update the appearance model accounting for the partial occlusion. Both qualitative and quantitative evaluations on challenging image sequence demonstrate that the proposed method outperforms several state-of-the-art trackers.

Competing Interests

The authors declare that they have no competing interests.

Acknowledgments

This project is supported by the Shandong Provincial Natural Science Foundation, China (no. ZR2015FL009).

References

[1] B. Babenko, M. H. Yang, and S. Belongie, "Visual tracking with online multiple instance learning," in *Proceedings of the 22th IEEE Conference on Computer Vision and Pattern Recognitionin*, pp. 983–990, San Francisco, Calif, USA, 2009.

[2] K. H. Zhang, L. Zhang, and M. H. Yang, "Real time compressive tracking," in *Proceedings of 12th European Conference on Computer Vision*, pp. 864–877, Florence, Italy, 2012.

[3] J. Wright, A. Y. Yang, A. Ganesh, S. S. Sastry, and Y. Ma, "Robust face recognition via sparse representation," *IEEE Transactions on Pattern Analysis and Machine Intelligence*, vol. 31, no. 2, pp. 210–227, 2009.

[4] J. Yang, J. Wright, T. S. Huang, and Y. Ma, "Image super-resolution via sparse representation," *IEEE Transactions on Image Processing*, vol. 19, no. 11, pp. 2861–2873, 2010.

[5] J. Mairal, M. Elad, and G. Sapiro, "Sparse representation for color image restoration," *IEEE Transactions on Image Processing*, vol. 17, no. 1, pp. 53–69, 2008.

[6] Y. Wu, J. Lim, and M.-H. Yang, "Online object tracking: a benchmark," in *Proceedings of the 26th IEEE Conference on Computer Vision and Pattern Recognition (CVPR '13)*, pp. 2411–2418, Portland, Ore, USA, June 2013.

[7] X. Jia, H. Lu, and M.-H. Yang, "Visual tracking via adaptive structural local sparse appearance model," in *Proceedings of the 2012 IEEE Conference on Computer Vision and Pattern Recognition (CVPR '12)*, pp. 1822–1829, Providence, RI, USA, June 2012.

[8] W. Zhong, H. Lu, and M.-H. Yang, "Robust object tracking via sparsity-based collaborative model," in *Proceedings of the 25th IEEE Conference on Computer Vision and Pattern Recognition (CVPR '12)*, pp. 1838–1845, Providence, RI, USA, June 2012.

[9] T. Zhang, B. Ghanem, S. Liu, and N. Ahuja, "Robust visual tracking via multi-task sparse learning," in *Proceedings of the 25th IEEE Conference on Computer Vision and Pattern Recognition*, pp. 2042–2049, Providence, RI, USA, 2012.

[10] X. Mei and H. Ling, "Robust visual tracking using ℓ_1 minimization," in *Proceedings of 12th IEEE International Conference on Computer Vision*, pp. 1436–1443, Kyoto, Japan, September 2009.

[11] D. A. Ross, J. Lim, R.-S. Lin, and M.-H. Yang, "Incremental learning for robust visual tracking," *International Journal of Computer Vision*, vol. 77, no. 1–3, pp. 125–141, 2008.

[12] B. H. Zhuang, H. Lu, Z. Y. Xiao, and D. Wang, "Visual tracking via discriminative sparse similarity map," *IEEE Transactions on Image Processing*, vol. 23, no. 4, pp. 1872–1881, 2014.

[13] C. Qian, Y. B. Zhuang, and Z. Z. Xu, "Visual tracking with structural appearance model based on extended incremental non-negative matrix factorization," *Neurocomputing*, vol. 136, pp. 327–336, 2014.

[14] D. Wang and H. Lu, "Object tracking via 2DPCA and *l*1-regularization," *IEEE Signal Processing Letters*, vol. 19, no. 11, pp. 711–714, 2012.

[15] D. Wang, H. Lu, and X. Li, "Two dimensional principal components of natural images and its application," *Neurocomputing*, vol. 74, no. 17, pp. 2745–2753, 2011.

[16] Z. Kalal, J. Matas, and K. Mikolajczyk, "P-N learning: bootstrapping binary classifiers by structural constraints," in *Proceedings of the IEEE Computer Society Conference on Computer Vision and Pattern Recognition (CVPR '10)*, pp. 49–56, San Francisco, Calif, USA, June 2010.

[17] J. Kwon and K. M. Lee, "Visual tracking decomposition," in *Proceedings of the IEEE Computer Society Conference on Computer Vision and Pattern Recognition (CVPR '10)*, pp. 1269–1276, IEEE, San Francisco, Calif, USA, June 2010.

[18] A. Adam, E. Rivlin, and I. Shimshoni, "Robust fragments-based tracking using the integral histogram," in *Proceedings of the 19th IEEE Conference on Computer Vision and Pattern Recognition*, pp. 798–805, New York, NY, USA, 2006.

17

Performance Estimation based Multicriteria Partitioning Approach for Dynamic Dataflow Programs

Małgorzata Michalska,[1] Nicolas Zufferey,[2] and Marco Mattavelli[1]

[1]*EPFL SCI-STI-MM, École Polytechnique Fédérale de Lausanne, 1015 Lausanne, Switzerland*
[2]*Geneva School of Economics and Management (GSEM), University of Geneva, 1211 Geneva 4, Switzerland*

Correspondence should be addressed to Małgorzata Michalska; malgorzata.michalska@epfl.ch

Academic Editor: Maxime Pelcat

The problem of partitioning a dataflow program onto a target architecture is a difficult challenge for any application design. In general, since the problem is NP-complete, it consists of looking for high quality solutions in terms of maximizing the achievable data throughput. The difficulty is given by the exploration of the design space which results in being extremely large for parallel platforms. The paper describes a heuristic partitioning methodology applicable to dynamic dataflow programs. The methodology is based on two elements: an execution model of the dynamic dataflow program which is used as estimation of the performance for the exploration of the large design space and several partitioning algorithms competing to lead to specific high quality solutions. Experimental results are validated with executions on a virtual platform.

1. Introduction

An interesting alternative to the classical sequential programming methods for signal processing system implementations is the approach based on dataflow programming. Dataflow programs are characterized by a high analyzability and platform independence and by providing an explicit exposition of the potential parallelism. For these reasons, they can be used for exploring a variety of parallel implementation options and to provide an extensive and systematic implementation analysis [1, 2]. Hence, they have been investigated in several research works [3–5].

Dataflow programs are in general structured as, possibly, hierarchical networks of communicating computational kernels, called actors. Actors are connected by directed, lossless, order preserving point-to-point communication channels, and data exchanges are only permitted by sending data packets (called tokens) over those channels. A general dataflow *Model of Computation* (MoC) is known in the literature as “Dataflow Process Network (DPN) with firings” [6]. A DPN network evolves as a sequence of discrete steps (called firings) corresponding to the executions of actions that may consume and/or produce a finite number of tokens and modify the internal actor state. At each step, according to the current actor internal state, only one action can be fired. The processing part of the actors is thus encapsulated in the atomic executions completely abstracting from time. Figure 1 illustrates the construction of a sample dataflow program and the underlying structure of an actor.

An important property of a dataflow program is the composability of its components (actors). Porting a program onto a target architecture requires determining three settings: the partitioning (the assignment of dataflow actors to the processing units), the scheduling (the execution order inside each unit, where, depending on the internal nature of actors, a static scheduling may or may not exist), and the dimensioning of buffers (a finite size for each communication channel in the network). An essential aspect of the problem is that although the configurations are applied at the level of actors, the program execution (i.e., in terms of data dependencies) is described at the level of action firings.

Depending on the architecture and on the dataflow program itself, the size of the solution space of admissible partitioning and scheduling configurations can be, in general, extremely large. Therefore, an important design challenge is to find a set of configurations that optimizes the desired

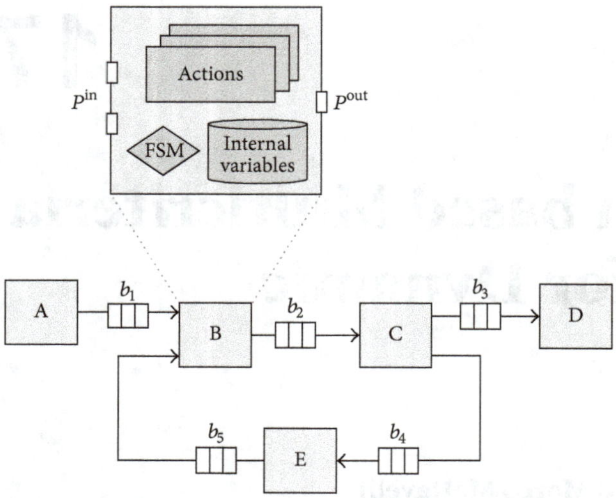

Figure 1: Construction of a sample dataflow program.

objective function. This problem has been proven to be NP-complete even for the case of only two processors [7]. Among different objective functions that might be considered, this work focuses on the maximization of the data throughput of a dataflow program. This particular objective function is often an appropriate choice for stream programs, since in many cases it contributes also to an optimization of other criteria, such as, for instance, resources utilization and energy consumption [8].

An efficient exploration of the multidimensional design space has two important applications. First, exploration of feasible regions leads to determining a close-to-optimal set of configurations according to the desired objective function. Second, it enables the identification of unreachable regions of the design space that could become reachable by applying refactorization stages to the considered design. For instance, a different implementation of an algorithm might be required for obtaining higher performances if its current exposed parallelism is lower than the potential parallelism offered by the processing platform.

Among different dimensions of the design space, more attention is usually paid to the partitioning, since its impact on the overall performance is considered to be dominant. Furthermore, the room for improvement left, for instance, to scheduling strongly depends on the quality of the applied partitioning (i.e., low-quality partitioning configurations cannot be improved much by applying efficient scheduling). This property has been confirmed also for other, nondataflow domains [9, 10]. The purpose of this work is to describe a new partitioning methodology to perform the design space exploration of dataflow program implementations.

The original contribution of this paper includes different stages of the partitioning process: starting from presenting and discussing different aspects of the problem, possible impact of a considered architecture, through the modeling approach, up to the description of several partitioning solution methods with different quality and complexity. An important property of the methodology is the construction of partitioning algorithms based on highly accurate models

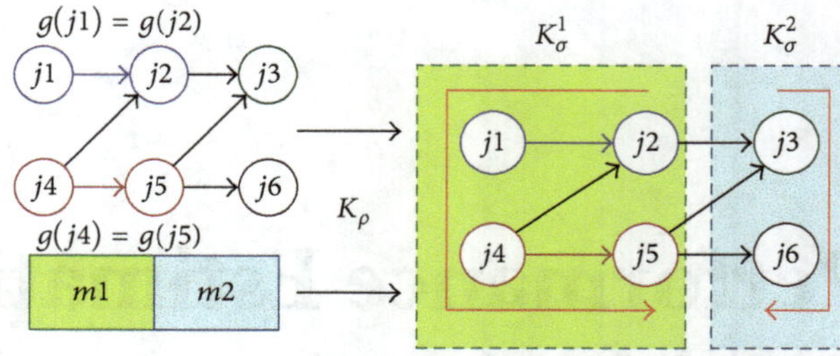

Figure 2: Partitioning and scheduling illustration.

of the program executions. The execution model is used in order to evaluate each solution and to extract some execution properties which are further considered as optimization criteria. The approach can be implemented with limited memory requirements (i.e., it can be implemented on a standard PC also for complex designs) and provides solutions according to quasi-constant evaluation times for different configurations considered. Unlike most of the existing approaches, the methodology can be used for analyzing static as well as fully dynamic dataflow programs.

The paper has the following structure. First, Section 2 presents and formulates the partitioning and scheduling problem explicitly for the case of dynamic dataflow programs. In this context, the related work is discussed in Section 3. Then, Section 4 describes the execution model used to derive the estimations and Section 5 reports in detail the proposed partitioning algorithms built upon this model. The algorithms are analyzed for complexity, quality of solution, and usability, based on the experimental results presented in Section 6. Finally, conclusions and future works are discussed in Section 7.

2. Dataflow Partitioning and Scheduling

The design problem considered here can be applied to any dataflow MoC.

2.1. Problem Presentation. Following the terminology commonly used in the production field [11], an objective is to find an assignment of n jobs (representing *action firings*) to a set of m parallel machines (corresponding to *processors*). The objective function to minimize is the overall makespan (i.e., the completion time of the last performed job among all processors). Assuming that the processors do not allow parallel execution, only one job can be performed at a time on each machine. Therefore, when all jobs are assigned to the machines, it must be decided in which order the jobs are going to be performed on each machine. The literature terminology often calls the first problem as mapping in the spatial domain (binding) and the second as mapping in the temporal domain (scheduling) [12]. In this work, partitioning (K_p) and scheduling (K_σ) refer to these problems, respectively, and are illustrated in Figure 2.

Each job j has an associated processing time (weight) p_j and a group (actor) g_j. There are k possible groups, and each one can be divided into subgroups where all jobs have the same processing time. This division can be easily identified with actions and their executions (firings). Between some pairs $\{j, j'\}$ of incompatible jobs (i.e., with $g_j \neq g_{j'}$) is associated a communication time $w_{jj'}$. The communication time is subject to a fixed quantity $q_{jj'}$ of information (or the number of tokens) that needs to be transferred. The size of this data is fixed for any subgroup (i.e., an action always produces/consumes the same amount of data). Due to the structure of dataflow programs, the partitioning and scheduling must take into account the following constraints:

(i) Group constraint: all jobs belonging to the same group have to be processed on the same machine (i.e., an actor must be entirely assigned to one processing unit). A fixed relative order is decided within each group. It can be assumed that this order is established based on the program's input data.

(ii) Precedence constraint: (j, j') means that a job j must be completed before job j' is allowed to start.

(iii) Setup constraint: it requires that, for each existing connection (j, j') involving jobs from different groups, a setup (or communication) time $w_{jj'}$ occurs.

(iv) Communication channel capacity constraint: the size of a communication channel (buffer) through which the information (tokens) is being transmitted is bounded by B. That is, the sum of $q_{jj'}$'s assigned to this buffer (q_b) cannot exceed B. If it occurs, it might affect the executability of j and j' and introduce serious delays in the overall makespan.

2.2. Architecture Impact. Executing the dataflow program on a given architecture requires introducing the notion of time to its originally time-abstracting definition. It is achieved by assigning a concrete value to the weight p_j of each job and to the communication time $w_{jj'}$. The types and ranges of values for p_j's and $w_{jj'}$'s fully depend on the properties of the targeted architecture. For the case of homogeneous platforms, p_j is constant no matter how a group (actor) is actually partitioned. It is not the case for heterogeneous platforms, where this value can vary according to the processor family that the processing unit belongs to (i.e., software or hardware).

Assuming that job j is assigned to the machine ρ, the actual value of communication time $w_{jj'}$ is a product of two elements: the number of tokens $q_{jj'}$ and the variable time $c_{jj'}(\rho_j, \rho_{j'})$ needed to transfer a single unit of information from ρ_j to $\rho_{j'}$. For two given jobs j and j', the largest $c_{jj'}$ can be significantly larger than the smallest $c_{jj'}$. In theory, every connection (j, j') can have as many different $c_{jj'}$'s as the number of different possible assignments to the machines. But in practice, this number can be usually reduced to few different values of latency, depending on the memory level serving a given communication (see Chapter 4 of [13]).

The case of heterogeneous platforms (typically hardware or software families of processors) may introduce some further constraints to the problem, such as the following:

(i) Eligibility constraint: not all jobs can be processed on each machine (i.e., each machine has a collection $d(\rho_i)$ of not supported operations).

(ii) Capacity constraint: the memory size is limited for a given family fl of processors. It involves that the memory requirements of all jobs assigned to processors of this family cannot exceed the given size.

Different figures of merit can be also introduced for communication time, especially for communication occurring between jobs partitioned to machines belonging to different families.

2.3. Problem Formulation. Following different aspects of the problem discussed in Sections 2.1 and 2.2, for each job it is required to find a partitioning configuration K_ρ and for each partition a scheduling configuration K_σ, so that the total execution time is minimized. The constraints include group, setup/communication, buffer capacity, eligibility, and capacity. These aspects can be summarized in the problem formulation provided below.

Decision Variables are as follows: $\forall_j K_\rho(j) = \rho_i$, $\forall_\rho K_\sigma = \{j, j', \ldots\}$.

Objective Function is $\min(T_{\text{end}}(j_{\text{last}}))$.

Constraints are as follows:

(i) $g_j = g_{j'} \Rightarrow K_\rho(j) = K_\rho(j')$,

(ii) $j \prec j' \Rightarrow T_{\text{start}}(j') \geq T_{\text{end}}(j)$,

(iii) $j_\sigma = j \Rightarrow \Sigma_{q_b} + \text{tokens}(j) \leq B$,

(iv) $K_\rho(j) = \rho_i \Rightarrow j \notin d(\rho_i)$,

(v) $\Sigma_{\rho \in \text{fl}} \Sigma_{g \in \rho} \Sigma_{j \in g} \text{mem}(j) \leq \text{size}(f)$.

3. Related Work

Due to the NP-completeness of the partitioning problem for realistic instances, it is only possible to develop methods providing close-to-optimal solutions [14]. They can be obtained by applying constructive heuristics, where a solution is generated from scratch by sequentially adding components to the current partial solution according to some criteria until the solution is complete [15]. Another possibility is metaheuristics, formally defined as an iterative generation process which guides a subordinate heuristic by combining intelligently different concepts for exploring and exploiting the search space [16]. Metaheuristics (e.g., simulated annealing, tabu search, variable neighborhood search, and guided local search) can usually lead to solutions of higher quality, but in general they require much longer computing times [17, 18].

There are several examples of the approaches based on metaheuristics used for partitioning or, more generally, for the design space exploration of dataflow programs. In [19], simulated annealing is employed for estimating the bounds of the partitioning program. Various optimization stages (including the selection of a target architecture, partitioning, scheduling, and designing space exploration) are applied in [20] in order to identify feasible solutions. The optimizations are performed using an evolutionary algorithm. Multiobjective evolutionary algorithms used for performing an automatic design space exploration are also an objective of work discussed in [21]. An interesting transition from simple heuristics to advanced metaheuristics (such as genetic algorithms) is also described in [22], where more advanced methods act as a refinement to the less advanced ones.

An important aspect of metaheuristic approaches for design space exploration, which requires explicitly providing an execution model, is performance estimation that can be used as an evaluation of a solution. The level of details considered in the model usually determines the accuracy of estimation as well as the evaluation time. The more detailed is the model, the more accurate estimation can be obtained, but employing a much longer time [23–25]. There are multiple frameworks which use the concept of performance estimation in order to build the optimization methods on them.

An interesting example, demonstrating some similarities with the methodology described in this paper, is the MPSoCs *Application Programming Studio* (MAPS) [26]. This design space exploration framework performs an estimation of a program execution, based on the analysis of the execution trace. Chapter 6 of [27] introduces a *trace replay module* employing a discrete-event simulator in order to simulate the scheduling of the segments of the trace. Several optimization heuristics are built on the estimation, with an emphasis on the advantages of light-weight heuristics over evolutionary methods, and the composability analysis for the purpose of executing simultaneously multiple applications on a given platform without interference [4]. In terms of modeling differences, the framework addresses only KPN dataflow programs and supports a limited set of dependencies. Another framework, supporting only the KPN programs, is *Sesame system-level simulation framework* [28]. Similar to the work presented in this paper, it introduces a concept of an application model which is independent from the architecture model and the partitioning configuration. The objective of the framework is to find the most suitable and efficient target MPSoC platform for a given program. Hence, the exploration process involves simulating and analyzing the candidate architectures.

Unlike the two approaches mentioned earlier, *Metropolis* is a framework that abandons the imposition of a specific language or flow model to the design [29]. It introduces a design description at different levels of abstraction and provides an infrastructure based on meta-modeling that remains generic enough to support existing MoCs, whereas it can also accommodate new MoCs. The meta-model allows capturing the architecture, the functionality, and the mapping between different abstraction levels. The simulation and formal analysis allow the user to determine how well an implementation satisfies the specified requirements. With its generic features, the framework responds only partially to the demands of DPN, because it does not support a simultaneous analysis at the level of partitioning, possibly dynamic scheduling and buffer dimensioning.

In the mentioned examples, the partitioning methodologies have been designed explicitly for the purpose of dataflow partitioning, which is a specific instance of a graph partitioning problem. The research field of graph partitioning is thoroughly covered by different algorithms proposed in the literature [30] as well as some software packages, such as METIS [31] or SCOTCH [32]. Such general purpose partitioning algorithms cannot be, however, easily applied for the case of dataflow programs, since they are not aware of the semantics related to the nodes and edges of a dataflow graph. An attempt of applying the METIS, mentioned earlier, for the purpose of run-time actor mapping of dataflow programs has been made in [33]. This approach explores the results of profiling and extracts some optimization criteria (connectivity between actors, critical path). It is, however, difficult to evaluate the obtained solutions in terms of being close-to-optimal, or point to possible optimizations in the design, since no execution model is provided. The considered partitioning graph is the program network itself and not the execution trace which can provide elements and measures of the execution properties of the dataflow program. Furthermore, such combinatorial approach, which might operate quite effectively for small instances of the problem, cannot be successfully applied to the exploration of design problems of larger size.

Summarizing, a comparison with existing approaches puts in evidence that there are some important differences that should be emphasized. First of all, the approach can be applied to all dataflow models of execution including fully dynamic dataflow programs [34]. Hence, it includes also other, less expressive dataflow variants such as, for instance, SDF and CSDF [35]. Furthermore, due to the demands of the dynamic programs, such as dynamic scheduling influenced by the applied buffer configuration, the methodology can be used directly for optimization of other design configurations, without imposing any particular order of optimizations. Finally, the long-term objective of the methodology is to find a close-to-optimal (close-to-the-potential-parallelism) design configuration on a given platform, in order to assess the maximum performance of a dataflow program, identify its bottlenecks, and hence identify new regions of the design space to be reached after applying some refactorization to the design.

4. Dataflow Program Execution Modeling

A general dataflow MoC, DPN with firings, is considered. It allows implementation of fully dynamic programs efficiently capturing all classes of signal processing applications (e.g., video and audio codecs, packet switching in communication networks). For this reason, a representation of a dataflow program must capture its entire dynamic behavior. Such a representation can be built by generating a directed, acyclic graph G, called *Execution Trace Graph* (ETG), where nodes of the graph represent a collection of action executions, called *firings*. Such representation has been already extensively studied to solve some optimization problems of dynamic dataflow implementations [36].

It is possible to explicitly characterize various intrinsic dependencies between two firings, which are used to describe the program execution. They can be grouped into two types. The first type includes *internal dependencies* related to FSM, internal variables, and ports (p^{in}/p^{out} in Figure 1) and describes the relations between two *firings* of the same actor. The second type (*token dependencies*) describes the relations between two *firings* of different actors that, respectively, produce and consume at least one token. Defining dependencies between executed firings establishes precedence orders: if firing f_2 depends on firing f_1, then f_1 has to be executed and completed (including the communication, if applicable) before f_2 can be started.

An ETG has to consider a sufficiently large and statistically meaningful set of input stimuli in order to cover the whole span of dynamic behavior of the application considered. The size of G varies according to the type of application and to the size of the input stimuli set. In fact, for some applications, it can be large. For instance, a very complex algorithm, such as HEVC video compression operating on an input signal of a few tenths of frames, may result in G containing a few billions of nodes. Still, it can be processed in reasonable time using standard PC platforms, as demonstrated by the examples described in Chapter 9 of [36].

Considering the program execution on a given architecture, the ETG is weighted at the nodes and at the edges, and the weights correspond to the values of p_j and $w_{jj'}$, respectively. These values are normally obtained by profiling the program on the target architecture. A weighted ETG allows simulating the program execution for a given partitioning, scheduling, and buffer dimensioning configuration. This task is accomplished by an event-driven performance estimation tool that processes the firings following the constraints and properties, as presented in Section 2.

5. Partitioning Algorithms

This section presents the algorithmic details of different partitioning solution methods, namely, greedy constructive procedures, the decent local search heuristics, and the tabu search metaheuristic. Each method takes the weighted ETG as input data.

5.1. Greedy Constructive Procedures. In order to construct a solution, the greedy procedures require specifying only the target number of processors. The solution generation succeeds in a negligible time frame, since no performance estimation needs to be performed.

5.1.1. Workload Balance (WB). The concept of balancing the workload in order to minimize the bottleneck of the program and hence maximize the throughput has been already successfully employed for partitioning purposes of systems of different types [37]. Inspired by such approaches, the very first constructive heuristic has been designed. The algorithm starts from calculating the total workload of each group throughout the program execution. It is expressed as the sum of the p_j's for all jobs (firings) belonging to one group (actor) g. The actors are then sorted decreasingly by the sum of weights (workload) $\mathrm{wl}(g) = \sum_{j\in g} p_j$. The partitioning decision is based on the sum of workloads of actors partitioned already in one processor ρ: $\mathrm{wl}(\rho) = \sum_{g\in\rho} \mathrm{wl}(g)$. The next actor on the list is always partitioned on the processor with the smallest sum of workloads $\mathrm{wl}(\rho)$. In this way, a balance of overall workload of each partition should be achieved and the bottleneck (understood as the workload of the most occupied processor) is likely to be minimized.

5.1.2. Balanced Pipeline (BP). The algorithm starts from giving each actor a dedicated processor. Next, the processors are being iteratively reduced and the members of the least occupied processors are attached to the remaining processors. The optimization criteria of the algorithm involve equalizing the average preceding workload (understood as the maximal sum of weights of each firing of each actor that precedes the given actor in the network in terms of topological order) and maximizing the number of common predecessors (understood as the number of actors appearing on the topological list of predecessors for a given pair of actors) within each partition. The simulation of the execution time might be incorporated to the algorithm in order to estimate the optimal number of processors from the perspective of processors utilization. The *Balanced Pipeline* algorithm, extended with some additional optimization procedures for communication volume and idle time, has been shown to outperform existing partitioning algorithms for dataflow programs. For a more detailed description of the algorithm, its metrics, and its results, the reader is referred to [38].

5.2. Decent Local Search Heuristics (DLS). As described in [39], a local search starts from an initial solution and then explores the solution space by moving from the current solution to a neighbor solution. A *neighbor* solution is usually obtained by making a slight modification of the current solution, called a *move*. The *neighborhood* $N(s)$ of a solution s is the set of solutions obtained from s by performing each possible move. In a *descent local search* (DLS), the best solution (according to the considered objective function f) of $s' \in N(s)$ is generated at each iteration. The main drawback of this method is that it stops in the first local optimum. Two DLS approaches are proposed below: the *Idle DLS* and the *Communication Frequency DLS*.

5.2.1. Idle DLS (IDLS). Representing the program execution with the ETG, and simulating it for a given partitioning, scheduling, and buffer dimensioning configuration using the performance estimation tool, may provide important information related to actor states throughout the execution. The following states may occur for an actor that is currently not processing and has not yet terminated.

(i) *Blocked reading* considers the situation where an actor has not yet received the required input tokens and therefore cannot be executed.

(ii) *Blocked writing* takes into account the situation where the buffer an actor is expecting to write to is full, so it has to wait for the available space.

(iii) *Idle* corresponds to the situation where although an actor has the necessary tokens and required space in the buffers, it cannot be fired because another actor is currently processing in the same processor (as previously mentioned, only one job can be executed on each processor at a time).

Assuming a design space exploration in the dimension of partitioning and scheduling, it is particularly important to minimize the occurrences of the idle state. In order to achieve that, in IDLS, all actors are sorted according to their idle times in decreasing order (*idle time list*). A newly created solution s is generated by moving a single actor to the most idle partition, where the *idleness* of a partition is defined as the overall time during the execution when none of its actors could be executed due to being blocked reading/writing or terminating already. In each iteration, the possible moves are prioritized according to the position of the considered actor on the *idle time list*. A move is evaluated by estimating the makespan of the new solution. For the case of a successful move, the statistics on the idle times of the actors and the corresponding *idle time list* are being regenerated. Since the moves are prioritized, there is a risk that if there is a move with a high priority that does not improve the solution, it will be unnecessarily repeated in each iteration. To prevent that from happening, a simple *release* mechanism is implemented: a (once unsuccessful) move, understood as an actor-partition pair, may be repeated only if the content of the target partition has been modified by applying another move.

5.2.2. Communication Frequency DLS (CFDLS). Another information that can be extracted from the ETG is the number of token dependencies between the firings of different actors. Accumulating these numbers for all firings leads to creating an actor-actor communication frequency map. This map is independent from the partitioning configuration, but taking the partitioning into consideration, it can be easily transformed into an actor-partition map. Indeed, this map is taken as an optimization criterion by another local search. For each actor, the algorithm calculates the internal communication frequency (token exchange with actors partitioned to the same processor) and external communication frequency (token exchange with actors partitioned to different processors). As described previously in Section 2.2, the partitioning of actors may strongly influence the values of communication cost and therefore the makespan.

If for any actor the external communication frequency with one processor exceeds the internal communication frequency, this actor-partition pair is considered as a move. The moves are prioritized according to the overall communication frequency of the actors and a *release* mechanism (similarly to IDLS) is implemented. The move can be evaluated in two ways: by estimating the execution time of a new solution or by analyzing if the overall external communication frequency (calculated collectively for all partitions) has decreased. In this work, for the purpose of consistency with other solution methods, only the first method is used.

5.3. Tabu Search (TS). Tabu search, as introduced by Glover [40], is still among the most cited and used local search metaheuristics for combinatorial optimization problems. It avoids the problem of getting stuck in the first local optimum by making use of recent memory with a *tabu list*. More precisely, it forbids performing the reverse of the moves done during the last *tab* (parameter) iterations, where *tab* is called *tabu tenure*. At each iteration of TS, the neighbor solution s' is obtained from the current solution s by performing on the latter the best nontabu move (ties are broken randomly). The process stops, for instance, when a time limit T (parameter) is reached. In most TS implementations, if the neighborhood size is too big, only a proportion is explored in each iteration. This proportion can be, for instance, a random sample involving e% (parameter) of the neighbor solutions.

TS has proven to have a good balance between intensification (i.e., the capability to focus on specific regions of the solution space) and diversification (i.e., the ability to visit diverse regions of the solution space). In addition, it has a good overall behavior according to the following measures [18]: (1) quality of the obtained results (according to a given objective function f that has to be optimized); (2) quickness (time needed to get competitive results); (3) robustness (sensitivity to variations in data characteristics); (4) simplicity (facility of adaptation); and (5) flexibility (possibility to integrate properties of the considered problem). To adapt TS to the studied problem, the following elements have to be designed: the representation of any solution s, the neighborhood structure (i.e., what is a move), the tabu list structure (i.e., what type of information is forbidden), and a stopping criterion (i.e., what is the most appropriate time limit).

5.3.1. Solution Encoding and Neighborhood Structure. A solution for partitioning is represented as a map of actors and processors, where the number of processors is fixed. Each actor can be mapped to only one processor at the time, and each processor must be mapped to at least one actor. Hence, leaving empty processors is not allowed. The following basic types of moves are possible: (1) REINSERT: moving an actor to another processor; (2) SWAP: two actors belonging to two different processors. For the purpose of swapping, the term *complementary move* is introduced. Assume that a move $m(j, \rho_i, \rho_{i'})$ consists of relocating an actor j from source partition ρ_i to target partition $\rho_{i'}$. A move $m(j', \rho_{i'}, \rho_i)$ is complementary to $m(j, \rho_i, \rho_{i'})$ if it involves moving any actor j' from source partition $\rho_{i'}$ to target partition ρ_i. In this work, the neighborhood structures are generated by performing REINSERT and SWAP moves according to four different criteria, presented below:

(1) $N^{(B)}$ (for balancing):

- (i) REINSERT: choose randomly an actor from the most occupied processor and move it to the least occupied processor.
- (ii) SWAP: choose randomly two actors in different partitions so that swapping the actors decreases the relative workload imbalance between the two partitions.

(2) $N^{(I)}$ (for idle):

- (i) REINSERT: for each actor which has a bigger idle time is than its processing time, find the most idle processor, different from the one currently mapped, where the definition of *idle* is as described in Section 5.2.1.

(ii) SWAP: generate a set of moves on the REINSERT basis, but allow actors to be moved to *any* partition except for the least idle one (search for complementary pairs of moves).

(3) $N^{(CF)}$ (for communication frequency):

(i) REINSERT: check the internal and external communication frequency of each actor and consider the moves, as described in Section 5.2.2.

(ii) SWAP: generate a set of moves on the REINSERT basis (search for complementary pairs of moves).

(4) $N^{(R)}$ (for random):

(i) REINSERT: choose randomly an actor and move it to a different processor (randomly chosen).

(ii) SWAP: generate a set of moves on the REINSERT basis (search for complementary pairs of moves).

5.3.2. Parameters. Any time an actor j is moved from a processor ρ to another processor, it is forbidden to put j back to ρ for *tab* iterations, where *tab* is an integer uniformly generated in interval $[a, b]$, and the values of parameters a and b were tuned to 5 and 15, based on the preliminary experiments. Smaller values do not allow escaping from local optima, whereas larger values do not allow intensifying the search around promising solutions. There are two other sensitive parameters that have to be tuned for TS, namely, e (the proportion of neighbor solutions explored during each iteration) and T (the time limit). Reaching the time limit T results in immediate termination of the search and returning of the best solution ever found. Usually, T is set so that the improvement potential is poor (i.e., the percentage of improvement is below a threshold during a predefined time interval) if the method is run for larger time limits. Next, the smaller is e, the more iterations are performed but the fewer neighbors are investigated in each iteration. A large value of e contributes to the intensification ability of the method (indeed, all the solutions around the current one are explored), whereas a small value plays a diversification role (indeed, no focus is put on the neighborhood of each solution). Finally, a small (resp., large) value of *tab* strengthens the intensification (resp., diversification) ability of the search.

5.3.3. Advanced Variants of Tabu Search. Since each of the used neighborhood structures relies on different properties, a more advanced version of the TS involves a consolidation of all neighborhood structures. It is applied in two different variants:

(i) *Joint Tabu Search* (JTS): at each iteration, the neighborhood structure includes moves obtained according to all types. Therefore, the used neighborhood structure is $N^{(J)} = N^{(B)} \cup N^{(I)} \cup N^{(CF)} \cup N^{(R)}$. This variant should have more flexibility, because it involves various types of moves. The proportion of the set sizes for different types of moves can be freely tuned.

(ii) *Probabilistic Tabu Search* (PTS): at each iteration, the search assigns a probability to the selection of each neighborhood of the set $\{N^{(B)}, N^{(I)}, N^{(CF)}, N^{(R)}\}$. This probability is tuned based on the history of the search during the considered run. As a result, the search is guided by the success rate of each type of move (where a success corresponds to an improvement of the current solution).

6. Experimental Results

6.1. Experimental Setup. The focus of this work is to explore the design space along the partitioning dimension. Hence, the other dimensions need to be fixed. The dynamic scheduling policy used in the experiments is *nonpreemptive*, which involves that the same actor is continuously scheduled, as long as its firing conditions (i.e., available input tokens and output space) are fulfilled. The nonsatisfaction of any of the conditions results in choosing the next actor (on the round-robin basis) among all actors with satisfied input/output conditions. Regarding the buffers, an infinite size would be, ideally, considered. Since it is not possible for the practical implementations, the buffer size must be sufficiently large, so that deadlocks are avoided and the overall *blocked writing* time of the actors, as defined in Section 5.2.1, is kept at a minimum level. In this way the influence of a buffer size which is too small remains negligible for the overall performance. For the dataflow programs considered here it has been sufficient to set all buffers to the size of 512 tokens.

6.1.1. Experimental Designs. The quality of a partitioning algorithm is, in principle, assessed by the quality of a solution it provides. For each solution it can be evaluated how close it is to the objective function and how it approaches the potential parallelism of a given dataflow program implementation, for instance, how far is a given throughput from the best throughput achievable by the dataflow program under consideration. Hence, it is essential to perform the experiments with an application that in principle can provide a sufficient level of parallelism. If this condition is not satisfied, it is likely that either none or *any* partitioning algorithm can provide any satisfactory performance. For this reason, finding appropriate dataflow programs for validating partitioning algorithms is not a trivial task. The application used in the experiments is an MPEG-4 SP decoder design that consists of 41 actors in total and provides the upper bound on the potential parallelism around 6.28. This upper bound on the potential parallelism is evaluated as a proportion of the critical executions in the overall execution time, assuming a full parallel execution and the unbounded buffer size configuration (as described in Chapter 5 of [36]).

MPEG-4 SP decoder network is an implementation of a full MPEG-4 4 : 2 : 0 Simple Profile decoder standard written in CAL Actor Language [41]. The main functional blocks include a parser, a reconstruction block, a 2D inverse discrete

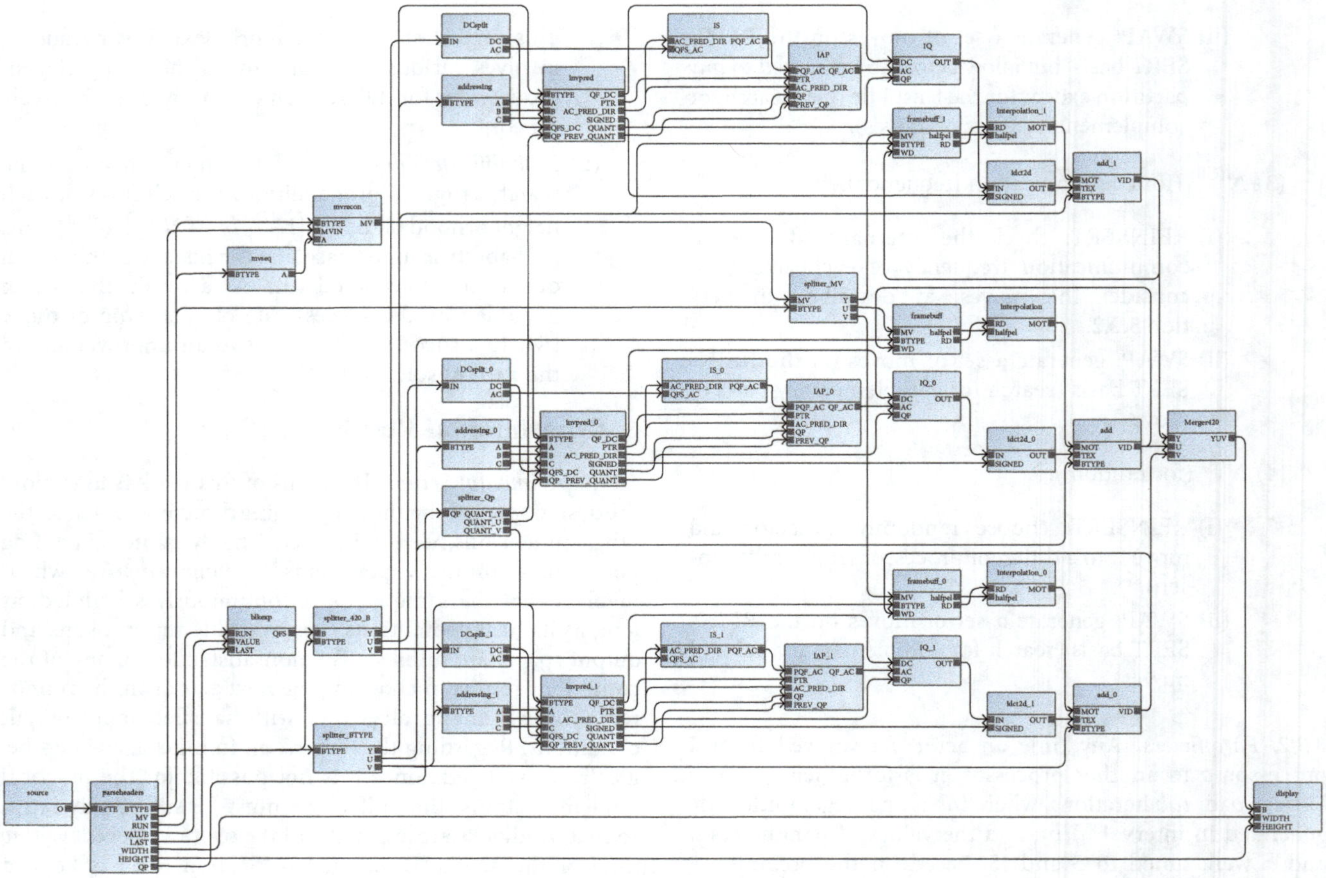

FIGURE 3: MPEG-4 SP decoder network.

cosine transform (IDCT) block, and a motion compensator. These functional units are hierarchical compositions of actors in themselves. The decoding starts from the parser (the most complicated actor in the network consisting of 71 actions) which extracts data from the incoming bitstream, through reconstruction blocks exploiting the correlation of pixels up to the motion compensator performing a selective adding of blocks. In order to give an overview of the complexity of the design, a schematic representation of the network is presented in Figure 3.

6.1.2. Target Platform. The platform used for the experiments is an array of *Transport Triggered Architecture* (TTA) processors [42]. There are several features of this platform that allow an easy modeling and a precise simulation for the purpose of partitioning and scheduling design space exploration. The most significant one is a cacheless, negligible interprocessor communication, and therefore an independence of the profiling results from the mapping configuration. In fact, as confirmed by our previous experiments on the platform, the results of a single profiling remain valid for multiple partitioning and scheduling configurations [43]. Hence, such a platform is a good choice for an initial validation of the effectiveness of the methodology.

6.1.3. Tuning of Parameters. As described in Section 5.3, apart from the length of the *tabu tenure*, TS is sensitive to two parameters that need to be properly tuned: the time limit T and the percentage e of explored neighbor solutions. For that purpose, 3 runs on a set of initial partitioning configurations have been performed for each: $N^{(B)}$, $N^{(I)}$, $N^{(\mathrm{CF})}$, and $N^{(R)}$. First, with a fixed value of $e = 0.5$, each TS variant was performed 3 times on the initial set of partitioning configurations. For each run, TS was stopped any time 5 minutes have elapsed without improving the best encountered solution (during the current run) by at least 1%. Parameter T has been set as the largest encountered stopping time (minus 5 minutes) among all these experiments.

Next, with the selected value of T, all TS variants were tested with different values of $e \in 0.2, 0.4, 0.6, 0.8$ in order to deduce the best value for each neighborhood type. The value of $e = 0.4$ has been chosen as the one providing the best average results among all instances in the test set. It was also observed that if a method is performed several times on the same instance, it gets similar results. This indicates the robustness of the proposed approach.

Proper tuning of parameters is important in order to reliably compare all of the iterative methods. All stages of parameters tuning in the proposed methodology have been performed automatically. The time limit T tuned for TS has been used also as a time limit for the DLS methods. Additionally, since these methods tend to get quickly stuck in local optima, a *restarting* procedure has been implemented.

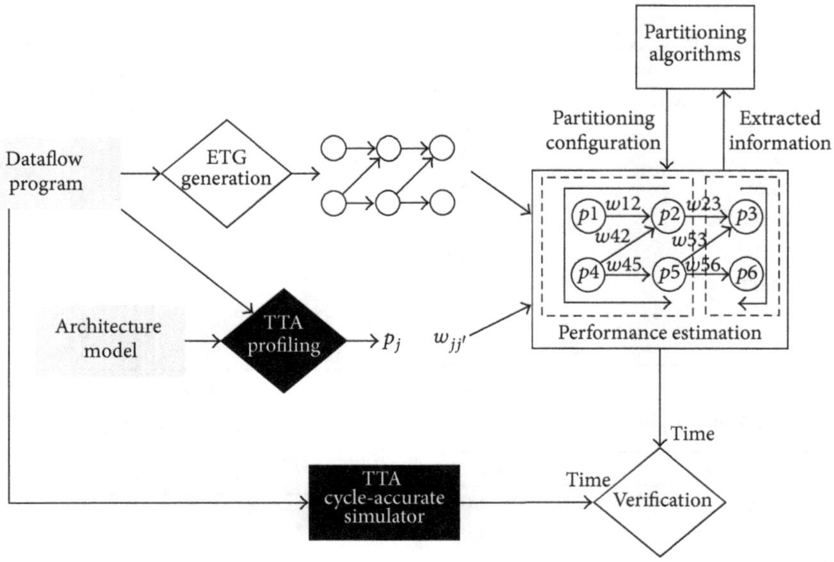

FIGURE 4: Experimental workflow.

If DLS finishes before elapsing the given time limit, it is restarted with a new random solution. At the end, the best found solution (among the restarts) is returned.

6.2. Methodology of Experiments. Most of the tools used for the experiments are the components of Turnus codesign framework [44]. They include the generation of an ETG for a given statistically meaningful input stimulus, the performance estimation tool exploiting the ETG, and the results of the platform profiling and the generation of partitioning configurations using different algorithms. Complementary units in this workflow are the profiling of the TTA architecture and a TTA cycle-accurate simulator [45] that allows a verification of the estimated results in terms of a real execution time obtained on the platform. The complete workflow is presented in Figure 4.

The partitioning configurations were generated using each of the described algorithm for the number of processors between 2 and 8. Considering the choice of application, 8 units should already approach their potential parallelism. For the local search methods that require specifying an initial solution, in each case, two sets were tested: the random one and the one generated by the WB algorithm. Such a choice was made in order to provide the algorithms with possibly good, as well as bad, initial configurations and also observe their sensitivity to the quality of an initial solution. The evaluation of the solutions generated by each algorithm was accomplished by the performance estimation tool that calculated the total execution time expressed in clock-cycles. Based on those values, the speed-up versus the mono-core execution was calculated in each case. The results presented in this section target the calculated values of speed-up in order to relate them easily to the potential parallelism of the application. Finally, the results obtained by estimation were also consistently verified by the platform execution.

6.3. Greedy and Decent Local Search Heuristics. Table 1 contains the speed-up values of partitioning configurations generated with the WB and BP algorithms along with the values estimated for a random set of configurations. Tables 2 and 3 contain the results obtained for the IDLS and CFDLS heuristics for the two sets of initial partitioning configurations.

TABLE 1: Speed-up: greedy constructive procedures.

Proc.	Workload balance	Balanced pipeline	Random
1	1.00	1.00	1.00
2	1.83	1.79	1.75
3	2.66	2.20	2.08
4	3.07	2.84	2.77
5	3.38	3.10	2.34
6	4.07	3.09	2.61
7	5.40	3.10	2.48
8	5.76	3.10	3.14

Since the purpose of a greedy constructive method is to build a solution from scratch, an important property is the scalability of the performance. In this case, both algorithms scale; however, the BP achieves a saturation already around 5 processors, unlike the WB that scales further. The maximal speed-up obtained for BP configurations is similar to the random configurations but is achieved on a smaller number of processors (5 versus 8). Applying the IDLS and CFDLS methods in all the cases improve the initial solution, but the improvement is bigger for CFDLS. The quality of the solution provided by the DLS heuristics depends also strongly on the quality of the solution provided as a starting configuration.

6.4. Tabu Search. The first experiment aimed at confirming the most beneficial types of moves. It was performed separately for each type of neighborhood structure. In the first execution, only the REINSERT moves were allowed,

TABLE 2: DLS speed-up: balanced start.

Proc.	IDLS	CFDLS
1	1.00	1.00
2	1.83	1.94
3	2.67	2.68
4	3.25	3.49
5	3.48	4.20
6	4.08	4.75
7	5.47	5.63
8	5.93	6.06

TABLE 3: DLS speed-up: random start.

Proc.	IDLS	CFDLS
1	1.00	1.00
2	1.77	1.95
3	2.09	2.32
4	2.78	3.19
5	2.35	4.27
6	2.95	3.98
7	3.02	3.93
8	3.85	4.23

whereas in the second one, the SWAP moves were also included. SWAP moves were not considered alone, since they do not lead to any change of the initial size of each partition (resulting in a nonconnected solution space). The results of this comparison for each neighborhood type are presented in Tables 4–11. Along with admitting the SWAP moves, a significant improvement has been brought only to $N^{(B)}$. In fact, the performance of $N^{(B)}$ based on REINSERT only was very poor and a slight improvement has been introduced only for certain initial configurations. It relies on the fact that the possible space of moves is very narrowed in this case (only actors from the most occupied partition are taken into consideration) and the tabu list can be very restrictive. Since it also aims at balancing the workload, for higher number of processors, it is not rare to encounter a solution where the heaviest bottleneck actor is placed alone on the processor. Due to the solution definition described in Section 5.3.1, the algorithm cannot proceed from that point. Relative balancing of the workload between two partitions instead of an overall balancing seems to be a much more effective approach.

For the other neighborhood structures, allowing SWAP moves decreased the quality of the final solution in the vast majority of the cases. It might be due to the fact that SWAP moves unnecessarily increased the set of neighbor solutions and reduce the diversification ability of the method. Comparing the neighborhood structures, there are some conclusions that can be made. First, $N^{(I)}$ seems to outperform the other variants, including $N^{(CF)}$. This observation is contrary to what has been previously observed for the DLS heuristics, where the search based on the *communication frequency* outperformed the *idle* optimization. It confirms that determining a local optimization criterion is challenging,

TABLE 4: Speed-up: $N^{(B)}$ with balanced start.

Proc.	REINSERT	SWAP
1	1.00	1.00
2	1.85	1.94
3	2.67	2.77
4	3.21	3.48
5	4.04	4.32
6	4.83	4.85
7	5.40	5.62
8	5.76	6.19

TABLE 5: Speed-up: $N^{(B)}$ with random start.

Proc.	REINSERT	SWAP
1	1.00	1.00
2	1.75	1.95
3	2.08	2.74
4	2.77	3.30
5	2.34	4.13
6	2.61	3.68
7	2.48	4.04
8	3.24	4.76

TABLE 6: Speed-up: $N^{(I)}$ with balanced start.

Proc.	REINSERT	SWAP
1	1.00	1.00
2	1.92	1.92
3	2.75	2.80
4	3.61	3.46
5	4.45	4.23
6	4.92	4.88
7	5.81	5.66
8	6.28	6.18

TABLE 7: Speed-up: $N^{(I)}$ with random start.

Proc.	REINSERT	SWAP
1	1.00	1.00
2	1.93	1.94
3	2.66	2.66
4	3.36	3.27
5	3.94	3.59
6	4.73	4.12
7	4.31	4.65
8	5.95	4.65

whereas employing an appropriate exploration strategy (i.e., the TS framework) is the other one. Finally, the results obtained for $N^{(I)}$ and $N^{(CF)}$ also prove that a *guided* choice of moves outperforms a random selection. In other words, the complete freedom of choice when choosing a move, as for $N^{(R)}$, does not necessarily lead to competitive solutions.

TABLE 8: Speed-up: $N^{(CF)}$ with balanced start.

Proc.	REINSERT	SWAP
1	1.00	1.00
2	1.88	1.93
3	2.79	2.78
4	3.49	3.56
5	4.30	4.29
6	4.95	4.97
7	5.79	5.74
8	6.26	6.18

TABLE 9: Speed-up: $N^{(CF)}$ with random start.

Proc.	REINSERT	SWAP
1	1.00	1.00
2	1.87	1.88
3	2.65	2.68
4	3.33	3.29
5	4.23	4.36
6	4.94	3.85
7	4.62	4.36
8	4.66	4.23

TABLE 10: Speed-up: $N^{(R)}$ with balanced start.

Proc.	REINSERT	SWAP
1	1.00	1.00
2	1.92	1.94
3	2.70	2.74
4	3.47	3.38
5	4.26	4.11
6	4.91	4.76
7	5.64	5.57
8	6.27	6.18

TABLE 11: Speed-up: $N^{(R)}$ with random start.

Proc.	REINSERT	SWAP
1	1.00	1.00
2	1.94	1.90
3	2.65	2.29
4	3.30	3.16
5	4.01	3.20
6	3.65	3.53
7	4.02	3.47
8	4.56	4.36

The final part of experiments with TS involves a comparison of its two advanced variants. Taking into consideration the previous observations, the analysis involved neighborhood $N^{(B)}$ based on the SWAP moves and neighborhoods $N^{(I)}$, $N^{(CF)}$, and $N^{(R)}$ based on REINSERT moves. Since different types provide different sizes of the neighborhood

TABLE 12: PTS and JTS speed-up: balanced start.

Proc.	PTS	JTS
1	1.00	1.00
2	1.93	1.96
3	2.83	2.80
4	3.58	3.57
5	4.44	4.37
6	5.13	5.03
7	5.81	5.72
8	6.22	6.22

TABLE 13: PTS and JTS speed-up: random start.

Proc.	PTS	JTS
1	1.00	1.00
2	1.96	1.93
3	2.77	2.71
4	3.62	3.39
5	4.42	4.26
6	4.98	4.31
7	5.10	4.86
8	5.72	5.12

sets, such sizes have been equalized according to the averaged values. For this reason, another parameter, namely, the *admission rate*, has been introduced for each neighborhood structure. Admission rate expresses the percentage of moves that is generated at each iteration. For $N^{(I)}$ and $N^{(CF)}$, a given percentage of moves is extracted according to the priorities (i.e., most idle or most communicative actors, resp.). For $N^{(B)}$ and $N^{(R)}$, since there are no priorities, the solutions are extracted randomly. The values of admission rate have been tuned as follows: 0.9 for $N^{(I)}$, 0.48 for $N^{(CF)}$, 0.16 for $N^{(R)}$, and 0.08 for $N^{(B)}$.

Tables 12 and 13 contain the results of the analysis of the advanced variants of TS. In almost all cases, PTS performed better than JTS and provided the results that, considering the previously mentioned upper bound on the potential parallelism of an application, can be considered as close-to-optimal. PTS and JTS were also less sensitive to the quality of the initial configuration. In fact, in few cases, a random initial solution leads to better results than a balanced initial configuration.

6.5. Consistency Verification. In order to verify the consistency of the obtained estimated results and their relation to the platform execution, a comparison of execution times was performed on a representative fraction of generated partitioning configurations spanned on each of the considered numbers of processors. Figure 5 presents a chart indicating the consistency of the estimation and platform execution results. Finally, Table 14 summarizes the best solutions obtained with PTS for each number of processors. Apart from the values of execution times expressed in clock-cycles and the speed-up, the distance between the execution

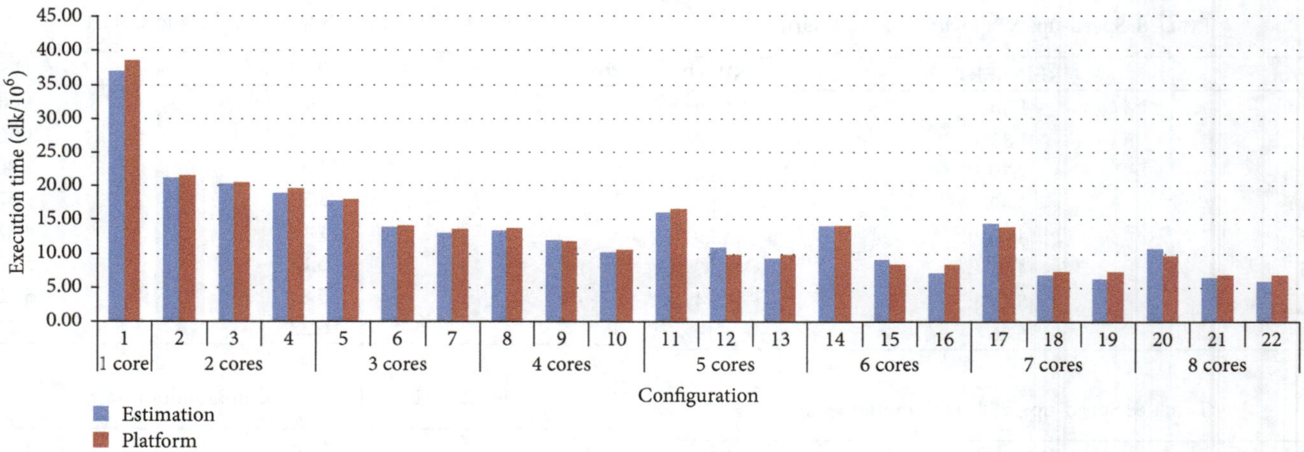

FIGURE 5: Consistency verification.

TABLE 14: Improvement summary.

Proc.	Time	Speed-up	CP dist [%]	Diff [%]
1	36938764	1.00	528	4.06
2	18839134	1.96	220	12.64
3	13045976	2.83	122	7.35
4	10200033	3.62	73	3.62
5	8321995	4.44	41	14.73
6	7194547	5.13	22	13.8
7	6354158	5.81	8	14.28
8	5941632	6.22	1	11.96

TABLE 15: Averaged time of the final improvement.

Algorithm	Time
WB	N/A
BP	N/A
IDLS	73 min
CFDLS	58 min
$N^{(B)}$	13 min
$N^{(I)}$	44 min
$N^{(CF)}$	84 min
$N^{(R)}$	318 min
JTS	308 min
PTS	276 min

time and the length of the critical path expressed in % is also highlighted. This value indicates how far a given solution is from the maximal parallelism of the application. It might be a particularly precious information for the application designer in terms of possible tracking the approach to the potential parallelism of an application. The last column contains the value of estimation discrepancy for this particular solution.

6.6. Discussion. Comparing the results obtained for all implemented algorithms, the first observation is that according to the decreasing quality of the output solutions, the algorithms can be ordered as follows: advanced TS variants, TS, DLS, and the greedy constructive procedures. This ranking is consistent, as a more refined approach outperforms a simpler one. It highlights that the specific ingredients belonging to a more refined method are relevant. Additionally, finding a good partitioning configuration for a small number of processors (i.e., 2 or 3) is relatively easy and the differences between the solutions provided by different algorithms are minor. For instance, for the case of two processors, the difference between the solutions provided by the best and the worst algorithms is less than 6%. With the increasing number of processors, the differences become more significant. For the case of the WB algorithm, the biggest difference of 30% with respect to PTS can be observed at around 5 processors, whereas for the BP algorithm, on 8 processors, the difference goes up to 100%.

The comparison of different variants of TS leads to a conclusion that the resulting solution might benefit from varying the definition of the neighborhood. In fact, both JTS and PTS outperformed the variants where only one type of neighborhood was taken into consideration. Among the advanced variants, the success of PTS over JTS might rely on two factors: (1) using the history of local search, which allows an adaptation of the search to the properties of the test case and (2) much smaller size of the neighborhood in each iteration that contributes to a diversification of the search.

An important aspect that must be also taken into account for evaluating the algorithms is the time required for their completion. It involves the evaluation time for all considered solutions in all iterations, extraction of the optimization criteria, and computation of new solutions. For DLS and TS, the upper bound on the time is defined by the user. However, for each algorithm, it was observed when the last improving move (before termination at the specified point) was performed. The averaged values among different instances are summarized in Table 15. For the TS, a big difference is visible between $N^{(R)}$ and the other variants. In fact, in this work, it is the time for $N^{(R)}$ that enforces the time limit for all other algorithms, but in the case of this particular variant, it does not necessarily correspond to the quality of the final solution. A promising observation can be made for the advanced variants of TS, since PTS not only provides with the best results, but also succeeds in ca. 10% shorter time than JTS. In all cases, the most impacting factor is the number of iterations performed, since the performance estimation and, at the same time, extraction of optimization criteria much outstrip the cost of computing the new solution.

Since the described partitioning methodology relies on an estimation of the actual performance, an important question is the estimation quality in terms of precision. This specific issue occurs in various fields [46]. For the case of experiments performed in this work, the average estimation discrepancy is between 3.35% (in total) and 10.31% (for the subset of best solutions). These values can be considered acceptable, since they still allow DLS and TS algorithms to iteratively improve an initial solution. In order to explain the reasons of this discrepancy, the first factor that must be taken into account is the general uncertainty of profiling [47]. The estimation of the execution time must rely on the limited set of measurements coming from the platform that can be burdened with some errors. On the other hand, there are some properties of the program execution not present in the proposed execution model, because of their hard tractability. A good example is the overhead introduced by a partition scheduler [48] that might depend on multiple factors, such as the number of actors in one partition, the properties of a scheduling policy, the number of conditions to be checked before an actor is executed, or even the order of appearance of actors on the list representing each partition.

7. Conclusion

This paper presents a partitioning methodology for dynamic dataflow programs that is based on a program execution model and uses multiple solution methods belonging to different classes of optimization algorithms: greedy constructive procedures, decent local search, and tabu search metaheuristic. The algorithms differ in terms of the time needed for generating a solution, the quality of the final solution, and the way they explore different properties of a dataflow application and its execution on the target platform. The best performing algorithms have been verified to approach the full potential parallelism of dataflow programs and hence to be capable of efficiently exploring the design space in all the partitioning dimensions.

The algorithms are based on the performance estimation of a program execution on a target platform. The estimation takes into account the partitioning, scheduling, and buffer size configuration and is capable of evaluating performances in terms of the total execution time. During the evaluation, the execution properties are also tracked and extracted in order to provide optimization criteria to the algorithms. The estimation has been experimentally verified to be highly accurate and consistent with real executions on the considered platform.

A direction for future improvements is to investigate the opportunities of extending the model and the methodology by further properties in order to minimize the estimation discrepancy. The methods might include tracking the variance for different executions of the same action in order to provide the model with individual weights for each job and profiling of scheduling cost that might be used for accurate scheduling and buffer size optimization. Further study is also required in order to determine the level of influence of these configurations on each other in order to effectively explore the design space in multiple dimensions. After validating the methodology on a simple platform with small level of uncertainty, an important objective is to consider more performing platforms which could turn to be more difficult to model. Studying the methodology of profiling communication cost occurring on such platform is currently an ongoing work.

Competing Interests

The authors declare that they have no competing interests.

Acknowledgments

This work is supported by the *Fonds National Suisse pour la Recherche Scientifique*, under Grant 200021.138214.

References

[1] J. B. Dennis, "First version of a data flow procedure language," in *Programming Symposium*, vol. 19 of *Lecture Notes in Computer Science*, pp. 362–376, Springer, Berlin, Germany, 1974.

[2] G. Kahn, *The Semantics of Simple Language for Parallel Programming*, IFIP Congress, 1974.

[3] S. Bhattacharyya, P. Murthy, and E. Lee, *Software Synthesis from Dataflow Graphs*, vol. 360, Springer Science & Business Media, 2012.

[4] J. Castrillon, R. Velasquez, A. Stulova et al., "Trace-based KPN composability analysis for mapping simultaneous applications to MPSoC platforms," in *Proceedings of the Design, Automation and Test in Europe Conference and Exhibition (DATE '10)*, pp. 753–758, Dresden, Germany, March 2010.

[5] M. Mattavelli, "MPEG reconfigurable video representation," in *The MPEG Representation of Digital Media*, L. Chiariglione, Ed., pp. 231–247, Springer, New York, NY, USA, 2012.

[6] E. A. Lee and T. M. Parks, "Dataflow process networks," *Proceedings of the IEEE*, vol. 83, no. 5, pp. 773–801, 1995.

[7] J. D. Ullman, "NP-complete scheduling problems," *Journal of Computer and System Sciences*, vol. 10, pp. 384–393, 1975.

[8] M. Hirzel, R. Soulé, S. Schneider, B. Gedik, and R. Grimm, "A catalog of stream processing optimizations," *ACM Computing Surveys*, vol. 46, no. 4, article 46, 2014.

[9] L. Benini, M. Lombardi, M. Milano, and M. Ruggiero, "Optimal resource allocation and scheduling for the CELL BE platform," *Annals of Operations Research*, vol. 184, pp. 51–77, 2011.

[10] S. Casale Brunet, E. Bezati, C. Alberti, M. Mattavelli, E. Amaldi, and J. W. Janneck, "Partitioning and optimization of high level stream applications for multi clock domain architectures," in *Proceedings of the IEEE Workshop on Signal Processing Systems (SiPS '13)*, pp. 177–182, October 2013.

[11] M. Pinedo, *Scheduling: Theory, Algorithms, and Systems*, Prentice Hall, New York, NY, USA, 2008.

[12] L. Thiele, I. Bacivarov, W. Haid, and K. Huang, "Mapping applications to tiled multiprocessor embedded systems," in *Proceedings of the 17th International Conference on Application of Concurrency to System Design (ACSD '07)*, pp. 29–40, IEEE, Bratislava, Slovakia, July 2007.

[13] M. Selva, *Performance monitoring of throughput constrained data flow programs executed on shared-memory multi-core architectures [Ph.D. thesis]*, INSA, Lyon, France, 2015.

[14] M. Garey and D. Johnson, *A Guide to the Theory of NP-Completeness*, W.H. Freeman, 1979.

[15] C. Blum and A. Roli, "Metaheuristics in combinatorial optimization: overview and conceptual comparison," *ACM Computing Surveys*, vol. 35, no. 3, pp. 268–308, 2003.

[16] I. H. Osman and G. Laporte, "Metaheuristics: a bibliography," *Annals of Operations Research*, vol. 63, pp. 513–623, 1996.

[17] M. Gendreau and J.-Y. Potvin, *Handbook of Metaheuristics*, Springer, Berlin, Germany, 2010.

[18] N. Zufferey, "Metaheuristics: some principles for an efficient design," *Computer Technology and Application*, vol. 3, pp. 446–462, 2012.

[19] M. A. Arslan, J. W. Janneck, and K. Kuchcinski, "Partitioning and mapping dynamic dataflow programs," in *Proceedings of the 46th Asilomar Conference on Signals, Systems and Computers (ASILOMAR '12)*, pp. 1452–1456, IEEE, Pacific Grove, Calif, USA, November 2012.

[20] J. Teich, T. Blickle, and L. Thiele, "An evolutionary approach to system-level synthesis," in *Proceedings of the 5th International Workshop on Hardware/Software Codesign (Codes/CASHE '97)*, pp. 167–171, Braunschweig, Germany, March 1997.

[21] T. Schlichter, M. Lukasiewycz, C. Haubelt, and J. Teich, "Improving system level design space exploration by incorporating sat-solvers into multi-objective evolutionary algorithms," in *Proceedings of the IEEE Computer Society Annual Symposium on VLSI (ISVLSI '06)*, pp. 309–316, Karlsruhe, Germany, March 2006.

[22] M. Pelcat, J. Piat, M. Wipliez, S. Aridhi, and J.-F. Nezan, "An open framework for rapid prototyping of signal processing applications," *EURASIP Journal on Embedded Systems*, vol. 2009, Article ID 598529, 2009.

[23] M. Obaidat and G. Papadimitriou, *Applied System Simulation: Methodologies and Applications*, Springer, New York, NY, USA, 2013.

[24] P. Bose and T. M. Conte, "Performance analysis and its impact on design," *Computer*, vol. 31, no. 5, pp. 41–49, 1998.

[25] S. Pllana, I. Brandic, and S. Benkner, "Performance modeling and prediction of parallel and distributed computing systems: a survey of the state of the art," in *Proceedings of the 1st International Conference on Complex, Intelligent and Software Intensive Systems (CISIS '07)*, pp. 279–284, Vienna, Austria, April 2007.

[26] J. Castrillon, R. Leupers, and G. Ascheid, "MAPS: mapping concurrent dataflow applications to heterogeneous MPSoCs," *IEEE Transactions on Industrial Informatics*, vol. 9, no. 1, pp. 527–545, 2013.

[27] J. Castrillon, *Programming Heterogeneous MPSoCs*, Springer, Berlin, Germany, 2013.

[28] A. D. Pimentel, C. Erbas, and S. Polstra, "A systematic approach to exploring embedded system architectures at multiple abstraction levels," *IEEE Transactions on Computers*, vol. 55, no. 2, pp. 99–111, 2006.

[29] F. Balarin, Y. Watanabe, H. Hsieh, L. Lavagno, C. Passerone, and A. Sangiovanni-Vincentelli, "Metropolis: an integrated electronic system design environment," *Computer*, vol. 36, no. 4, pp. 45–52, 2003.

[30] A. Buluc, H. Meyerhenke, I. Safro, P. Sanders, and C. Schulz, "Recent advances in graph partitioning," in *Algorithm Engineering: Selected Results and Surveys*, vol. 9220 of *Lecture Notes in Computer Science*, 2015.

[31] G. Karypis and V. Kumar, "A fast and high quality multilevel scheme for partitioning irregular graphs," *SIAM Journal on Scientific Computing*, vol. 20, no. 1, pp. 359–392, 1998.

[32] F. Pellegrini and J. Roman, "A software package for static mapping by dual recursive bipartitioning of process and architecture graphs," in *High-Performance Computing and Networking: International Conference and Exhibition HPCN EUROPE 1996 Brussels, Belgium, April 15–19, 1996 Proceedings*, vol. 1067 of *Lecture Notes in Computer Science*, pp. 493–498, Springer, Berlin, Germany, 1996.

[33] H. Yviquel, E. Casseau, M. Raulet, P. Jaaskelainen, and J. Takala, "Towards run-time actor mapping of dynamic dataflow programs onto multi-core platforms," in *Proceedings of the 8th International Symposium on Image and Signal Processing and Analysis (ISPA '13)*, pp. 732–737, September 2013.

[34] S. Bhattacharyya, E. Deprettere, and B. Theelen, "Dynamic dataflow graphs," in *Handbook of Signal Processing Systems*, pp. 905–944, Springer, Berlin, Germany, 2013.

[35] T. Parks, J. Pino, and E. A. Lee, "A comparison of synchronous and cyclo-static dataflow," in *Proceedings of the Asilomar Conference on Signals, Systems and Computers*, Pacific Grove, Calif, USA, November 1995.

[36] S. Casale-Brunet, *Analysis and optimization of dynamic dataflow programs [Ph.D. thesis]*, EPFL, Lausanne, Switzerland, 2015.

[37] T. Tabirca, S. Tabirca, L. Freeman, and L. T. Yang, "A static workload balance scheduling algorithm," in *Proceedings of the International Conference on Parallel Processing Workshops (ICPPW '02)*, pp. 235–239, August 2002.

[38] M. Michalska, S. Casale-Brunet, E. Bezati, and M. Mattavelli, "Execution trace graph based multi-criteria partitioning of stream programs," *Procedia Computer Science*, vol. 51, pp. 1443–1452, 2015.

[39] S. Thevenin, N. Zufferey, and M. Widmer, "Metaheuristics for a scheduling problem with rejection and tardiness penalties," *Journal of Scheduling*, vol. 18, no. 1, pp. 89–105, 2015.

[40] F. Glover, "Tabu search—part I," *ORSA Journal on Computing*, vol. 1, pp. 190–205, 1989.

[41] J. W. Janneck and J. Eker, "CAL language report," Tech. Memo UCB/ERL M03/48, UC Berkeley, 2003.

[42] TTA-Based Co-Design Environment, http://tce.cs.tut.fi/tta.html.

[43] M. Michalska, J. Boutellier, and M. Mattavelli, "A methodology for profiling and partitioning stream programs on many-core architectures," *Procedia Computer Science*, vol. 51, pp. 2962–2966, 2015.

[44] S. Casale-Brunet, C. Alberti, M. Mattavelli, and J. W. Janneck, "Turnus: a unified dataflow design space exploration framework for heterogeneous parallel systems," in *Proceedings of the 7th Conference on Design and Architectures for Signal and Image Processing (DASIP '13)*, pp. 47–54, Cagliari, Italy, October 2013.

[45] H. Yviquel, A. Sanchez, P. Jääskeläinen, J. Takala, M. Raulet, and E. Casseau, "Embedded multi-core systems dedicated to dynamic dataflow programs," *Journal of Signal Processing Systems*, vol. 80, no. 1, pp. 121–136, 2014.

[46] E. A. Silver and N. Zufferey, "Inventory control of an item with a probabilistic replenishment lead time and a known supplier shutdown period," *International Journal of Production Research*, vol. 49, no. 4, pp. 923–947, 2011.

[47] V. M. Weaver, D. Terpstra, and S. Moore, "Non-determinism and overcount on modern hardware performance counter implementations," in *Proceedings of the IEEE International Symposium on Performance Analysis of Systems and Software (ISPASS '13)*, pp. 215–224, Austin, Tex, USA, April 2013.

[48] R. I. Davis and A. Burns, "A survey of hard real-time scheduling for multiprocessor systems," *ACM Computing Surveys*, vol. 43, no. 4, article 35, 2011.

18

Improved Collaborative Representation Classifier based on l_2-Regularized for Human Action Recognition

Shirui Huo,[1] Tianrui Hu,[2] and Ce Li[3,4]

[1] *City University of Hong Kong, Kowloon Tong, Hong Kong*
[2] *Beijing University of Posts and Telecommunications, Beijing, China*
[3] *State Key Laboratory of Coal Resources and Safe Mining, China University of Mining & Technology, Beijing, China*
[4] *University of Chinese Academy of Sciences, Beijing, China*

Correspondence should be addressed to Ce Li; licekong@gmail.com

Academic Editor: Naiyang Guan

Human action recognition is an important recent challenging task. Projecting depth images onto three depth motion maps (DMMs) and extracting deep convolutional neural network (DCNN) features are discriminant descriptor features to characterize the spatiotemporal information of a specific action from a sequence of depth images. In this paper, a unified improved collaborative representation framework is proposed in which the probability that a test sample belongs to the collaborative subspace of all classes can be well defined and calculated. The improved collaborative representation classifier (ICRC) based on l_2-regularized for human action recognition is presented to maximize the likelihood that a test sample belongs to each class, then theoretical investigation into ICRC shows that it obtains a final classification by computing the likelihood for each class. Coupled with the DMMs and DCNN features, experiments on depth image-based action recognition, including MSRAction3D and MSRGesture3D datasets, demonstrate that the proposed approach successfully using a distance-based representation classifier achieves superior performance over the state-of-the-art methods, including SRC, CRC, and SVM.

1. Introduction

Human action recognition has been studied in the computer vision community for decades, due to its applications in video surveillance [1], human computer interaction [2], and motion analysis [3]. Prior to the Microsoft Kinect, the conventional research focused on human action recognition from RGB, but Kinect sensors provide an affordable technology to capture RGB and depth (D) images in real time, which can offer better geometric cues and less sensitivity to illumination changes for action recognition. In [1], a bag of 3D points and graphical model are obtained to characterize spatial and temporal information from depth images. In [3], three depth motion maps (DMMs) are projected to capture body shape and motion, which is a discriminant feature to describe the spatiotemporal information of a specific action from a sequence of depth images. Seen from the literature review, although depth based methods appear to be compelling toward a practical application, even if there are a few of deep-learned features for depth based action recognition, the performance is still far from satisfactory due to the large variations of the motion. In this paper, we focus on leveraging one kind of the structure of representative model to improve performance in multiclass classification with handcrafted DMMs descriptor. In [4], three channel deep convolutional neural networks are trained to extract features of depth map sequences after projecting weighted DMMs on three orthogonal planes at several temporal scales. It was verified that the method using DCNN features can achieve almost the same state-of-the-art results on the MSRAction3D and MSRGesture3D dataset. DCNNs have been demonstrated as an effective kind of models for performing state-of-the-art results in the tasks of image recognition, segmentation, detection, and retrieval. With the success of DCNN, we also take it as feature extraction and apply it in our classifier model.

As for representative models, many achievements based on space representation include image restoration [5], compressive sensing [6, 7], morphological component analysis [8], and super-resolution [9, 10]. As the advances of classifiers based on representation, several pattern recognition problems in the field of computer vision can be effectively solved by sparse coding or sparse representation methods in recent decades. In particular, the linear models can be represented as $y = A\alpha$ [11], where y, α, and A represent the data, a sparse vector, and a given matrix with the overcomplete sample set, respectively. Because of the great success of sparse coding algorithms on image processing, the sparse representation based classifiers, such as sparse representation classification (SRC) and collaborative representation classification (CRC), have gained more attention nowadays.

The basic idea of SRC/CRC is to code the test sample over a set of samples with sparsity constraints, which can be calculated by l_1-minimization. In [12], Wright proposed a basic SRC model for classification by the discriminative nature of sparse representation, which is based on the theory that newly signals are recognized as the linear combinations of previously observed ones. Based on SRC, Yang and Zhang proposed a Gabor occlusion dictionary based SRC, which can significantly reduce the computational cost [13]. In [14], the authors combined sparse representation with linear pyramid matching for image classification. Rather than using the entire training set, Zhang and Li [15] proposed learned dictionary using SRC. In [16], l_1-graph is constructed by a sparse representation subspace over the other samples. Yang et al. [14] also proposed a method to preserve the l_1-graph for image classification by using a subspace to solve misalignment problems in image classification task. Besides, SRC is used for robust illumination [17], image-plane transformation [18], and so on. However, Zhang argued that the good performance of SRC should be largely attributed to the collaborative representation of a test sample by training samples across all classes and proposed more effective CRC. In summary, SRC/CRC simply uses the reconstruction error or residual by each class-specific subspace to determine the class label, and many modified models and solution algorithms to SRC/CRC are also proposed for visual recognition tasks, including Augmented Lagrange Multiplier, Proximal Gradient, Gradient Projection, Iterative Shrinkage-Thresholding, and Homotopy [19]. Recently, some researchers [20, 21] have pointed out the purpose of l_1-regularized based sparsity in pattern classification. On the contrary, using l_2-regularized based representation for classification can do a similar job to l_1-regularized but the computational cost will reduce a lot.

Motivated by the work of modifications of CRC, in this paper, we mainly present the improved collaborative representation classifier (ICRC) based on l_2-regularized for human action recognition. Based on three DMMs' descriptor feature, the ICRC approach is to jointly maximize the likelihood that a test sample belongs to each of the multiple classes, then the final classification is performed by computing the likelihood for each class. The experiments on human action classification tasks, including MSRAction3D and MSRGesture3D datasets, are demonstrated and analyzed on the superior performance of this algorithm over the state-of-the-art methods, including SRC, CRC, and SVM. The rest of the paper is organized as follows. In Section 2, we introduce related feature descriptors using DMM. Section 3 details the action classifier based on ICRC, and Section 4 shows the experimental results of our approach on relevant datasets. The conclusion and acknowledgment are drawn in Section 5 and Acknowledgments section.

2. Feature Descriptors

2.1. Using Depth Motion Maps. In this section, we explain the extracted feature descriptor using depth motion maps (DMMs) from depth images, which is generated by selecting and stacking motion energy of depth maps projected onto three orthogonal Cartesian planes, aligning with front (f), side (s), and top (t) views (i.e., p^f, p^s, and p^t, resp.). As for each projected map, its motion energy is computed by thresholding the difference between consecutive maps. The binary map of motion energy provides a strong clue of the action category being performed and indicates motion regions or where movement happens in each temporal interval. We suggest that all frames should be deployed to calculate motion information instead of selecting frames. Considering both the discriminability and robustness of feature descriptors, we use the l_1-norm of the absolute difference of a frame to define the salient information on depth sequences. Because l_1-norm is invariant to the length of a depth sequence, and l_1-norm contains more salient information than other norms (e.g., l_2), we have

$$\mathrm{DMM}^{\{f,s,t\}} = \sum_{i=1}^{N-v} \left| p_{i+v}^{\{f,s,t\}} - p_i^{\{f,s,t\}} \right|, \quad (1)$$

where v is the frame interval, i represents the frame index, and N is the total number of frames in a depth sequence. In the case that the sum operation in (1) is only used given a threshold satisfied, the scale of v affects little the local pattern histogram on the DMMs.

2.2. Using Deep Convolutional Neural Networks. In this section, we introduce three deep convolutional neural networks (DCNNs) to train the features on three projected planes of DMMs and perform fusion of three nets by combining the softmax in fully connected layer. The layer configuration of our three CNNs is schematically shown in Figure 1, in which there are five convolutional layers and three fully connected layers in each net. The detail of our implementation is illustrated in Section 4.1.2.

3. Action Classifier Based on ICRC

Based on depth motion maps, to incorporate the feature descriptors into a powerful classifier, an improved collaborative representation classifier (ICRC) is presented for human action recognition.

3.1. l_2-Regularized Collaborative Representation Classifier. The basic idea of SRC is to get a test sample by sparsely choosing a small number of atoms from an overcomplete dictionary

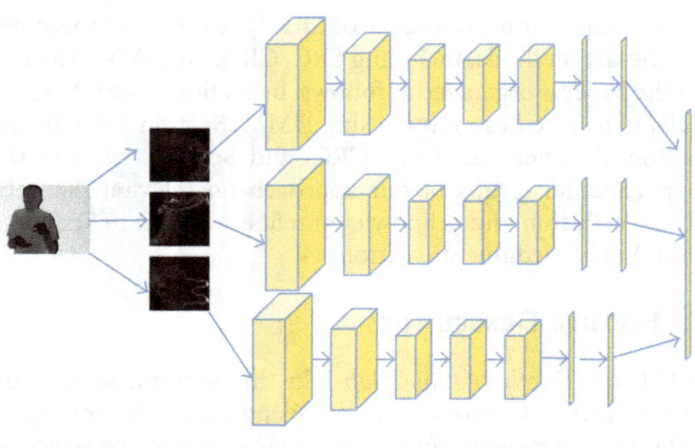

FIGURE 1: Three DCNNs architecture for a depth action sequence to extract features.

that contain all training samples [12]. Denoted by $A_j \in \mathbb{R}^{d\times n}$, the set of training samples form class j, and suppose we have C class of subjects. So $A = [A_1, A_2, \ldots, A_C] \in \mathbb{R}^{d\times n}$ involves many samples from all classes, and A_j ($j = 1, 2, \ldots, C$) is the individual class of training samples, n is the total number of training samples, and d is the dimension of training samples. A query sample $y \in \mathbb{R}^d$ can be presented by $y = A\alpha$, where y, α, and A represent the data, a sparse vector, and a given matrix with the overcomplete training samples, respectively.

To be specific, in the mechanism of collaborative representation classifier (CRC), each data point in the collaborative subspace can be represented as a linear combination of samples in A, where $\alpha = [\alpha_1, \alpha_2, \ldots, \alpha_C]$ is an $n \times 1$ representation vector associated with training sample and α_j ($j = 1, 2, \ldots, C$) is the subvector corresponding to A_j. Generally, it is formulized as a l_1-norm minimization problem with a convex objective and solved by

$$\min_{\alpha} \left\{\|y - A\alpha\|_2^2 + \theta \|\alpha\|_1\right\}, \quad \text{s.t.} \quad y = A\alpha, \tag{2}$$

where θ is a positive scalar to balance the sparsity term and the residual. The residual can be computed as

$$e_j = \left\|y - A_j\alpha_j\right\|_2, \tag{3}$$

where α_j is the coefficient vector corresponding to class j. And then the output of the identity of y can be obtained by the lowest residual as

$$\text{class}(y) = \arg\min_j \left\{e_j\right\}. \tag{4}$$

For more details of SRC/CRC, one can refer to [12]. Because of the computational time consuming in l_1-regularized minimization, (1) is approximated as

$$\min_{\alpha} \left\{\|y - A\alpha\|_2^2 + \lambda \|L\alpha\|_2^2\right\}, \quad \text{s.t.} \quad y = A\alpha, \tag{5}$$

where $y \in \mathbb{R}^d$ is representative by α and A if the l_2-norm of α is smaller. In (5), L is the Tikhonov regularization [27] to calculate the coefficient vector, and λ is the regularization parameter. $\|L\alpha\|_2$ is the l_2-regularization term to add a certain amount of sparsity to α, which is weaker than l_1-norm minimization. The diagonal matrix L and the coefficient vector α are calculated as follows [21]:

$$L = \begin{pmatrix} \|y - h_1\|_2 & \cdots & 0 \\ \vdots & \ddots & \vdots \\ 0 & \cdots & \|y - h_k\|_2 \end{pmatrix}, \quad \alpha = Py, \tag{6}$$

where $P = (A^T A + \lambda L^T L)^{-1} A^T$ is independent of y and precalculated. With (3) and (4), the data y is assigned different identities based on α.

3.2. The Proposed ICRC Method. Based on the training sample set, we propose an improved collaborative representation classifier based on l_2-regularized term, which assigns the data points with different probabilities based on α by adding a term $\sum_j^C \|A\alpha - A_j\alpha_j\|_2^2$ that attempts to find a point $A_j\alpha_j$ close to the common point y inside each subspace of class j. The first two terms $\|y - A\alpha\|_2^2 + \lambda\|L\alpha\|_2^2$ still form a l_2-regularized collaborative representation term, which encourages to find a point $A\alpha$ close to y in the collaborative subspace. Therefore, (5) is rewritten as

$$\min_{\alpha} \left\{\|y - A\alpha\|_2^2 + \lambda \|L\alpha\|_2^2 + \frac{\gamma}{C}\sum_j^C \left\|A\alpha - A_j\alpha_j\right\|_2^2\right\}, \quad \text{s.t.} \quad y = A\alpha. \tag{7}$$

Obviously, the parameters λ and γ balance three terms, which can be set from the training data. Accordingly, a new solution of representative vector α is obtained from (7).

In the condition of $\gamma = 0$, (7) will degenerate to CRC with the first two terms, and $\|y - A\alpha\|_2^2 + \lambda\|L\alpha\|_2^2$ will play an important role in determining α. When $\gamma > 0$, these two terms $\|y - A\alpha\|_2^2 + \lambda\|L\alpha\|_2^2$ will be the same for all classes, and thus the term $\sum_j^C \|A\alpha - A_j\alpha_j\|_2^2$ will be dominant to further fine-tune α_j by A_j yielding to a precise α. That is, the last newly added term is introduced to further adjust α_j by A_j, resulting in a more stable solution to representative vector α.

We can omit the first two same terms for all classes, make the classifier rule by the last term, and formulize it as a probability exponent:

$$\text{class}(y) = \arg\max_j \left\{\exp\left(-\left\|A\alpha - A_j\alpha_j\right\|_2^2\right)\right\}. \tag{8}$$

The proposed l_2-regularized method for human action recognition is presented to maximize the likelihood that a test sample belongs to each class, then the experiments in the following section show that it obtains a final classification by checking which class has the maximum likelihood. So far, the abovementioned classifier model in (7) and (8) is named as the improved collaboration representation classifier (ICRC).

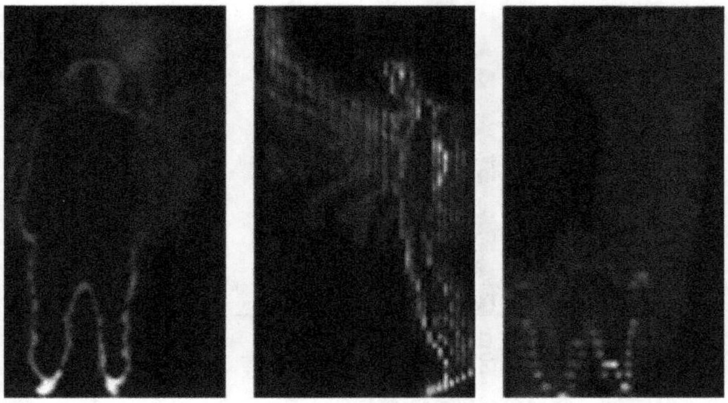

FIGURE 2: Three DMMs of a depth action sequence "ASL Z" from the front (f) view, side (s) view, and top (t) view, respectively.

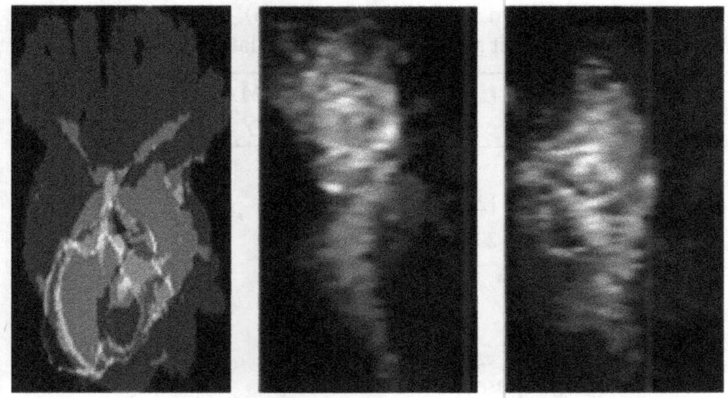

FIGURE 3: Three DMMs of a depth action sequence "Swipe left" from the front (f) view, side (s) view, and top (t) view, respectively.

4. Experimental Results

Based on depth motion maps, to incorporate the feature descriptors into a powerful classifier, an ICRC is presented for human action recognition. To verify the effectiveness of the proposed ICRC algorithm on action recognition applications using DMM descriptors of depth sequences, we carry out experiments on challenging depth based action datasets MSRAction3D [1] and MSRGesture3D [1] for human action recognition.

4.1. Feature Descriptors

4.1.1. DMMs. The MSRAction3D [1] dataset is composed of depth images sequence captured by the Microsoft Kinect-V1 camera. It includes 20 actions performed by 10 subjects facing the camera. Each subject performed each action 2 or 3 times. There are 20 action types: high arm wave, horizontal arm wave, hammer, hand catch, forward punch, high throw, draw X, draw tick, draw circle, hand clap, two-hand wave, side boxing, bend, forward kick, side kick, jogging, tennis swing, tennis serve, golf swing, and pick up and throw. The size of each depth image is 240×320 pixels. The background information has been removed in the depth data.

The MSRGesture3D [1] dataset is for continuous online human action recognition from a Kinect device. It consists of 12 gestures defined by American Sign Language (ASL). Each person performs each gesture 2 or 3 times. There are 333 depth sequences. For action recognition on the MSRAction3D and MSRGesture3D dataset, we use the feature computed from the DMMs, and each depth action sequence generates three DMMs corresponding to three projection views. The DMMs of high arm wave class from the MSRAction dataset are shown in Figure 2, and the DMMs of ASL Z class from the MSRGesture3D dataset are shown in Figure 3.

4.1.2. DCNNs. Furthermore, our implementation of DCNN features is based on the publicly available MatConvNet toolbox [28] using one Nvidia Titan X card. The network weights are learned by mini-batch stochastic gradient descent. Similar to [4], the momentum is set to 0.9 and weight decay is set to 0.0005, and all hidden weight layers use the rectification activation function. At each iteration, 256 samples in each batch are constructed and resized to 256×256, then 224×224 patches are randomly cropped from the center of the selected image to artificial data augmentation. The dropout regularization ratio is 0.5 in the nets. Besides, the initial learning rate is set to 0.01 with pretrained model on ILSVRC-2012 to fine-tune our model, and the learning rate decreases every 20 epochs. Finally, we concatenate three 4096 dimensional feature vectors in 7th fully connected layer to input the subsequent classifier.

4.2. Experiment Setting. The same experimental setup in [1] was adopted, and the actions in MSRAction3D dataset were divided into three subsets as follows: AS1: horizontal wave, hammer, forward punch, high throw, hand clap, bend, tennis serve, and pickup throw; AS2: high wave, hand catch, draw x, draw tick, draw circle, two-hand wave, forward kick, and side boxing; AS3: high throw, forward kick, side kick, jogging, tennis swing, tennis serve, golf swing, and pickup throw. We performed three experiments with 2/3 training samples and 1/3 testing samples in AS1, AS2, and AS3, respectively. Thus, the performance on MSRAction3D is evaluated by the average accuracy (Accu., unit: %) on three subsets. On the other hand, the same experimental setting reported in [26, 29, 30] was followed. 12 gestures were tested by leave-one-subject-out cross-validation to evaluate the performance of the proposed method.

4.3. Recognition Results with DMMs and ICRC. We concatenate the sign, magnitude, and center features to form the feature based on DMMs as the final feature representation. The compared methods are similar to [29, 30]. The same parameters reported in [26] were used here for the sizes of SI and block. A total of 20 actions are employed and one-half of the subjects (1, 3, 5, 7, and 9) are used for training and the remaining subjects are used for testing. The recognition performance of our method and existing approaches are listed in Table 1. It is clear that our method achieves better performance than other competing methods.

To show the outcome of our method, Figures 4 and 5 illustrate the recognition rates of each class in two datasets. It is stated that there are 14 classes obtaining 100% recognition rates in the MSRAction3D dataset, and the performance of 3

Table 1: Recognition accuracies (unit: %) comparison on the MSRAction3D dataset and MSRGesture3D dataset.

Method	Accu. (%) on two datasets	
	MSRAction3D	MSRGesture3D
DMM-HOG [3]	85.5	89.2
Random Occupancy [22]	86.5	88.5
Actionlet Ensemble [23]	88.2	88.2
Depth Cuboid [24]	89.3	90.5
Vemulapalli [25]	89.5	87.7
DMM + CRC [26]	92.3	92.5
DMM + ICRC (proposed)	*93.8*	*94.8*

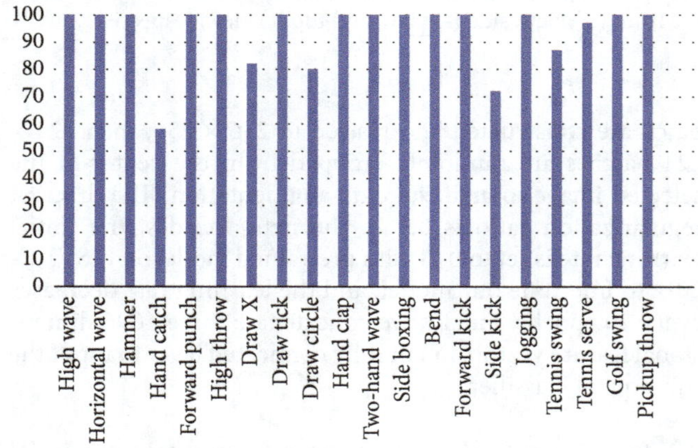

Figure 4: Recognition rates (unit: %) of 20 classes in MSRAction3D dataset (average results of three subsets).

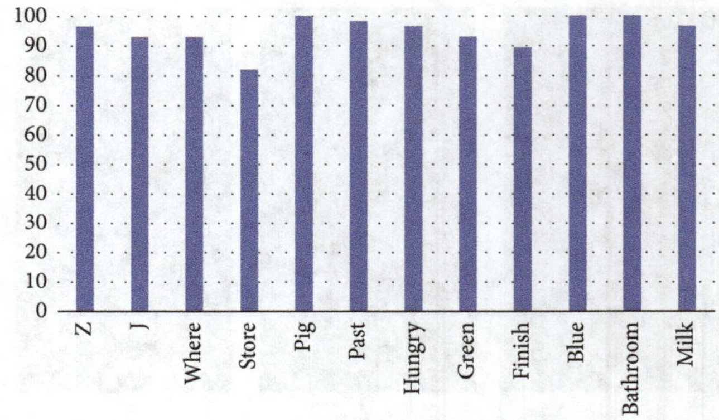

Figure 5: Recognition rates (unit: %) of 12 classes in MSRGesture3D dataset.

classes reaches up to best in the MSRGesture3D dataset. All experiments are carried out using MATLAB 2016b on an Intel i7-6500U desktop with 8 GB RAM, and the average time of video processing gets about 26 frames per second, meeting a real-time processing demand basically.

4.4. Comparison with DCNN Features and ICRC. Furthermore, in order to evaluate our proposed classifier method, we also extract the deep features by the abovementioned conventional CNN model and then input the 12288 dimensional vectors to the proposed ICRC for action recognition. Table 2 shows that DCNN algorithm indeed has advances as good as in other popular tasks of image classification and object detection, and it can improve the accuracy greatly up to 6% in MSRAction3D and MSRGesture3D. This would also explain the importance of effective feature to ICRC classifier.

5. Conclusion

In this paper, we propose improved collaborative representation classifier (ICRC) based on l_2-regularized for human action recognition. The DMMs and DCNN feature descriptors are involved as an effective action representation. For the action classifier, ICRC is proposed based on collaborative representation with the additional regularization term. The new insight focuses on a subspace constraints on the solution. The experimental results on MSRAction3D and MSRGesture3D

Table 2: Recognition accuracies (unit: %) comparison of DMM + ICRC and DCNN + ICRC on the MSRAction3D dataset and MSRGesture3D dataset.

Method	Accu. (%) on two datasets	
	MSRAction3D	MSRGesture3D
DMM + ICRC (proposed)	93.8	94.8
DCNN + ICRC (proposed)	99.99	100.0

show that the proposed algorithm performs favorably against the state-of-the-art methods, including SRC, CRC, and SVM. Future work will focus on involving the deep-learned network in the depth image representation and evaluating more complex datasets such as MSR3DActivity, UTKinect-Action, and NTU RGB+D, for the action recognition task.

Conflicts of Interest

The authors declare that they have no conflicts of interest.

Acknowledgments

The work was supported by the State Key Laboratory of Coal Resources and Safe Mining under Contracts SKLCRSM16KFD04 and SKLCRSM16KFD03, in part by the Natural Science Foundation of China under Contract 61601466, and in part by the Fundamental Research Funds for the Central Universities under Contract 2016QJ04.

References

[1] W. Li, Z. Zhang, and Z. Liu, "Action recognition based on a bag of 3D points," in *Proceedings of the IEEE Computer Society Conference on Computer Vision and Pattern Recognition Workshops (CVPRW '10)*, pp. 9–14, San Francisco, Calif, USA, June 2010.

[2] S. Wang, Z. Ma, Y. Yang, X. Li, C. Pang, and A. G. Hauptmann, "Semi-supervised multiple feature analysis for action recognition," *IEEE Transactions on Multimedia*, vol. 16, no. 2, pp. 289–298, 2014.

[3] X. Yang, C. Zhang, and Y. Tian, "Recognizing actions using depth motion maps-based histograms of oriented gradients,"

in *Proceedings of the 20th ACM International Conference on Multimedia, MM 2012*, pp. 1057–1060, jpn, November 2012.

[4] P. Wang, W. Li, Z. Gao, J. Zhang, C. Tang, and P. Ogunbona, "Deep convolutional neural networks for action recognition using depth map sequences," *Computer Vision and Pattern Recognition*, arXiv preprint, https://arxiv.org/abs/1501.04686, 2015.

[5] W. E. Vinje and J. L. Gallant, "Sparse coding and decorrelation in primary visual cortex during natural vision," *Science*, vol. 287, no. 5456, pp. 1273–1276, 2000.

[6] E. Candès, "Compressive sampling," in *Proceedings of the International Congress of Mathematics*, vol. 3, pp. 1433–1452, 2006.

[7] N. Guan, D. Tao, Z. Luo, and B. Yuan, "NeNMF: an optimal gradient method for nonnegative matrix factorization," *IEEE Transactions on Signal Processing*, vol. 60, no. 6, pp. 2882–2898, 2012.

[8] J.-L. Starck, M. Elad, and D. Donoho, "Redundant multiscale transforms and their application for morphological component separation," *Advances in Imaging and Electron Physics*, vol. 132, pp. 287–348, 2004.

[9] J. Yang, J. Wright, T. Huang, and Y. Ma, "Image super-resolution as sparse representation of raw image patches," in *Proceedings of the 26th IEEE Conference on Computer Vision and Pattern Recognition*, pp. 1–8, June 2008.

[10] W. Dong, L. Zhang, G. Shi, and X. Wu, "Image deblurring and super-resolution by adaptive sparse domain selection and adaptive regularization," *IEEE Transactions on Image Processing*, vol. 20, no. 7, pp. 1838–1857, 2011.

[11] K. Huang and S. Aviyente, "Sparse representation for signal classification," in *Proceedings of the NIPS*, pp. 609–616, Vancouver, Canada, December 2006.

[12] J. Wright, A. Y. Yang, A. Ganesh, S. S. Sastry, and Y. Ma, "Robust face recognition via sparse representation," *IEEE Transactions on Pattern Analysis and Machine Intelligence*, vol. 31, no. 2, pp. 210–227, 2009.

[13] M. Yang and L. Zhang, "Gabor feature based sparse representation for face recognition with gabor occlusion dictionary," in *Proceedings of the 11th European Conference on Computer Vision (ECCV '10)*, pp. 448–461, Springer, Crete, Greece, 2010.

[14] J. Yang, K. Yu, Y. Gong, and T. Huang, "Linear spatial pyramid matching using sparse coding for image classification," in *Proceedings of the IEEE Computer Society Conference on Computer Vision and Pattern Recognition (CVPR '09)*, pp. 1794–1801, June 2009.

[15] Q. Zhang and B. Li, "Discriminative K-SVD for dictionary learning in face recognition," in *Proceedings of the IEEE Computer Society Conference on Computer Vision and Pattern Recognition (CVPR '10)*, pp. 2691–2698, June 2010.

[16] B. Cheng, J. Yang, S. Yan, Y. Fu, and T. S. Huang, "Learning with l1-graph for image analysis," *IEEE Transactions on Image Processing*, vol. 19, no. 4, pp. 858–866, 2010.

[17] A. Wagner, J. Wright, A. Ganesh, Z. Zhou, and Y. Ma, "Towards a practical face recognition system: Robust registration and illumination by sparse representation," in *Proceedings of the 2009 IEEE Computer Society Conference on Computer Vision and Pattern Recognition Workshops, CVPR Workshops 2009*, pp. 597–604, usa, June 2009.

[18] J. Z. Huang, X. L. Huang, and D. Metaxas, "Simultaneous image transformation and sparse representation recovery," in *Proceedings of the 26th IEEE Conference on Computer Vision and Pattern Recognition (CVPR '08)*, pp. 1–8, Anchorage, Alaska, USA, June 2008.

[19] A. Y. Yang, A. Genesh, Z. Zhou, S. Sastry, and Y. Ma, "A Review of Fast L1-Minimization Algorithms for Robust Face Recognition," Defense Technical Information Center, 2010.

[20] L. Zhang, M. Yang, and X. Feng, "Sparse representation or collaborative representation: Which helps face recognition?" in *Proceedings of the IEEE International Conference on Computer Vision (ICCV '11)*, pp. 471–478, Barcelona, Spain, November 2011.

[21] P. Berkes, B. L. White, and J. Fiser, "No evidence for active sparsification in the visual cortex," in *Proceedings of the 23rd Annual Conference on Neural Information Processing Systems, NIPS 2009*, pp. 108–116, can, December 2009.

[22] J. Wang, Z. Liu, J. Chorowski, Z. Chen, and Y. Wu, "Robust 3D action recognition with random occupancy patterns," in *Computer Vision—ECCV 2012: 12th European Conference on Computer Vision, Florence, Italy, October 7–13, 2012, Proceedings, Part II*, pp. 872–885, Springer, Berlin, Germany, 2012.

[23] J. Wang, Z. Liu, Y. Wu, and J. Yuan, "Mining actionlet ensemble for action recognition with depth cameras," in *Proceedings of the IEEE Conference on Computer Vision and Pattern Recognition (CVPR '12)*, pp. 1290–1297, Providence, RI, USA, June 2012.

[24] L. Xia and J. K. Aggarwal, "Spatio-temporal depth cuboid similarity feature for activity recognition using depth camera," in *Proceedings of the 26th IEEE Conference on Computer Vision and Pattern Recognition (CVPR '13)*, pp. 2834–2841, IEEE, Portland, Ore, USA, June 2013.

[25] R. Vemulapalli, F. Arrate, and R. Chellappa, "Human action recognition by representing 3D skeletons as points in a lie group," in *Proceedings of the 27th IEEE Conference on Computer Vision and Pattern Recognition (CVPR '14)*, pp. 588–595, Columbus, Ohio, USA, June 2014.

[26] C. Chen, M. Liu, B. Zhang, J. Han, J. Junjun, and H. Liu, "3D action recognition using multi-temporal depth motion maps and fisher vector," in *Proceedings of the 25th International Joint Conference on Artificial Intelligence, IJCAI 2016*, pp. 3331–3337, usa, July 2016.

[27] A. N. Tikhonov and V. Y. Arsenin, *Solutions of Ill-Posed Problems*, John Wiley & Sons, New York, NY, USA, 1977.

[28] http://www.vlfeat.org/matconvnet/.

[29] Y. Yang, B. Zhang, L. Yang, C. Chen, and W. Yang, "Action recognition using completed local binary patterns and multiple-class boosting classifier," in *Proceedings of the 3rd IAPR Asian Conference on Pattern Recognition, ACPR 2015*, pp. 336–340, mys, November 2016.

[30] A. W. Vieira, E. R. Nascimento, G. L. Oliveira, Z. Liu, and M. F. M. Campos, "On the improvement of human action recognition from depth map sequences using space-time occupancy patterns," *Pattern Recognition Letters*, vol. 36, no. 1, pp. 221–227, 2014.

19

Score-Informed Source Separation for Multichannel Orchestral Recordings

Marius Miron, Julio J. Carabias-Orti, Juan J. Bosch, Emilia Gómez, and Jordi Janer

Music Technology Group, Universitat Pompeu Fabra, Barcelona, Spain

Correspondence should be addressed to Marius Miron; marius.miron@upf.edu

Academic Editor: Alexander Petrovsky

This paper proposes a system for score-informed audio source separation for multichannel orchestral recordings. The orchestral music repertoire relies on the existence of scores. Thus, a reliable separation requires a good alignment of the score with the audio of the performance. To that extent, automatic score alignment methods are reliable when allowing a tolerance window around the actual onset and offset. Moreover, several factors increase the difficulty of our task: a high reverberant image, large ensembles having rich polyphony, and a large variety of instruments recorded within a distant-microphone setup. To solve these problems, we design context-specific methods such as the refinement of score-following output in order to obtain a more precise alignment. Moreover, we extend a close-microphone separation framework to deal with the distant-microphone orchestral recordings. Then, we propose the first open evaluation dataset in this musical context, including annotations of the notes played by multiple instruments from an orchestral ensemble. The evaluation aims at analyzing the interactions of important parts of the separation framework on the quality of separation. Results show that we are able to align the original score with the audio of the performance and separate the sources corresponding to the instrument sections.

1. Introduction

Western classical music is a centuries-old heritage traditionally driven by well-established practices. For instance, large orchestral ensembles are commonly tied to a physically closed place, the concert hall. In addition, Western classical music is bounded by established customs related to the types of instruments played, the presence of a score, the aesthetic guidance of a conductor, and compositions spanning a large time frame. Our work is conducted within the PHENICX project [1], which aims at enriching the concert experience through technology. Specifically, this paper aims at adapting and extending score-informed audio source separation to the inherent complexity of orchestral music. This scenario involves challenges like changes in dynamics and tempo, a large variety of instruments, high reverberance, and simultaneous melodic lines but also opportunities as multichannel recordings.

Score-informed source separation systems depend on the accuracy of different parts which are not necessarily integrated in the same parametric model. For instance, they rely on a score alignment framework that yields a coarsely aligned score [2–5] or, in a multichannel scenario, they compute a panning matrix to assess the weight of each instrument in each channel [6, 7]. To account for that, we adapt and improve the parts of the system within the complex scenario of orchestral music. Furthermore, we are interested in establishing a methodology for this task for future research and we propose a dataset in order to objectively assess the contribution of each part of the separation framework to the quality of separation.

1.1. Relation to Previous Work. Audio source separation is a challenging task when sources corresponding to different instrument sections are strongly correlated in time and frequency [8]. Without any previous knowledge, it is difficult to separate two sections which play, for instance, consonant notes simultaneously. One way to approach this problem is to introduce into the separation framework information about the characteristics of the signals such as a well-aligned score [2–5, 9, 10]. Furthermore, previous research relates the accuracy of the alignment to the quality of source separation [4, 11]. For Western classical music, a correctly aligned score yields the exact time where each instrument is playing. Thus, an important step in score-informed source separation is obtaining a correctly aligned score, which can be done automatically with an audio-to-score alignment system.

Audio-to-score alignment deals with the alignment of a symbolic representation such as the score with the audio of the rendition. In a live scenario, this task deals with following the musical score while listening to the live performance, and it is known as score-following. To our knowledge, with the exception of [12], audio-to-score alignment systems have not been rigorously tested in the context of orchestral music. However, with respect to classical music, limited experimental scenarios comprising Bach chorales played by a four instruments have been discussed in [4, 13]. Furthermore, in [14], a subset of RWC classical music database [15] is used for training and testing, though no details are given regarding the instrumental complexity of the pieces. Moreover, a Beethoven orchestral piece from the same database is tested in [16], obtaining lower accuracy than the other evaluated pieces. These results point out the complexity of an orchestral scenario, underlined in [12]. Particularly, a larger number of instruments, many instruments playing concurrently different melody lines [17], prove to be a more difficult problem than tracking a limited number of instruments as in pop music or, for instance, string quartets and piano pieces. Although, in this paper, we do not propose a new system for orchestral music audio-to-score alignment, the task being a complex and extensive itself, we are interested in analyzing the relation between the task and the quality of score-informed source separation in such a complex scenario.

Besides the score, state-of-the-art source separation methods take into account characteristics of the audio signal which can be integrated into the system, thus achieving better results. For instance, the system can learn timbre models for each of the instruments [7, 18]. Moreover, it can rely on the assumption that the family of featured instruments is known, and their spectral characteristics are useful to discriminate between different sections playing simultaneously, when the harmonics of the notes overlap. In a more difficult case, when neither the score or the timbre of the instrument is available, temporal continuity and frequency sparsity [19] help in distributing the energy between sources in a musically meaningful way. Furthermore, an initial pitch detection can improve the results [6, 20–22], if the method assumes a predominant source. However, our scenario assumes multichannel recordings with distant microphones, in contrast to the close-microphone approach in [6], and we cannot assume that a source is predominant. As a matter a fact, previous methods deal with a limited case: the separation between small number of harmonic instruments or piano [3, 4, 6, 18], leaving the case of orchestral music as an open issue. In this paper, we investigate a scenario characterized by large reverberation halls [23], a large number of musicians in each section, a large diversity of instruments played, abrupt tempo changes, and many concurrent melody lines, often within the same instrument section.

Regarding the techniques used for source separation, matrix decomposition has been increasingly popular for source separation during the recent years [2, 3, 6, 18–20]. Nonnegative matrix factorization (NMF) is a particular case of decomposition which restricts the values of the factor matrices to be nonnegative. The first-factor matrix can be seen as a dictionary representing spectral templates. For audio signals, a dictionary is learned for each of the sources and stored into the basis matrix as a set of spectral templates. The second-factor matrix holds the temporal activation or the weights of the templates. Then, the resulting factorized spectrogram is calculated as a linear combination of the template vectors with a set of weight vectors forming the activation matrix. This representation allows for parametric models such as the source-filter model [3, 18, 20] or the multiexcitation model [9], which can easily capture important traits of harmonic instruments and help separate between them, as it is the case with orchestral music. The multiexcitation model has been evaluated in a restricted scenario of Bach chorales played by a quartet [4] and for this particular database has been extended in the scope of close-microphone recordings [6] and score-informed source separation [11]. From a source separation point of view, in this article, we extend and evaluate the work in [6, 11, 18] for orchestral music.

In order to obtain a better separation with any NMF parametric model, the sparseness of the gains matrix is increased by initializing it with time and frequency information obtained from the score [3–5, 24]. The values between the time frames where a note template is not activated are set to zero and will remain this way during factorization, allowing for the energy from the spectrogram to be redistributed between the notes and the instruments which actually play during that interval. A better alignment leads to better gains initialization and better separation. Nonetheless, audio-to-score alignment mainly fixes global misalignments, which are due to tempo variations, and does not deal with local misalignments [21]. To account for local misalignments, score-informed source separation systems include onset and offset information into the parametric model [3, 5, 24] or use image processing in order to refine the gains matrix so that it closely matches the actual time and frequency boundaries of the played notes [11]. Conversely, local misalignments can be fixed explicitly [25–27]. To our knowledge, none of these techniques have been explored for orchestral music, although there is a scope for testing their usefulness, if we take into account several factors as the synchronization of musicians in large ensembles, concurrent melody lines, and reverberation. Furthermore, the alignment systems are monaural. However, in our case, the separation is done on multichannel recordings and the delays between the sources and microphones might yield local misalignments. Hence, towards a better separation and more precise alignment, we propose a robust method to refine the output of a score alignment system with respect to each audio channel of the multichannel audio.

In the proposed distant-microphone scenario, we normally do not have microphones close to a particular instrument or soloist and, moreover, in an underdetermined case,

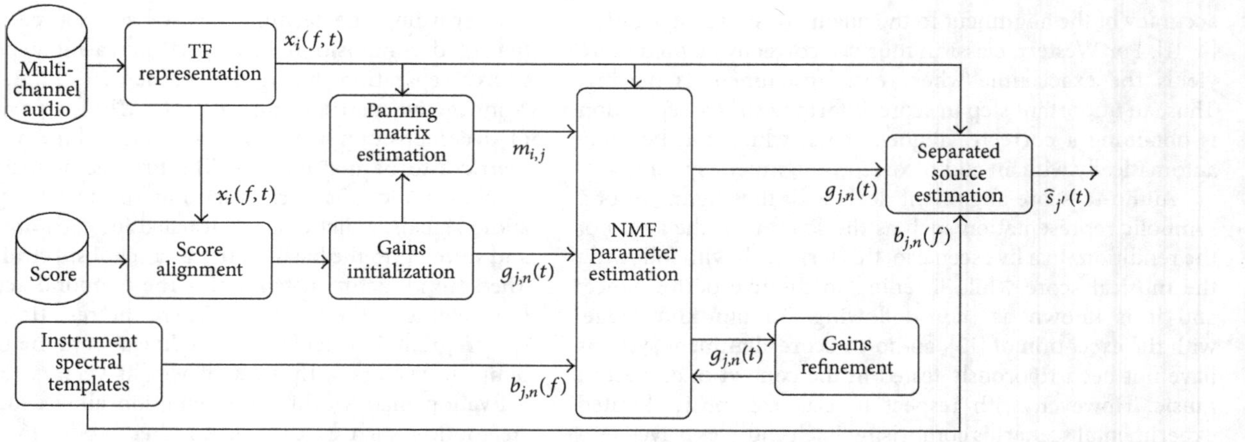

FIGURE 1: The diagram representing the flow of operations in the system.

the number of sources surpasses the number of microphones. Therefore, recording the sound of an entire section also captures interference from other instruments and the reverberation of the concert hall. To that extent, our task is different from interference reduction in close-microphone recordings [6, 7, 23], these approaches being evaluated for pop concerts [23] or quartets [6]. Additionally, we do not target a blind case source separation as the previous systems [6, 7, 23]. Subsequently, we adapt and improve the systems in [6, 11], by using information from all the channels, similar to parallel factor analysis (PARAFAC) in [28, 29].

With respect to the evaluation, to our knowledge, this is the first time score-informed source separation is objectively evaluated on such a complex scenario. An objective evaluation provides a more precise estimation of the contribution of each part of the framework and their influence on the separation. Additionally, it establishes a methodology for future research and eases the research reproducibility. We annotated the database proposed in [30], comprising four pieces of orchestral music recorded in an anechoic room, in order to obtain a score which is perfectly aligned with the anechoic recordings. Then, using the Roomsim software in [31], we simulate a concert hall in order to obtain realistic multitrack recordings.

1.2. Applications. The proposed framework for score-informed source separation has been used to separate recordings by various orchestras. The recordings are processed automatically and stored in the multimodal repository Repovizz [32]. The repository serves the data through its API for several applications. The first application is called instrument emphasis or orchestra focus and allows for emphasizing a particular instrument over the full orchestra. The second application relates to spatialization of the separated musical sources in the case of virtual reality scenarios and it is commonly known as Acoustic Rendering. Third, we propose an application to estimating the spatial locations of the instruments on the stage. All the three applications are detailed in Section 7.

1.3. Outline. We introduce the architecture of the framework with its main parts in Section 2. Then, in Section 3, we give an outline of the baseline source separation system. Furthermore, we present the extension of the baseline system: the initialization of the gains with score information (Section 4) and the note refinement (Section 4.1). Additionally, the proposed extension to the multichannel case is introduced in Section 5. We present the dataset and the evaluation procedures and discuss the results in Section 6. The demos and applications are described in Section 7.

2. Proposed Approach Overview

The diagram of the proposed framework is presented in Figure 1. The baseline system relies on training spectral templates for the instruments we aim to separate (Section 3.3). Then, we compute the spectrograms associated with the multichannel audio. The spectrograms along with the score of the piece are used to align the score to the audio. From the aligned score, we derive gains matrix that serves as an input for the NMF parameter estimation stage (Section 4), along with the learned spectral templates. Furthermore, the gains and the spectrogram are used to calculate a panning matrix (Section 3.1) which yields the contribution of each instrument in each channel. After the parameter estimation stage (Section 3.2), the gains are refined in order to improve the separation (Section 4.1). Then, the spectrograms of the separated sources are estimated using Wiener filtering (Section 3.5).

For the score alignment step, we use the system in [13] which aligns the scores to a chosen microphone and achieved the best results in MIREX score-following challenge (http://www.music-ir.org/mirex/wiki/2015:Real-time_Audio_to_Score_Alignment_(a.k.a._Score_Following)_Results). However, other state-of-the-art alignment systems can be used at this step, since our final goal is to refine a given score with respect to each channel, in order to minimize the errors in separation (Section 5.2). Accounting for that, we extend the model proposed by [6] and the gains refinement for monaural recordings in [11] to the case of score-informed multichannel source separation in the more complex scenario of orchestral music.

3. Baseline Method for Multichannel Source Separation

According to the baseline model in [6], the short-term complex valued Fourier transform (STFT) in time frame t and frequency f for channel $i = 1, \dots, I$, where I is the total number of channels, is expressed as

$$\underline{x}_i(f,t) \approx \underline{\hat{x}}_i(f,t) = \sum_{j=1}^{J} \underline{m}_{i,j} s_j(f,t), \quad (1)$$

where $s_j(f,t)$ represents the estimation of the complex valued STFT computed for the source $j = 1, \dots, J$, with J the total number of sources. Note that, in this paper, we consider a source or instrument, one or more instruments of the same kind (e.g., a section of violins). Additionally, $\underline{m}_{i,j}$ is a mixing matrix of size $I \times J$ that accounts for the contribution of source i to channel j. In addition, we denote $x_i(f,t)$ as the magnitude spectrogram and $m_{i,j}$ as the real-valued panning matrix.

Under the NMF model described in [18], each source $s_j(f,t)$ is factored as a product of two matrices: $g_{j,n}(t)$, the matrix which holds the gains or activation of the basis function corresponding to pitch n at frame t, and $b_{j,n}(f)$, $n = 1, \dots, N$, the matrix which holds bases, where $n = 1, \dots, N$ is defined as the pitch range for instrument j. Hence, source j is modeled as

$$s_j(f,t) \approx \sum_{n=1}^{N} b_{j,n}(f)\, g_{j,n}(t). \quad (2)$$

The model represents a pitch for each source j as a single template stored in the basis matrix $b_{j,n}(f)$. The temporal activation of a template (e.g., onset and offset times for a note) is modeled using the gains matrix $g_{j,n}(t)$. Under harmonicity constraints [18], the NMF model for the basis matrix is defined as

$$b_{j,n}(f) = \sum_{h=1}^{H} a_{j,n}(h)\, G(f - hf_0(n)), \quad (3)$$

where $h = 1, \dots, H$ is the number of harmonics, $a_{j,n}(h)$ is the amplitude of harmonic h for note n and instrument j, $f_0(n)$ is the fundamental frequency of note n, $G(f)$ is the magnitude spectrum of the window function, and the spectrum of a harmonic component at frequency $hf_0(n)$ is approximated by $G(f - hf_0(n))$.

Considering the model given in (3), the initial equation (2) for the computation of the magnitude spectrogram for the source j is expressed as

$$s_j(f,t) \approx \sum_{n=1}^{N} g_{j,n}(t) \sum_{h=1}^{H} a_{j,n}(h)\, G(f - hf_0(n)), \quad (4)$$

and (1) for the factorization of magnitude spectrogram for channel i is rewritten as

$$\hat{x}_i(f,t) = \sum_{j=1}^{J} m_{i,j} \sum_{n=1}^{N} g_{j,n}(t) \sum_{h=1}^{H} a_{j,n}(h)\, G(f - hf_0(n)). \quad (5)$$

3.1. Panning Matrix Estimation. The panning matrix gives the contribution of each instrument in each channel and as seen in (1) influences directly the separation of the sources. The panning matrix is estimated by calculating an overlapping mask which discriminates the time-frequency zones for which the partials of a source are not overlapped with the partials of other sources. Then, using the overlapping mask, a panning coefficient is computed for each pair of sources at each channel. The estimation algorithm in the baseline framework is described in [6].

3.2. Augmented NMF for Parameter Estimation. According to [33], the parameters of the NMF model are estimated by minimizing a cost function which measures the reconstruction error between the observed $x_i(f,t)$ and the estimated $\hat{x}_i(f,t)$. For flexibility reasons, we use the beta-divergence [34] cost function, which allows for modeling popular cost functions for different values of β, such as Euclidean (EUC) distance ($\beta = 2$), Kullback-Leibler (KL) divergence ($\beta = 1$), and the Itakura-Saito (IS) divergence ($\beta = 0$).

The minimization procedures assure that the distance between $x_i(f,t)$ and $\hat{x}_i(f,t)$ does not increase with each iteration, thus accounting for the nonnegativity of the basis and the gains. By these means, the magnitude spectrogram of a source is explained solely by additive reconstruction.

3.3. Timbre-Informed Signal Model. An advantage of the harmonic model is that templates can be learned for various instruments, if the appropriate training data is available. The RWC instrument database [15] offers recordings of solo instrument playing isolated notes along all their corresponding pitch range. The method in [6] uses these recordings along with the ground truth annotation to learn instrument spectral templates for each note of each instrument. More details on the training procedure can be found in the original paper [6].

Once the basis functions $b_{j,n}(f)$ corresponding to the spectral templates are learned, they are used at the factorization stage in any orchestral setup which contains the targeted instruments. Thus, after training the basis $b_{j,n}(f)$ are kept fixed, while the gains $g_{j,n}(t)$ are estimated during the factorization procedure.

3.4. Gains Estimation. The factorization procedure to estimate the gains $g_{j,n}(t)$ considers the previously computed panning matrix $m_{i,j}$ and the learned basis $b_{j,n}(f)$ from the training stage. Consequently, we have the following update rules:

$$g_{j,n}(t) \longleftarrow g_{j,n}(t) \frac{\sum_{f,i} m_{i,j} b_{j,n}(f)\, x_i(f,t)\, \hat{x}_i(f,t)^{\beta-2}}{\sum_{f,i} m_{i,j} b_{j,n}(f)\, \hat{x}(f,t)^{\beta-1}}. \quad (6)$$

3.5. From the Estimated Gains to the Separated Signals. The reconstruction of the sources is done by estimating the complex amplitude for each time-frequency bin. In the case of binary separation, a cell is entirely associated with a single source. However, when having many instruments as in orchestral music, it is more advantageous to redistribute

(1) Initialize $b_{j,n}(f)$ with the values learned in Section 3.3.
(2) Initialize the gains $g_{j,n}(t)$ with score information.
(3) Initialize the mixing matrix $m_{i,j}$ with the values learned in Section 3.1.
(4) Update the gains using equation (6).
(5) Repeat Step (2) until the algorithm converges (or maximum number of iterations is reached).

ALGORITHM 1: Gain estimation method.

energy proportionally over all sources as in the Wiener filtering method [23].

This model allows for estimating each separated source $s_j(t)$ from mixture $x_i(t)$ using a generalized time-frequency Wiener filter over the short-time Fourier transform (STFT) domain as in [3, 34].

Let $\alpha_{j'}$ be the Wiener filter of source j', representing the relative energy contribution of the predominant source with respect to the energy of the multichannel mixed signal $x_i(t)$ at channel i:

$$\alpha_{j'}(t,f) = \frac{\left|A_{i,j'}\right|^2 \left|\underline{s}_{j'}(f,t)\right|^2}{\sum_j \left|A_{i,j}\right|^2 \left|\underline{s}_j(f,t)\right|^2}. \tag{7}$$

Then, the corresponding spectrogram of source j' is estimated as

$$\widehat{\underline{s}}_{j'}(f,t) = \frac{\alpha_{j'}(t,f)}{\left|A_{i,j'}\right|^2} \underline{x}_i(f,t). \tag{8}$$

The estimated source $\widehat{s}_{j'}(f,t)$ is computed with the inverse overlap-add STFT of $\widehat{\underline{s}}_{j'}(f,t)$.

The estimated source magnitude spectrogram is computed using the gains $g_{j,n}(t)$ estimated in Section 3.4 and $b_{j,n}(f)$ the fixed basis functions learned in Section 3.3: $\widehat{s}_j(t,f) = g_{n,j}(t)b_{j,n}(f)$. Then, if we replace $s_j(t,f)$ with $\widehat{s}_j(t,f)$ and if we consider the mixing matrix coefficients computed in Section 3.1, we can calculate the Wiener mask from (7):

$$\alpha_{j'}(t,f) = \frac{m_{i,j'}^2 \widehat{s}_j(f,t)^2}{\sum_j m_{i,j}^2 \widehat{s}_j(f,t)^2}. \tag{9}$$

Using (8), we can apply the Wiener mask to the multichannel signal spectrogram, thus obtaining $\widehat{s}_{j'}(f,t)$, the estimated predominant source spectrogram. Finally, we use the phase information from the original mixture signal $\underline{x}_i(t)$, and, through inverse overlap-add STFT, we obtain the estimated predominant source $\widehat{s}_{j'}(t)$.

4. Gains Initialization with Score Information

In the baseline method [6], the gains are initialized following a transcription stage. In our case, the automatic alignment system yields a score which offers an analogous representation to the one obtained by the transcription. To that extent, the output of the alignment is used to initialize the gains for the NMF based methods for score-informed source separation.

Although the alignment algorithm aims at fixing global misalignments, it does not account for local misalignments. In the case of score-informed source separation, having a better aligned score leads to better separation [11], since it increases the sparseness of the gains matrix by setting to zero the activation for a time frame in which a note is not played (e.g., the corresponding spectral template of the note in the basis matrix is not activated outside this time boundary). However, in a real-case scenario, the initialization of gains derived from the MIDI score must take into account the local misalignments. This has been traditionally done by setting a tolerance window around the onsets and offsets [3, 5, 24] or by refining the gains after a number of NMF iterations and then reestimating the gains [11]. While the former integrates note refinement into the parametric model, the latter detects contours in the gains using image processing heuristics and explicitly associates them with meaningful entities as notes. In this paper, we present two methods for note refinement: in Section 4.1, we detail the method in [11] which is used as a baseline for our framework and, in Section 5.2, we adapt and improve this baseline to the multichannel case.

On these terms, if we account for errors up to d frames in the audio-to-score alignment, we need to increase the time interval around the onset and the offset for a MIDI note when we initialize the gains. Thus, the values in $g_{j,n}(t)$ for instrument j and pitch corresponding to a MIDI note n are set to 1 for the frames where the MIDI note is played, as well as the neighboring d frames. The other values in $g_{j,n}(t)$ are set to 0 and do not change during computation, while the values set to 1 evolve according to the energy distributed between the instruments.

Having initialized the gains, the classical augmented NMF factorization is applied to estimate the gains corresponding to each source j in the mixture. The process is detailed in Algorithm 1.

4.1. Note Refinement. The note refinement method in [11] aims at associating the values in the gains matrix $g_{j,n}(t)$ with notes. Therefore, it is applied after a certain number of iterations of Algorithm 1, when the gains matrix yields a meaningful distribution of the energy between instruments.

The method is applied on each note separately, with the scope of refining the gains associated with the targeted note. The gains matrix $g_{j,n}(t)$ can be understood as a greyscale image with each element in the matrix representing a pixel in the image. A set of image processing heuristics are deployed to detect shapes and contours in this image commonly known

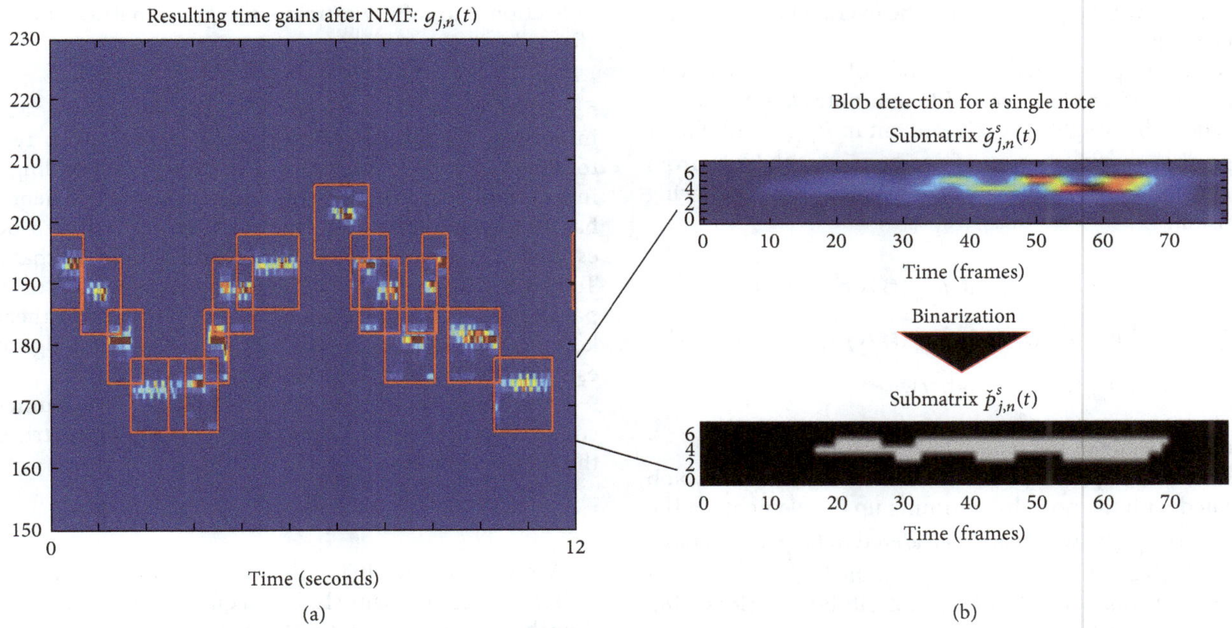

FIGURE 2: After NMF, the resulting gains (a) are split in submatrices (b) and used to detect blobs [11].

as blobs [35, p. 248]. As a result, each blob is associated with a single note, giving the onset and offset times for the note and its frequency contour. This representation further increases the sparsity of the gains $g_{j,n}(t)$, yielding less interference and better separation.

As seen in Figure 2, the method considers an image patch around the pixels corresponding to the pitch of the note and the onset and offset of the note given by the alignment stage, plus the additional d frames accounting for the local misalignments. In fact, the method works with the same submatrices of $g_{j,n}(t)$ which are set to 1 according to their corresponding MIDI note during the gains initialization stage, as explained in Section 3.4. Therefore, for a given note $k = 1, \ldots, K_j$, we process submatrix $\breve{g}^{k}_{j,n}(t)$ of the gains matrix $g_{j,n}(t)$, where K_j is the total number of notes for instrument j.

The steps of the method in [11], image preprocessing, binarization, and blob selection, are explained in Sections 4.1.2, 4.1.1, and 4.1.3.

4.1.1. Image Preprocessing. The preprocessing stage ensures through smoothing that there are no energy discontinuities within an image patch. Furthermore, it gives more weight to the pixels situated closer to the central bin in the blob in order to eliminate interference from the neighboring notes (close in time and frequency), but still preserving vibratos or transitions between notes.

First, we convolve with a smoothing Gaussian filter [35, p. 86] each row of the submatrix $\breve{g}^{k}_{j,n}(t)$. We choose a one-dimension Gaussian filter:

$$w(t) = \frac{1}{\sqrt{2\pi}\phi} e^{-(-t^2/2\phi^2)}, \tag{10}$$

where t is the time axis and ϕ is the standard deviation. Thus, each row vector of $\breve{g}^{k}_{j,n}(t)$ is convolved with $w(t)$, and the result is truncated in order to preserve the dimensions of the initial matrix by removing the mirrored frames.

Second, we penalize values in $\breve{g}^{k}_{j,n}(t)$ which are further away from the central bin by multiplying each column vector of this matrix with a 1-dimensional Gaussian centered in the central frequency bin, represented by vector $v(n)$:

$$v(n) = \frac{1}{\sqrt{2\pi}v} e^{-(n-\kappa)^2/2v^2}, \tag{11}$$

where n is the frequency axis, κ is the position of the central frequency bin, and v is the standard deviation. The values of the parameters above are given in Section 6.4.2 as a part of the evaluation setup.

4.1.2. Image Binarization. Image binarization sets to zero the elements of the matrix $\breve{g}^{k}_{j,n}(t)$ which are lower than a threshold and to one the elements larger than the threshold. This involves deriving a submatrix $\breve{p}^{k}_{j,n}(t)$, associated with note k:

$$\breve{p}^{k}_{j,n}(t) = \begin{cases} 0, & \text{if } \breve{g}^{k}_{j,n}(t) < \text{mean}\left(\breve{g}^{k}_{j,n}(t)\right), \\ 1, & \text{if } \breve{g}^{k}_{j,n}(t) \geq \text{mean}\left(\breve{g}^{k}_{j,n}(t)\right). \end{cases} \tag{12}$$

4.1.3. Blob Selection. First, we detect blobs in each binary submatrix $\breve{p}^{k}_{j,n}(t)$, using the connectivity rules described in [35, p. 248] and [27]. Second, from the detected blob candidates, we determine the best blob for each note in a similar way to [11]. We assign a value to each blob, depending on its area and the overlap with the blobs corresponding to adjacent notes, which will help us penalize the overlap between blobs of adjacent notes.

As a first step, we penalize parts of the blobs which overlap in time with other blobs from different notes $k-1, k, k+1$. This is done by weighting each element in $\breve{g}_{j,n}^{k}(t)$ with factor γ, depending on the amount of overlapping with blobs from adjacent notes. The resulting score matrix has the following expression:

$$\breve{q}_{j,n}^{k}(t) = \begin{cases} \gamma * \breve{g}_{j,n}^{k}(t), & \text{if } \breve{p}_{j,n}^{k}(t) \wedge \breve{p}_{j,n}^{k-1}(t) = 1, \\ \gamma * \breve{g}_{j,n}^{k}(t), & \text{if } \breve{p}_{j,n}^{k}(t) \wedge \breve{p}_{j,n}^{k+1}(t) = 1, \\ \breve{g}_{j,n}^{k}(t), & \text{otherwise,} \end{cases} \quad (13)$$

where γ is a value in the interval $[0, 1]$.

Then, we compute a score for each note l and for each blob associated with the note, by summing up the elements in the score matrix $\breve{q}_{j,n}^{k}(t)$ which are considered to be part of a blob. The best blob candidate is the one with the highest score and further on; it is associated with the note, its boundaries giving the note onset and offsets.

4.2. Gains Reinitialization and Recomputation. Having associated a blob with each note (Section 4.1.3), we discard obtrusive energy from the gains matrix $g_{j,n}(t)$, by eliminating the pixels corresponding to the blobs which were not selected, making the matrix sparser. Furthermore, the energy excluded from instrument's gains is redistributed to other instruments, contributing to better source separation. Thus, the gains are reinitialized with the information obtained from the corresponding blobs and we can repeat the factorization Algorithm 1 to recompute $g_{j,n}(t)$. Note that the energy which is excluded by note refinement is set to zero $g_{j,n}(t)$ and will remain zero during the factorization.

In order to refine the gains $g_{j,n}(t)$, we define a set of matrices $p_{j,n}^{k}(t)$ derived from the matrices corresponding to the best blobs $\breve{p}_{j,n}^{k}(t)$ which contain 1 only for the elements associated with the best blob and 0 otherwise. We rebuild the gains matrix $g_{j,n}(t)$ with the set of submatrices $p_{j,n}^{k}(t)$. For the corresponding bins n and time frames t of note k, we initialize the values in $g_{j,n}(t)$ with the values in $p_{j,n}^{k}(t)$. Then, we reiterate the gains estimation with Algorithm 1. Furthermore, we obtain the spectrogram of the separated sources with the method described in Section 3.5.

5. PARAFAC Model for Multichannel Gains Estimation

Parallel factor analysis methods (PARAFAC) [28, 29] are mostly used under the nonnegative tensor factorization paradigm. By these means, the NMF model is extended to work with 3-valence tensors, where each slice of the sensor represents the spectrogram for a channel. Another approach is to stack up spectrograms for each channel in a single matrix [36] and perform a joint estimation of the spectrograms of sources in all channels. Hence, we extend the NMF model in Section 3 to jointly estimate the gains matrices in all the channels.

5.1. Multichannel Gains Estimation. The algorithm described in Section 3 estimates gains $g_{n,j}$ for source j with respect to single channel i determined as the corresponding row in column j of the panning matrix where element $m_{i,j}$ has the maximum value. However, we argue that a better estimation can benefit from the information in all channels. To this extent, we can further include update rules for other parameters such as mixing matrix $m_{i,j}$ which were otherwise kept fixed in Section 3, because the factorization algorithm estimates the parameters jointly for all the channels.

We propose to integrate information from all channels by concatenating their corresponding spectrogram matrices on the time axis, as in

$$x(f,t) = [x_1(f,t) \;\; x_1(f,t) \;\; \cdots \;\; x_I(f,t)]. \quad (14)$$

We are interested in jointly estimating the gains $g_{n,j}$ of the source j in all the channels. Consequently, we concatenate in time the gains corresponding to each channel i for $i = 1, \ldots, I$, where I is the total number of channels, as seen in (15). The new gains are initialized with identical score information obtained from the alignment stage. However, during the estimation of the gains for channel i, the new gains $g_{n,j}^{i}(t)$ evolve accordingly, taking into account the corresponding spectrogram $x_i(f,t)$. Moreover, during the gains refinement stage, each gain is refined separately with respect to each channel:

$$g_{n,j}(t) = \left[g_{n,j}^{1}(t) \;\; g_{n,j}^{2}(t) \;\; \cdots \;\; g_{n,j}^{I}(t)\right]. \quad (15)$$

In (5), we describe the factorization model for the estimated spectrogram, considering the mixing matrix, the basis, and the gains. Since we estimate a set of I gains for each source $j = 1, \ldots, J$, this will result in J estimations of the spectrograms corresponding to all the channels $i = 1, \ldots, I$, as seen in

$$\hat{x}_i^{j}(f,t) = \sum_{j=1}^{J} m_{i,j} \sum_{n=1}^{N} g_{j,n}^{i}(t) \sum_{h=1}^{H} a_{j,n}(h)\, G(f - hf_0(n)). \quad (16)$$

Each iteration of the factorization algorithm yields additional information regarding the distribution of energy between each instrument and each channel. Therefore, we can include in the factorization update rules for mixing matrix $m_{i,j}$ as in (17). By updating the mixing parameters at each factorization step, we can obtain a better estimation for $\hat{x}_i^{j}(f,t)$:

$$m_{i,j} \longleftarrow m_{i,j} \frac{\sum_{f,t} b_{j,n}(f)\, g_{n,j}(t)\, x_i(f,t)\, \hat{x}_i(f,t)^{\beta-2}}{\sum_{f,t} b_{j,n}(f)\, g_{n,j}(t)\, \hat{x}(f,t)^{\beta-1}}. \quad (17)$$

Considering the above, the new rules to estimate the parameters are described in Algorithm 2.

(1) Initialize $b_{j,n}(f)$ with the values learned in Section 3.3.
(2) Initialize the gains $g_{j,n}(t)$ with score information.
(3) Initialize the panning matrix $m_{i,j}$ with the values learned in Section 3.1.
(4) Update the gains using equation (6).
(5) Update the panning matrix using equation (17).
(6) Repeat Step (2) until the algorithm converges (or maximum number of iterations is reached).

ALGORITHM 2: Gain estimation method.

Note that the current model does not estimate the phases for each channel. In order to reconstruct source j, the model in Section 3 uses the phase of the signal corresponding to channel i where it has the maximum value in the panning matrix, as described in Section 3.5. Thus, in order to reconstruct the original signals, we can solely rely on the gains estimated in single channel i, in a similar way to the baseline method.

5.2. Multichannel Gains Refinement. As presented in Section 5.1, for a given source, we obtain an estimation of the gains corresponding to each channel. Therefore, we can apply note refinement heuristics in a similar manner to Section 4.1 for each of the gains $[g^1_{n,j}(t), \ldots, g^I_{n,j}(t)]$. Then, we can average out the estimations for all the channel, making the blob detection more robust to the variances between the channels:

$$g'_{n,j}(t) = \frac{\sum_{i=1}^{I} g^i_{n,j}(t)}{I}. \tag{18}$$

Having computed the mean over all channels as in (18), for each note $k = 1, \ldots, K_j$, we process submatrix $\overline{g}^k_{j,n}(t)$ of the new gains matrix $g'_{j,n}(t)$, where K_j is the total number of notes for an instrument j. Specifically, we apply the same steps: preprocessing (Section 4.1.1), binarization (Section 4.1.2), and blob selection (Section 4.1.3), to each matrix $\overline{g}^k_{j,n}(t)$ and we obtain a binary matrix $\overline{p}^k_{j,n}(t)$ having 1 s for the elements corresponding to the best blob and 0 s for the rest.

Our hypothesis is that averaging out the gains between all channels makes blob detection more robust. However, when performing the averaging, we do not account for the delays between the channels. In order to compute the delay for a given channel, we can compute the best blob separately with the method in Section 4.1 (matrix $\check{p}^k_{j,n}(t)$) and compare it with the one calculated with the averaged estimation ($\overline{p}^k_{j,n}(t)$). This step is equivalent to comparing the onset times of the two best blobs for the two estimations. Subtracting these onset times, we get the delay between the averaged estimation and the one obtained for a channel and we can correct this in matrix $\overline{p}^k_{j,n}(t)$. Accordingly, we zero-pad the beginning of $\overline{p}^k_{j,n}(t)$ with the amount of zeros corresponding to the delay, or we remove the trailing zeros for a negative delay.

6. Materials and Evaluation

6.1. Dataset. The audio material used for evaluation was presented by Pätynen et al. [30] and consists of four passages of symphonic music from the Classical and Romantic periods. This work presented a set of anechoic recordings for each of the instruments, which were then synchronized between them so that they could later be combined to a mix of the orchestra. Musicians played in an anechoic chamber, and, in order to be synchronous with the rest of the instruments, they followed a video featuring a conductor and a pianist playing each of the four pieces. Note that the benefits of having isolated recordings comes at the expense of ignoring the interactions between musicians which commonly affect intonation and time-synchronization [37].

The four pieces differ in terms of number of instruments per instrument class, style, dynamics, and size. The first passage is a soprano aria of Donna Elvira from the opera Don Giovanni by W. A. Mozart (1756–1791), corresponding to the Classical period, and traditionally played by a small group of musicians. The second passage is from L. van Beethoven's (1770–1827) Symphony no. 7, featuring big chords and string crescendo. The chords and pauses make the reverberation tail of a concert hall clearly audible. The third passage is from Bruckner's (1824–1896) Symphony no. 8, and represents the late Romantic period. It features large dynamics and size of the orchestra. Finally, G. Mahler's Symphony no. 1, also featuring a large orchestra, is another example of late romanticism. The piece has a more complex texture than the one by Bruckner. Furthermore, according to the musicians which recorded the dataset, the last two pieces were also more difficult to play and record [30].

In order to keep the evaluation setup consistent between the four pieces, we focus in the following instruments: violin, viola, cello, double bass, oboe, flute, clarinet, horn, trumpet, and bassoon. All tracks from a single instrument were joined into a single track for each of the pieces.

For the selected instruments, we list the differences between the four pieces in Table 1. Note that in the original dataset the violins are separated into two groups. However, for brevity of evaluation and because in our separation framework we do not consider sources sharing the same instrument templates, we decided to merge the violins into a single group. Note that the pieces by Mahler and Bruckner have a divisi in the groups of violins, which implies a larger number of instruments playing different melody lines simultaneously. This results in a scenario which is more challenging for source separation.

We created a ground truth score, by hand annotating the notes played by the instruments. In order to facilitate this process, we first gathered the scores in MIDI format and automatically computed a global audio-score alignment, using the method from [13] which has won the MIREX

TABLE 1: Anechoic dataset [30] characteristics.

Piece	Duration	Period	Instrument sections	Number of tracks	Max. tracks/instrument
Mozart	3 min 47 s	Classical	8	10	2
Beethoven	3 min 11 s	Classical	10	20	4
Mahler	2 min 12 s	Romantic	10	30	4
Bruckner	1 min 27 s	Romantic	10	39	12

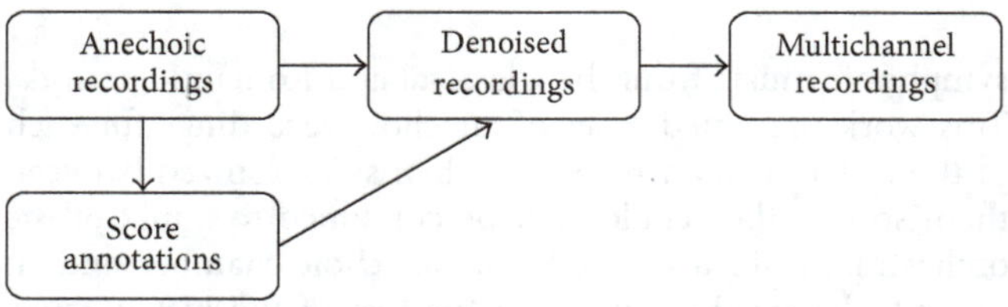

FIGURE 3: The steps to create the multichannel recordings dataset.

score-following challenge for the past years. Then, we locally aligned the notes of each instrument by manually correcting the onsets and offsets to fit the audio. This was performed using Sonic Visualiser, with the guidance of the spectrogram and the monophonic pitch estimation [38] computed for each of the isolated instruments. The annotation was performed by two of the authors, which cross-checked the work of their peer. Note that this dataset and the proposed annotation are useful not only for our particular task but also for the evaluation of multiple pitch estimation and automatic transcription algorithms in large orchestral settings, a context which has not been considered so far in the literature. The annotations can be found at the associated page (http://mtg.upf.edu/download/datasets/phenicx-anechoic).

During the recording process detailed in [30], the gain of the microphone amplifiers was fixed to the same value for the whole process, which reduced the dynamic range of the recordings of the quieter instruments. This led to noisier recordings for most of the instruments. In Section 6.2, we describe the score-informed denoising procedure we applied to each track. From the denoised isolated recordings, we then used Roomsim to create a multichannel image, as detailed in Section 6.3. The steps necessary to pass from the anechoic recordings to the multichannel dataset are represented in Figure 3. The original files can be obtained from the Acoustic Group at Aalto Univeristy (http://research.cs.aalto.fi/acoustics/). For the denoising algorithm, please refer to http://mtg.upf.edu/download/datasets/phenicx-anechoic.

6.2. Dataset Denoising. The noise related problems in the dataset were presented in [30]. We remove the noise in the recordings with the score-informed method in [39], which relies on learned noise spectral patters. The main difference is that we rely on a manually annotated score, while in [39] the score is assumed to be misaligned, so further regularization is included to ensure that only certain note combinations in the score occur.

The annotated score yields the time interval where an instrument is not playing. Thus, the noise pattern is learned only within that interval. In this way, the method assures that

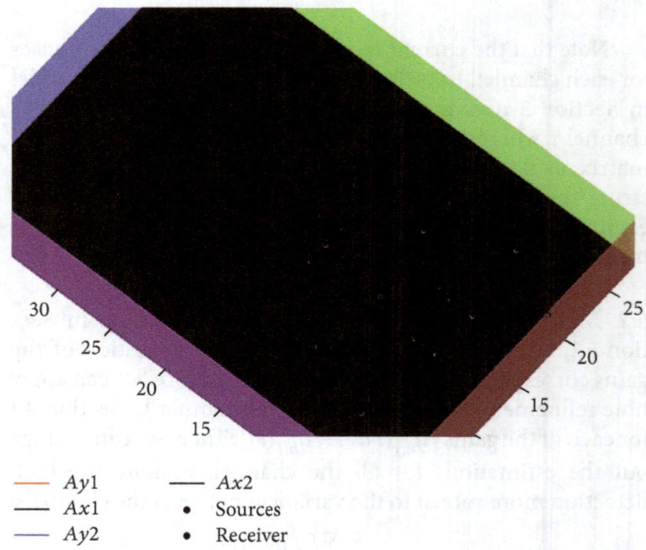

FIGURE 4: The sources and the receivers (microphones in the simulated room).

the desired noise, which is a part of the actual sound of the instrument, is preserved in the denoised recording.

The algorithm takes each anechoic recording of a given instrument and removes the noise for the time interval where an instrument is playing, while setting to zero the frames where an instrument is not playing.

6.3. Dataset Spatialization. To simulate a large reverberant hall, we use the software Roomsim [31]. We define a configuration file which specifies the characteristics of the hall and, for each of the microphones, their position relative to each of the sources. The simulated room has similar dimensions to the Royal Concertgebouw in Amsterdam, one of the partners in the PHENICX project, and represents a setup in which we tested our framework. The simulated room's width, length, and height are 28 m, 40 m, and 12 m. The absorption coefficients are specified in Table 2.

The positions of the sources and microphones in the room are common for orchestral concerts (Figure 4). A configuration file is created for each microphone which contains its coordinates (e.g., (14, 17, 4) for the center microphone). Then, each source is defined through polar coordinates relative to the microphone (e.g., (11.4455, −95.1944, −15.1954) radius, azimuth, and elevation for the bassoon relative to the center microphone). We selected all the microphones to be cardioid, in order to match the realistic setup of Concertgebouw Hall.

TABLE 2: Room surface absorption coefficients.

Standard measurement frequencies (Hz)	125	250	500	1000	2000	4000
Absorption of wall in $x = 0$ plane	0.4	0.3	0.3	0.3	0.2	0.1
Absorption of wall in $x = Lx$ plane	0.4	0.45	0.35	0.35	0.45	0.3
Absorption of wall in $y = 0$ plane	0.4	0.45	0.35	0.35	0.45	0.3
Absorption of wall in $y = Ly$ plane	0.4	0.45	0.35	0.35	0.45	0.3
Absorption of floor, that is, $z = 0$ plane	0.5	0.6	0.7	0.8	0.8	0.9
Absorption of ceiling, that is, $z = Lz$ plane	0.4	0.45	0.35	0.35	0.45	0.3

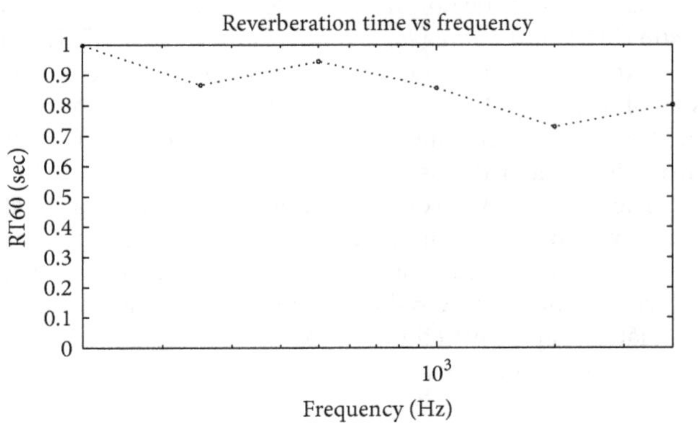

FIGURE 5: The reverberation time versus frequency for the simulated room.

Using the configuration file and the anechoic audio files corresponding to the isolated sources, Roomsim generates the audio files for each of the microphones along with the impulse responses for each pair of instruments and microphones. The impulse responses and the anechoic signals are used during the evaluation to obtain the ground truth spatial image of the sources in the corresponding microphone. Additionally, we plot Roomsim the reverberation time RT60 [31] across the frequencies in Figure 5.

We need to adapt ground truth annotations to the audio generated with Roomsim, as the original annotations were done on the isolated audio files. Roomsim creates an audio for a given microphone by convolving each of the sources with the corresponding impulse response and then summing up the results of the convolution. We compute a delay for each pair of microphones and instruments by taking the position of the maximum value in the associated impulse response vector. Then, we generate a score for each of the pairs by adding the corresponding delay to the note onsets. Additionally, since the offset time depends on the reverberation and the frequencies of the notes, we add 0.8 s to each note offset to account for the reverberation, besides the added delay.

6.4. Evaluation Methodology

6.4.1. Parameter Selection. In this paper, we use a low-level spectral representation of the audio data which is generated from a windowed FFT of the signal. We use a Hanning window with the size of 92 ms and a hop size of 11 ms. Here, a logarithmic frequency discretization is adopted. Furthermore, two time-frequency resolutions are used. First, to estimate the instrument models and the panning matrix, a single semitone resolution is proposed. In particular, we implement the time-frequency representation by integrating the STFT bins corresponding to the same semitone. Second, for the separation task, a higher resolution of 1/4 of semitone is used, which has proven to achieve better separation results [6]. The time-frequency representation is obtained by integrating the STFT bins corresponding to 1/4 of semitone. Note that in the separation stage, the learned basis functions $b_{j,n}(f)$ are adapted to the 1/4 of semitone resolution by replicating 4 times the basis at each semitone to the 4 samples of the 1/4 of semitone resolution that belong to this semitone. For image binarization, we pick for the first Gaussian $\phi = 3$ as the standard deviation and for the second Gaussian $\kappa = 4$ as the position of the central frequency bin and $\nu = 4$ as the standard deviation, corresponding to one semitone.

We picked 10 iterations for the NMF, and we set the beta-divergence distortion, $\beta = 1.3$, as in [6, 11].

6.4.2. Evaluation Setup. We perform three different kind of evaluations: audio-to-score alignment, panning matrix estimation, and score-informed source separation. Note that, for the alignment, we evaluate the state-of-the-art system in [13]. This method does not align notes but combinations of notes in the score (a.k.a states). Here, the alignment is performed with respect to a single audio channel, corresponding to the microphone situated in the center of the stage. On the other hand, the offsets are estimated by shifting the original duration for each note in the score [13] or by assigning the offset time as the onset for the next state. We denote these two cases as INT or NEX.

Regarding the initialization of the separation framework, we can use the raw output of the alignment system. However, as stated in Section 4 and [3, 5, 24], a better option is to extend the onsets and offsets along a tolerance window to account for the errors of the alignment system and for the delays between center channel (on which the alignment is performed) and the other channels and for the possible errors in the alignment itself. Thus, we test two hypotheses regarding the tolerance window for the possible errors. In the first case, we extend the boundaries with 0.3 s for onsets and 0.6 s for offsets (T1) and in the second with 0.6 s for onsets and 1 s for offsets (T2). Note that the value for the onset times of 0.3 s is not arbitrary but the usual threshold for onsets in the score-following in

Table 3: Score information used for the initialization of score-informed source separation.

Tolerance window size	Offset estimation
T1: onsets, 0.3 s; offsets, 0.6 s	INT: interpolation of the offset time
T2: onsets, 0.6 s; offsets, 0.9 s	NEX: the offset is the onset of the next note

MIREX evaluation of real-time score-following [40]. Two different tolerance windows were tested to account for the complexity of this novel scenario. The tolerance window is slightly larger for offsets due to the reverberation time and because the ending of the note is not as clear as its onset. A summary of the score information used to initialize the source separation framework is found in Table 3.

We label the test case corresponding to the initialization with the raw output of the alignment system as Ali. Conversely, the test case corresponding to the tolerance window initialization is labeled as Ext. Furthermore, within the tolerance window, we can refine the note onsets and offsets with the methods in Section 4.1 (Ref1) and Section 5.2 (Ref2), resulting in other two test cases. Since method Ref1 can only refine the score to a single channel, the results are solely computed with respect to that channel. For the multichannel refinement Ref2, we report the results of the alignment of each instrument with respect to each microphone. A graphic of the initialization of the framework with the four test cases listed above (Ali, Ext, Ref1, and Ref2), along with the ground truth score initialization (GT), is found in Figure 7, where we present the results for these cases in terms of source separation.

In order to evaluate the panning matrix estimation stage, we compute an ideal panning matrix based on the impulse responses generated by Roomsim during the creation of the multichannel audio (see Section 6.3). The ideal panning matrix gives the ideal contribution of each instrument in each channel and it is computed by searching the maximum in the impulse response vector corresponding to each instrument-channel pair, as in

$$m_{\text{ideal}}(i,j) = \max(\text{IR}(i,j)(t)), \qquad (19)$$

where $\text{IR}(i,j)(t)$ is the impulse response of source i in channel j. By comparing the estimated matrix $m(i,j)$ with the ideal one $m_{\text{ideal}}(i,j)$, we can determine if the algorithm picked a wrong channel for separation.

6.4.3. Evaluation Metrics. For score alignment, we are interested in a measure which relates to source separation and accounts for the audio frames which are correctly detected, rather than an alignment rate computed per note onset, as found in [13]. Thus, we evaluate the alignment at the frame level rather than at a note level. A similar reasoning on the evaluation of score alignment is found in [4].

We consider 0.011 s the temporal granularity for this measure and the size of a frame. Then, a frame of a musical note is considered a true positive (tp) if it is found in the ground truth score and in the aligned score in the exact time boundaries. The same frame is labeled as a false positive (fp) if it is found only in the aligned score and a false negative (fn) if it is found only in the ground truth score. Since the gains initialization is done with score information (see Section 4), lost frames (recall), and incorrectly detected frames (precision) impact the performance of the source separation algorithm, precision is defined as $p = \text{tp}/(\text{tp}+\text{fp})$ and recall as $r = \text{tp}/(\text{tp}+\text{fn})$. Additionally, we compute the harmonic mean of precision and recall to obtain F-measure as $F = 2\cdot((p\cdot r)/(p+r))$.

The source separation evaluation framework and metrics employed are described in [41, 42]. Correspondingly, we use *Source to Distortion Ratio* (SDR), *Source to Interference Ratio* (SIR), and *Source to Artifacts Ratio* (SAR). While SDR measures the overall quality of the separation and ISR the spatial reconstruction of the source, SIR is related to rejection of the interferences and SAR to the absence of forbidden distortions and artifacts.

The evaluation of source separation is a computationally intensive process. Additionally, to process the long audio files in the dataset would require large memory to perform the matrix calculations. To reduce the memory requirements, the evaluation is performed for blocks of 30 s with 1 s overlap to allow for continuation.

6.5. Results

6.5.1. Score Alignment. We evaluate the output of the alignment Ali, along the estimation of the note offsets: INT and NEX in terms of F-measure (see Section 6.4.3), precision, and recall, ranging from 0 to 1. Additionally, we evaluate the optimal size for extending the note boundaries along the onsets and offsets, T1 and T2, for the refinement methods, Ref1 and Ref2, and the baseline, Ext. Since there are lots of differences between pieces, we report the results individually per song in Table 4.

Methods Ref1 and Ref2 depend on a binarization threshold which determines how much energy is set to zero. A lower threshold will result in the consolidation of larger blobs in blob detection. In [11], this threshold is set to 0.5 for a dataset of monaural recordings of Bach chorales played by four instruments. However, we are facing a multichannel scenario where capturing the reverberation is important, especially when we consider that offsets were annotated with a low energy threshold. Thus, we are interested in losing the least energy possible and we set lower values for the threshold: 0.3 and 0.1. Consequently, when analyzing the results, a lower threshold achieves better performance in terms of F-measure for Ref1 (0.67 and 0.72) and for Ref2 (0.71 and 0.75).

According to Table 4, extending note offsets (NEX), rather than interpolating them (INT), gives lower recall in all pieces, and the method leads to losing more frames which cannot be recovered even by extending the offset times in T2: NEX T2 yields always a lower recall when compared to INT T2 (e.g., $r = 0.85$ compared to $r = 0.97$ for Mozart).

The output of the alignment system Ali is not a good option to initialize the gains of source separation system. It has a high precision and a very low recall (e.g., the case INT Ali has $p = 0.95$ and $r = 0.39$ compared to case INT Ext

TABLE 4: Alignment evaluated in terms of F-measure, precision, and recall, ranging from 0 to 1.

			Mozart			Beethoven			Mahler			Bruckner		
			F	p	r	F	p	r	F	p	r	F	p	r
INT		Ali	0.69	0.93	0.55	0.56	0.95	0.39	0.61	0.85	0.47	0.60	0.94	0.32
	T1	Ref1	0.77	0.77	0.76	0.79	0.81	0.77	0.63	0.74	0.54	0.72	0.73	0.70
		Ref2	**0.83**	0.79	0.88	0.82	0.82	0.81	0.67	0.77	0.60	0.81	0.77	0.85
		Ext	0.82	0.73	0.94	**0.84**	0.84	0.84	**0.77**	0.69	0.87	**0.86**	0.78	0.96
	T2	Ref1	0.77	0.77	0.76	0.76	0.75	0.77	0.63	0.74	0.54	0.72	0.70	0.71
		Ref2	**0.83**	0.78	0.88	**0.82**	0.76	0.87	0.67	0.77	0.59	**0.79**	0.73	0.86
		Ext	0.72	0.57	0.97	0.79	0.70	0.92	**0.69**	0.55	0.93	0.77	0.63	0.98
NEX		Ali	0.49	0.94	0.33	0.51	0.89	0.36	0.42	0.90	0.27	0.48	0.96	0.44
	T1	Ref1	0.70	0.77	0.64	0.72	0.79	0.66	0.63	0.74	0.54	0.68	0.72	0.64
		Ref2	**0.73**	0.79	0.68	**0.71**	0.79	0.65	0.66	0.77	0.58	0.73	0.74	0.72
		Ext	**0.73**	0.77	0.68	**0.71**	0.80	0.65	**0.69**	0.75	0.64	**0.76**	0.79	0.72
	T2	Ref1	0.74	0.78	0.71	0.72	0.74	0.70	0.63	0.74	0.54	0.72	0.79	0.72
		Ref2	**0.80**	0.80	0.80	**0.75**	0.75	0.75	0.67	0.77	0.59	**0.79**	0.73	0.86
		Ext	0.73	0.65	0.85	0.73	0.69	0.77	**0.72**	0.64	0.82	**0.79**	0.69	0.91

Table 5: Instruments for which the closest microphone was incorrectly determined for different score information (GT, Ali, T1, T2, INT, and NEX) and two room setups.

	GT	INT			NEX		
		Ali	T1	T2	Ali	T1	T2
Setup 1	Clarinet	Clarinet, double bass	Clarinet, flute	Clarinet, flute, horn	Clarinet, double bass		Clarinet, flute
Setup 2	Cello	Cello, flute	Cello, flute	Cello, flute	Bassoon	Flute	Cello, flute

which has $p = 0.84$ and $r = 0.84$ for Beethoven). For the case of Beethoven, the output is particularly poor compared to other pieces. However, by extending the boundaries (Ext) and applying note refinement (Ref1 or Ref2), we are able to increase the recall and match the performance on the other pieces.

When comparing the size for the tolerance window for the onsets and offsets, we observe that the alignment is more accurate with detecting the onsets within 0.3 s and offsets within 0.6 s. In Table 4, T1 achieves better results than T2 (e.g., $F = 0.77$ for T1 compared to $F = 0.69$ for T2, Mahler). Relying on a large window retrieves more frames but also significantly damages the precision. However, when considering source separation we might want to lose as less information as possible. It is in this special case that the refinement methods Ref1 and Ref2 show their importance. When facing larger time boundaries as T2, Ref1 and especially Ref2 are able to reduce the errors by achieving better precision with the minimum amount of loss in recall.

The refinement Ref1 has a worse performance than Ref2, the multichannel refinement (e.g., $F = 0.72$ compared to $F = 0.81$ for Bruckner, INT T1). Note that, in the original version [11], Ref1 was assuming monophony within a source as it was tested in the simple case of Bach10 dataset [4]. To that extent, it was relying on a graph computation to determine the best distribution of blobs. However, due to the increased polyphony within an instrument (e.g., violins playing divisi), with simultaneous melodic lines, we disabled this feature and in this case Ref1 has lower recall, it loses more frames. On the other hand, Ref2 is more robust because it computes a blob estimation per channel. Averaging out these estimations yields better results.

The refinement works worse for more complex pieces (Mahler and Bruckner) than for simple pieces (Mozart and Beethoven). Increasing the polyphony within a source and the number of instruments, having many interleaving melodic lines, a less sparse score, also makes the task more difficult.

6.5.2. Panning Matrix. Estimating correctly the panning matrix is an important step in the proposed method, since Wiener filtering is performed on the channel where the instrument has the most energy. If the algorithm picks a different channel for this step, in the separated audio files we can find more interference between instruments.

As described in Section 3.1, the estimation of the panning matrix depends on the number of nonoverlapping partials of the notes found in the score and their alignment with the audio. To that extent, the more nonoverlapping partials we have, the more robust the estimation.

Initially, we experimented with computing the panning matrix separately for each piece. In the case of Bruckner the piece is simply too short, and there are few nonoverlapping partials to yield a good estimation, resulting in errors in the panning matrix. Since the instrument setup is the same for Bruckner, Beethoven, and Mahler pieces (10 sources in the same position on the stage), we decided to jointly estimate the matrix for the concatenated audio pieces and the associated scores. We denote Setup 1 as the Mozart piece played by 8 sources and Setup 2 the Beethoven, Mahler, and Bruckner pieces played by 10 sources.

Since the panning matrix is computed using the score, different score information can yield very different estimation of the panning matrix. To that extent, we evaluate the influence of audio-to-score alignment, namely, the cases INT, NEX, Ali, T1, and T2, and the initialization with the ground truth score information, GT.

In Table 5, we list the instruments for which the algorithm picked the wrong channel. Note that, in the room setup generated with Roomsim, most of the instruments in Table 5 are placed close to other sources from the same family of instruments: for example, cello and double bass, flute with clarinet, bassoon, and oboe. In this case, the algorithm makes more mistakes when selecting the correct channel to perform source separation.

In the column GT of Table 5, we can see that having a perfectly aligned score yields less errors when estimating the panning matrix. Conversely, in a real-life scenario, we cannot rely on hand annotated score. In this case, for all columns of the Table 5 excluding GT, the best estimation is obtained by the combination of NEX and T1: taking the offset time as the onsets of the next note and then extending the score with a smaller window.

Furthermore, we compute the SDR values for the instruments in Table 5, column GT (clarinet and cello), if the separation was to be done in the correct channel or in the estimated channel. For Setup 1, the channel for clarinet is wrongly mistaken to WWL (woodwinds left), the correct one being WWR (woodwinds right), when we have a perfectly aligned score (GT). However, the microphones WWL and WWR are very close (see Figure 4), and they do not capture significant energy from other instrument sections and the SDR difference is less than 0.01 dB. However, in Setup 2, the cello is wrongly separated in the WWL channel, and the SDR difference between this audio and the audio separated in the correct channel is around −11 dB for each of the three pieces.

6.5.3. Source Separation. We use the evaluation metrics described in Section 6.4.3. Since there is a lot of variability

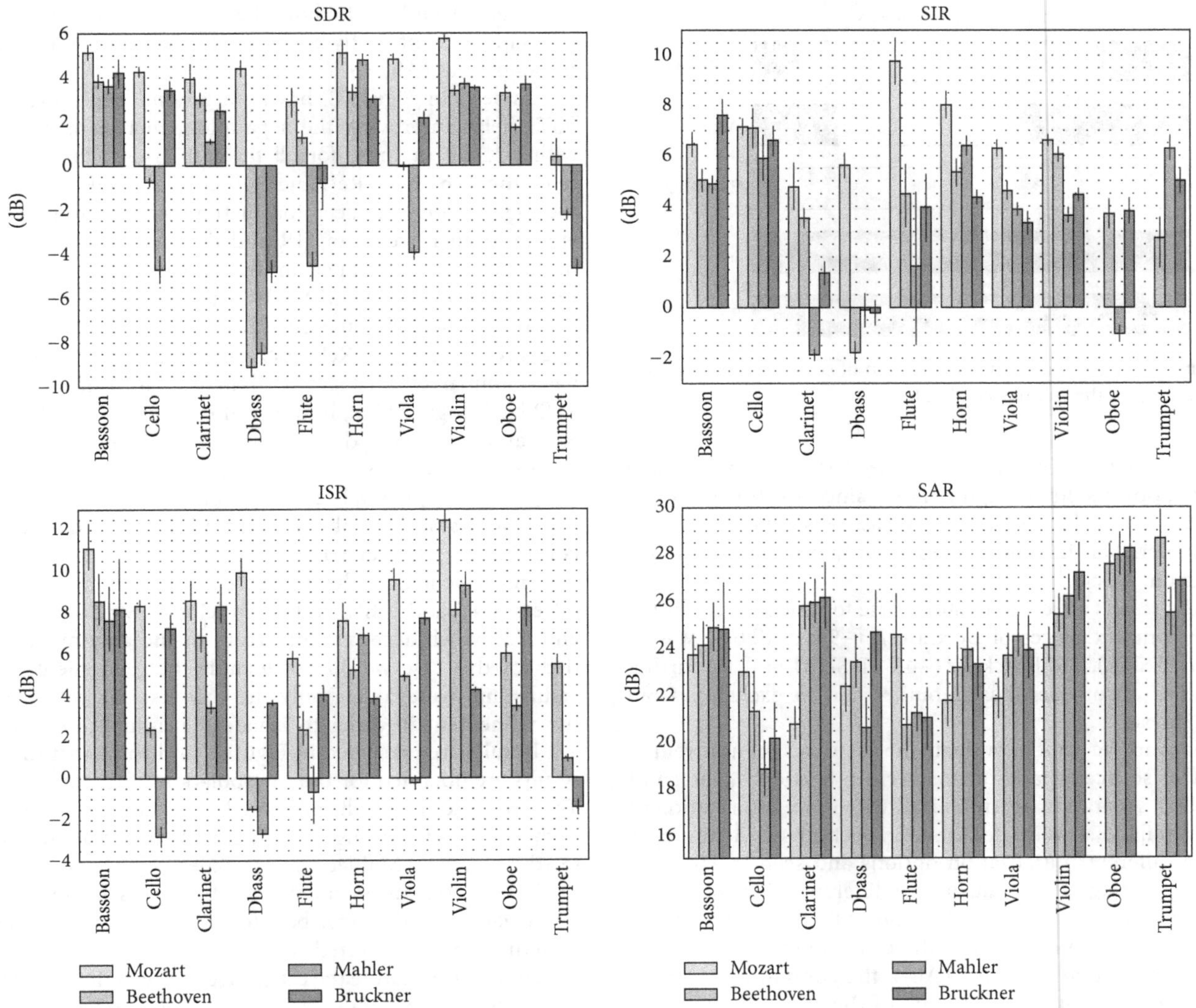

FIGURE 6: Results in terms of SDR, SIR, SAR, and ISR for the instruments and the songs in the dataset.

between the four pieces, it is more informative to present the results per piece rather than aggregating them.

First, we analyze the separation results per instruments in an ideal case. We assume that the best results for score-informed source separation are obtained in the case of a perfectly aligned score (GT). Furthermore, for this case, we calculate the separation in the correct channel for all the instruments, since, in Section 6.5.2, we could see that picking a wrong channel could be detrimental. We present the results as a bar plot in Figure 6.

As described in Section 6.1 and Table 1, the four pieces had different levels of complexity. In Figure 6, we can see that the more complex the piece is, the more difficult it is to achieve a good separation. For instance, note that cello, clarinet, flute, and double bass achieve good results in terms of SDR on Mozart piece but significantly worse results on other three pieces (e.g., 4.5 dB for cello in Mozart, compared to −5 dB in Mahler). Cello and double bass are close by in both of the setups, similarly for clarinet and flute, and we expect interference between them. Furthermore, these instruments usually share the same frequency range which can result in additional interference. This is seen in lower SIR values for double bass (5.5 dB SIR for Mozart, but −1.8, −0.1, and −0.2 dB SIR for the others) and flute.

An issue for the separation is the spatial reconstruction, measured by the ISR metric. As seen in (9), when applying the Wiener mask, the multichannel spectrogram is multiplied with the panning matrix. Thus, wrong values in this matrix can yield wrong amplitude values of the resulting signals.

This is the case for trumpet, which is allocated a close microphone in the current setup, and for which we expect a good separation. However, trumpet achieves a poor ISR (5.5, 1, and −1 dB) but has a good separation in terms of SIR and SAR. Similarly, other instruments as cello, double bass, flute, and viola face the same problem, particularly for the piece by Mahler. Therefore, a good estimation of the panning matrix is crucial for a good ISR.

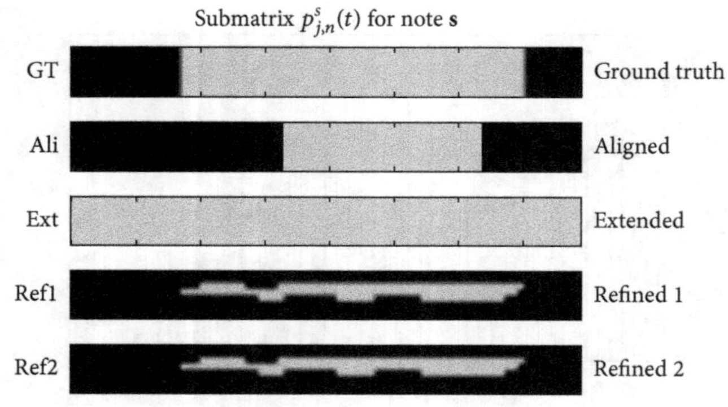

FIGURE 7: The test cases for initialization of score-informed source separation, for the submatrix $p^s_{j,n}(t)$.

A low SDR is obtained in the case of Mahler that is related to the poor results in alignment obtained for this piece. As seen in Table 4 for INT case, F-measure is almost 8% lower in Mahler than in other pieces, mainly because of the bad precision.

The results are considerably worse for double bass for the more complex pieces of Beethoven (−9.1 dB SDR), Mahler (−8.5 dB SDR), and Bruckner (−4.7 dB SDR), and, for further analysis, we consider it as an outlier, and we exclude it from the analysis.

Second, we want to evaluate the usefulness of note refinement in source separation. As seen in Section 4, the gains for NMF separation are initialized with score information or with a refined score as described in Section 4.2. A summary of the different initialization options and seen in Figure 7. Correspondingly, we evaluate five different initializations of the gains: the perfect initialization with the ground truth annotations (Figure 7 (GT)), the direct output of the score alignment system (Figure 7 (Ali)), the common practice of NMF gains initialization in state-of-the-art score-informed source separation [3, 5, 24] (Figure 7 (Ext)), and the refinement approaches (Figure 7 (Ref1 and Ref2)). Note that Ref1 is the refinement done with the method in Section 4.1 and Ref2 with the multichannel method described in Section 5.2.

We test the difference between the binarization thresholds 0.5 and 0.1, used in the refinement methods Ref1 and Ref2. One-way ANOVA on SDR results gives F-value $= 0.0796$ and p value $= 0.7778$, which shows no significant difference between both binarization thresholds.

The results for the five initializations, GT, Ref1, Ref2, Ext, and Ali, are presented in Figure 8, for each of the four pieces. Note that, for Ref1, Ref2, and Ext, we aggregate information across all possible outputs of the alignment: INT, NEX, T1, and T2. Analyzing the results, we note that the more complex the piece, the more difficult to separate between the instruments, the piece by Mahler having the worse results, and the piece by Bruckner a large variance, as seen in the error bars. For these two pieces, other factors as the increased polyphony within a source, the number of instruments (e.g., 12 violins versus 4 violins in a group), and the synchronization issues we described in Section 6.1 can increase the difficulty of separation up to the point that Ref1, Ref2, and Ext have a minimal improvement. To that extent, for the piece by Bruckner, extending the boundaries of the notes (Ext) does not achieve significantly better results than the raw output of the alignment (Ali).

As seen in Figure 8, having a ground truth alignment (GT) helps improving the separation, increasing the SDR with 1–1.5 dB or more for all the test cases. Moreover, the refinement methods Ref1 and Ref2 increase SDR for most of the pieces with the exception of the piece by Mahler. This is due to an increase of SIR and decrease of interferences in the signal. For instance, in the piece by Mozart, Ref1 and Ref2 increase the SDR with 1 dB when compared to Ext. For this piece, the difference in SIR is around 2 dB. Then, for Beethoven, Ref1 and Ref2 increase 0.5 dB in terms of SDR when compared to Ext and 1.5 dB in SIR. For Bruckner, solely Ref2 has a higher SDR; however SIR increases with 1.5 dB in Ref1 and Ref2. Note that not only do Ref1 and Ref2 refine the time boundaries of the notes, but also the refinement happens in frequency, because the initialization is done with the contours of the blobs, as seen in Figure 7. This can also contribute to a higher SIR.

Third, we look at the influence of the estimation of note offsets: INT and NEX, and the tolerance window sizes, T1 and T2, which accounts for errors in the alignment. Note that for this case we do not include the refinement in the results and we evaluate only the case Ext, as we leave out the refinement in order to isolate the influence of T1 and T2. Results are presented in Figure 9 and show that the best results are obtained for the interpolation of the offsets INT. This relates to the results presented in Section 6.5.1. Similarly to the analysis regarding the refinement, the results are worse for the pieces by Mahler and Bruckner, and we are not able to draw a conclusion on which strategy is better for the initialization, as the error bars for the ground truth overlap with the ones of the tested cases.

Fourth, we analyze the difference between the PARAFAC model for multichannel gains estimation as proposed in Section 5.1, compared with the single channel estimation of the gains in Section 3.4. We performed a one-way ANOVA on SDR and we obtain a F-value $= 1.712$ and a p-value $= 0.1908$. Hence, there is no significant difference between single channel and multichannel gain estimation, when we are not performing postprocessing of the gains using grain refinement. However, despite the new updates rule do not help, in the multichannel case we are able to better refine the gains. In this case, we aggregate information all over the channels, and blob detection is more robust, even to variations of the binarization threshold. To account for that, for the piece by Bruckner, Ref2 outperforms Ref1 in terms of SDR and SIR. Furthermore, as seen in Table 4 the alignment is always better for Ref2 than Ref1.

The audio excerpts from the dataset used for evaluation, as well as tracks separated with the ground truth annotated score are made available (http://repovizz.upf.edu/phenicx/anechoic_multi/).

7. Applications

7.1. Instrument Emphasis. The first application of our approach for multichannel score-informed source separation

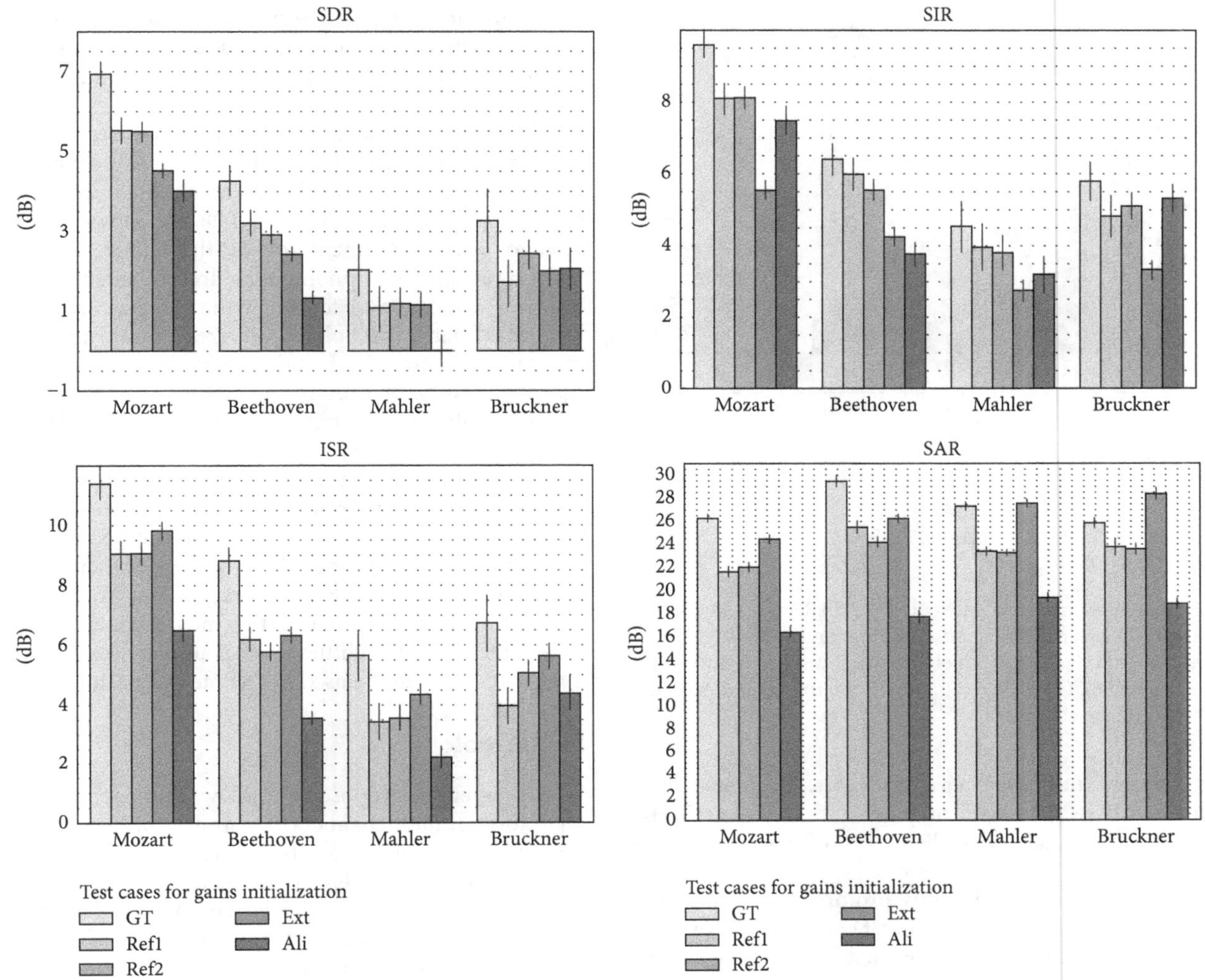

FIGURE 8: Results in terms of SDR, SIR, AR, and ISR for the NMF gains initialization in different test cases.

is *Instrument Emphasis*, which aims at processing multitrack orchestral recordings. Once we have the separated tracks, it allows emphasizing a particular instrument over the full orchestra downmix recording. Our workflow consists in processing multichannel orchestral recording from which the musical score is available. The multitrack recordings are obtained from a typical on-stage setup in a concert hall, where multiple microphones are placed on stage at certain distance of the sources. The goal is to reduce leakage of other sections, obtaining enhanced signal for the selected instrument.

In terms of system integration, this application has two parts. The front-end is responsible for interacting with the user in the uploading of media content and present the results to the user. The back-end is responsible for managing the audio data workflow between the different signal processing components. We process the audio files in batch estimating the signal decomposition for the full length. For long audio files, as in the case of symphonic recordings, the memory requirements can be too demanding even for a server infrastructure. Therefore, to overcome this limitation, audio files are split into blocks. After the separation has been performed, the blocks associated with each instrument are concatenated, resulting in the separated tracks. The separation quality is not degraded if the blocks have sufficient duration. In our case, we set the block duration to 1 minute. Examples of this application are found online (http://repovizz.upf.edu/phenicx/) and are integrated into the PHENICX prototype (http://phenicx.com/).

7.2. Acoustic Rendering. The second application for the separated material is augmented or virtual reality scenarios, where we apply a spatialization of the separated musical sources. Acoustic Rendering aims at recreating acoustically the recorded performance from a specific listening location and orientation, with a controllable disposition of instruments on the stage and the listener.

We have considered binaural synthesis as the most suitable spatial audio technique for this application. Humans locate the direction of incoming sound based on a number of cues: depending on the angle and distance between listener and source, the sound will arrive with a different intensity

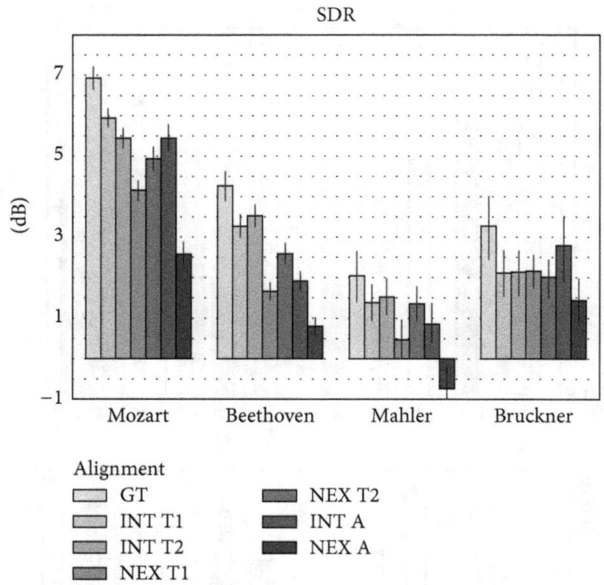

FIGURE 9: Results in terms of SDR, SIR, SAR, and ISR for the combination between different offset estimation methods (INT and Ext) and different sizes for the tolerance window for note onsets and offsets (T1 and T2). GT is the ground truth alignment and Ali is the output of the score alignment.

and at different time instances at both ears. The idea behind binaural synthesis is to artificially generate these cues to be able to create an illusion of directivity of a sound source when reproduced over headphones [43, 44]. For a rapid integration into the virtual reality prototype, we have used the noncommercial plugin for Unity3D of binaural synthesis provided by 3DCeption (https://twobigears.com/index.php).

Specifically, this application opens new possibilities in the area of virtual reality (VR), where video companies are already producing orchestral performances specifically recorded for VR experiences (e.g., the company WeMakeVR has collaborated with the London Symphony Orchestra and the Berliner Philharmoniker https://www.youtube.com/watch?v=ts4oXFmpacA). Using a VR headset with headphones, the Acoustic Rendering application is able to perform an acoustic zoom effect when pointing at a given instrument or section.

7.3. Source Localization. The third application aims to estimate the spatial location of musical instruments on stage. This application is useful for recordings where the orchestra layout is unknown (e.g., small ensemble performances) for the instrument visualization and Acoustic Rendering use-cases introduced above.

As for inputs for the source localization method we need the multichannel recordings and the approximate position of the microphones on stage. In concert halls, the recording setup consists typically of a grid structure of hanging overhead mics. The position of the overhead mics is therefore kept as metadata of the performance recording.

Automatic Sound Source Localization (SSL) methods make use of microphone arrays and complex signal processing techniques; however, undesired effects such as acoustic reflections and noise make this process difficult, being currently a hot-topic task in acoustic signal processing [45].

Our approach is a novel time difference of arrival (TDOA) method based on note-onset delay estimation. It takes the refined score alignment obtained prior to the signal separation (see Section 4.1). It follows two steps: first, for each instrument source, the relative time delays for the various microphone pairs are evaluated, and, then, the source location is found as the intersection of a pair of a set of half-hyperboloids centered around the different microphone pairs. Each half-hyperboloid determines the possible location of a sound source based on the measure of the time difference of arrival between the two microphones for a specific instrument.

To determine the time delay for each instrument and microphone pair, we evaluate a list of time delay values corresponding to all note onsets in the score and take the maximum of the histogram. In our experiment, we have a time resolution of 2.8 ms, corresponding to the Fourier transform hop size. Note that this method does not require time intervals in which the sources to play isolated as SRP-PHAT [45] and can be used in complex scenarios.

8. Outlook

In this paper, we proposed a framework for score-informed separation of multichannel orchestral recordings in distant-microphone scenarios. Furthermore, we presented a dataset which allows for an objective evaluation of alignment and separation (Section 6.1) and proposed a methodology for future research to understand the contribution of the different steps of the framework (Section 6.5). Then, we introduced several applications of our framework (Section 7). To our knowledge, this is the first time the complex scenario of orchestral multichannel recordings is objectively evaluated for the task of score-informed source separation.

Our framework relies on the accuracy of an audio-to-score alignment system. Thus, we assessed the influence of the alignment on the quality of separation. Moreover, we proposed and evaluated approaches to refining the alignment which improved the separation for three of the four pieces in the dataset, when compared to other two initialization options for our framework: the raw output of the score alignment and the tolerance window which the alignment relies on.

The evaluation shows that the estimation of panning matrix is an important step. Errors in the panning matrix can result into more interference in separated audio, or to problems in recovery of the amplitude of a signal. Since the method relies on finding nonoverlapping partials, an estimation done on a larger time frame is more robust. Further improvements on determining the correct channel for an instrument can take advantage of our approach for source localization in Section 7, provided that the method is reliable enough in localizing a large number of instruments. To that extent, the best microphone to separate a source is the closest one determined by the localization method.

When looking at separation across the instruments, viola, cello, and double bass were more problematic in the more complex pieces. In fact, the quality of the separation in our experiment varies within the pieces and the instruments, and future research could provide more insight on this problem. Note that increasing degree of consonance was related to a more difficult case for source separation [17]. Hence, we could expect a worse separation for instruments which are harmonizing or accompanying other sections, as the case of viola, cello, and double bass in some pieces. Future research could find more insight on the relation between the musical characteristics of the pieces (e.g., tonality and texture) and source separation quality.

The evaluation was conducted on the dataset presented in Section 6.1. The creation of the dataset was a very laborious task, which involved annotating around 12000 pairs of onsets and offsets, denoising the original recordings and testing different room configurations in order to create the multichannel recordings. To that extent, annotations helped us to denoise the audio files, which could then be used in score-informed source separation experiments. Furthermore, the annotations allow for other tasks to be tested within this challenging scenario, such as instrument detection, or transcription.

Finally, we presented several applications of the proposed framework related to Instrument Emphasis or Acoustic Rendering, some of which are already at the stage of functional products.

Competing Interests

The authors declare that they have no competing interests.

Acknowledgments

The authors would like to thank all the partners of the PHENICX consortium for a very fruitful collaboration. Particularly, they would like to thank Agustin Martorell for his insights on the musicology side and correcting the scores and Oscar Mayor for his help regarding Repovizz and his contribution on the applications. This work is partially supported by the Spanish Ministry of Economy and Competitiveness under CASAS Project (TIN2015-70816-R).

References

[1] E. Gómez, M. Grachten, A. Hanjalic et al., "PHENICX: performances as highly enriched aNd Interactive concert experiences," in *Proceedings of the SMAC Stockholm Music Acoustics Conference 2013 and SMC Sound and Music Computing Conference*, 2013.

[2] S. Ewert and M. Müller, "Using score-informed constraints for NMF-based source separation," in *Proceedings of the IEEE International Conference on Acoustics, Speech, and Signal Processing (ICASSP '12)*, pp. 129–132, Kyoto, Japan, March 2012.

[3] J. Fritsch and M. D. Plumbley, "Score informed audio source separation using constrained nonnegative matrix factorization and score synthesis," in *Proceedings of the 38th IEEE International Conference on Acoustics, Speech, and Signal Processing (ICASSP '13)*, pp. 888–891, Vancouver, Canada, May 2013.

[4] Z. Duan and B. Pardo, "Soundprism: an online system for score-informed source separation of music audio," *IEEE Journal on Selected Topics in Signal Processing*, vol. 5, no. 6, pp. 1205–1215, 2011.

[5] R. Hennequin, B. David, and R. Badeau, "Score informed audio source separation using a parametric model of non-negative spectrogram," in *Proceedings of the 36th IEEE International Conference on Acoustics, Speech, and Signal Processing (ICASSP '11)*, pp. 45–48, May 2011.

[6] J. J. Carabias-Orti, M. Cobos, P. Vera-Candeas, and F. J. Rodriguez-Serrano, "Nonnegative signal factorization with learnt instrument models for sound source separation in close-microphone recordings," *EURASIP Journal on Advances in Signal Processing*, vol. 2013, article 184, 2013.

[7] T. Pratzlich, R. M. Bittner, A. Liutkus, and M. Muller, "Kernel Additive Modeling for interference reduction in multi-channel music recordings," in *Proceedings of the 40th IEEE International Conference on Acoustics, Speech, and Signal Processing (ICASSP '15)*, pp. 584–588, Brisbane, Australia, April 2014.

[8] S. Ewert, B. Pardo, M. Mueller, and M. D. Plumbley, "Score-informed source separation for musical audio recordings: an overview," *IEEE Signal Processing Magazine*, vol. 31, no. 3, pp. 116–124, 2014.

[9] F. J. Rodriguez-Serrano, Z. Duan, P. Vera-Candeas, B. Pardo, and J. J. Carabias-Orti, "Online score-informed source separation with adaptive instrument models," *Journal of New Music Research*, vol. 44, no. 2, pp. 83–96, 2015.

[10] F. J. Rodriguez-Serrano, S. Ewert, P. Vera-Candeas, and M. Sandler, "A score-informed shift-invariant extension of complex matrix factorization for improving the separation of overlapped partials in music recordings," in *Proceedings of the IEEE International Conference on Acoustics, Speech and Signal Processing (ICASSP '16)*, pp. 61–65, Shanghai, China, 2016.

[11] M. Miron, J. J. Carabias, and J. Janer, "Improving score-informed source separation for classical music through note refinement," in *Proceedings of the 16th International Society for Music Information Retrieval (ISMIR '15)*, Málaga, Spain, October 2015.

[12] A. Arzt, H. Frostel, T. Gadermaier, M. Gasser, M. Grachten, and G. Widmer, "Artificial intelligence in the concertgebouw," in *Proceedings of the 24th International Joint Conference on Artificial Intelligence*, pp. 165–176, Buenos Aires, Argentina, 2015.

[13] J. J. Carabias-Orti, F. J. Rodriguez-Serrano, P. Vera-Candeas, N. Ruiz-Reyes, and F. J. Cañadas-Quesada, "An audio to score alignment framework using spectral factorization and dynamic time warping," in *Proceedings of the 16th International Society for Music Information Retrieval (ISMIR '15)*, Malaga, Spain, 2015.

[14] Ö. Izmirli and R. Dannenberg, "Understanding features and distance functions for music sequence alignment," in *Proceedings of the 11th International Society for Music Information Retrieval Conference (ISMIR '10)*, pp. 411–416, Utrecht, The Netherlands, 2010.

[15] M. Goto, "Development of the RWC music database," in *Proceedings of the 18th International Congress on Acoustics (ICA '04)*, pp. 553–556, Kyoto, Japan, April 2004.

[16] S. Ewert, M. Müller, and P. Grosche, "High resolution audio synchronization using chroma onset features," in *Proceedings of the IEEE International Conference on Acoustics, Speech, and Signal Processing (ICASSP '09)*, pp. 1869–1872, IEEE, Taipei, Taiwan, April 2009.

[17] J. J. Burred, *From sparse models to timbre learning: new methods for musical source separation [Ph.D. thesis]*, 2008.

[18] F. J. Rodriguez-Serrano, J. J. Carabias-Orti, P. Vera-Candeas, T. Virtanen, and N. Ruiz-Reyes, "Multiple instrument mixtures source separation evaluation using instrument-dependent NMF models," in *Proceedings of the 10th international conference on Latent Variable Analysis and Signal Separation (LVA/ICA '12)*, pp. 380–387, Tel Aviv, Israel, March 2012.

[19] A. Klapuri, T. Virtanen, and T. Heittola, "Sound source separation in monaural music signals using excitation-filter model and EM algorithm," in *Proceedings of the IEEE International Conference on Acoustics, Speech, and Signal Processing (ICASSP '10)*, pp. 5510–5513, March 2010.

[20] J.-L. Durrieu, A. Ozerov, C. Févotte, G. Richard, and B. David, "Main instrument separation from stereophonic audio signals using a source/filter model," in *Proceedings of the 17th European Signal Processing Conference (EUSIPCO '09)*, pp. 15–19, Glasgow, UK, August 2009.

[21] J. J. Bosch, K. Kondo, R. Marxer, and J. Janer, "Score-informed and timbre independent lead instrument separation in real-world scenarios," in *Proceedings of the 20th European Signal Processing Conference (EUSIPCO '12)*, pp. 2417–2421, August 2012.

[22] P. S. Huang, M. Kim, M. Hasegawa-Johnson, and P. Smaragdis, "Singing-voice separation from monaural recordings using deep recurrent neural networks," in *Proceedings of the International Society for Music Information Retrieval Conference (ISMIR '14)*, Taipei, Taiwan, October 2014.

[23] E. K. Kokkinis and J. Mourjopoulos, "Unmixing acoustic sources in real reverberant environments for close-microphone applications," *Journal of the Audio Engineering Society*, vol. 58, no. 11, pp. 907–922, 2010.

[24] S. Ewert and M. Müller, "Score-informed voice separation for piano recordings," in *Proceedings of the 12th International Society for Music Information Retrieval Conference (ISMIR '11)*, pp. 245–250, Miami, Fla, USA, October 2011.

[25] B. Niedermayer, *Accurate audio-to-score alignment-data acquisition in the context of computational musicology [Ph.D. thesis]*, Johannes Kepler University Linz, Linz, Austria, 2012.

[26] S. Wang, S. Ewert, and S. Dixon, "Compensating for asynchronies between musical voices in score-performance alignment," in *Proceedings of the 40th IEEE International Conference on Acoustics, Speech, and Signal Processing (ICASSP '15)*, pp. 589–593, April 2014.

[27] M. Miron, J. Carabias, and J. Janer, "Audio-to-score alignment at the note level for orchestral recordings," in *Proceedings of the 15th International Society for Music Information Retrieval Conference (ISMIR '14)*, Taipei, Taiwan, 2014.

[28] D. Fitzgerald, M. Cranitch, and E. Coyle, "Non-negative tensor factorisation for sound source separation. acoustics, speech and signal processing," in *Proceedings of the 2006 IEEE International Conference on Acoustics, Speech, and Signal (ICASSP '06)*, Toulouse, France, 2006.

[29] C. Févotte and A. Ozerov, "Notes on nonnegative tensor factorization of the spectrogram for audio source separation: statistical insights and towards self-clustering of the spatial cues," in *Exploring Music Contents: 7th International Symposium, CMMR 2010, Málaga, Spain, June 21–24, 2010. Revised Papers*, vol. 6684 of *Lecture Notes in Computer Science*, pp. 102–115, Springer, Berlin, Germany, 2011.

[30] J. Pätynen, V. Pulkki, and T. Lokki, "Anechoic recording system for symphony orchestra," *Acta Acustica United with Acustica*, vol. 94, no. 6, pp. 856–865, 2008.

[31] D. Campbell, K. Palomäki, and G. Brown, "A Matlab simulation of shoebox room acoustics for use in research and teaching," *Computing and Information Systems*, vol. 9, no. 3, p. 48, 2005.

[32] O. Mayor, Q. Llimona, M. Marchini, P. Papiotis, and E. Maestre, "RepoVizz: a framework for remote storage, browsing, annotation, and exchange of multi-modal data," in *Proceedings of the 21st ACM International Conference on Multimedia (MM '13)*, pp. 415–416, Barcelona, Spain, October 2013.

[33] D. D. Lee and H. S. Seung, "Learning the parts of objects by non-negative matrix factorization," *Nature*, vol. 401, no. 6755, pp. 788–791, 1999.

[34] C. Févotte, N. Bertin, and J.-L. Durrieu, "Nonnegative matrix factorization with the Itakura-Saito divergence: with application to music analysis," *Neural Computation*, vol. 21, no. 3, pp. 793–830, 2009.

[35] M. Nixon, *Feature Extraction and Image Processing*, Elsevier Science, 2002.

[36] R. Parry and I. Essa, "Estimating the spatial position of spectral components in audio," in *Independent Component Analysis and Blind Signal Separation: 6th International Conference, ICA 2006, Charleston, SC, USA, March 5–8, 2006. Proceedings*, J. Rosca, D. Erdogmus, J. C. Príncipe, and S. Haykin, Eds., vol. 3889 of *Lecture Notes in Computer Science*, pp. 666–673, Springer, Berlin, Germany, 2006.

[37] P. Papiotis, M. Marchini, A. Perez-Carrillo, and E. Maestre, "Measuring ensemble interdependence in a string quartet through analysis of multidimensional performance data," *Frontiers in Psychology*, vol. 5, article 963, 2014.

[38] M. Mauch and S. Dixon, "PYIN: a fundamental frequency estimator using probabilistic threshold distributions," in *Proceedings of the IEEE International Conference on Acoustics, Speech, and Signal Processing (ICASSP '14)*, pp. 659–663, Florence, Italy, May 2014.

[39] F. J. Cañadas-Quesada, P. Vera-Candeas, D. Martínez-Muñoz, N. Ruiz-Reyes, J. J. Carabias-Orti, and P. Cabanas-Molero, "Constrained non-negative matrix factorization for score-informed piano music restoration," *Digital Signal Processing*, vol. 50, pp. 240–257, 2016.

[40] A. Cont, D. Schwarz, N. Schnell, and C. Raphael, "Evaluation of real-time audio-to-score alignment," in *Proceedings of the 8th International Conference on Music Information Retrieval (ISMIR '07)*, pp. 315–316, Vienna, Austria, September 2007.

[41] E. Vincent, R. Gribonval, and C. Févotte, "Performance measurement in blind audio source separation," *IEEE Transactions on Audio, Speech and Language Processing*, vol. 14, no. 4, pp. 1462–1469, 2006.

[42] V. Emiya, E. Vincent, N. Harlander, and V. Hohmann, "Subjective and objective quality assessment of audio source separation," *IEEE Transactions on Audio, Speech and Language Processing*, vol. 19, no. 7, pp. 2046–2057, 2011.

[43] D. R. Begault and E. M. Wenzel, "Headphone localization of speech," *Human Factors*, vol. 35, no. 2, pp. 361–376, 1993.

[44] G. S. Kendall, "A 3-D sound primer: directional hearing and stereo reproduction," *Computer Music Journal*, vol. 19, no. 4, pp. 23–46, 1995.

[45] A. Martí Guerola, *Multichannel audio processing for speaker localization, separation and enhancement*, Universitat Politècnica de València, 2013.

20

Iterative Signal Processing for Mitigation of Analog-to-Digital Converter Clipping Distortion in Multiband OFDMA Receivers

Markus Allén, Toni Levanen, Jaakko Marttila, and Mikko Valkama

Department of Communications Engineering, Tampere University of Technology, P.O. Box 553, 33101 Tampere, Finland

Correspondence should be addressed to Markus Allén, markus.allen@tut.fi

Academic Editor: Tzi-Dar Chiueh

In modern wideband communication receivers, the large input-signal dynamics is a fundamental problem. Unintentional signal clipping occurs, if the receiver front-end with the analog-to-digital interface cannot respond to rapidly varying conditions. This paper discusses digital postprocessing compensation of such unintentional clipping in multiband OFDMA receivers. The proposed method iteratively mitigates the clipping distortion by exploiting the symbol decisions. The performance of the proposed method is illustrated with various computer simulations and also verified by concrete laboratory measurements with commercially available analog-to-digital hardware. It is shown that the clipping compensation algorithm implemented in a turbo decoding OFDM receiver is able to remove almost all the clipping distortion even under significant clipping in fading channel circumstances. That is to say, it is possible to nearly recover the receiver performance to the level, which would be achieved in the equivalent nonclipped situation.

1. Introduction

Modern wideband radio receivers, such as cognitive radios, are setting significant challenges for the design of the receiver front-end. One of the key issues is to have enough dynamic range [1, 2]. Current communication waveforms tend to have a high peak-to-average-power ratio (PAPR), which makes it challenging to optimize the usage of power amplifier on the transmitter side, and it can also increase the required dynamic range in the receiver. Moreover, the reception becomes even more challenging, if a wide frequency band with several independent signals is received with a single receiver front-end and A/D interface. Those independent signals can have considerably different power levels and hence the required dynamic range is not determined only by the PAPR of a single-user signal but by the dynamic range of the overall received waveform [3, 4]. The power difference between two independent signals in mobile environment can be several tens of dBs, which is a particularly important issue to take into account in cognitive radios. Due to the continuously changing conditions, the automatic gain control (AGC) of the receiver might not be able to follow the signal dynamics. Especially the suddenly appearing strong adjacent-channel signals, with no power control from secondary (cognitive) radio point of view will cause problems. The AGC failure causes too high input level for the A/D converter and therefore the amplitude of the digitized waveform is saturated, that is, the highest signal peaks are clipped. This causes considerable signal distortion especially at the weak signal bands due to intermodulation of strong input components.

This paper discusses the compensation of unintentional receiver clipping occurring in A/D converters using digital postprocessing. More specifically, the focus is on orthogonal frequency division multiplexing- (OFDM-) based multiple access scheme, noted in the literature as OFDMA. The reception of a weak signal can be seriously compromised in the presence of a stronger adjacent-channel signal, because clipping causes relatively high amount of nonlinear distortion to the weak signal band of interest. The current literature has only a very limited number of publications proposing digital signal processing methods for reducing unintentional clipping taking place *on the receiver side* [5–7]. The aforementioned references contain rather general methods, which can be applied to different kinds of communication systems. Nevertheless, a compensation method, which is designed for a certain system, can perform better from the accuracy or computational complexity point of view. Therefore, this paper specifically focuses on OFDM, which is widely used in emerging radio systems, and the receiver clipping compensation methods are likely to be needed. Currently, most of the OFDM-related clipping compensation algorithms in the literature concentrate on deliberate clipping to reduce signal PAPR *on the transmitter side* [8–10]. These methods are not directly applicable to mitigate clipping occurring on the receiver since the exact clipping level is not known in the case of unintentional clipping. In addition, the receiver-side clipping case is more challenging, because it happens after the radio channel and hence interferes with the channel estimation.

The proposed receiver clipping compensation method relies on the idea that most of the bit decisions made on the receiver are still correct and hence they can be exploited in clipping distortion estimation. This kind of decision-based digital compensation methods have been proposed in the literature for mitigation of various RF impairments, for example, phase noise [11, 12] and power amplifier nonlinearity [13]. This paper extends the work initiated in [14], which uses the PAPR reduction scheme of [9, 10] as a starting point and then modifies it to be suitable for the challenging task of removing unintentional receiver-side clipping. This paper provides more thorough signal modeling, performance analysis, and considers the use of turbo codec as well as more advanced channel estimation than what is presented in [14]. Additionally, the performance of the receiver clipping compensation method is verified with laboratory measurements employing commercially available A/D hardware. The results show that the proposed receiver clipping compensation method effectively mitigates the clipping distortion, is robust against various estimation errors, and is implementable with reasonable hardware costs in practice.

The remainder of the paper is organized as follows. Section 2 provides background information about the modeling of A/D converter clipping and also defines the notation used in the paper to express OFDM(A) signals. Section 3 then introduces the postprocessing algorithm for compensating clipping distortion and discusses its implementation details. The performance of the proposed algorithm is carefully studied in Section 4 using computer simulations. In addition, practical laboratory measurements with real A/D hardware are used to verify the functionality of the algorithm. After that, the computational complexity of the algorithm in general is discussed. Section 5 finishes the paper by drawing conclusions.

2. Received OFDM Signal Model and Clipping Phenomenon

This paper considers the traditional OFDM signal model, where the transmitted baseband OFDM symbol in time domain is first expressed as

$$x(t) = \frac{1}{\sqrt{N}} \sum_{k=-N/2}^{N/2-1} X_k e^{j2\pi kt/T_s}, \quad 0 \le t \le T_s, \tag{1}$$

where X_k is the kth complex data symbol (subcarrier k) and T_s is the OFDM symbol duration. This OFDM symbol

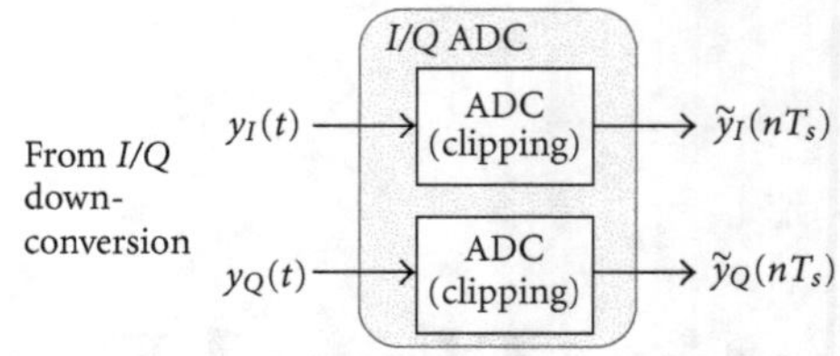

FIGURE 1: Illustration of an I/Q A/D converter, which contains separate converters for the I and Q branches of the received downconverted signal $y(t) = y_I(t) + jy_Q(t)$.

contains N_A active subcarriers (SC) and has an oversampling factor J. Therefore, the total amount of SCs is $N = JN_A$ and the SC indices are expressed as a set $\Omega = \{-N/2, -N/2+1, \dots, N/2-1\}$. The active SCs carry the data symbol sequence X_k, $k \in \Omega_A = \{-N_A/2, \dots, -1, 1, \dots, N_A/2\}$, whereas the nonactive SCs contain zeros, that is, $X_k = 0$, $k \in \Omega \setminus \Omega_A$. It is worth noticing that the presented OFDM signal model does not include cyclic prefix (CP), which is required in practice to avoid intersymbol interference in frequency-selective mobile environments. This paper omits the CP from the mathematical notation in order to make it simpler. This simplification does not affect the presented clipping analysis in any way, because clipping is a memoryless phenomenon. On the other hand, the CP is taken into account in the proposed receiver structures of Section 3 and in the simulations of Section 4 due to the CP's essential role in implementations of practical communication systems.

A zero-symmetric hard-clipping model is a simple but accurate way to characterize signal clipping in receiver A/D converters. Let us first define a general received waveform as $y(t) = y_I(t) + jy_Q(t) = h(t) * x(t) + w(t)$, where the noise $w(t)$ is added to the convolution of the channel impulse response $h(t)$ and the transmitted signal $x(t)$. After an I/Q A/D converter (see Figure 1), clipping occurring separately in both branches, the discrete-time received signal is $\tilde{y}(nT_s) = \tilde{y}_I(nT_s) + j\tilde{y}_Q(nT_s)$, where the zero-symmetric hard-clipping model determines that

$$\tilde{y}_I(t) = \begin{cases} y_I(t), & |y_I(t)| < V_{0,I} \\ V_{0,I}, & y_I(t) \ge V_{0,I} \\ -V_{0,I}, & y_I(t) \le -V_{0,I}, \end{cases}$$
$$\tilde{y}_Q(t) = \begin{cases} y_Q(t), & |y_Q(t)| < V_{0,Q} \\ V_{0,Q}, & y_Q(t) \ge V_{0,Q} \\ -V_{0,Q}, & y_Q(t) \le -V_{0,Q}, \end{cases} \tag{2}$$

where $V_{0,I}$ and $V_{0,Q}$ denote the maximum output levels of the A/D converters of I and Q branches, respectively. Often in practice, a convenient way to express the clipping level of the signal is needed, but $V_{0,I}$ ($V_{0,Q}$) by itself is not very descriptive if nothing about the signal is known. Therefore, the common practice is to define the clipping level using a ratio of $V_{0,I}$ ($V_{0,Q}$) and the average branch power level of the nonclipped signal $P_{\text{in},I}$ ($P_{\text{in},Q}$). In other words, the clipping level for I branch γ_I can be written in the form of

$$\gamma_I = \frac{V_{0,I}}{\sqrt{P_{\text{in},I}}}, \tag{3}$$

and similarly

$$\gamma_Q = \frac{V_{0,Q}}{\sqrt{P_{\text{in},Q}}}, \tag{4}$$

for the Q branch. These variables give a precise understanding of how intensely the signal branch is actually clipped. In addition, it can be concluded from (3) and (4) that the clipping level γ is different for I and Q branches, if the maximum input levels of the converters are different ($V_{0,I} \neq V_{0,Q}$) or the branch signal powers are not equal ($P_{\text{in},I} \neq P_{\text{in},Q}$). Notice that even though $V_{0,I}$ and $V_{0,Q}$ are most likely designed to be equal, unavoidable implementation tolerances of two physical separate converter circuits are indeed likely to cause small relative deviations between them.

The clipping phenomenon removes the highest signal amplitudes and hence causes nonlinear distortion. This process defined in (2) is not directly invertible. Therefore, another kind of approach is required for compensating the clipping distortion, especially at weak signal bands. This paper exploits Bussgang's theorem [15] for the compensation purpose. It has already been proven to be an adequate way to model the clipping distortion [14, 16–18]. Basically, Bussgang's theorem states that the clipping attenuates the received signal $y(nT_s)$ with a factor α and also causes additive clipping distortion $d(nT_s)$, which is uncorrelated with the nonclipped $y(nT_s)$. Therefore, the clipped signal branches can be expressed as

$$\begin{aligned} \widetilde{y}_I(nT_s) &= \alpha_I y_I(nT_s) + d_I(nT_s), \\ \widetilde{y}_Q(nT_s) &= \alpha_Q y_Q(nT_s) + d_Q(nT_s), \end{aligned} \tag{5}$$

when for modeling purposes $y_I(nT_s)$ and $y_Q(nT_s)$ are assumed to be Gaussian. Here, α_I and α_Q depend on γ_I and γ_Q, respectively. For the corresponding complex signal $\widetilde{y}(nT_s)$, it can be written that

$$\begin{aligned} \widetilde{y}(nT_s) &= \alpha_I y_I(nT_s) + d_I(nT_s) + j[\alpha_Q y_Q(nT_s) + d_Q(nT_s)] \\ &= \alpha_1 y(nT_s) + \alpha_2 y^*(nT_s) + d(nT_s) \\ &= \alpha_1 y(nT_s) + z(nT_s), \end{aligned} \tag{6}$$

where $\alpha_1 = (\alpha_I + \alpha_Q)/2$, $\alpha_2 = (\alpha_I - \alpha_Q)/2$, complex clipping distortion $d(nT_s) = d_I(nT_s) + jd_Q(nT_s)$, and total interference $z(nT_s) = \alpha_2 y^*(nT_s) + d(nT_s)$. In the general case with unequal clipping levels in the I and Q branches, the total interference $z(nT_s)$ contains both the nonlinear clipping distortion $d(nT_s)$ and the conjugate signal interference $\alpha_2 y^*(nT_s)$, which in the frequency domain corresponds to mirror-frequency interference. In the special case of identical clipping levels for both branches, $\gamma_I = \gamma_Q = \gamma$, the attenuation factor is $\alpha = \alpha_I = \alpha_Q$ and, therefore, $\alpha_1 = \alpha$ and $\alpha_2 = 0$. This means that the general complex signal model presented in (6) reduces to

$$\widetilde{y}(nT_s) = \alpha y(nT_s) + d(nT_s), \tag{7}$$

which is in accordance with the traditional Bussgang's theorem for a complex signal.

Even a clipping model as simple as (6) or (7) is justified due to the memoryless nature of the clipping phenomenon. In addition, the model is linear and hence straightforward to utilize in clipping compensation purposes as described in Section 3. In short, if α_1 and $z(nT_s)$ can be estimated, recovery of the nonclipped signal $y(t)$ is simple using (6). The Gaussianity assumption for $y_I(nT_s)$ and $y_Q(nT_s)$ also makes it possible to derive concise equations for α_I and α_Q, namely,

$$\begin{aligned} \alpha_I &= \frac{E[y_I(nT_s)\widetilde{y}_I(nT_s)]}{E[y_I(nT_s)y_I(nT_s)]} = \operatorname{erf}\left(\frac{\gamma_I}{\sqrt{2}}\right), \\ \alpha_Q &= \frac{E[y_Q(nT_s)\widetilde{y}_Q(nT_s)]}{E[y_Q(nT_s)y_Q(nT_s)]} = \operatorname{erf}\left(\frac{\gamma_Q}{\sqrt{2}}\right). \end{aligned} \tag{8}$$

Here the error function $\operatorname{erf}(x)$ is defined as

$$\operatorname{erf}(x) = \frac{2}{\sqrt{\pi}} \int_0^x e^{-t^2} dt. \tag{9}$$

The details related to the derivation of (8) are presented in [19].

3. Iterative Decision-Aided Method for Compensating Unintentional Receiver Clipping

This section proposes a compensation algorithm for removing signal distortion caused by unintentional clipping in the A/D interface of an OFDM receiver. First, the basic compensation idea and its justification through mathematical analysis are described in Section 3.1 with the help of a block diagram presented in Figure 2. After that, the most essential details from the implementation point of view are discussed in Section 3.2. Unintentional clipping is most likely to happen in multiuser/multiband situations. Therefore, Section 3.3 deals with the multiuser aspects of the compensation algorithm implementation from the uplink (UL) and downlink (DL) point of view.

3.1. The Compensation Algorithm. On top part of Figure 2, a traditional OFDM transmitter exploiting turbo coding is illustrated. After channel coding and interleaving, the coded bits $c_{1,\ldots,M}^{(X_k)}$, $k \in \Omega_A$, are mapped to data symbols $X_k, k \in \Omega_A$, and then the time-domain signal with a CP is formed to be sent on the radio channel. Here, M denotes the number of bits in one data symbol. Figure 2 also illustrated two different receiver options for implementing the proposed clipping compensation algorithm. In other words, the purpose of Figure 2 is to illustrate the proposed receiver structures in a simplified manner. It does not present system-level hierarchy, where there are several users and, therefore, several transmitters and receivers. However, the proposed clipping

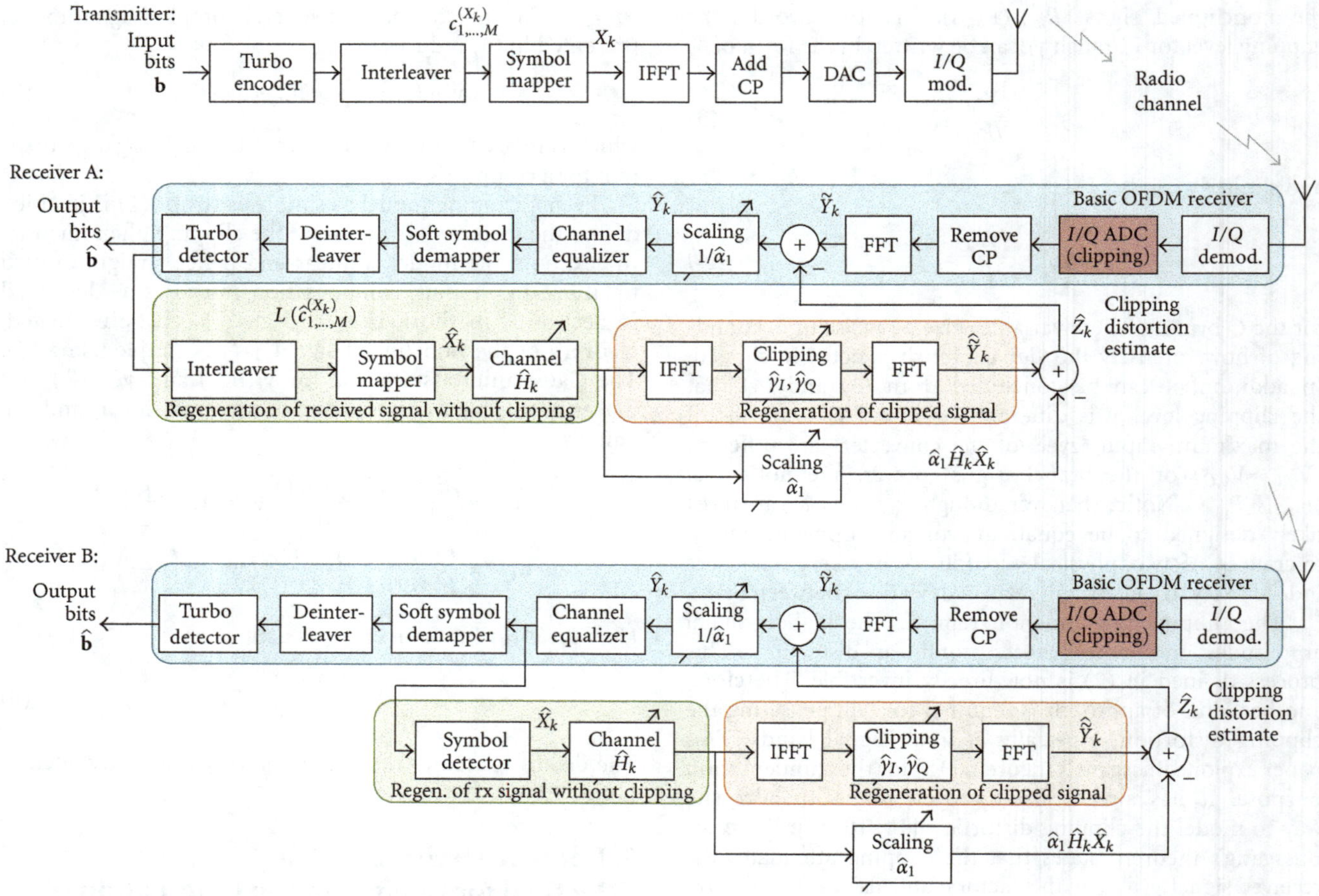

FIGURE 2: Undesired clipping distortion is caused by the A/D converter on the receiver side. Block diagrams of traditional OFDM transmitter and two alternative receivers implementing the proposed clipping compensation algorithm are illustrated. In Receiver A, bits are decoded on every compensation iteration round whereas Receiver B first iterates the clipping compensation on symbol level and only in the end decodes the information bits.

compensation algorithm is able to jointly process multiuser data inside one receiver as is discussed in Section 3.3.

Receiver A in Figure 2 presents the first proposition of this paper on how to remove the clipping distortion caused by the A/D interface of the receiver. In brief, the received and digitized signal is first detected in a conventional manner, that is, transformed to frequency domain, equalized, mapped to bits, deinterleaved, and decoded. Now, the decoded bits are exploited in the clipping compensation feedback loop by generating an estimate of the received signal with and without clipping. Based on (6), an estimate of the clipping distortion in the original received signal can be obtained by subtracting the scaled nonclipped estimate of the received signal from the clipped one. After that, the clipping distortion estimate can be subtracted from the original received signal in order to enhance it and obtain better detection results. Then, the whole process is iterated, because a better clipping distortion estimate can be obtained based on the enhanced detection results. Receiver B in Figure 2 implements this same clipping compensation idea, but the symbol demapper, deinterleaver, and turbo decoder are now located outside the clipping compensation loop. Hence, the computational complexity is greatly reduced since the data is kept on symbol level until the end of the clipping compensation iterations and after that the turbo decoder is used only once. The trade-off is that without turbo decoding, there are likely to be more symbol errors and, therefore, the clipping distortion estimate is not as accurate as it could be.

The following paragraphs describe the proposed clipping compensation algorithm in details by providing the exact mathematical notation. The key steps are also summarized in Algorithms 1 and 2 for Receiver A and B, respectively. It is worth noticing that only one OFDM symbol duration is considered in the equations in order to simplify the notation. The clipping compensation can be done separately for each OFDM symbol and hence the simplified notation does not omit any important details. However, in practice, the processing block length may be longer than one OFDM symbol due to the other receiver functions such as channel decoding, deinterleaving, and equalization, but this does not affect the execution of the clipping compensation, which is performed subcarrierwise.

Initialization: for all $k \in \Omega_A$

(1) Set $i = 0$, $\hat{\alpha}_1^{(0)} = 1$ and $\hat{Z}_k^{(0)} = 0$

(2) Obtain channel estimates $\hat{H}_k^{(i)}$ using signal $\hat{Y}_k^{(i)} = (1/\hat{\alpha}_1^{(i)})\,(\tilde{Y}_k - \hat{Z}_k^{(i)})$

(3) Channel equalization: $\hat{Y}_k^{(i)}/\hat{H}_k^{(i)}$

(4) Demap symbols $\hat{Y}_k^{(i)}/\hat{H}_k^{(i)}$ to bits, deinterleave and decode to obtain output data bits $\hat{\mathbf{b}}^{(i)}$

Iteration: Increase i by one and do the following steps for all $k \in \Omega_A$

(5) Create $\hat{X}_k^{(i)}$ based on the coded bits $\hat{c}_{1,\ldots,M}^{(X_k^{(i-1)})}$ (hard) or $L(\hat{c}_{1,\ldots,M}^{(X_k^{(i-1)})})$ (soft)

(6) Estimate clipping levels $\hat{\gamma}_I^{(i)}$ and $\hat{\gamma}_Q^{(i)}$ as well as attenuation factor $\hat{\alpha}_1^{(i)}$

(7) Clip $\hat{H}_k^{(i-1)}\hat{X}_k^{(i)}$ according to $\hat{\gamma}_I^{(i)}$ and $\hat{\gamma}_Q^{(i)}$ in *time domain* to produce $\hat{\tilde{Y}}_k^{(i)}$

(8) Calculate clipping distortion estimate $\hat{Z}_k^{(i)} = \hat{\tilde{Y}}_k^{(i)} - \hat{\alpha}_1^{(i)}\hat{H}_k^{(i-1)}\hat{X}_k^{(i)}$

(9) Remove the distortion from the received signal, that is, $\hat{Y}_k^{(i)} = (1/\hat{\alpha}_1^{(i)})(\tilde{Y}_k - \hat{Z}_k^{(i)})$

(10) Obtain new channel estimates $\hat{H}_k^{(i)}$ using signal $\hat{Y}_k^{(i)}$

(11) Channel equalization: $\hat{Y}_k^{(i)}/\hat{H}_k^{(i)}$

(12) Demap symbols $\hat{Y}_k^{(i)}/\hat{H}_k^{(i)}$ to bits, deinterleave and decode to obtain output data bits $\hat{\mathbf{b}}^{(i)}$

ALGORITHM 1: Key steps of the proposed receiver clipping compensation algorithm for Receiver A.

Initialization: for all $k \in \Omega_A$

(1) Set $i = 0, \hat{\alpha}_1^{(0)} = 1$ and $\hat{Z}_k^{(0)} = 0$

(2) Obtain channel estimates $\hat{H}_k^{(i)}$ using signal $\hat{Y}_k^{(i)} = (1/\hat{\alpha}_1^{(i)})(\tilde{Y}_k - \hat{Z}_k^{(i)})$

(3) Channel equalization: $\hat{Y}_k^{(i)}/\hat{H}_k^{(i)}$

Iteration: Increase i by one and do the following steps for all $k \in \Omega_A$

(4) Detect $\hat{Y}_k^{(i-1)}/\hat{H}_k^{(i-1)}$ to obtain symbols $\hat{X}_k^{(i)}$

(5) Estimate clipping levels $\hat{\gamma}_I^{(i)}$ and $\hat{\gamma}_Q^{(i)}$ as well as attenuation factor $\hat{\alpha}_1^{(i)}$

(6) Clip $\hat{H}_k^{(i-1)}\hat{X}_k^{(i)}$ according to $\hat{\gamma}_I^{(i)}$ and $\hat{\gamma}_Q^{(i)}$ in *time domain* to produce $\hat{\tilde{Y}}_k^{(i)}$

(7) Calculate clipping distortion estimate $\hat{Z}_k^{(i)} = \hat{\tilde{Y}}_k^{(i)} - \hat{\alpha}_1^{(i)}\hat{H}_k^{(i-1)}\hat{X}_k^{(i)}$

(8) Remove the distortion from the received signal, that is, $\hat{Y}_k^{(i)} = (1/\hat{\alpha}_1^{(i)})(\tilde{Y}_k - \hat{Z}_k^{(i)})$

(9) Obtain new channel estimates $\hat{H}_k^{(i)}$ using signal $\hat{Y}_k^{(i)}$

(10) Channel equalization: $\hat{Y}_k^{(i)}/\hat{H}_k^{(i)}$

After iteration:

(11) Demap symbols $\hat{Y}_k^{(i)}/\hat{H}_k^{(i)}$ to bits, deinterleave and decode to obtain output data bits $\hat{\mathbf{b}}^{(i)}$

ALGORITHM 2: Key steps of the proposed receiver clipping compensation algorithm for Receiver B.

If the signal is received perfectly without clipping in the A/D interface, it can be written as

$$Y_k = H_k X_k + W_k, \quad k \in \Omega \tag{10}$$

in frequency domain after removing CP and taking FFT. In (10), H_k is the channel gain and W_k is the additive white Gaussian noise component for the kth SC. In other words, this equation represents the ideal case for reference. However, if the A/D converter clips the signal, the frequency-domain version of it is then

$$\begin{aligned}\widetilde{Y}_k &= \alpha_1 Y_k + \alpha_2 Y_{-k}^* + D_k \\ &= \alpha_1 (H_k X_k + W_k) + \alpha_2 \left(H_{-k}^* X_{-k}^* + W_{-k}^*\right) + D_k \\ &= \alpha_1 H_k X_k + Z_k + \alpha_1 W_k + \alpha_2 W_{-k}^*, \quad k \in \Omega,\end{aligned} \tag{11}$$

as can be derived from (6). This signal is also illustrated in Figure 2. In (11), $Z_k = \alpha_2 H_{-k}^* X_{-k}^* + D_k$, $k \in \Omega$, which essentially means that unequal clipping levels γ_I and γ_Q cause mirror-subcarrier interference ($\alpha_2 H_{-k}^* X_{-k}^*$, $k \in \Omega$) in addition to the clipping distortion (D_k, $k \in \Omega$). However, if $\gamma_I = \gamma_Q$, then $\alpha_2 = 0$ and thus $Z_k = D_k$, $k \in \Omega$. Because all the interference in Z_k, $k \in \Omega$, stems from the clipping phenomenon, it is called clipping distortion in the rest of the paper. As shown in Figure 2, prior to the clipping compensation, the data symbols $\widetilde{Y}_k$, $k \in \Omega_A$ are equalized, mapped to soft bits, deinterleaved, and then decoded. In the channel equalization stage, the channel gain estimates $\widehat{H}_k$, $k \in \Omega_A$, for active SCs are obtained based on pilot SCs. Section 3.2 contains more discussion about the channel estimation.

In the first step of the clipping compensation, Receiver A uses the bit estimates after the turbo decoder to regenerate the originally transmitted OFDM signal. All the decoded bits may not be correct and, therefore, it is possible that the regenerated signal does not exactly match the transmitted signal. Hence, the regenerated signal is called the estimate of the transmitted signal and it is denoted as $\widehat{X}_k$, $k \in \Omega$, in Figure 2. The data bit estimates $\widehat{\mathbf{b}}$ are not re-encoded in the clipping compensation feedback loop, but rather the soft-coded bit estimates $L(\widehat{c}_{1,\ldots,M}^{(X_k)})$, $k \in \Omega_A$, from the turbo decoder output (after interleaving) are used. Here, $\widehat{c}_{1,\ldots,M}^{(X_k)}$ are hard bit estimates related to the symbol X_k and $L(\widehat{c}_{1,\ldots,M}^{(X_k)})$ are the log-likelihood ratio (LLR) values for the bit estimates $\widehat{c}_{1,\ldots,M}^{(X_k)}$. It is possible to generate either hard or soft symbol estimates and hence both options are studied in Section 4. In Receiver B, it is enough to obtain hard symbol estimates using directly the channel equalizer output. This way, the symbol demapping, channel decoding, and symbol mapping process can be avoided during the clipping compensation process. Since channel decoding is usually the most complex process in the receiver chain, significant savings in computational complexity and latency can be achieved if the channel decoder is located outside the clipping compensation loop. From this point onward, both the Receiver A and B proceed in the same way.

The estimate of the transmitted signal is exposed to the estimated channel and then to the estimated clipping process (separately for I and Q branches) in order to obtain a regenerated version of the clipped received signal $\widehat{\widetilde{Y}}_k$, $k \in \Omega_A$. The clipping process is performed in time domain using the estimated clipping levels $\widehat{\gamma}_I$ and $\widehat{\gamma}_Q$, which should be as close as possible to the clipping levels originally occurred in the A/D converter during the reception of the signal. The knowledge of the exact clipping levels γ_I and γ_Q is not available, but the estimates $\widehat{\gamma}_I$ and $\widehat{\gamma}_Q$ can be calculated as shown in (3). Here, the maximum output levels of the A/D converter $V_{0,I}$ and $V_{0,Q}$ are approximated by following the absolute maximum level of the corresponding signal branch after the A/D converter. An estimate of $P_{\text{in},I}$ and $P_{\text{in},Q}$ can be calculated from the hard symbol estimates $\widehat{X}_k$, $k \in \Omega$. The clipping process performed in time domain essentially produces a signal, which according to Bussgang's theorem can be written in frequency domain as $\widehat{\widetilde{Y}}_k = \widehat{\alpha}_1 \widehat{H}_k \widehat{X}_k + \widehat{Z}_k$, $k \in \Omega_A$, where $\widehat{\alpha}_1$ is the estimate of the attenuation factor. Now, the clipping distortion estimate $\widehat{Z}_k$, $k \in \Omega_A$ can be extracted from $\widehat{\widetilde{Y}}_k$, $k \in \Omega_A$, by removing the attenuated desired signal part $\widehat{\alpha}_1 \widehat{H}_k \widehat{X}_k$, $k \in \Omega_A$, from it. This estimated desired signal can easily be produced from the nonclipped version of the regenerated received signal as illustrated with the lower branch in the clipping compensation loop in Figure 2 (both in Receivers A and B). That is to say, the estimate of the clipping distortion $\widehat{Z}_k$, $k \in \Omega_A$, is calculated as

$$\widehat{Z}_k = \widehat{\widetilde{Y}}_k - \widehat{\alpha}_1 \widehat{H}_k \widehat{X}_k, \quad k \in \Omega_A, \tag{12}$$

where $\widehat{\alpha}_1 = (\widehat{\alpha}_I + \widehat{\alpha}_Q)/2$ is calculated from (8) using $\widehat{\gamma}_I$ and $\widehat{\gamma}_Q$. The out-of-band clipping distortion can be ignored and this is easy to do by processing only the active SCs Ω_A, as indicated in (12). Now, the last step in clipping compensation is to improve the received signal $\widetilde{Y}_k$, $k \in \Omega_A$, by removing the estimated clipping distortion $\widehat{Z}_k$, $k \in \Omega_A$, from it. The compensated signal is expressed as

$$\widehat{Y}_k = \frac{1}{\widehat{\alpha}_1}\left(\widetilde{Y}_k - \widehat{Z}_k\right), \quad k \in \Omega_A, \tag{13}$$

and it is used as a starting point for a new processing round, which begins with the channel estimation and equalization.

One way to show the validity of the clipping compensation approach is to substitute (11) into (13). This leads to the equation that expresses the estimate of the received signal without clipping as

$$\begin{aligned}\widehat{Y}_k &= \frac{\alpha_1}{\widehat{\alpha}_1}(H_k X_k + W_k) \\ &\quad + \frac{1}{\widehat{\alpha}_1}\left(Z_k - \widehat{Z}_k\right) + \frac{\alpha_2}{\widehat{\alpha}_1} W_{-k}^*, \quad k \in \Omega_A.\end{aligned} \tag{14}$$

If exactly correct estimates $\widehat{\alpha}_1$ and $\widehat{Z}_k$, $k \in \Omega_A$, can be achieved, all essential clipping distortion is thus removed. This is straightforward to see from (14), because with the correct estimates, it reduces to the form of (10) plus additional scaled mirror-subcarrier noise $(\alpha_2/\widehat{\alpha}_1) W_{-k}^*$, $k \in \Omega_A$. However, it is impossible to actually produce exactly correct clipping estimate, because the noise W_k, $k \in \Omega$, gets also clipped in the A/D converter, but the model cannot reproduce the clipping distortion term of the noise.

In practice, it is usually the case that there are errors in the symbol estimates $\hat{X}_k$, $k \in \Omega_A$, on the first processing round. This leads to the inaccurate clipping distortion estimate and, therefore, all the distortion is not removed from the received signal on the first round. However, on every processing iteration round the symbol estimates get better since more and more accurate estimate of the clipping distortion can be obtained. In other words, the compensation algorithm works in the desired manner, if high enough portion of the initial symbol estimates are correct. In order to make the proposed algorithm work in the best possible manner, all the estimated parameters should be recalculated on every iteration. This means the estimates of the channel $\hat{H}_k$, $k \in \Omega_A$, the clipping levels $\hat{\gamma}_I$ and $\hat{\gamma}_Q$ as well as the attenuation factor $\hat{\alpha}_1$.

The receiver clipping compensation discussed in this paper has similarities with the deliberate transmitter-clipping compensation presented, for example, in [9] and [10] as well as to the power amplifier nonlinearity compensation in [13]. However, there are several fundamental differences, which can be analyzed comparing the aforementioned references to the equations shown in Sections 2 and 3 of this paper. Using consistent notation, the received signal under deliberate *transmitter clipping* can be expressed as $\tilde{Y}_k = H_k(\alpha X_k + D_k) + W_k$, $k \in \Omega$. When compared to (11), it is evident that the role of the channel is different. In the transmitter clipping case, the clipping is applied before transmission and hence the channel estimation in the receiver is less challenging compared to the receiver clipping case. It is shown in Section 4 that the channel estimation is severely interfered in the receiver clipping case, if clipping compensation is not used. Another essential difference is the clipping model itself. In the deliberate transmitter clipping, the amplitude of a complex signal is limited, that is,

$$|\tilde{x}(t)| = \begin{cases} |x(t)|, & |x(t)| \leq V_0, \\ V_0, & |x(t)| > V_0, \end{cases} \tag{15}$$

where as the receiver clipping means that the I and Q signal branches are clipped separately as expressed in (2). This means that it is possible in the receiver clipping case that different clipping levels take place in I and Q branches. Nevertheless, since our signal modeling takes it into account, the clipping compensation with unequal clipping levels is possible as verified in Section 4.

3.2. Implementation Details. In this paper, a zero-forcing channel equalizer is used for simplicity. The channel estimates $\hat{H}_k$, $k \in \Omega_A$, are obtained using part of the active SCs Ω_A, as pilots, which are denoted here as Ω_P. These are evenly distributed among the data SCs $\Omega_D = \Omega_A \setminus \Omega_P$ so that every Lth SC is a pilot. The channel estimates of pilot SCs are $\hat{H}_{P,k} = \tilde{Y}_k/X_k$, $k \in \Omega_P$, which are exploited to obtain channel estimates for the data SCs using low-pass interpolation. This means that, first, $L-1$ zeros are inserted between every pilot SC channel estimate. Then, a properly designed FIR lowpass interpolation filter is used to obtain the channel estimates for data SCs. The width of the passband of the interpolation filter corresponds to the original data bandwidth and, therefore, the mean-squared error (MSE) of the interpolated SCs is minimized assuming that the pilot SC channel estimates are correct. As the channel estimation is performed in frequency domain, the use of interpolation filter means calculating convolution between the impulse response of the interpolation filter and the zero-padded channel estimates of the pilots in frequency domain.

From the second clipping compensation iteration round onward, also the symbol estimates from the previous iteration round can be exploited in the channel estimation task. Due to the symbol errors, purely data-symbol-based channel estimation may give poorer results than the estimation using pilots. Therefore, a heuristic approach, which combines these two methods, is proposed in this paper. In the combining, the channel estimates are weighted with the corresponding symbol reliabilities $p(X_k)$, $k \in \Omega_A$, so that the channel estimates after combining are

$$\hat{H}_k = \frac{p(X_k)}{2}\hat{H}_{D,k} + \left(1 - \frac{p(X_k)}{2}\right)\hat{H}_{P,k}, \quad k \in \Omega_A, \tag{16}$$

where $\hat{H}_{D,k} = \tilde{Y}_k/\hat{X}_k$, $k \in \Omega_D$ are the channel estimates obtained from the latest data symbol estimates and $\hat{H}_{P,k}$, $k \in \Omega_A$, are the latest channel estimates obtained using pilots. The estimates $\hat{H}_{D,k}$, $k \in \Omega_P$ for pilot SCs are not needed since (16) implicates that $\hat{H}_k = \hat{H}_{P,k}$, $k \in \Omega_P$. This is guaranteed by setting $p(X_k) = 0$, $k \in \Omega_P$. For the data symbols, the reliabilities are calculated as

$$p(X_k) = \prod_{j=1}^{M} \frac{\exp\left[\hat{c}_j^{(X_k)} L\left(\hat{c}_j^{(X_k)}\right)\right]}{1 + \exp\left[L\left(\hat{c}_j^{(X_k)}\right)\right]}, \quad k \in \Omega_D. \tag{17}$$

This data-symbol-based estimation is used in Receiver A. Since the LLR values for the bits are not calculated in Receiver B in every iteration, it uses only the pilot-based channel estimation. One variation of the data-symbol-based channel estimation proposed above would be such that a certain threshold is used for the symbol reliabilities. In other words, only the symbols which are reliable enough, let us say $p(X_k) > 0.95$, are used in the channel estimation. However, this variation of the proposed channel estimation technique is omitted from the performance analysis of Section 4 and is left for future studies.

The output of the zero-forcing channel equalizer should be limited so that the maximum amplitude is

$$V_{\max,\mathrm{eq}} = \max_r \|A_r\| + \frac{3.29}{\sqrt{2}}\sigma_n, \tag{18}$$

where A_r, $r = 1, \ldots, 2^M$ denotes all the possible symbols from the used alphabet and σ_n is the standard deviation of the complex noise $w(t)$, which is assumed to be Gaussian. The factor of 3.29 corresponds to the 99.9% confidence level of one-dimensional Gaussian distribution and, therefore, the standard deviation of the complex noise has to be scaled with $\sqrt{2}$. The amplitude limit of the equalizer output does not have to be exactly what is proposed in (18), but significantly too tight or loose limit affects decoding results. The reason to use such an amplitude limiter after the channel equalizer is the soft symbol demapper defined based on an assumption of Gaussian noise. The clipping distortion and the channel equalizer cause interference peaks, which degrade the performance of the soft symbol demapper. In this paper, it is chosen to set the LLR values of the bits related to the amplitude-limited symbols to zero. This is justified, because if there is a significant distortion term present in a symbol, all the usable information carried by the symbol has already been lost. The main target in the limiting of the equalizer output is to tell the decoder which symbols are very unreliable due to the channel fading or unintentional clipping, but in the same time avoid affecting the symbols which are spread only because of the Gaussian noise.

The channel codec used in this paper is a turbo codec [20] with a generator matrix:

$$G = \begin{bmatrix} 1 & \frac{15}{13} \end{bmatrix}. \tag{19}$$

The used interleavers are bit-wise S-interleavers [21] for which the distance parameter is defined as $S = \sqrt{U/2}$, where U is the length of the unit to be interleaved. In channel interleaving, the unit is the whole coded block (U_{out}). Inside the turbo encoder/decoder, the length of the interleaved unit is equal to one uncoded data block (U_{in}). This is defined as $U_{\text{in}} = \lfloor R(U_{\text{out}} - 2m) \rfloor$, where $m = 3$, is the memory length of the component encoder and the term $2m$ is due to the unpunctured termination bits [22]. Different coding rates are achieved by puncturing the parity bits, which is done based on [22]. The turbo decoder uses the Max-Log-MAP algorithm for decoding, and it does not have a correction function for the max-operator [23]. The iteration of turbo decoding is stopped based on the hard-data-aided (HDA) criterion presented in [24].

3.3. Multiuser Considerations. Since the proposed receiver clipping compensation method is very general, it can be used both in UL and DL directions. However, there are some essential differences when a multiuser scenario is concerned. In DL direction, the data symbol sequence X_k, $k \in \Omega_A$, contains symbols, which belong to several individual mobile users. A single mobile receiver should be able to process, in addition to its own data, also the symbols of neighboring users in order to exploit the clipping compensation method in an optimal manner. This is because of the nonlinear behavior of clipping, which causes that part of the clipping distortion is stemming from outside the frequency band of the user. If the signal power for the neighboring users is higher than for the interested user, the amount of clipping distortion can be very high. Therefore, joint clipping compensation processing of all neighboring users is preferable. The challenge in the implementation of the clipping compensation in mobile receiver is that the amount of signal processing is significantly increased when the symbols of neighboring users are processed and the channel estimation is performed over the whole frequency band instead of concentrating on one user only as the mobile receiver usually does. The proposed Receiver B model is especially suitable for DL direction due to its considerably lower computational complexity but still relatively good performance, as shown in Section 4.1. One way to decrease the computational burden even more is to exploit cooperative interference cancellation among the users so that they exchange information of their own data and interference using short-range communications. For example, spatial correlation of the channels can be exploited, if users are close to each other.

Similarly as in the DL direction, also in the UL scenario the data symbol sequence X_k, $k \in \Omega_A$, contains data of multiple users. Furthermore, in UL, a band-wise channel estimation for each user must be performed individually, since the different mobile signals have propagated through different channels. Therefore, the channel estimate sequence $\hat{H}_k$, $k \in \Omega_A$, is not continuous. In general, the implementation of the clipping compensation method is more viable in the UL scenario since the base station has to receive and process the signals of all users anyway. Hence, including the clipping compensation processing is not going to excessively increase the computational complexity. On the other hand, modern base stations can also have vast amount of computing power, which makes complex digital signal processing algorithms feasible. Due to its higher computational complexity, Receiver A model is proposed for the UL direction. More discussion of the receiver complexity and performance can be found in Section 4.1.

4. Performance Results for the Clipping Compensation Approach

In this section, the performance of the proposed receiver-side clipping compensation algorithm is analyzed and discussed in details. This is done using a UL scenario with two individual users in the block-fading extended ITU-R Vehicular A channel [25]. First, Section 4.1 analyzes the performance of the clipping compensation algorithm through computer simulations. After that, Section 4.2 considers the same scenario using laboratory measurements with real A/D hardware. Finally, Section 4.3 presents analysis of computational complexity for Receivers A and B in the considered scenario.

Since an UL scenario is considered, both users experience different realizations of the fading channel. Here, it is also assumed that the channels stays static for ten consecutive OFDM symbols, which makes possible to average obtained channel estimates over this period. The channel conditions are set so that the average received power level for User 1 is 15 dB less than for User 2. In this section, it is assumed that there is perfect delay and frequency alignment between the users. The most essential parameters of the simulated scenario are collected into Table 1. Both users have 512 active SCs of which every eighth is a pilot ($L = 8$). Therefore, $N_A = 1024$ in the receiver side, when the composite signal of two users is processed. The pilot power is adjusted so that it corresponds to the average power of the data symbols.

TABLE 1: Parameters for the simulated uplink multiuser scenario.

Parameter	Value
Number of users	2
FFT size N	2048
Active subcarriers per user	512
Scattered pilots per user	64
CP length	128 samples
Sampling rate	30.72 MHz
Subcarrier spacing	15 kHz
Subcarrier modulation	16-QAM ($M = 4$)
Channel profile	Extended ITU-R Veh. A [25]
Channel codec	Turbo codec (Max-Log-MAP)
Coding rate R	1/2
Coded block length U_{out}	3584 bits (2 OFDM symbols)
Decoding iterations	Max. 5
Clipping comp. iterations I	5 (or 10)
Clipping level γ	0–10 dB
SNR for User 1 SNR_1	16–36 dB

Both users have 16-QAM as an SC modulation and hence $M = 4$. The oversampling factor used is $J = 2$ and the overall sampling rate is 30.72 MHz, which results in the SC spacing of 15 kHz. The length of CP is $(1/16)N = 128$ samples. On average, the PAPR of the composite signal of two users is 10.1 dB. For the channel coding, the turbo codec is used and its coded block length is $U_{out} = 3584$ bits, meaning that two consecutive OFDM symbols form one coded block. Coding rate of $R = 1/2$ is used and hence $U_{in} = 1789$ bits. In every clipping compensation iteration, the turbo decoder is allowed to do five decoding iterations, if not stopped earlier based on the HDA criterion. In the following, five clipping compensation iterations ($I = 5$) have been performed in all cases, if not specifically stated otherwise. In practical receiver, unnecessary iterations can be avoided, to decrease the amount of computation, by stopping iterations for the user, if its bit-error ratio (BER) goes to zero before the maximum amount of iterations has been performed. This can be checked, for example, by exploiting the cyclic redundancy check (CRC).

4.1. Simulation Results. Figure 3 gives an overall illustration of the above-defined scenario, when the clipping level $\gamma = 6$ dB occurs. For simplicity, equal clipping levels $\gamma_I = \gamma_Q = \gamma$ are assumed in this section, if not otherwise stated. Figure 3(a) shows a piece of the real branch signal before and after clipping. Correspondingly, Figure 3(b) illustrates the received waveform in frequency domain before and after clipping. It is clearly visible that the weak User 1 suffers from the nonlinear clipping distortion stemming from the band of the considerably stronger User 2.

The progress of the clipping compensation process for User 1 is illustrated in Figure 4 when Receiver A is used for the case shown in Figure 3. The BER results are averaged over 20,000 OFDM symbols, that is, 2000 independent channel

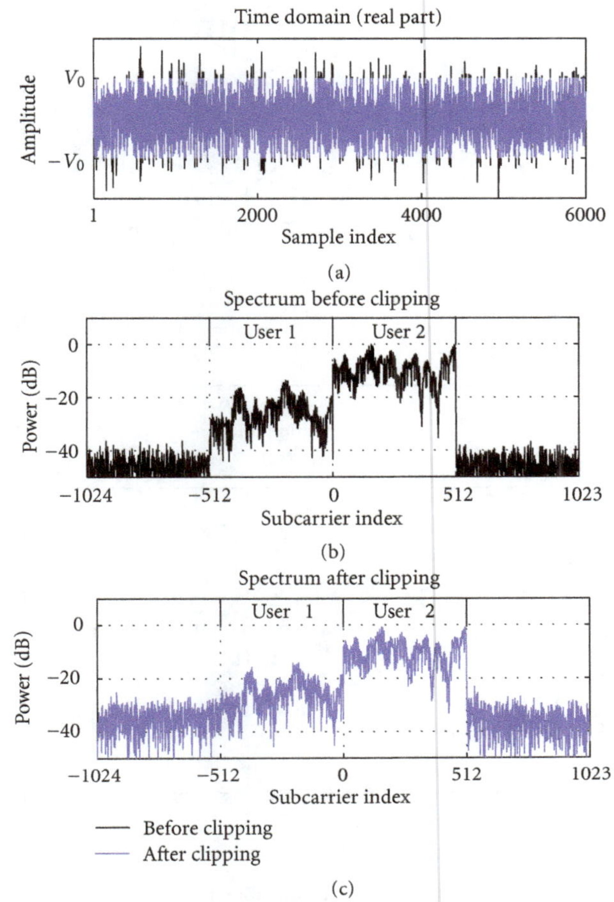

FIGURE 3: (a) Time domain illustration of the real part of the received signal before and after clipping, $\gamma = 6$ dB (b) spectrum of the received signal before clipping (c) spectrum of the signal after clipping.

realizations, as is done also for all the other performance figures in this Section. In this paper, BER is calculated for the data bits **b**. Figure 4 shows that the proposed clipping compensation algorithm is able to almost reach the BER of the nonclipped signal, which represents here the error floor. For the nonclipped signal, perfect channel knowledge is used in order to present the ideal reference case whereas the channel estimation proposed in Section 3.2 is used in the clipping compensation. Figure 4 also illustrates the information flow inside the receiver. Decoding arrow (D) refers to the turbo decoding process and feedback arrow (F) denotes the process where bit decisions are exploited to remove clipping distortion from the received signal. It can be seen from Figure 4 that the turbo decoder cannot reduce the BER of the noncompensated signal very much due to the significant amount of clipping distortion. Nonetheless, the clipping compensation method is still able to reduce the amount of distortion, and in the next iteration, the turbo decoder reduces the BER more. This means that the clipping compensation performs well even without turbo decoding and thus the Receiver B model is justified.

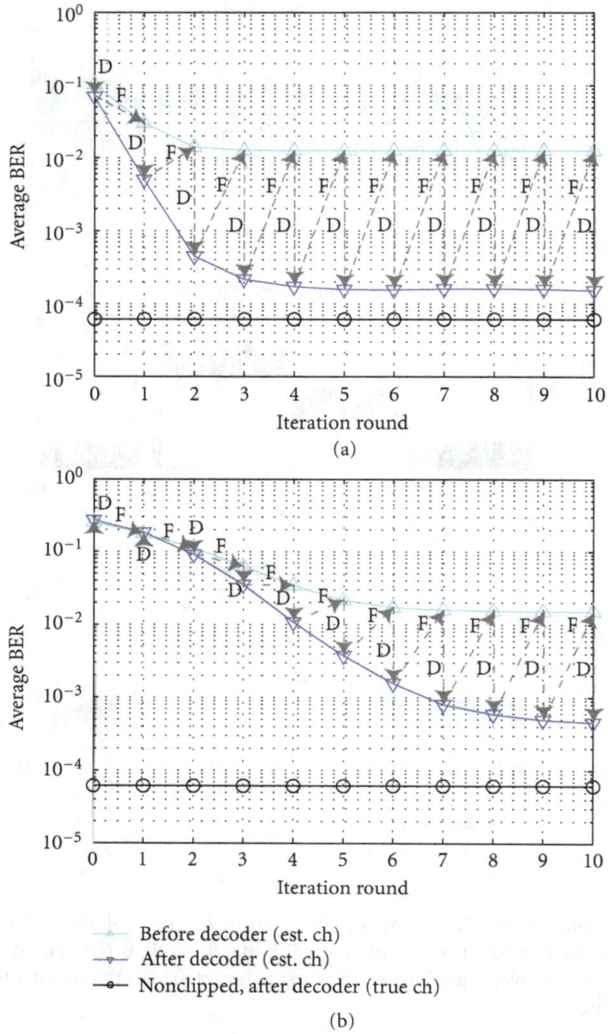

FIGURE 4: Average BER as a function of clipping compensation iterations for User 1 using Receiver A when $SNR_1 = 21.9$ dB and (a) $\gamma = 6$ dB or (b) $\gamma = 2$ dB. Iteration round 0 corresponds to the noncompensated signal. Arrows refers to the decoding (D) and compensation feedback (F) stages.

A detailed performance comparison between Receiver A and B is given later in this Section. For the results shown in Figure 4, the average received signal-to-noise ratio (SNR) for User 1 is 21.9 dB whereas for User 2 it is 36.9 dB. Here, the average received SNR is defined using the average signal and noise powers on the user frequency band, that is,

$$SNR_u = \frac{E\left[\sum_{k\in\Omega_{A,u}} |H_k X_k|^2\right]}{E\left[\sum_{k\in\Omega_{A,u}} |W_k|^2\right]}, \tag{20}$$

where u is the user index, $\Omega_{A,1} = \{-N_A/2, \ldots, -1\}$ for User 1 and $\Omega_{A,2} = \{1, \ldots, N_A/2\}$ for User 2. By comparing Figures 4(a) and 4(b), it can be concluded that stronger clipping requires more compensation iterations rounds, but eventually most of the clipping distortion is removed.

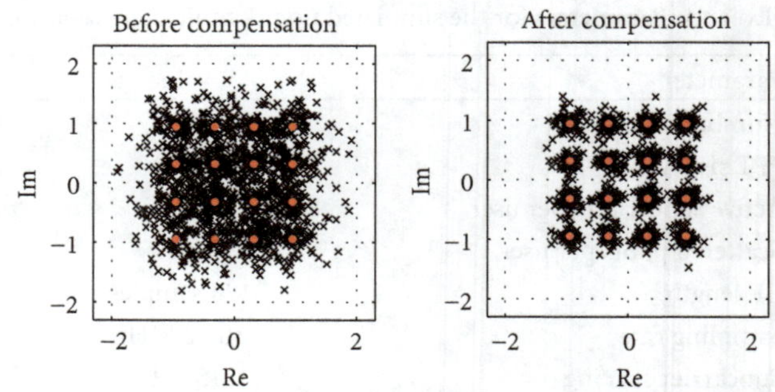

FIGURE 5: Example realization of the received constellation for User 1 before and after clipping compensation in Receiver A when $\gamma = 6$ dB and $SNR_1 = 21.9$ dB.

An example of the received constellation before and after the clipping compensation for User 1 is provided in Figure 5 to visually illustrate the reduction of clipping distortion due to the applied compensation method. This matches with the cases shown in Figures 3 and 4. In addition to BER results, one way to illustrate the performance of the clipping compensation method is the subcarrier-wise signal-to-clipping-distortion ratio ($SCDR$), which is defined as

$$SCDR_k = \frac{E\left[|Y_k|^2\right]}{E\left[\left|\hat{Y}_k - Y_k\right|^2\right]}, \quad k \in \Omega_A. \tag{21}$$

It describes how much there is clipping distortion compared to the ideal nonclipped situation. Figure 6 shows SCDR results for both users when $\gamma = 6$ dB, which is the same case that is considered in Figures 3–5. It can be seen from Figure 6 that in this particular case the amount of clipping distortion is decreased over 20 dB for both users. It should be kept in mind that SCDR values are relative. User 1 has considerably more clipping distortion than User 2 in the beginning and hence the improvement in SCDR has much more significant effect for User 1. Naturally, the employed channel estimation approach has a finite accuracy, but this is a minor matter compared to the fact that the receiver-side clipping affects the pilots and, therefore, notably distorts the channel estimates. On the other hand, the clipping compensation also effectively enhances the channel estimates. The subcarrier-wise channel estimation MSE is given in Figure 7 for the currently discussed case of $\gamma = 6$ dB. This figure verifies that the clipping compensation without a doubt improves the channel estimates. The channel estimation MSE for the nonclipped signal in Figure 7 illustrates the error floor for this particular channel estimation method and noise level. The relatively high error on the sides of the user bands is stemming from the interpolation filter transients.

In order to provide a wider view of the clipping compensation performance, Figure 8 illustrates the simulated BER results for User 1 as a function of occurred clipping level when $SNR_1 = 21.9$ dB. The same figure also provides a comparison between the proposed channel estimation and perfect channel knowledge. Although the channel estimation works well when $\gamma > 4$ dB, the channel estimates gets

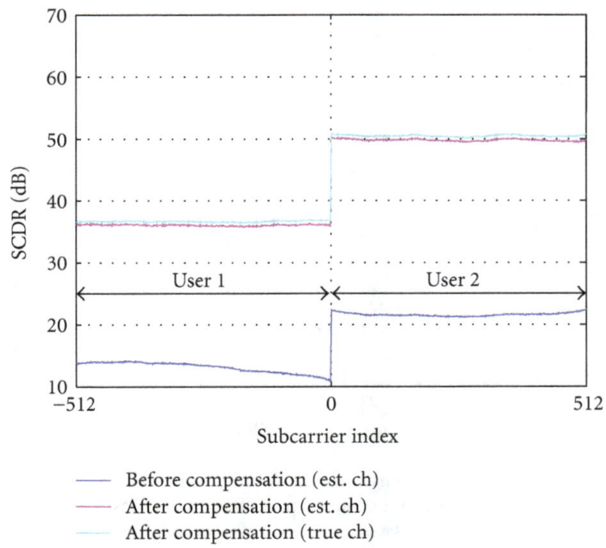

FIGURE 6: Subcarrier-wise signal-to-clipping-distortion ratio when $\gamma = 6$ dB, $SNR_1 = 21.9$ dB and $SNR_2 = 36.9$ dB. The results are obtained after 5 clipping compensation iterations using Receiver A.

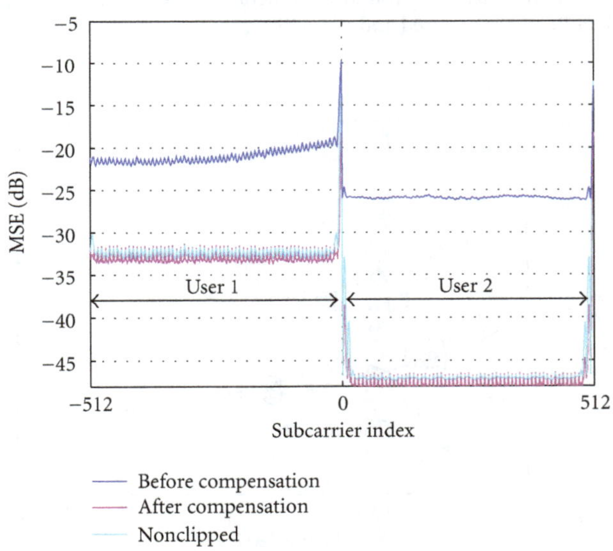

FIGURE 7: Mean squared error of channel estimates when $\gamma = 6$ dB, $SNR_1 = 21.9$ dB and $SNR_2 = 36.9$ dB. The results are obtained after 5 clipping compensation iterations using Receiver A.

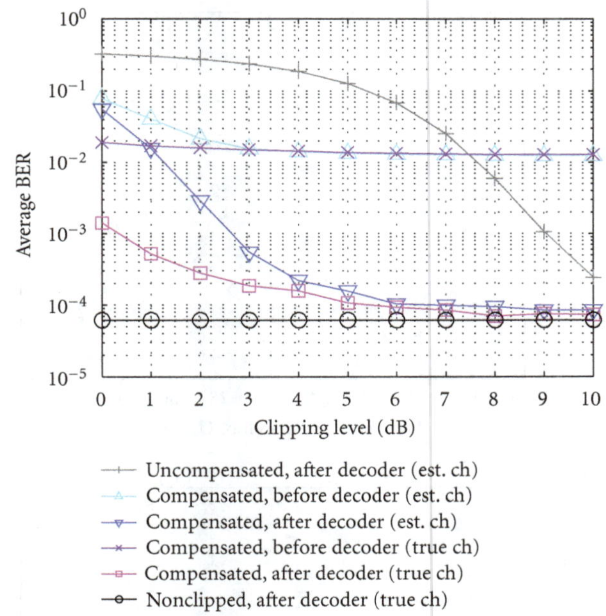

FIGURE 8: Performance of the clipping compensation for User 1 as a function of clipping level when $SNR_1 = 21.9$ dB. The results are obtained after 5 clipping compensation iterations using Receiver A.

significantly poorer if heavier clipping occurs. Because the input SNR is the same for all clipping levels, it can be concluded that the channel estimation accuracy is limited due to the clipping distortion and not because of the noise. These results were obtained after five clipping compensation iterations, which does not guarantee that BER is always minimized. Under heavy clipping, additional iterations would provide lower BER to some extent, but there is always a certain error floor despite the number of iterations. Another support for this issue is provided in Figure 9, where the BER of User 1 is illustrated as a function SNR_1 for clipping levels $\gamma = 6$ dB and $\gamma = 2$ dB. From here, it can be seen that the clipping compensation performs decently, if the noise level and the clipping level are reasonable. The limited channel estimation accuracy due to the heavy clipping can be seen in Figure 9(b) by comparing the BER curves in cases of estimated channel and perfect channel knowledge. This means that the clipping compensation method is able to recover the BER almost to the level of nonclipped signal even under heavy clipping, if the channel estimation could be performed accurately. The limitations of the channel estimation can be relieved to some extend by increasing the number of compensation iterations, but in order to minimize the number of iterations, the optimal channel estimation approach under clipping is an interesting topic for future research. Figure 10 shows the BER of User 1 separately for every clipping compensation iteration round when $\gamma = 6$ dB. From here, it can be concluded that almost the maximum performance is achieved within only three iterations. The rapid saturation of the performance is also observable from Figure 4(a) whereas stronger clipping would require a few iterations more. It is desirable to minimize the number of iterations since it directly affects the processing time of the clipping compensation per code block and thus also the overall latency of the receiver.

One interesting question: does the performance of the clipping compensation method change if soft symbol estimates are used instead of the hard ones. Figures 11 and 12 show that the performance for User 1 using Receiver A is practically the same for hard and soft symbol estimates. It is worth clarifying that the soft symbol estimates are used only for the regeneration of the transmitted signal

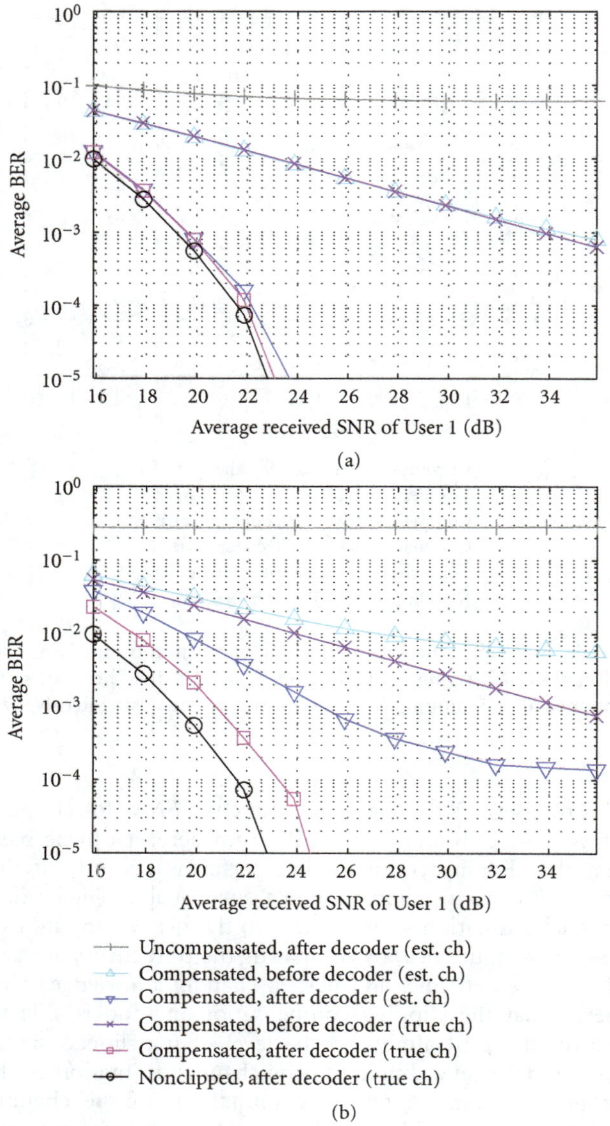

FIGURE 9: Performance of the clipping compensation for User 1 as a function of average received SNR for clipping levels (a) 6 dB and (b) 2 dB. The results are obtained after 5 clipping compensation iterations using Receiver A.

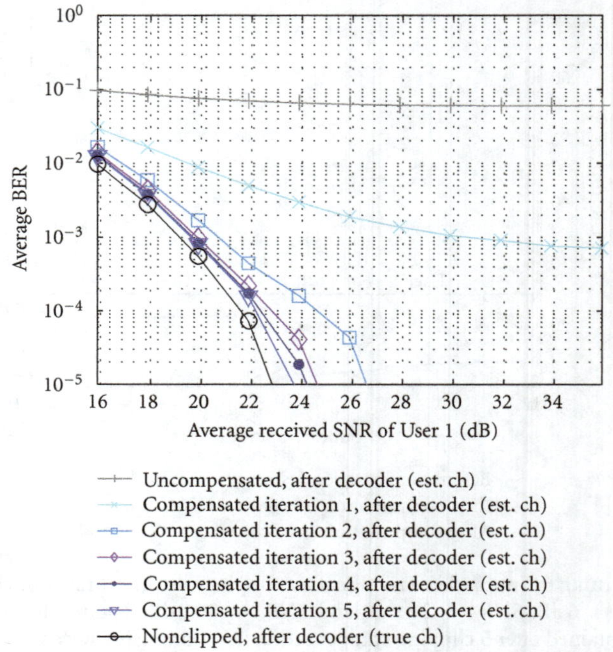

FIGURE 10: Average BER of User 1 separately for every clipping compensation iteration round as a function of average received SNR when Receiver A is used and $\gamma = 6$ dB.

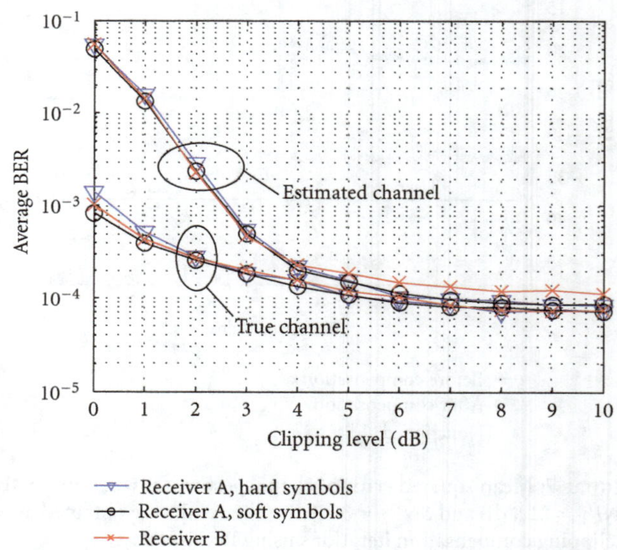

FIGURE 11: Performance comparison between Receivers A and B for User 1 as a function of clipping level when $SNR_I = 21.9$ dB and 5 compensation iterations is performed.

in the compensation feedback loop. The channel estimates $\hat{H}_{D,k}$, $k \in \Omega_D$, and the estimate of P_{in} are always obtained using the hard symbol estimates, because the soft symbol estimates would give misleading results. For a deliberate transmitter clipping compensation, it is reported in [10] that the soft symbol estimates lead to equal or even worse performance compared to the hard symbol estimates. Based on Figures 11 and 12, it can be stated that the hard symbol estimates should be used in Receiver A, since the soft symbol estimates would only unnecessarily increase the computational load. Another comparison made in Figures 11 and 12 is between Receivers A and B. Even though the computational complexity of Receiver B is significantly lower, it performs only marginally worse than Receiver A with all the clipping levels and SNRs studied. Therefore, the Receiver B model is suggested especially for mobile receivers. In practice, one option would be to consider a hybrid of Receiver A and B. This means first iterating the clipping compensation on the symbol level until the performance saturates, then perform the turbo decoding, and finally do

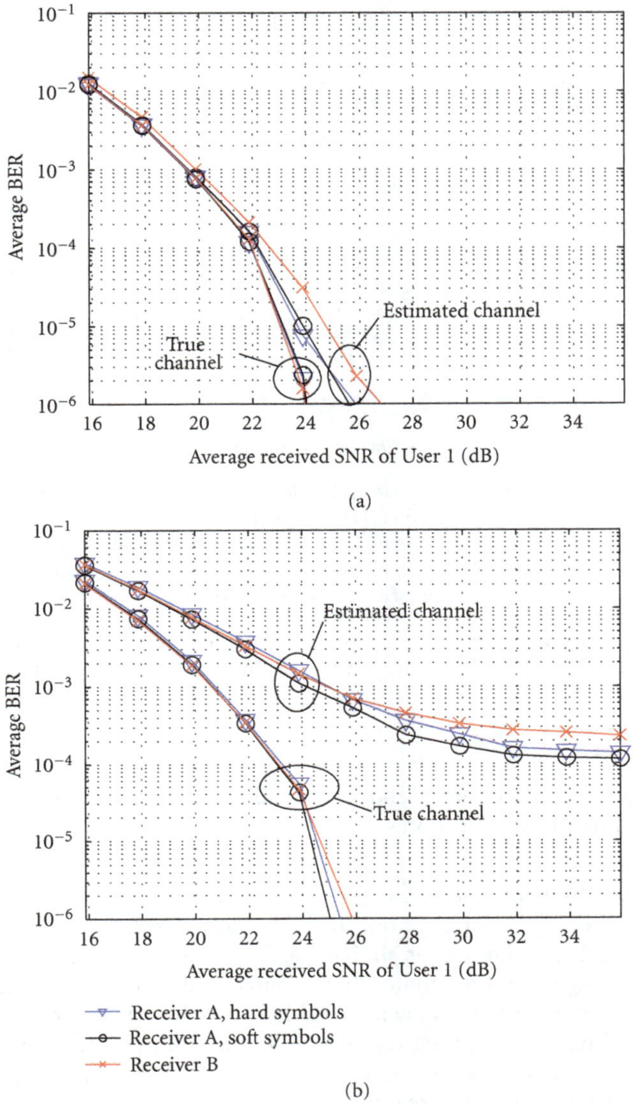

FIGURE 12: Performance comparison between Receivers A and B for User 1 as a function of average received SNR when 5 compensation iterations is performed and clipping level is (a) 6 dB or (b) 2 dB.

one more clipping compensation iteration as well as the final turbo decoding. This would provide a compromise between the performance and computational complexity.

Generally in this section, equal clipping levels in I and Q branches are assumed. However, the proposed clipping compensation algorithm completely supports unequal clipping levels as discussed in Sections 2 and 3. This feature is desired, since in practice there can be slight differences between the A/D converters of I and Q branch. In addition, other imbalances of analog circuitry can also affect the clipping levels. Figure 13 presents BER results for a clipping compensation example, where $\gamma_I = 6$ dB and $\gamma_Q = 3$ dB. The 3-dB difference in clipping levels can be considered to be a rather extreme situation but is used for demonstration purposes. However, the clipping compensation algorithm is

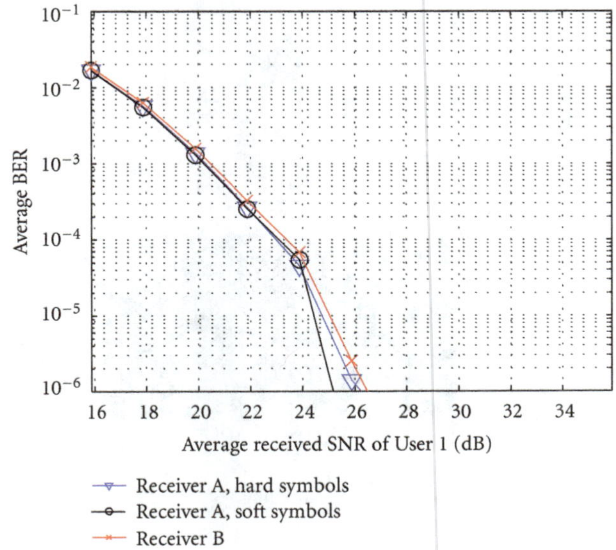

FIGURE 13: Performance comparison between Receivers A and B with channel estimation for User 1 as a function of average received SNR when 5 compensation iterations is performed and the clipping levels for I and Q branches are $\gamma_I = 6$ dB and $\gamma_Q = 3$ dB.

able to perform very well as can be seen from Figure 13. There is only a slight performance decrease compared to the Figure 12(a), where equal clipping levels ($\gamma = 6$ dB) are assumed. The small performance decrease is expected, because stronger clipping is more challenging to compensate as was shown in Figure 8.

4.2. Laboratory Measurement Experiment. The same two-user scenario used in the simulations of the previous Section is also used in laboratory measurements in order to provide comparable results. Only difference is that 10,000 OFDM symbols (instead of 20,000) are used in averaging the BER results due to the slowness of performing the measurements. The complete measurement setup is illustrated in Figure 14. First, a composite waveform of two users is created with a PC. Also the fading channel model is applied to the waveform already on the PC and noise is added so that the SNR of the overall waveform is 31 dB. This corresponds to the case used in the simulations, where $SNR_1 = 21.9$ dB and $SNR_2 = 36.9$ dB were calculated. This digital waveform is then sent to the baseband signal generator [26] in order to convert it to the analog domain. The communication interface between the PC and the signal generator is implemented with TCP/IP. The typical spurious free dynamic range of the used signal generator with a 14-bit D/A converter is 83 dBc [26]. Therefore, it can be assumed that the signal generator does not significantly change the generated waveform. The analog waveform is then fed to the 14-bit A/D converter [27] using balanced coaxial cable connection. The output voltage of the signal generator is adjusted so that it exceeds the maximum voltage level of the A/D converter input and the waveform gets clipped. The digitized clipped waveform is then read from the buffer memory of the A/D converter

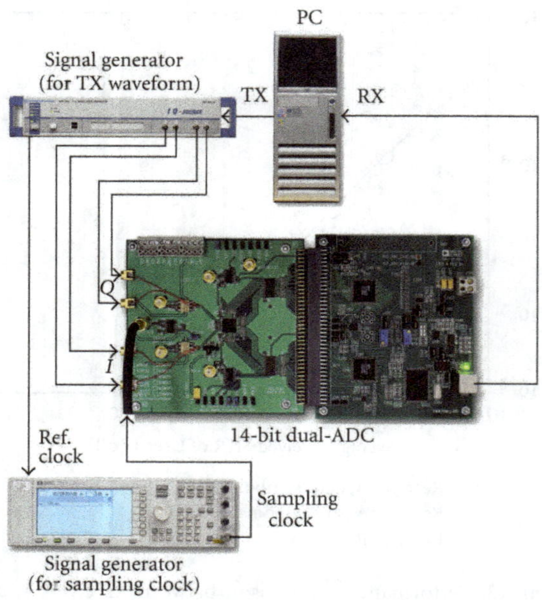

FIGURE 14: Laboratory measurement set-up. A signal waveform with fading channel is generated using the PC and fed to the baseband signal generator [26] to create an analog waveform. The output power of the signal generator is adjusted so that the signal is clipped in the A/D converter [27]. The digitized signal is then collected to the PC for clipping compensation.

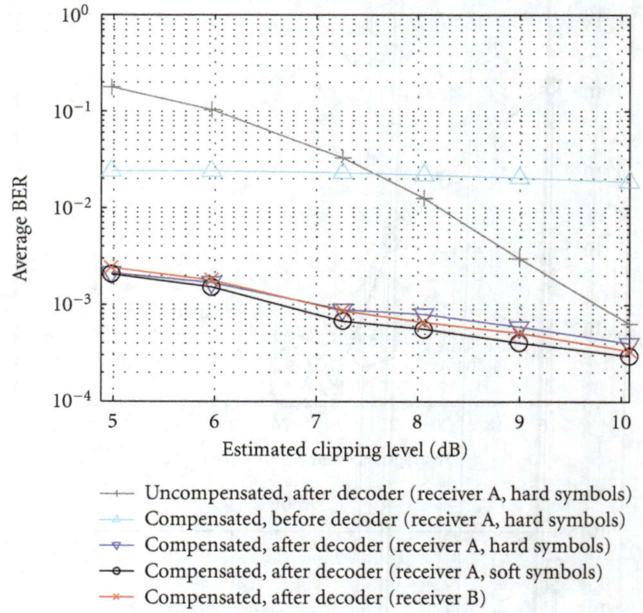

FIGURE 15: Laboratory measurement results for User 1 when $SNDR_1 = 21.7$ dB and 5 clipping compensation iterations is performed.

board to the PC using USB interface. Finally, the PC applies the proposed iterative clipping compensation method to the clipped waveform. The other signal generator shown on the lower left corner of Figure 14 is used to feed sampling clock signal for the A/D converter. The sampling rate is 30.72 MHz as specified in Table 1. There is also a reference clock signal between the signal generators to synchronize their internal oscillators.

The amount of noise and distortion the measurement setup creates can be obtained by running the measurements with a low voltage level so that the waveform does not clip in the A/D converter. The estimated SNR after measurements for the overall nonclipped waveform was 30.87 dB. In order to take also the distortion into account, the signal-to-noise-and-distortion ratio ($SNDR$) for the frequency band of user u is defined as

$$SNDR_u = \frac{E\left[\sum_{k\in\Omega_{A,u}} |H_k X_k|^2\right]}{E\left[\sum_{k\in\Omega_{A,u}} \left|\tilde{Y}_k - H_k X_k\right|^2\right]}. \quad (22)$$

Using the measured nonclipped waveform, it was observed that $SNDR_1 = 21.7$ dB and $SNDR_2 = 36.5$ dB. Figure 15 illustrates the clipping compensation results for User 1 in the laboratory measurement cases in which the occurred clipping is 5–10 dB. These are average estimated clipping levels for certain output voltage levels of the signal generator since the exact knowledge of clipping levels is not available. By comparing the BERs before and after the clipping compensation in Figure 15, it can be concluded that considerable gain is obtained. In addition, the performance difference between Receivers A and B is small as was indicated in the simulations of Section 4.1.

4.3. Computational Complexity of Clipping Compensation. Preceding Sections show that, in many cases, Receiver A and B provide rather similar performance. However, there is a significant difference in computational complexity. This is more formally shown in Table 2, which provides rough estimates of required real additions and multiplications per data symbol in the scenario discussed in Section 4. The numbers of operations are only suggestive since they are strongly depending on implementation platform in practice and the way how the particular functions are carried out. Only the usage of hard symbols estimates in clipping compensation is considered in this Section, because the usage of soft symbols does not change any main conclusions drawn here.

The channel estimation in Receiver A is more complex than the one of Receiver B, because it has to calculate the symbol reliabilities in (17) and combine estimates as defined in (16). For pilot-based channel estimates, the length of the used interpolation filter is 65 and $L = 8$ is considered in Table 2. The complexity of the turbo decoder is defined in [28] using the so-called equivalent additions meaning that computational costs of all mathematical and logical operations are expressed as multiples of one real addition operation. The same principle is adapted to all parts of Table 2 where needed. The complexity of FFT and IFFT operations are assumed to be as defined in [29].

As can be concluded from Table 2, the most complex part of the receivers is the turbo decoder. Receiver A uses the symbol demapper and turbo decoder in every clipping compensation iteration, that is, I times in total, whereas

TABLE 2: Rough estimate for computational complexity of the clipping compensation per data symbol expressed with real addition and real multiplication operations.

Operation	Additions	Multiplications
Scaling with $1/\widehat{\alpha}_1$	0	2
Channel estimation (Receiver A)	$13M + 17$	$20 + M$
Channel estimation (Receiver B)	14	16
Channel equalization	0	4
Soft symbol demapping	$27M \cdot 2^M$	$3M \cdot 2^M$
Turbo decoding	$2130MR$	0
Apply channel estimate to symbol estimate ($\widehat{H}_k\widehat{X}_k$)	0	4
IFFT	$(2N\log_2 N)/((1 - 1/L)N_A)$	$(N\log_2 N)/((1 - 1/L)N_A)$
Clipping	$(2N)/((1 - 1/L)N_A)$	0
Scaling with $\widehat{\alpha}_1$	0	2
FFT	$(2N\log_2 N)/((1 - 1/L)N_A)$	$(N\log_2 N)/((1 - 1/L)N_A)$
Calculating $\widehat{Z}_k$ and subtracting it from $\widetilde{Y}_k$	4	0
Estimating $\widehat{\alpha}_1$	$(3N - 1)/((1 - 1/L)N_A)$	$(N + 7)/((1 - 1/L)N_A)$
Total (Receiver A)	$[((5 + 4\log_2 N)N - 1)/((1 - 1/L)N_A)$ $+(27 \cdot 2^M + 2130R + 13)M + 21]I$	$[((1 + 2\log_2 N)N + 7)/((1 - 1/L)N_A)$ $+(1 + 3 \cdot 2^M)M + 32]I$
Total (Receiver B)	$[((5 + 4\log_2 N)N - 1)/((1 - 1/L)N_A) + 18]I$ $+(27 \cdot 2^M + 2130R)M$	$[((1 + 2\log_2 N)N + 7)/((1 - 1/L)N_A) + 28]I$ $+3M \cdot 2^M$

Receiver B uses them only once. Therefore, Receiver B can be considered to be noticeably less complex. When using the parameter values from Table 1 and five clipping compensation iterations, it can be calculated from Table 2 that Receiver A requires 30865 additions and 1403 multiplications per data symbol whereas Receiver B needs only 6638 additions and 595 multiplications.

5. Conclusions

This paper discussed the compensation of unintentional clipping occurring in the A/D converter of a radio receiver. A digital postprocessing compensation method iteratively exploiting the symbol decisions and clipping distortion regeneration was proposed for multiband OFDMA receivers. The performance of the method was carefully studied in fading channel circumstances using computer simulations and also verified using concrete laboratory measurements with real A/D hardware. It was shown that it is possible to recover the system performance under significant clipping almost to the level of a nonclipped situation. The performance is mostly limited by the accuracy of the channel estimation, which is compromised under heavy clipping. Although it cannot be avoided that the clipping interferes with the channel estimation, it should be studied in the future what is the optimal channel estimation scheme under heavy unintentional clipping. This paper also discussed the complexity of the proposed clipping compensation. It was proven that it is possible, without considerably decreasing the performance, to iterate the compensation process on symbol level and perform the symbol demapping and channel decoding only once after finishing the clipping compensation. This way the computational complexity of the receiver signal processing can be kept at reasonable level.

Acknowledgments

This work was supported by the Academy of Finland (under the project "Digitally-Enhanced RF for Cognitive Radio Devices"), the Finnish Funding Agency for Technology and Innovation (Tekes, under the projects "Advanced Techniques for RF Impairment Mitigation in Future Wireless Radio Systems" and "Enabling Methods for Dynamic Spectrum Access and Cognitive Radio"), Austrian Center of Competence in Mechatronics (ACCM), Jenny and Antti Wihuri Foundation, Finnish Foundation for Technology Promotion, and Tampere University of Technology Graduate School.

References

[1] J. Yang, R. W. Brodersen, and D. Tse, "Addressing the dynamic range problem in cognitive radios," in *Proceedings of the IEEE International Conference on Communications (ICC '07)*, pp. 5183–5188, Glasgow, UK, June 2007.

[2] B. Razavi, "Cognitive radio design challenges and techniques," *IEEE Journal of Solid-State Circuits*, vol. 45, no. 8, Article ID 5518490, pp. 1542–1553, 2010.

[3] C. Svensson, "The blocker challenge when implementing software defined radio receiver RF frontends," *Analog Integrated Circuits and Signal Processing*, vol. 64, no. 2, pp. 81–89, 2010.

[4] A. Rusu, D. Rodríguez de Llera González, and M. Ismail, "Reconfigurable ADCs enable smart radios for 4G wireless

connectivity," *IEEE Circuits and Devices Magazine*, vol. 22, no. 3, pp. 6–11, 2006.

[5] T. Tomioka, R. Sakata, T. Horiguchi, T. Tomizawa, and K. Inoue, "A/D converter clipping noise suppression for high-sensitivity carrier-sensing of cognitive radio transceiver," in *Proceedings of the 50th Annual IEEE Global Telecommunications Conference (GLOBECOM '07)*, pp. 4170–4174, Washington, DC, USA, November 2007.

[6] M. Allén, J. Marttila, and M. Valkama, "Modeling and mitigation of nonlinear distortion in wideband A/D converters for cognitive radio receivers," *International Journal of Microwave and Wireless Technologies*, vol. 2, no. 2, pp. 183–192, 2010.

[7] M. Allén, J. Marttila, and M. Valkama, "Digitally-enhanced wideband analog-digital interfaces for future cognitive radio devices," in *Proceedings of the 8th IEEE International NEWCAS Conference (NEWCAS '10)*, pp. 361–364, Montréal, Canada, June 2010.

[8] S. H. Han and J. H. Lee, "An overview of peak-to-average power ratio reduction techniques for multicarrier transmission," *IEEE Wireless Communications*, vol. 12, no. 2, pp. 56–65, 2005.

[9] H. Chen and A. M. Haimovich, "Iterative estimation and cancellation of clipping noise for OFDM signals," *IEEE Communications Letters*, vol. 7, no. 7, pp. 305–307, 2003.

[10] H. Zhang, X.-G. Xia, Q. Zhang, and W. Zhu, "Iterative decision-aided clipping compensation and its application to scalable video transmission with multiband OFDM," *IEEE Transactions on Vehicular Technology*, vol. 56, no. 2, pp. 756–765, 2007.

[11] D. Petrovic, W. Rave, and G. Fettweis, "Effects of phase noise on OFDM systems with and without PLL: characterization and compensation," *IEEE Transactions on Communications*, vol. 55, no. 8, pp. 1607–1616, 2007.

[12] V. Syrjälä and M. Valkama, "Receiver DSP for OFDM systems impaired by transmitter and receiver phase noise," in *Proceedings of the IEEE International Conference on Communications (ICC '11)*, Kyoto, Japan, June 2011.

[13] J. Šterba, J. Gazda, M. Deumal, and D. Kocur, "Iterative algorithm for nonlinear noise cancellation and channel re-estimation in nonlinearly distorted OFDM system," in *Proceedings of the 8th International Symposium on Applied Machine Intelligence and Informatics (SAMI '10)*, pp. 65–70, Herl'any, Slovakia, January 2010.

[14] M. Allén, J. Marttila, and M. Valkama, "Iterative signal processing for mitigation of wideband ADC nonidealities in cognitive radio receiver," in *Proceedings of the 19th European Signal Processing Conference (EUSIPCO '11)*, pp. 2279–2283, Barcelona, Spain, August 2011.

[15] J. J. Bussgang, "Crosscorrelation functions of amplitude distorted Gaussian signals," Tech. Rep. 216, MIT Research Laboratory of Electronics, Cambridge, Mass, USA, 1952.

[16] H. Ochiai and H. Imai, "Performance analysis of deliberately clipped OFDM signals," *IEEE Transactions on Communications*, vol. 50, no. 1, pp. 89–101, 2002.

[17] H. E. Rowe, "Memoryless nonlinearities with gaussian inputs: elementary results," *The Bell System Technical Journal*, vol. 61, no. 7, pp. 1519–1525, 1982.

[18] H. Ochiai and H. Imai, "Performance of the deliberate clipping with adaptive symbol selection for strictly band-limited OFDM systems," *IEEE Journal on Selected Areas in Communications*, vol. 18, no. 11, pp. 2270–2277, 2000.

[19] L. Cheded, "Invariance property of gaussian signals: a new interpretation, extension and applications," *Circuits, Systems, and Signal Processing*, vol. 16, no. 5, pp. 523–536, 1997.

[20] C. Berrou, A. Glavieux, and P. Thitimajshima, "Near Shannon limit error-correcting coding and encoding: turbo-codes," in *IEEE International Conference on Communications (ICC '93)*, pp. 1064–1070, Geneva, Switzerland, May 1993.

[21] D. Divsalar and F. Pollara, "Turbo codes for PCS applications," in *Proceedings of the IEEE International Conference on Communications (ICC '95)*, pp. 54–59, Seattle, Wash, USA, June 1995.

[22] 3rd Generation Partnership Project, "3GPP TS 25.212 V7.2.0, technical specification group radio access network, multiplexing and channel coding (FDD), Release 7," 2006, http://www.3gpp.org/ftp/Specs/2006-09/Rel-7/25_series/25212-720.zip.

[23] P. Robertson, E. Villebrun, and P. Hoeher, "Comparison of optimal and sub-optimal MAP decoding algorithms operating in the log domain," in *Proceedings of the IEEE International Conference on Communications (ICC '95)*, pp. 1009–1013, Seattle, Wash, USA, June 1995.

[24] C. L. Kei and W. H. Mow, "Improved stopping criteria for iterative decoding of short-frame multi-component turbo codes," in *Proceedings of the IEEE International Conference on Communications, Circuits and Systems and West Sino Expositions (ICCCAS & WeSino Expo '02)*, pp. 42–45, Chengdu, China, June 2002.

[25] T. B. Sørensen, P. E. Mogensen, and F. Frederiksen, "Extension of the ITU channel models for wideband (OFDM) systems," in *Proceedings of the IEEE 62nd Vehicular Technology Conference (VTC-2005-Fall)*, pp. 392–396, Dallas, Tex, USA, September 2005.

[26] Rohde and Schwarz, "R&S AFQ100A/B I/Q modulation generator—data sheet," 2008, http://www2.rohde-schwarz.com/file_10716/AFQ100A_AFQ100B_dat_sw_en.pdf.

[27] Analog Devices Inc., "AD9248 data sheet, ver. B," 2010, http://www.analog.com/static/imported-files/data_sheets/AD9248.pdf.

[28] P. H.-Y. Wu, "On the complexity of turbo decoding algorithms," in *Proceedings of the IEEE VTS 53rd Vehicular Technology Conference (VTC-2001-Spring)*, pp. 1439–1443, Rhodes, Greece, May 2001.

[29] S. Winograd, *Arithmetic Complexity of Computations*, Society for Industrial and Applied Mathematics, Philadelphia, Pa, USA, 1980.

21

Signal Processing for Digital Beamforming FMCW SAR

Qin Xin,[1] Zhihong Jiang,[1] Pu Cheng,[1] and Mi He[2]

[1] *College of Electronic Science and Engineering, National University of Defense Technology, ChangSha, Hunan 410073, China*
[2] *School of Biomedical Engineering, Third Military Medical University, Chongqing 400038, China*

Correspondence should be addressed to Pu Cheng; hagooddemon@163.com and Mi He; hmcherry@126.com

Academic Editor: Gianluca Ranzi

According to the limitations of single channel Frequency Modulation Continuous Wave (FMCW) Synthetic Aperture Radar (SAR), Digital Beamforming (DBF) technology is introduced to improve system performance. Combined with multiple receive apertures, DBF FMCW SAR can obtain high resolution in low pulse repetition frequency, which can increase the processing gain and decrease the sampling frequency. The received signal model of DBF FMCW SAR is derived. The continuous antenna motion which is the main characteristic of FMCW SAR received signal is taken into account in the whole signal processing. The detailed imaging diagram of DBF FMCW SAR is given. A reference system is also demonstrated in the paper by comparing with a single channel FMCW SAR. The validity of the presented diagram is demonstrated with a point target simulation results.

1. Introduction

The combination of Frequency Modulated Continuous Wave (FMCW) technology and Synthetic Aperture Radar (SAR) paves a way to light-weight, cost-effective, high-resolution active microwave remote sensing instrument. This system is suitable for small platforms such as Unmanned Aerial Vehicles (UAV). In the last years, several FMCW SAR systems have been constructed and have been successfully used in ice measurement [1], environment monitoring [2], three-dimensional application [3], and so on [4, 5].

However, these systems are currently applied in the short range and narrow swath case. According to the single channel FMCW SAR radar equation [6], the most immediate method for improving the system performance is to increase the transmission power. Nevertheless, high power transmission is not a suitable solution because of the isolation problem between transmitter and receiver and the increasing system complexity. Moreover, when operating in high resolution such as 0.1 m mode, the aperture length of antenna in azimuth is very small. This will decrease the antenna gain of single channel FMCW SAR systems, and then it will decrease the Signal Noise Radio (SNR) of SAR images.

In pulsed SAR systems, wide unambiguous swath coverage and high azimuth resolution are also contradicting requirements. Several proposals, such as multiple beam SAR system operating in a squinted imaging geometry [7], displaced phase center antenna technique [8], quad array system [9], and High-Resolution Wide-Swath (HRWS) SAR system [10], have been presented to resolve the aforementioned dilemma. Those systems have their own advantages and disadvantages. However, their basic principle is ambiguity suppression using Digital Beamforming (DBF) technology combined with multiple antennas.

The HRWS SAR concept relies on separated transmit and receive antennas, which is also the characteristic of FMCW SAR systems. So the HRWS concept may be a suitable solution for improving the system performance of FMCW SAR. Now, multichannel FMCW SAR is becoming a hot area of miniature SAR research. And many research institutions have announced their research projects on multichannel FMCW SAR [11–14].

According to the imaging algorithm of pulsed DBF SAR, the basic steps are firstly the azimuth reconstruction and then single channel imaging. However, the multichannel azimuth reconstruction algorithm [15] is based on "stop and go" hypothesis. Compared with pulsed SAR received signal, the main characteristic of FMCW SAR received signal is that the continuous antenna motion must be taken into account; that is, the "stop and go" hypothesis is not valid for FMCW SAR [16]. So the impact of continuous antenna motion to azimuth reconstruction must be analyzed. Furthermore, there are many imaging algorithms for single channel FMCW SAR. The best algorithm for DBF FMCW SAR when considering some nonideal factors such as sweep frequency nonlinearity should also be chosen carefully.

The remaining sections are organized as follows. Compared with single channel FMCW SAR, the performance improvement of DBF FMCW SAR is given in Section 2. The signal model, the impact of continuous antenna motion, and the complete imaging diagram of DBF FMCW SAR are all presented in Section 3. Next, an example system is designed and the numerical simulation results are given in Section 4. Finally, Section 5 concludes the whole paper.

2. Performance Improvement

The noise equivalent sigma zero (NESZ) is a quantity directly related to SAR imaging performance, which is defined as the target Radar Cross Section (RCS) when the SNR of the image is equal to one [17]. The NESZ of single channel FMCW SAR is expressed by [6]

$$\text{NESZ} = \frac{(4\pi)^3 r^4 \sin\phi K T_0 F B_N L_s}{P_t G^2 \lambda^2 \rho_a \rho_r G_a G_r}, \quad (1)$$

where r is the slant range, ϕ is the incidence angle, K is the Boltzmann constant, T_0 is the system noise temperature, F is the receiver noise figure, B_N is the noise bandwidth, L_s is the loss factor, P_t is the transmitted signal power, G is the antenna gain, ρ_a and ρ_r are the azimuth and range resolution, respectively, and G_a and G_r are the azimuth and range processing gain, respectively.

2.1. Antenna Gain. From (1), we know that the transmit and receive antenna gains in single channel FMCW SAR system are the same. However, in DBF FMCW SAR, the receive antenna gain is not equal to that of the transmit antenna. Assum that MN receive channels are used, M is the antenna number in azimuth, and N is the antenna number in elevation. The multiple receive apertures are shown in Figure 1.

The DBF in azimuth is applied to suppress the azimuth ambiguity. This does not increase the antenna gain. In elevation, the channels are coherently combined. Performed by the SCan-On-REceive (SCORE) technique, the DBF in elevation can obtain full array gain over a wide area. Thus, under the same resolution and swath width, the maximum receive antenna gain of the DBF system is

$$G_{\text{re}} = NG. \quad (2)$$

The DBF technique increases the antenna gain by N times, compared to the single channel system.

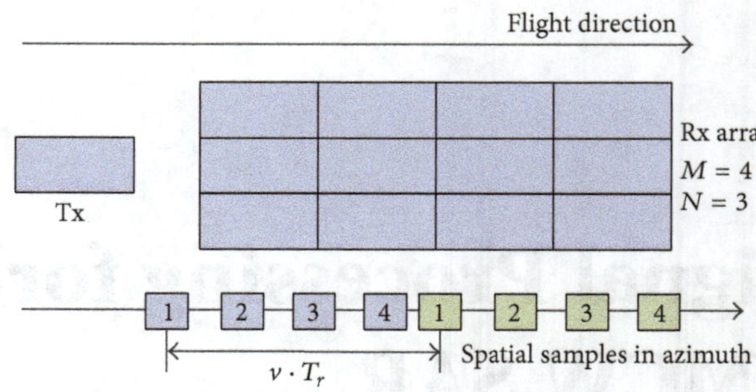

FIGURE 1: Illustration of multiple receive apertures.

2.2. Range Processing Gain. Range processing of FMCW SAR is analog with pulsed SAR. They are all based on pulse compressing using matched filter. Theoretically, range processing gain is about the following:

$$G_r = T_r B_r, \quad (3)$$

where T_r is the pulse width and B_r is the signal bandwidth. For FMCW SAR, the value of T_r is equal to $1/f_p$, where f_p is the pulse repetition frequency (PRF). If we decrease the system PRF using M receive channels in azimuth, we can get M times signal processing gain compared with single channel system. The signal processing gain is increased at the cost of long processing time.

2.3. Sampling Frequency. In order to maintain the low-power and lightweight characteristics of FMCW SAR, the sampling frequency should be chosen as lowly as possible. According to the Nyquist theorem, the lowest sampling frequency for FMCW SAR is

$$f_s = k_r \cdot \frac{2(r_{\max} - r_{\min})}{c_0} = B_r f_p \cdot \frac{2(r_{\max} - r_{\min})}{c_0}, \quad (4)$$

where k_r is the chirp modulation rate and $r_{\max}$ and $r_{\min}$ are the maximum and minimum slant range, respectively. The sampling frequency is also directly proportional to the PRF. That is to say, the sampling frequency of multichannel FMCW SAR is $1/M$ times of that of single channel FMCW SAR.

The DBF FMCW SAR operates with a low PRF to increase the swath width. To decrease the PRF, the significant and challenging processing is the azimuth ambiguity suppression. In the next section, we focus on the multichannel signal reconstruction in azimuth.

3. Signal Processing

3.1. Received Signal Model. Let the transmitted signal of FMCW SAR be written in complex form as follows:

$$ss_T(n,t) = \exp\left[j2\pi\left(f_c \cdot t + \frac{1}{2}k_r \cdot (t - nT_r)^2\right)\right], \quad (5)$$

where f_c is the center frequency, n is the sweep number, and t is the time variable. Because signal amplitude has little effect in SAR signal processing, so for simplicity, antenna gain, amplitude effects of propagation on the signal and any additional time delays due to atmospheric effects are ignored in the following discussions.

The received signal is a delayed version of the transmitted one. The M channels of received signals in azimuth are processed. The mth received signal is

$$ss_{m,R}(n,t) = \exp\left[j2\pi\left(f_c\cdot\left(t-\frac{r_{t,m}}{c_0}\right)+\frac{1}{2}k_r \cdot\left(t-nT_r-\frac{r_{t,m}}{c_0}\right)^2\right)\right], \quad (6)$$

where c_0 is the velocity of light and $r_{t,m}$ is the roundtrip instantaneous slant range which can be written as

$$r_{t,m}(t;r_0) = r_{tx}(t;r_0) + r_{rx,m}(t;r_0), \quad (7)$$

where $r_{tx}(t;r_0)$ and $r_{rx,m}(t;r_0)$ are transmit range and receive range, respectively. In the single channel case, $r_{tx}(t;r_0)$ is equal to $r_{rx,m}(t;r_0)$. However, it cannot be valid in the multichannel case. This is the main difference between the multichannel and the single channel FMCW SAR received signal.

Dechirp on receive is often used in FMCW SAR. In order to further decrease the sampling frequency, and then to decrease the data in range dimension, the radar demodulates the received signal by mixing it with a replica of the transmitted waveform delayed by a time $\tau_{\text{ref}} = 2\cdot r_c/c_0$, where r_c is the slant range distance of the swath center. The reference signal is

$$ss_{\text{ref}}(n,t;r_c) = \exp\left[j2\pi\left(f_c\cdot(t-\tau_{\text{ref}})+\frac{1}{2}k_r\cdot(t-nT_r-\tau_{\text{ref}})^2\right)\right]. \quad (8)$$

The intermediate frequency signal that results from mixing (8) with (6) is

$$\begin{aligned} ss_m(n,t;r_0) &= \exp\left(-j\frac{2\pi f_c}{c_0}\cdot r_{t,m}\right)\cdot\exp\left[j\frac{\pi k_r}{c_0^2}\cdot(r_{t,m}-2r_c)^2\right] \\ &\quad\cdot\exp\left[-j\frac{2\pi k_r}{c_0}\cdot(r_{t,m}-2r_c)\cdot\left(t-nT_r-\frac{2r_c}{c_0}\right)\right]. \end{aligned} \quad (9)$$

Using Taylor series, the round-trip range $r_{t,m}(t;r_0)$ can be expanded as

$$\begin{aligned} r_{t,m}(t;r_0) &= \sqrt{r_0^2+v^2t^2}+\sqrt{r_0^2+(vt-\Delta x_m)^2} \\ &\approx \sqrt{r_0^2+v^2\cdot t_a^2}+\sqrt{r_0^2+(vt_a-\Delta x_m)^2}-\lambda f_d t_r \\ &\approx 2r_0+\frac{\Delta x_m^2}{4r_0}+\frac{(vt_a-\Delta x_m/2)^2}{r_0}-\lambda f_d t_r, \end{aligned} \quad (10)$$

where $t_a = nT_r$ and $t_r = t - nT_r$ represent azimuth time and range time, respectively. v is the velocity of the platform, r_0 is the closest distance between antenna and the scatter. Δx_m is the distance between the transmit antenna and the mth receive antenna, $\lambda = c_0/f_c$ is the wavelength, and f_d is the approximate Doppler frequency shift during the transmission of one sweep as follows:

$$f_d = -\frac{1}{\lambda}\left[\frac{v^2t_a}{\sqrt{r_0^2+v^2\cdot t_a^2}}+\frac{v(vt_a-\Delta x_m)}{\sqrt{r_0^2+(vt_a-\Delta x_m)^2}}\right]. \quad (11)$$

Different from pulsed SAR, f_d introduced by the continuous antenna motion should be taken into consideration, for the "stop and go" hypothesis is not valid for FMCW SAR. This can be seen from the simulated results in Figure 5.

After substituting (10) into (9), the received signal of channel m can be rewritten as

$$\begin{aligned} ss_m(t_a,t_r;r_0) &= \exp\left(-j\frac{4\pi}{\lambda}\cdot r_0\right)\cdot\exp\left[-j\frac{2\pi}{\lambda}\cdot\frac{\Delta x_m^2}{4r_0}\right] \\ &\quad\cdot\exp\left[-j\frac{2\pi}{\lambda}\cdot\frac{(vt_a-\Delta x_m/2)^2}{r_0}\right]\cdot\exp(j2\pi f_d t_r) \\ &\quad\cdot\exp\left[-j\frac{4\pi k_r}{c_0}\cdot(r(t_a;r_0)-r_c)\cdot\left(t_r-\frac{2r_c}{c_0}\right)\right] \\ &\quad\cdot\exp\left[j\frac{4\pi k_r}{c_0^2}\cdot(r(t_a;r_0)-r_c)^2\right], \end{aligned} \quad (12)$$

where $r(t_a;r_0)$ is the azimuth dependent distance to the scatter which can be expressed by

$$r(t_a;r_0) \approx \sqrt{r_0^2+v^2\cdot t_a^2}. \quad (13)$$

In (12), the first to third exponential terms are the azimuth phase history of multichannel FMCW SAR. The fourth term represents the range cell migration due to continuous antenna motion, that is, the Doppler impact in one sweep. The fifth term represents the range signal, which is a sinusoidal signal with a constant frequency value corresponding to the azimuth dependent distance to the point target. The sixth exponential term represents the residual video phase (RVP) term.

3.2. The Impact of Continuous Antenna Motion. As mentioned above, the main characteristic of FMCW SAR is the continuous antenna motion. The range cell migration (RCM) introduced by the continuous antenna motion may result in reconstruction mismatch. Even worse, the extraphase term introduced by the continuous antenna motion may degrade the quality of the focus image.

Multichannel azimuth reconstruction algorithm [15] is a suited approach to process pulsed DBF SAR data in azimuth.

According to the algorithm [15] and the received signal model of DBF FMCW SAR described as (12), the reconstruction filters $\mathbf{P}(f_a)$ for DBF FMCW SAR can be calculated by the following equation:

$$\mathbf{P}(f_a)=\mathbf{H}^{-1}(f_a)=\begin{bmatrix} P_{11}(f_a) & P_{12}(f_a+f_p) & \cdots & P_{1M}(f_a+(M-1)f_p) \\ P_{21}(f_a) & P_{22}(f_a+f_p) & \cdots & P_{2M}(f_a+(M-1)f_p) \\ \vdots & \vdots & \vdots & \vdots \\ P_{M1}(f_a) & P_{M2}(f_a+f_p) & \cdots & P_{MM}(f_a+(M-1)f_p) \end{bmatrix}, \tag{14}$$

where $\mathbf{H}(f_a)$ is

$$\mathbf{H}(f_a)=\begin{bmatrix} \exp\left(-j\frac{\pi\Delta x_1^2}{2\lambda r_0}\right)\exp\left(-j\frac{\pi\Delta x_1}{v}f_a\right) & \cdots & \exp\left(-j\frac{\pi\Delta x_M^2}{2\lambda r_0}\right)\exp\left(-j\frac{\pi\Delta x_M}{v}f_a\right) \\ \exp\left(-j\frac{\pi\Delta x_1^2}{2\lambda r_0}\right)\exp\left(-j\frac{\pi\Delta x_1}{v}(f_a+f_p)\right) & \cdots & \exp\left(-j\frac{\pi\Delta x_M^2}{2\lambda r_0}\right)\exp\left(-j\frac{\pi\Delta x_M}{v}(f_a+f_p)\right) \\ \vdots & \vdots & \vdots \\ \exp\left(-j\frac{\pi\Delta x_1^2}{2\lambda r_0}\right)\exp\left(-j\frac{\pi\Delta x_1}{v}(f_a+(M-1)f_p)\right) & \cdots & \exp\left(-j\frac{\pi\Delta x_M^2}{2\lambda r_0}\right)\exp\left(-j\frac{\pi\Delta x_M}{v}(f_a+(M-1)f_p)\right) \end{bmatrix}. \tag{15}$$

Using basic matrix theory, the system matrix $\mathbf{H}$ can be decomposed as a matrix $\mathbf{Q}$ multiplied by a diagonal matrix $\mathbf{D}$, where only the elements of $\mathbf{D}$ are dependent on the slant range r_0. And the diagonal elements of $\mathbf{D}$ can be written as

$$(\mathbf{D})_{m,m}=\exp\left[-j\pi\frac{\Delta x_m^2}{2\lambda r_0}\right]. \tag{16}$$

In the SAR case, the range history is not a line but spans several range cells, which results in mismatch between the signal and the reconstruction. In FMCW SAR, the additional mismatch introduced by the continuous antenna motion is

$$\Delta\varphi_m=\frac{\pi\Delta x_m^2}{2\lambda}\cdot\frac{\Delta r_{\mathrm{dfm}}}{r_0\cdot(r_0+\Delta r_{\mathrm{dfm}})}, \tag{17}$$

in which Δr_{dfm} represents the maximum RCM introduced by the continuous antenna motion of channel m. The mismatch can be derived from (14) to (16).

Figure 2 shows a simulated RCM introduced by the continuous antenna motion using the parameter listed in Table 1. In this case, Δr_{dfm} is about 1 m. Calculated by (17), the mismatch introduced by this RCM is about 4×10^{-5} rad, which is negligible. Here, $\Delta x_m = 5$ m is assumed. On the other side, Δr_{dfm} is about 10 times of the range resolution. This RCM must be compensated according to SAR imaging theory [17]. In a word, for DBF FMCW SAR, the impact of continuous motion antenna is negligible during multichannel azimuth reconstructing [15]; however, it must be considered during RCM correction.

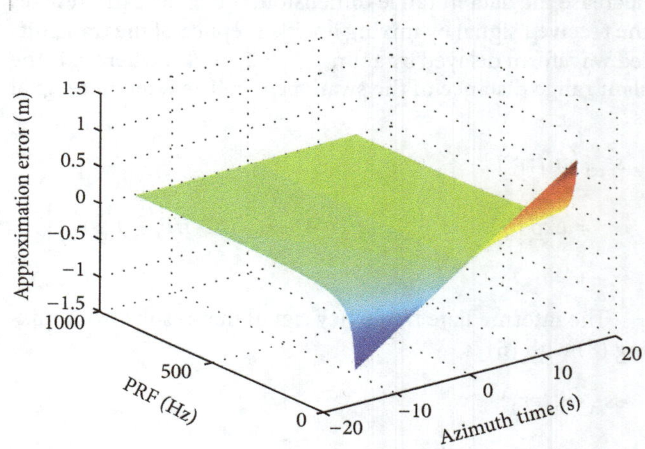

FIGURE 2: RCM introduced by the continuous antenna motion.

3.3. Signal Processing Diagram. The first part of signal processing for DBF FMCW SAR is the azimuth ambiguity suppression. All the recorded subaperture signals whose azimuth frequencies are aliasing become an equivalent single channel signal without azimuth ambiguity. To focus the signal, the second part of the diagram is the classic single channel SAR imaging.

In the first part, multichannel azimuth reconstruction algorithm [15] is chosen because the algorithm can relax the imposed constraints on system design and the algorithm performance is rarely affected by the continuous antenna

TABLE 1: System parameters.

Parameter	Multi-Channel System	Single Channel System
RF center frequency	15 GHz	15 GHz
Band Width	1.5 GHz	1.5 GHz
Operational Altitude	5000 m	5000 m
Operational Velocity	70 m/s	70 m/s
Swath Width	7 km	7 km
Resolution	0.1 m × 0.1 m	0.1 m × 0.1 m
Radar Losses	2 dB	2 dB
Incidence Angle	5–70°	5–70°
Number of Receive Apertures	2	1
Transmitted Power	5 W	5 W
Transmit Antenna Gain	19.5 dB	19.5 dB
Receive Antenna Gain	19.5 dB	19.5 dB
PRF	350 Hz	700 Hz
Range Processing Gain	66.3 dB	63.3 dB
Sampling Frequency	16 MHz	32 MHz

motion. In the second part, frequency scaling algorithm [16, 18–20] is combined with the DBF technology. This algorithm has not only efficient and accurate processing but also real-time integrated motion compensation [19] and easy sweep frequency nonlinear correction [16, 20].

The entire signal processing block diagram of DBF FMCW SAR is shown in Figure 3 including the following steps.

Step 1. Apply Azimuth Fast Fourier Transform (FFT) to the received signal of each channel. Then, the Doppler domain signal $Ss_m(f_a, t_r; r_0)$ is obtained, where f_a is the azimuth frequency.

Step 2. Design the azimuth reconstruction filter $\mathbf{P}(f_a)$ by using (14) and (15). The reconstruction filter of channel m can be written as

$$P_m(f_a) = [P_{m1}(f_a) \;\; P_{m2}(f_a) \;\; \cdots \;\; P_{mM}(f_a)]. \qquad (18)$$

Step 3. Multiply each channel signal to the corresponding reconstruction filter. At this time, the signal becomes

$$Ss_{\mathrm{AR},m}(f_a, t_r; r_0) = Ss_m(f_a, t_r; r_0) \cdot P_m(f_a). \qquad (19)$$

Step 4. Add each channel signal after passing the corresponding reconstruction filter. Then, we can obtain the equivalent single channel signal $Ss(f_a, t_r; r_0)$, which has nonazimuth frequency aliasing. And it can be written in convolution [16] as follows:

$$\begin{aligned} Ss(f_a, t_r; r_0) &= \left\{\exp\left(-j\frac{4\pi r_0 \beta}{\lambda}\right)\exp(j2\pi f_d t_r)\right. \\ &\quad \cdot \exp\left[-j\frac{4\pi k_r}{c_0}\left(\frac{r_0}{\beta} - r_c\right)\left(t_r - \frac{2r_c}{c_0}\right)\right] \\ &\quad \left.\times \mathrm{src}(f_a, t_r; r_0)\right\} \\ &\quad \otimes \exp\left(-j\pi k_r t_r^2\right), \end{aligned} \qquad (20)$$

where β is the RCM factor defined by

$$\beta(f_a) = \sqrt{1 - \frac{f_a^2\lambda^2}{4v^2}}, \qquad (21)$$

where $\mathrm{src}(f_a, t_r; r_0)$ is the secondary range compression as follows:

$$\begin{aligned} \mathrm{src}(f_a, t_r; r_0) &= \exp\left[-j\frac{2\pi r_0 k_r^2 \lambda}{c_0^2}\frac{(\beta^2 - 1)}{\beta^3}\left(t_r - \frac{2r_c}{c_0}\right)^2\right] \\ &\quad \cdot \exp\left[j\frac{2\pi r_0 k_r^3 \lambda^2}{c_0^3}\frac{(\beta^2 - 1)}{\beta^5}\left(t_r - \frac{2r_c}{c_0}\right)^3\right]. \end{aligned} \qquad (22)$$

Step 5 (frequency scaling transform). Multiplying the signal $Ss(f_a, t_r; r_0)$ with the frequency scaling factor, we have the following:

$$H_{\mathrm{FS}}(f_a, t_r) = \exp\left[j\pi k_r (1 - \beta) t_r^2\right]. \qquad (23)$$

Step 6 (continuous antenna motion correction). Multiplying the signal with the Doppler frequency correction factor, we have the following:

$$H_{\mathrm{DFC}}(f_a, t_r) = \exp(-j2\pi f_d \beta \cdot t_r). \qquad (24)$$

In actual implementation, Steps 5 and 6 can be completed at the same time. The signal $Ss_{\mathrm{FS}}(f_a, t_r; r_0)$ is obtained as follows:

$$Ss_{\mathrm{FS}}(f_a, t_r; r_0) = Ss(f_a, t_r; r_0) \cdot H_{\mathrm{FS}} \cdot H_{\mathrm{DFS}}. \qquad (25)$$

After an equivalent transform, (25) can be expressed as

$$\begin{aligned} Ss_{\mathrm{FS}}(f_a, t_r; r_0) &= \left\{\exp\left(-j\frac{4\pi r_0 \beta}{\lambda}\right)\right. \\ &\quad \times \exp\left[-j\frac{4\pi k_r}{c_0}\left(\frac{r_0}{\beta} - r_c\right)\left(\beta t_r - \frac{2r_c}{c_0}\right)\right] \\ &\quad \cdot \exp\left[-j\pi k_r \beta(\beta - 1) t_r^2\right] \\ &\quad \left.\times \mathrm{src}(f_a, \beta t_r; r_0)\right\} \otimes \exp\left(-j\pi k_r \beta t_r^2\right). \end{aligned} \qquad (26)$$

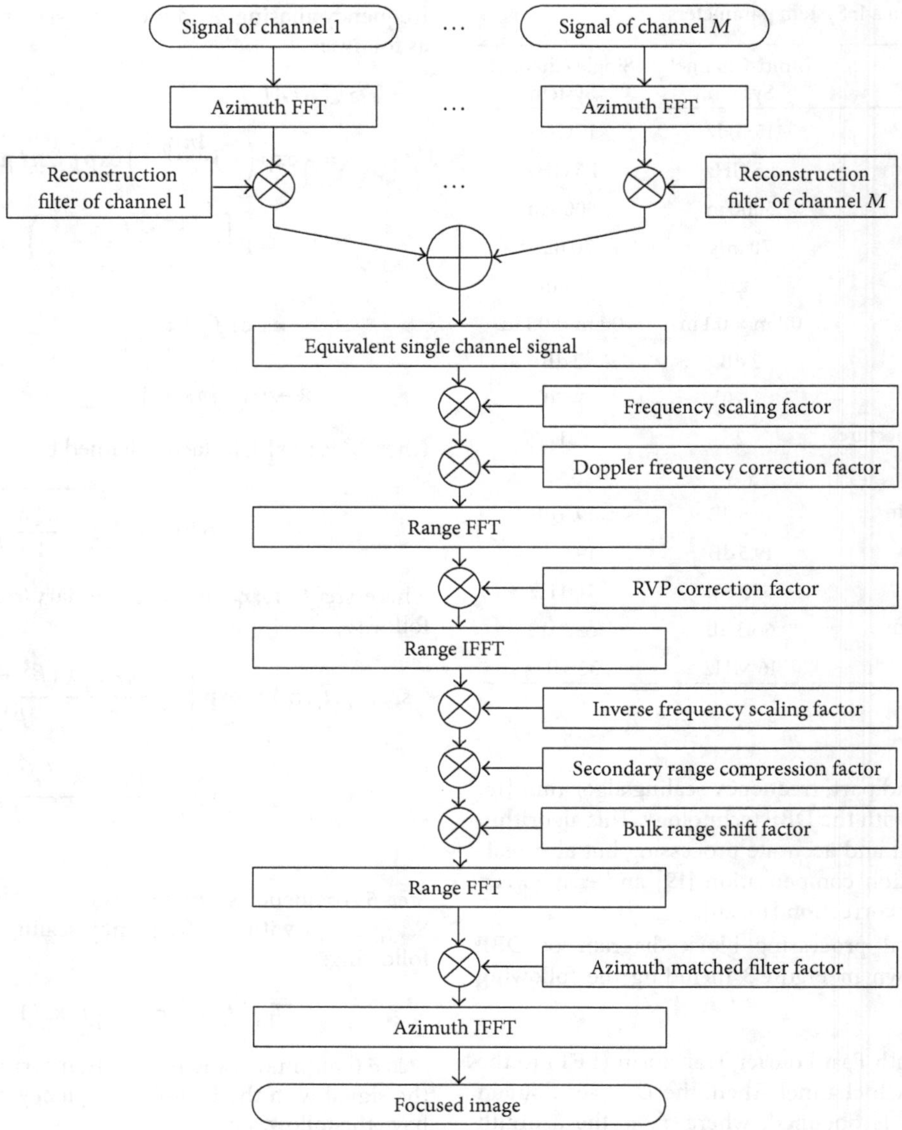

FIGURE 3: Signal processing block diagram for DBF FMCW SAR.

Step 7 (RVP correction). The correction is applied in 2D frequency domain. It is to eliminate the last exponential term in (26). After FFT in range dimension and multiplying with the RVP correction factor $H_{\mathrm{RVPC}}(f_a, f_r)$, IFFT in range domain is performed. The signal $Ss_{\mathrm{RVPC}}(f_a, t_r; r_0)$ is obtained as follows:

$$Ss_{\mathrm{RVPC}}(f_a, t_r; r_0) = \exp\left(-j\frac{4\pi r_0\beta}{\lambda}\right)\exp\left[-j\frac{4\pi k_r}{c_0}\left(\frac{r_0}{\beta}-r_c\right)\left(\beta t_r-\frac{2r_c}{c_0}\right)\right] \cdot \exp\left[-j\pi k_r\beta(\beta-1)t_r^2\right]\mathrm{src}(f_a, \beta t_r; r_0). \tag{27}$$

The RVP correction factor is expressed as

$$H_{\mathrm{RVPC}}(f_a, f_r) = \exp\left(-j\frac{\pi f_r^2}{k_r\beta}\right). \tag{28}$$

Step 8 (inverse frequency scaling transform). Multiply $Ss_{\mathrm{RVPC}}(f_a, t_r; r_0)$ with the inverse frequency scaling factor $H_{\mathrm{IFS}}(f_a, t_r)$ which is

$$H_{\mathrm{IFS}}(f_a, t_r) = \exp\left[j\pi k_r\beta(\beta-1)t_r^2\right]. \tag{29}$$

Step 9. Secondary range compression: multiply the signal with the secondary range compression factor $H_{\mathrm{SRC}}(f_a, t_r; r_c)$ which is described as

$$H_{\mathrm{SRC}}(f_a, t_r; r_c) = \mathrm{src}(f_a, \beta t_r; r_c)^*, \tag{30}$$

where the approximation $r_c \approx r_0$ is required.

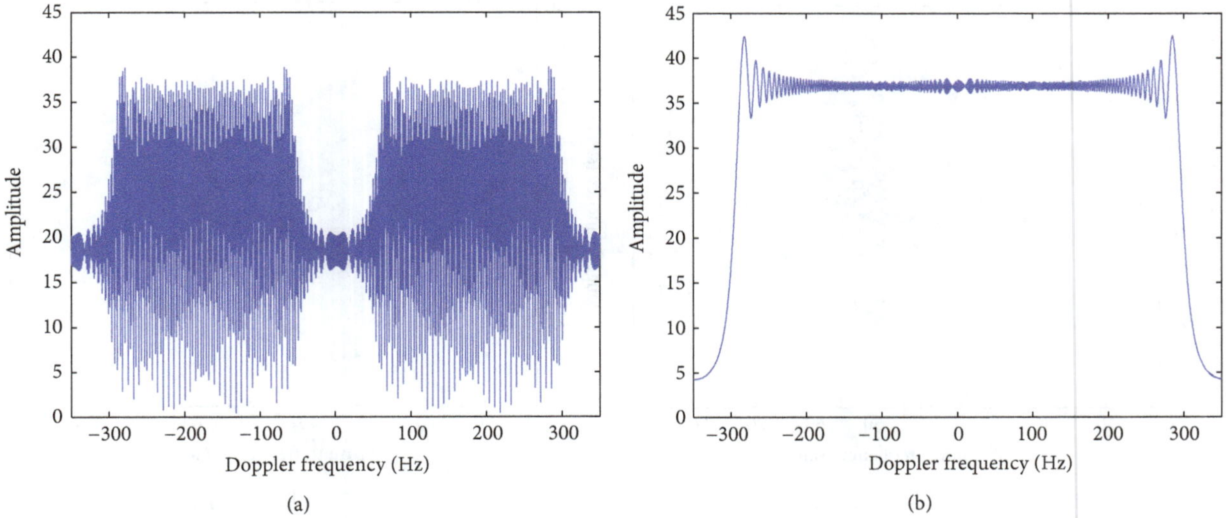

FIGURE 4: The azimuth aliasing spectrum before reconstruction filters (a) and the aliasing-removal spectrum after reconstruction filters (b).

Step 10. Bulk range shift: multiply the signal with the bulk range shift factor $H_{\mathrm{BV}}(f_a, t_r; r_c)$ which is described as

$$H_{\mathrm{BV}}(f_a, t_r; r_c) = \exp\left[j\frac{4\pi k_r r_c}{c_0}\left(\frac{1}{\beta} - 1\right)\left(\beta t_r - \frac{2r_c}{c_0}\right)\right]. \tag{31}$$

In actual implementation, Step 8 to Step 10 can also be completed at the same time. We obtain

$$\begin{aligned} Ss_{\mathrm{RCMC}}(f_a, t_r; r_0) &= Ss_{\mathrm{RVPC}}(f_a, t_r; r_0) \cdot H_{\mathrm{IFS}} \cdot H_{\mathrm{SRC}} \cdot H_{\mathrm{BV}} \\ &\approx \exp\left(-j\frac{4\pi r_0 \beta}{\lambda}\right) \\ &\quad \times \exp\left[-j\frac{4\pi k_r (r_0 - r_c)}{c_0 \beta}\left(\beta t_r - \frac{2r_c}{c_0}\right)\right]. \end{aligned} \tag{32}$$

Step 11. Apply FFT to $Ss_{\mathrm{RCMC}}(f_a, t_r; r_0)$ in range dimension. At this time, the range compression is completed as follows:

$$\begin{aligned} &Ss_{\mathrm{RCMC}}(f_a, f_r; r_0) \\ &\quad = \frac{T}{\beta}\exp\left\{-j\frac{4\pi f_r r_c}{\beta c_0}\right\}\exp\left(-j\frac{4\pi r_0 \beta}{\lambda}\right) \\ &\quad \cdot \operatorname{sinc}\left[\pi\frac{T}{\beta}\left(f_r + \frac{2k_r(r_0 - r_c)}{c_0}\right)\right]. \end{aligned} \tag{33}$$

Step 12. The 2D focused image of DBF FMCW SAR is obtained after azimuth matched filtering [18].

4. Simulation Results

The parameters of an example FMCW SAR system using dual channels combined with DBF are listed in Table 1. The corresponding parameters of single channel FMCW SAR, which is very similar to MicroBSAR described in [6], are also given. There are two parts in the simulation. One is imaging simulation, the other is performance simulation.

4.1. System Parameters Design. To achieve the required resolution in azimuth, the Doppler bandwidth must be greater than 700 Hz. The transmit antenna length in azimuth should be less than 0.2 m. According to the swath width, the transmit antenna length in elevation can be chosen to be 2.9 cm, which corresponds to 40° beamwidth. The transmit antenna gain is 19.5 dB with the antenna efficiency 0.5.

The receive antenna is the same with the transmit antenna for single channel FMCW SAR. In the two-channel system, the receive antennas are placed in azimuth. The receive antenna gain is still the same. However the overall receive antenna length in azimuth reaches 0.4 m.

According to the Nyquist theorem, the PRF of classic FMCW SAR should be greater than the Doppler bandwidth. However, the PRF of DBF FMCW SAR can be chosen to be 700/2 Hz using two receive channels. To ensure a slant range resolution of 0.1 m, a chirp bandwidth of 1.5 GHz is necessary. The range processing gain 66.3 dB is obtained for DBF FMCW SAR. Nevertheless, that value of single channel FMCW SAR is only 63.3 dB. The sampling frequency of DBF FMCW SAR and single channel system are 16 MHz and 32 MHz, respectively. All the parameters are listed in Table 1.

4.2. Imaging Simulation. The target raw data are simulated using the two-channel system parameters listed in Table 1 by (9). To investigate the method of DBF technology and the effects of the continuous antenna motion, the raw data are processed by using the algorithm described in Figure 3.

The azimuth reconstruction is shown in Figure 4. Figure 4(a) shows the azimuth aliasing spectrum of the first channel signal in the two-channel system before reconstruction filter. After filters of $P_1(f_a)$ and $P_2(f_a)$, respectively, add the first channel signal and the second

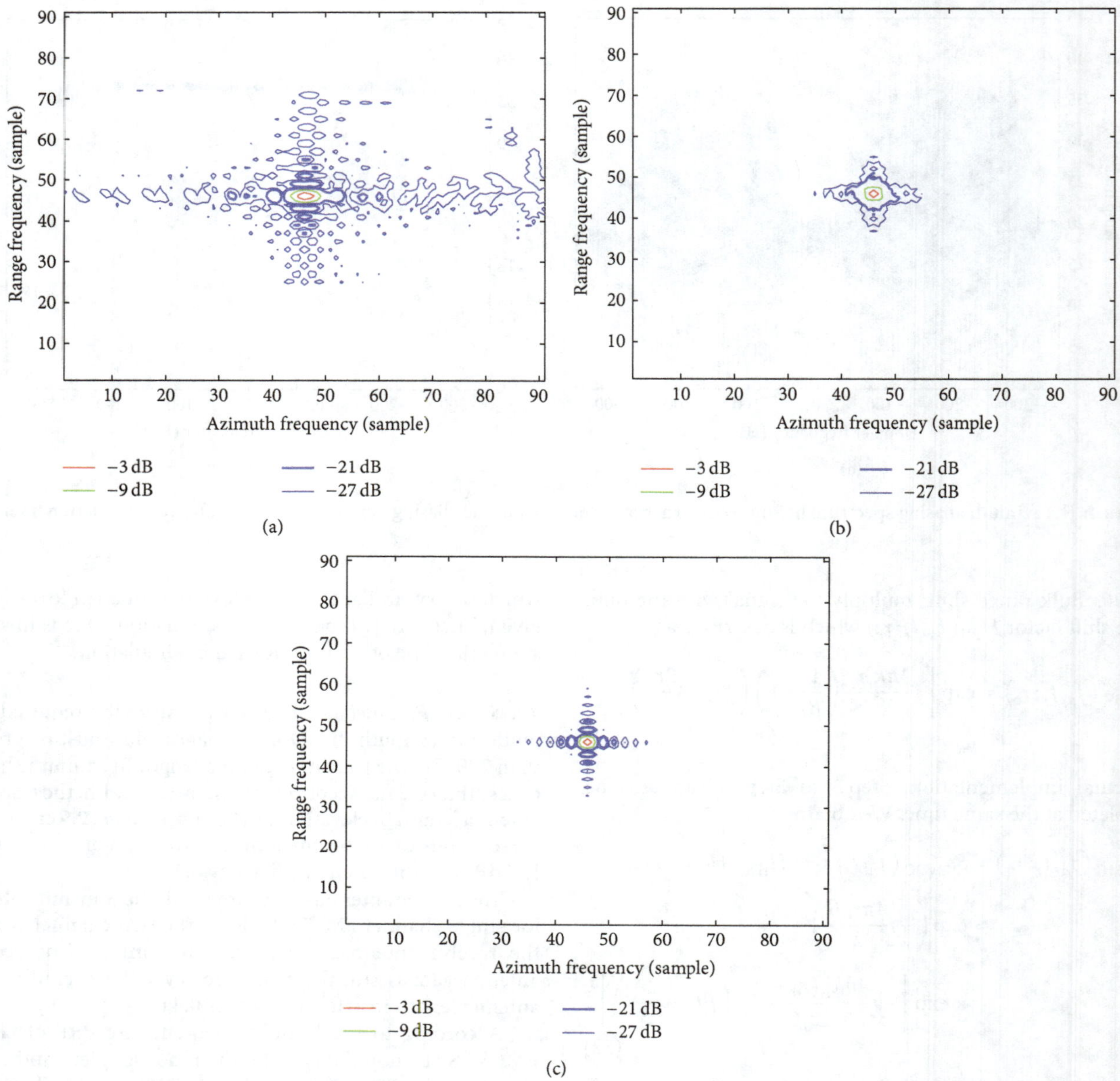

FIGURE 5: Contour plots of impulse response function of a point target without azimuth reconstruction (a), without Doppler frequency correction (b), and with azimuth reconstruction and Doppler frequency correction (c).

channel signal together. Then the azimuth spectrum is completely reconstructed. The reconstruction result is shown in Figure 4(b). The DBF coherently combines the two channel signals in azimuth and reconstructs a wide Doppler spectrum about 700 Hz.

Using the parameters of the two-channel example system, the signal processing presented in Section 3.3 is simulated. Figure 5 plots the 2D contour maps of the impulse response function of a point target. Only the values above −27 dB are plotted. The result of the proposed processing algorithm is plotted at the bottom. Figure 5(a) is the result without azimuth reconstruction. The performance is seriously degraded because the azimuth spectrum is overlapped as Figure 4(a) shows. To explore the characteristic of FMCW SAR distinguished with pulse SAR, Figure 5(b) illustrates the result without Doppler frequency correction factor which is expressed as (24). In terms of the resolution and shape of the response, it is shown that the reconstruction is successful and the impact of the motion within one sweep in FMCW SAR needs to be compensated.

4.3. Performance Simulation. Using the parameters in Table 1, NESZ for DBF FMCW SAR has been calculated over a range of incidence angles, which is shown in Figure 6. The result for single channel system is also shown in Figure 6. When calculating NESZ, the receiver noise figure is chosen

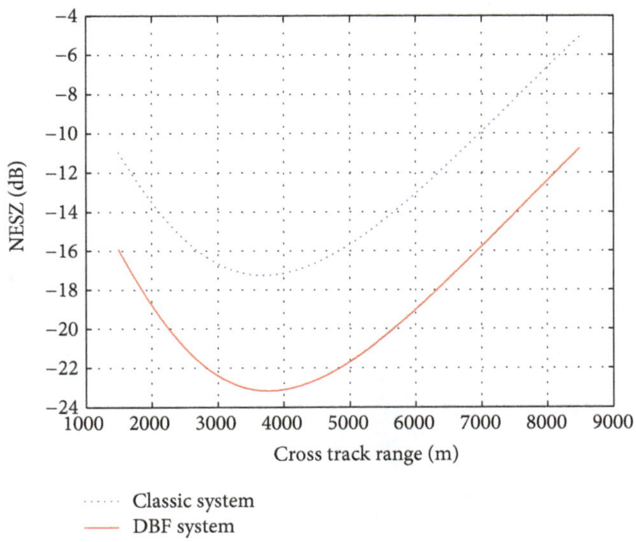

FIGURE 6: NESZ versus cross track range for DBF (solid) and single channel FMCW SAR (dotted).

as 3 dB. The DBF FMCW SAR achieves a NESZ below −10 dB along the entire swath. 5 W of transmitted power is sufficient to image vegetation and farmland. However, the result of classic FMCW SAR in the edge of the swath is only −5 dB, which is not sufficient to obtain good far-range imagery. The reason is that there is higher signal gain for DBF FMCW SAR.

5. Conclusions

A multichannel receive antenna FMCW SAR system combined with DBF is introduced. DBF FMCW SAR overcomes the limitation of high PRF caused by high azimuth resolution. It has higher receive antenna gain and range processing gain than single channel system. The received signal of DBF FMCW SAR is modeled. The impact of continuous antenna motion which is the main characteristic of FMCW SAR is negligible in multichannel azimuth reconstruction for DBF FMCW SAR, but it must be considered in the process during RCM correction. The whole signal processing diagram is given. Simulated point target experiments have been performed to verify the processing diagram successfully. Future works will include performance analysis thoroughly, moving target indication, and the miniaturization of the system.

Conflict of Interests

The authors declare that there is no conflict of interests regarding the publication of this paper.

Acknowledgment

This work was supported by the National Natural Science Foundation of China (nos. 41301397).

References

[1] E. Zaugg, D. Long, M. Edwards et al., "Using the MicroASAR on the NASA SIERRA UAS in the Characterization of Arctic Sea Ice Experiment," in *Proceedings of the IEEE International Radar Conference (RADAR '10)*, pp. 271–276, Washington, DC, USA, May 2010.

[2] M. Edrich, "Ultra-lightweight synthetic aperture radar based on a 35 GHz FMCW sensor concept and online raw data transmission," *IEE Proceedings: Radar, Sonar and Navigation*, vol. 153, no. 2, pp. 129–134, 2006.

[3] M. Weiss, O. Peters, and J. Ender, "A three dimensional SAR system on an UAV," in *Proceedings of the IEEE International Geoscience and Remote Sensing Symposium (IGARSS '07)*, pp. 5315–5318, Barcelona, Spain, July 2007.

[4] E. Zaugg, M. Edwards, D. Long, and C. Stringham, "Developments in compact high-performance synthetic aperture radar systems for use on small unmanned aircraft," in *Proceedings of the IEEE Aerospace Conference (AERO '11)*, pp. 1–14, Big Sky, Mont, USA, March 2011.

[5] A. Meta, J. J. M. de Wit, and P. Hoogeboom, "Development of a high resolution airborne millimeter wave FM-CW SAR," in *Proceedings of the 1st European Radar Conference (EuRAD '04)*, pp. 209–212, Amsterdam, The Netherlands, October 2004.

[6] M. C. Edwards, *Design of a continuous-wave synthetic aperture radar system with analog dechirp [M.S. thesis]*, Brigham Young University, Provo, Utah, USA, 2009.

[7] B. R. Jean and J. W. Rouse Jr., "A multiple beam synthetic aperture radar design concept for geoscience applications," *IEEE Transactions on Geoscience and Remote Sensing*, vol. 21, no. 2, pp. 201–207, 1983.

[8] A. Currie and M. A. Brown, "Wide-swath SAR," *IEE Proceedings F: Radar and Signal Processing*, vol. 139, no. 2, pp. 122–135, 1992.

[9] G. D. Callaghan and I. D. Longstaff, "Wide-swath space-borne SAR using a quad-element array," *IEE Proceedings: Radar, Sonar and Navigation*, vol. 146, no. 3, pp. 159–165, 1999.

[10] M. Suess, B. Grafmueller, and R. Zahn, "A novel high resolution, wide swath SAR system," in *Proceedings of the International Geoscience and Remote Sensing Symposium (IGARRS '01)*, pp. 1013–1015, Sydney, Australia, July 2001.

[11] W. van Rossum, M. Otten, and P. van Dorp, "Multichannel FMCW SAR," in *Proceedings of the 9th European Conference on Synthetic Aperture Radar (EUSAR '12)*, pp. 279–282, Nuremberg, Germany, April 2012.

[12] A. Meta, E. Imbembo, C. Trampuz, A. Coccia, and G. de Luca, "A selection of MetaSensing airborne campaigns at L-, X- and ku band," in *Proceedings of the IEEE International Geoscience and Remote Sensing Symposium (IGARSS '12)*, pp. 4571–4574, Munich, Germany, July 2012.

[13] M. C. Edwards and E. C. Zaugg, "Design of a compact, modular, multi-frequency band, multi-mode, multi-channel synthetic aperture radar," in *Proceedings of the 9th European Conference on Synthetic Aperture Radar (EUSAR '12)*, pp. 44–47, Nuremberg, Germany, April 2012.

[14] X. Qin, W. Zhan, Z.-H. Jiang, and H.-F. Kan, "Improving FMCW SAR system performance by digital beamforming," in *Proceedings of the IEEE International Geoscience and Remote Sensing Symposium (IGARSS '12)*, pp. 4557–4560, Munich, Germany, July 2012.

[15] N. Gebert, G. Krieger, and M. A. Moreira, "Digital beamforming on receive: techniques and optimization strategies for high-resolution wide-swath SAR imaging," *IEEE Transactions on*

Aerospace and Electronic Systems, vol. 45, no. 2, pp. 564–592, 2009.

[16] Z.-H. Jiang and K. Huang-Fu, "Squint LFMCW SAR data processing using doppler-centroid-dependent frequency scaling algorithm," *IEEE Transactions on Geoscience and Remote Sensing*, vol. 46, no. 11, pp. 3535–3543, 2008.

[17] J. C. Curlander and R. N. McDonough, *Synthetic Aperture Radar: Systems and Signal Processing*, John Wiley & Sons, New York, NY, USA, 1991.

[18] J. Mittermayer, A. Moreira, and O. Loffeld, "Spotlight SAR data processing using the frequency scaling algorithm," *IEEE Transactions on Geoscience and Remote Sensing*, vol. 37, no. 5, pp. 2198–2214, 1999.

[19] E. C. Zaugg and D. G. Long, "Theory and application of motion compensation for LFM-CW SAR," *IEEE Transactions on Geoscience and Remote Sensing*, vol. 46, no. 10, pp. 2990–2998, 2008.

[20] A. Meta, P. Hoogeboom, and L. P. Ligthart, "Signal processing for FMCW SAR," *IEEE Transactions on Geoscience and Remote Sensing*, vol. 45, no. 11, pp. 3519–3532, 2007.

Permissions

All chapters in this book were first published in JECE, by Hindawi Publishing Corporation; hereby published with permission under the Creative Commons Attribution License or equivalent. Every chapter published in this book has been scrutinized by our experts. Their significance has been extensively debated. The topics covered herein carry significant findings which will fuel the growth of the discipline. They may even be implemented as practical applications or may be referred to as a beginning point for another development.

The contributors of this book come from diverse backgrounds, making this book a truly international effort. This book will bring forth new frontiers with its revolutionizing research information and detailed analysis of the nascent developments around the world.

We would like to thank all the contributing authors for lending their expertise to make the book truly unique. They have played a crucial role in the development of this book. Without their invaluable contributions this book wouldn't have been possible. They have made vital efforts to compile up to date information on the varied aspects of this subject to make this book a valuable addition to the collection of many professionals and students.

This book was conceptualized with the vision of imparting up-to-date information and advanced data in this field. To ensure the same, a matchless editorial board was set up. Every individual on the board went through rigorous rounds of assessment to prove their worth. After which they invested a large part of their time researching and compiling the most relevant data for our readers.

The editorial board has been involved in producing this book since its inception. They have spent rigorous hours researching and exploring the diverse topics which have resulted in the successful publishing of this book. They have passed on their knowledge of decades through this book. To expedite this challenging task, the publisher supported the team at every step. A small team of assistant editors was also appointed to further simplify the editing procedure and attain best results for the readers.

Apart from the editorial board, the designing team has also invested a significant amount of their time in understanding the subject and creating the most relevant covers. They scrutinized every image to scout for the most suitable representation of the subject and create an appropriate cover for the book.

The publishing team has been an ardent support to the editorial, designing and production team. Their endless efforts to recruit the best for this project, has resulted in the accomplishment of this book. They are a veteran in the field of academics and their pool of knowledge is as vast as their experience in printing. Their expertise and guidance has proved useful at every step. Their uncompromising quality standards have made this book an exceptional effort. Their encouragement from time to time has been an inspiration for everyone.

The publisher and the editorial board hope that this book will prove to be a valuable piece of knowledge for researchers, students, practitioners and scholars across the globe.

List of Contributors

Yong Wang and Wei Liu
Faculty of Mechanical and Electronic Information, China University of Geosciences, Wuhan 430074, China

Dianhong Wang
Faculty of Mechanical and Electronic Information, China University of Geosciences, Wuhan 430074, China
Department of Geological Science and Engineering, Wuhan University of Engineering Sciences, Wuhan 430200, China

Qian Lu
Department of Geological Science and Engineering, Wuhan University of Engineering Sciences, Wuhan 430200, China

Chengchang Zhang
Department of Electrical Engineering, College of Electronic Engineering, Chongqing University of Posts and Telecommunications, Chongqing, China
Electrical and Computer Engineering, Faculty of Engineering and Applied Science, Memorial University, St. John's, NL, Canada

Lihong Zhang
Electrical and Computer Engineering, Faculty of Engineering and Applied Science, Memorial University, St. John's, NL, Canada

Shuhan Chen and Xiaorun Li
College of Electrical Engineering, Zhejiang University, Hangzhou 310027, China

Liaoying Zhao
Institute of Computer Application Technology, Hangzhou Dianzi University, Hangzhou 310018, China

Li Yang
School of Computer Science and Engineering, Beifang University of Nationalities, Yinchuan 750021, China

Chang Xia
The Institute of Information and System Science, School of Mathematics and Information Science, Beifang University of Nationalities, Yinchuan 750021, China

Chang Juan
Graduate School, Ningxia University, Yinchuan 750021, China

Zahid Akhtar and Gian Luca Foresti
Department of Mathematics and Computer Science, University of Udine, Via delle Scienze 206, 33100 Udine, Italy

Hong Tang and Tianshuang Qiu
Department of Biomedical Engineering, Dalian University of Technology, Dalian, China

Ting Li
College of Information and Communication Engineering, Dalian Minzu University, Dalian, China

Yongwan Park
Department of Information and Communication Engineering, Yeungnam University, Gyeongsan, Republic of Korea

Juha Partala, Angelos Fylakis, Anu Pramila, Anja Keskinarkaus and Tapio Seppänen
Physiological Signal Analysis Team, Center for Machine Vision and Signal Analysis, University of Oulu, Oulu, Finland

A. M. Dehnavi and H. Rabbani
Department of Bioelectrics and Biomedical Engineering, Medical Image & Signal Processing Research Center, School of Advanced Technologies in Medicine, Isfahan University of Medical Sciences, Isfahan, Iran

M. Esmaeili
Department of Bioelectrics and Biomedical Engineering, Medical Image & Signal Processing Research Center, School of Advanced Technologies in Medicine, Isfahan University of Medical Sciences, Isfahan, Iran
Department of Advanced Technologies in Medicine, Faculty of Medicine, Tabriz University of Medical Sciences, Tabriz, Iran

Huayang Li, Dehui Kong and Shaofan Wang
Beijing Key Laboratory of Multimedia & Intelligent Software Technology, College of Metropolitan Transportation, Beijing University of Technology, Beijing 100124, China

Baocai Yin
Faculty of Electronic Information and Electrical Engineering, Dalian University of Technology, Dalian 116024, China

Qiong Li, Tingting Zhao, Lingchao Zhang, Wenhui Sun and Xi Zhao
College of Computer Science and Information Engineering, Tianjin University of Science and Technology, Tianjin 300457, China

Yasunori Sugita
Nagaoka University of Technology, 1603-1 Kamitomioka, Nagaoka City, Niigata 940-2188, Japan

Huidong He and Xiaoqian Mao
School of Electrical and Information Engineering, Tianjin University, Tianjin 300072, China

Wei Li
Department of Computer & Electrical Engineering and Computer Science, California State University, Bakersfield, CA 93311, USA

Linwei Niu
Department of Math and Computer Science, West Virginia State University, Institute, WV 25112, USA

Genshe Chen
Intelligent Fusion Technology, Inc., Germantown, MD 20876, USA

Hui Ren, Jun Wang and Qiong-Hua Wang
School of Electronics and Information Engineering, Sichuan University, Chengdu 610065, China

Christian Gentner, Siwei Zhang and Thomas Jost
German Aerospace Center (DLR), Institute of Communications and Navigation, Oberpfaffenhofen, 82234Wessling, Germany

Johannes Jordan, Elli Angelopoulou and Andreas Maier
Pattern Recognition Lab, University of Erlangen-Nuremberg, Erlangen, Germany

Hongjuan Ge
College of Civil Aviation, Nanjing University of Aeronautics and Astronautics, Nanjing 211106, China

Haijun Wang
College of Civil Aviation, Nanjing University of Aeronautics and Astronautics, Nanjing 211106, China
Aviation Information Technology R & D Center, Binzhou University, Binzhou 256603, China

Shengyan Zhang
Aviation Information Technology R & D Center, Binzhou University, Binzhou 256603, China

Małgorzata Michalska and Marco Mattavelli
EPFL SCI-STI-MM, École Polytechnique Fédérale de Lausanne, 1015 Lausanne, Switzerland

Nicolas Zufferey
Geneva School of Economics and Management (GSEM), University of Geneva, 1211 Geneva 4, Switzerland

Shirui Huo
City University of Hong Kong, Kowloon Tong, Hong Kong

Tianrui Hu
Beijing University of Posts and Telecommunications, Beijing, China

Ce Li
State Key Laboratory of Coal Resources and Safe Mining, China University of Mining & Technology, Beijing, China
University of Chinese Academy of Sciences, Beijing, China

Marius Miron, Julio J. Carabias-Orti, Juan J. Bosch, Emilia Gómez and Jordi Janer
Music Technology Group, Universitat Pompeu Fabra, Barcelona, Spain

Qin Xin, Zhihong Jiang and Pu Cheng
College of Electronic Science and Engineering, National University of Defense Technology, ChangSha, Hunan 410073, China

Mi He
School of Biomedical Engineering,Third Military Medical University, Chongqing 400038, China

Index

www.ingramcontent.com/pod-product-compliance
Lightning Source LLC
LaVergne TN
LVHW060515210926
848485LV00005B/1084